# 고온 원자력 재료의 크리프 수명 평가 및 예측

Creep Life Assessment and
Prediction of High-Temperature Nuclear Materials

한국원자력연구원 김우곤 저

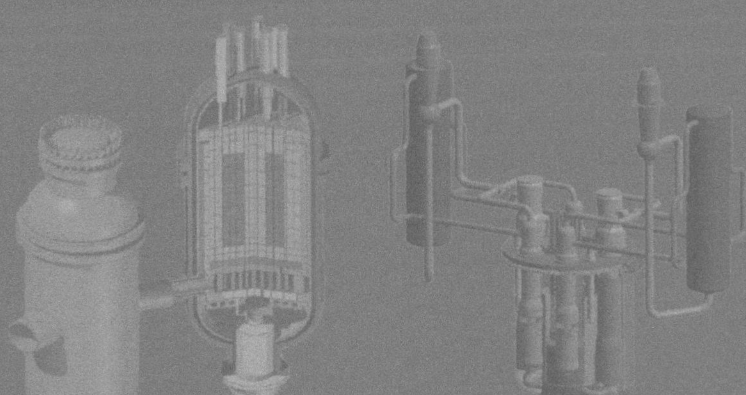

GS인터비전

# 목 차

# 머 리 말

제4세대 (Gen-IV) 원자로 및 핵융합로 등의 혁신 원자력 시스템은 높은 중성자 조사량은 물론 효율을 증대시키기 위해서 더 높은 고온에서 40-60년간의 장기 가동을 목표로 설계되기 때문에 원자로의 안전성은 크리프 (creep) 성능에 의해 좌우된다 해도 과언이 아닐 것이다. 고온 하에서의 금속의 기계적 성질은 시간과 관계없는 상온과는 다르게 시간 변화에 따라서 크게 달라진다. 고온에서는 원자들의 활동이 활발해지고 확산의 과정을 통하여 전위들의 움직임이 빨라지므로 상온에서 설명할 수 없는 새로운 변형기구에 의해 작동되기 때문일 것이다. 고온에서는 항복강도보다 매우 낮은 하중에서도 지속적으로 장기간 받게 되면 크리프 현상에 의해 파괴가 일어나므로 혁신 원자력 시스템의 부품의 안전을 위해서는 크리프 성질이 반드시 평가되어야 한다. 또한 이러한 부품의 설계 시에는 크리프 응력과 파단시간 시험 자료가 마련되어야 하고 장시간 파단시험 자료가 없을 시에는 단시간 시험 자료로부터 적어도 $10^5$ 시간 이상의 장기간에 대하여 정확한 수명예측 (외삽)을 하여야 한다.

한편, 고온 크리프 조건하의 재료 파괴는 균열의 생성과 성장의 두 과정에 의해 진행된다. 두꺼운 후육부재 (thick section component)의 경우 심한 온도 및 응력 구배가 존재하고 열응력과 산화 등에 의하여 취약부에서 균열이 생성되기 쉽다. 초기에 생성된 균열은 크리프 조건하에서 성장하게 되며, 균열성장이 부재 수명의 대부분을 차지하게 되므로 부재의 건전성 유지를 위해서는 크리프 균열성장 (creep crack growth) 거동이 평가되어야 한다. 시간-의존 파괴역학 매개변수를 사용하는 크리프 균열성장 평가 방법은 제4세대 원자로의 구조재 뿐만 아니라 화력 발전소의 터빈, 보일러 헤더, 항공기 엔진 등의 고온 부품에도 적용될 수 있을 것이다.

본 책자에 수록된 29편의 논문들은 저자가 한국원자력연구원에서 원자력 재료의 고온 강도 평가 및 해석에 관한 시험과 연구를 30여 년간 수행하면서 제1저자로 SCI (Science Citation Index) 급 학술지에 게재한 논문만을 발췌한 것이다. 주요 내용은 제4세대 미래형 원자력 소재들에 대하여 크리프 거동 평가와 장기 수명예측, 그리고 크리프 균열성장 속도를 평가하기 위한 고온 파괴역학 매개변수의 적용과 균열성장 속도의 신뢰성 예측 방법 등에 관한 내용들을 담고 있다. 아무쪼록 수록된 각 논문의 내용들이 동 분야를 연구하는 후배 연구자 및 공학도들에게 조금이라도 도움이 되는 유용한 자료로 활용되었으면 좋겠다.

2021년 11월 김 우 곤

# 논 문 목 록

이하에 수록된 29편의 논문은 본 저자가 한국원자력연구원에 근무하면서 제4세대 원자력 재료의 고온 크리프 분야에 대하여 제1저자로 SCI 국제 학술지에 게재된 논문만을 발췌하여 담은 것이다. 각 논문들의 주요내용은 크리프 거동 평가/해석, 장시간 크리프 수명예측 방법, 그리고 크리프 균열성장 거동 평가/해석, 균열성장속도의 신뢰성 예측 방법 등에 관한 내용을 포함하고 있다.

Nuclear Engineering and Technology 53 (2021) 572–582

Contents lists available at ScienceDirect

Nuclear Engineering and Technology

journal homepage: www.elsevier.com/locate/net

Original Article

# Creep and creep crack growth behaviors for base, weld, and heat affected zone in a grade 91 weldment

Woo-Gon Kim [a, *], Injin Sah [a], Seon-Jin Kim [b], Hyeong-Yeon Lee [a], Eung-Seon Kim [a]

[a] Korea Atomic Energy Research Institute, 989-111 Daedeokdaero, Yuseong-gu, Daejeon, 305-353, Republic of Korea
[b] Pukyong National University, 365 Shinsunro, Nam-gu, Busan, 608-739, Republic of Korea

ARTICLE INFO

Article history:
Received 16 April 2020
Received in revised form
6 July 2020
Accepted 12 July 2020
Available online 28 July 2020

Keywords:
Creep
Creep crack growth
Grade 91 steel
Weld metal
Heat affected zone

ABSTRACT

This study investigated the creep and creep crack growth (CCG) behavior of the base metal (BM), weld metal (WM), and heat affected zone (HAZ) in a Gr. 91 weldment, which was made by a shield metal arc weld process. A series of tensile, creep, and CCG tests were performed for the BM, WM, and HAZ at 550 °C. Creep behavior of the BM, WM, and HAZ was analyzed in terms of various creep laws; Norton's power-law, Monkman-Grant relation and damage tolerance factor (λ), and their constants were determined. In addition, each CCGR law for the BM, WM, and HAZ was proposed and compared in terms of a C*-fracture parameter. The WM and HAZ revealed faster creep rate, lower rupture ductility, and faster CCGRs than the BM, but they showed a similar behavior in the creep and CCG. The CCGRs obtained in the present study exhibited a marginal difference when compared with those of RCC-MRx of currently elevated design code in France. A creep crack path in the HAZ plane progressed towards a weak fine-grained HAZ adjacent to the BM.



## 1. Introduction

Modified 9Cr−1Mo−VNb steel (Grade 91, hereafter Gr. 91) was developed to improve thermal efficiency and life span, and to reduce $CO_2$ emission in a fossil-fuel fired power plant. This steel was used for the pressure vessels and piping at the higher temperature above 500 °C in the fossil power plant because of good creep strength and corrosion/erosion resistance. Particular, Gr. 91 steel is being expected to use for the reactor pressure vessels (RPV) in a very high temperature reactor (VHTR) system and the steam generator in sodium-cooled fast reactor (SFR) systems at the heart in Generation-IV (Gen-IV) nuclear reactors [1,2]. The Gen-IV reactors will be designed to operate 60 years under much higher temperatures than those of currently operating reactors. These conditions may induce creep damage at a localized stress concentration, initiate and develop creep cracks, and eventually a failure. A most portion of total life will be spent for crack growth (or propagation) duration [3–5].

In addition, good creep strength of Gr. 91 steel arises from the addition of strong carbides and nitrides forming elements of V and Nb along with controlled addition of N in the plain 9Cr−1Mo steel. However, the creep strength of its weld joint reveals a significant reduction from that of the base metals owing to Type-IV failure, which is cracking observed in the heat affected zone (HAZ) during service at high temperatures. Creep damage in the HAZ become life limiting factor for the welded components during long-term service life. Type-IV failure in the HAZ of Gr. 91 steel is approximately reduced with 20% of the base metal (BM) [6]. A fracture location of Type-IV cracking is commonly reported to be observed in the three HAZ subzones of the coarse-grained HAZ, fine-grained HAZ, and intercritical HAZ displaying considerable structural changes [7,8]. Therefore, it is essential to accurately assess the reductions of the creep and creep crack growth (CCG) behaviors of the weld metal (WM) and HAZ from the base metal of Gr. 91 steel.

In the present study, the creep and CCG behaviors for the BM, WM, and HAZ of a Gr. 91 weldment at 550 °C were comparatively examined, and the creep behavior was analyzed using useful creep laws of Norton's power law, the Monkman-Grant (M-G) relations, and damage tolerance factor (λ). Material constants $D$, $m$, $A$, and $n$ to calculate a C*-fracture parameter were determined. Their CCGR laws were proposed and compared in terms of an empiral relationship of the $da/dt$ vs. C*. In addition, the CCGR laws obtained in the present tests were compared with those in RCC-MRx code. A crack propagation path between the BM and HAZ locations was

* Corresponding author.
E-mail address: wgkim@kaeri.re.kr (W.-G. Kim).

https://doi.org/10.1016/j.net.2020.07.015
1738-5733/

described using an evident photo.

## 2. Experimental procedures

### 2.1. Welding method

Gr. 91 steel used in the BM has a 32 mm thickness as a hot-rolled plate type of a commercial grade. The BM steel had been normalized and tempered (N + T) at 1050 °C/1min per mm and 770 °C/3min per mm. The chemical compositions are given in Table 1. The filler metal of CM-9Cb (brand name) was used for the AWS Class, E9016-G (3.2–4.0 mm) manufactured in Kobe steel. A shielded metal arc weld (SMAW) process was used to manufacture welded blocks. A single V-groove shape of 60° was designed to join two plates. A weld-block size has a 32 mm thickness, 320 mm width, and 375 mm length. Post-weld heat treatment was held for 255 min at 750 °C. A photo of a welded block is shown in Fig. 1. The three regions of the base, weld, and HAZ in the photo were clearly observed from macro-etching technique.

In the author's previous study [1], micro-Vickers hardness (Hv) values were measured along the middle plane of the weld joint. In the hardness values, the WM region was measured to be higher for Hv = 40 as about Hv = 290 than Hv = 250 in the BM. In particular, the hardness of the HAZ region located between the BM and WM was steeply dropped, as shown in Fig. 2. In a fine-grained heat affect zone (FGHAZ) adjacent to the BM, the hardness value was lower than Hv = 250 in the BM as Hv = 220. This FGHAZ will be very weak from cracks, and a final failure will occur in the FGHAZ. A failure location under the long-term service of a low stress most commonly occurs in the FGHAZ [9]. The directions of Type-IV cracking resulted from the FGHAZ region appear to be parallel or offset from the fusion line.

**Fig. 1.** A weld block manufactured by SMAW method.

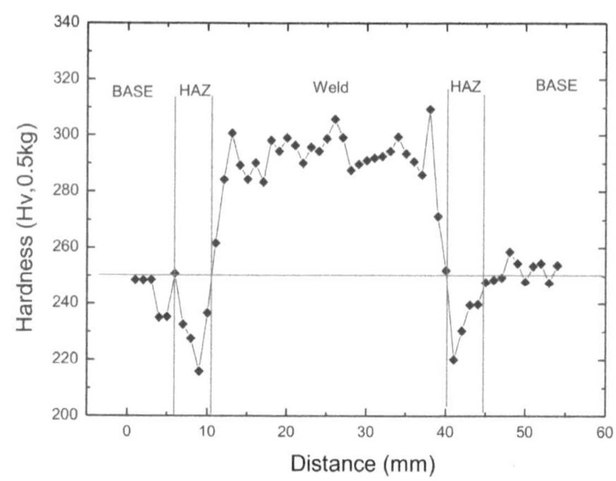


**Fig. 2.** Variations of Vickers hardness in Gr. 91 weld joint [1].

### 2.2. Tensile, creep, and CCG tests

The tensile and creep tests were conducted for the BM, WM, and HAZ at the identical temperature of 550 °C. The HAZ specimens were taken for a cross-weld direction in the welded blocks. The rectangular-cross section specimens for tensile tests had a 2 mm thickness and 6.25 mm width for a 25 mm gauge length. A strain rate in the tensile tests was constant as $1 \times 10^{-4}$/s at 550 °C. The round-type creep specimens had a 6 mm diameter for a 30 mm gauge length. The HAZ specimens position the HAZ location at the centre of the gauge length. Constant-load creep machines with lever ratio of 20:1 were used, and the creep tests were conducted under various applied stresses at 550 °C. Experimental procedures for the creep tests referred to the ASTM standard E139 [10]. Each constant value of *D*, *m*, *A*, and *n* for the BM, WM, and HAZ was obtained from the tensile and creep tests, respectively.

An apparatus for the CCG tests is illustrated in Fig. 3. The CCG tests were conducted under different applied load levels at 550 °C. Load-line displacement (LLD) data were continuously acquainted through a data acquisition system using a linear gauge assembly attached to the compact tension (CT) specimens. Crack length data with time variations were measured by a direct current potential drop (DCPD) technique. The CT specimens had a 25.4 mm width (W), a 12.7 mm thickness (B), and 10% depth side groove. An initial crack ratio (*a/W*) in the CT sample was for about 0.5. To make a sharp crack-tip starter in front of notch for the BM, WM, and HAZ specimens, a pre-cracking size was machined for 2.0 mm by an electric discharge machining (EDM). The pre-cracking for the WM was made within fully welded metal and the pre-cracking for the HAZ was made to conform to one of the HAZ location, as shown in Fig. 4a and b. All procedures in the CCG tests referred to the recommendations of the ASTM standard E1457 [11].

To measure an actual crack length after CCG testing, a CT specimen was split into two parts at its centre after enough cooling in a solution of liquid nitrogen. Typical fracture photos of CT specimens after the CCG tests for the WM and HAZ are illustrated in Fig. 5. An actually measured final crack length, $a_{mf}$ can be obtained by calculating the measurements on a fracture surface at nine equally spaced points, that is, named as "nine-points method" using a high-magnification photo for the fractured surfaces. The reason for this is that the individual crack measurements on the fracture surface are different due to irregularities of cracks.

The predicted crack length, $a_p$ by the DCPD method can be obtained by calculating Johnson's equation. Herein, the difference between $a_{mf}$ and $a_{pf}$ can be examined to determine whether their values are satisfied within the ASTM requirements, or not. Namely, if$0.85 \le \left(\frac{a_{pf}-a_o}{a_{mf}-a_o}\right) \le 1.15$, the CCG data are identified to be valid for further processing. In this equation, the predicted crack extension,

**Table 1**
Chemical composition of the Gr. 91 steel (wt. %).

| C | Si | Mn | P | S | Ni | Cr | Mo | Cu | V | Al | N | Nb |
|---|---|---|---|---|---|---|---|---|---|---|---|---|
| 0.115 | 0.23 | 0.415 | 0.012 | 0.0014 | 0.22 | 8.9 | 0.869 | 0.038 | 0.194 | 0.020 | 0.0513 | 0.073 |

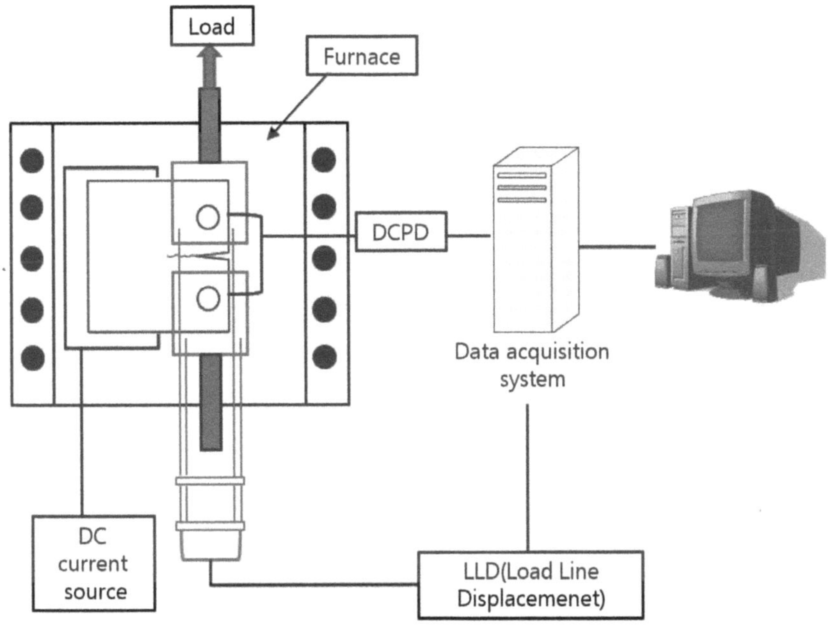


**Fig. 3.** A schematic diagram for the CCG tests.

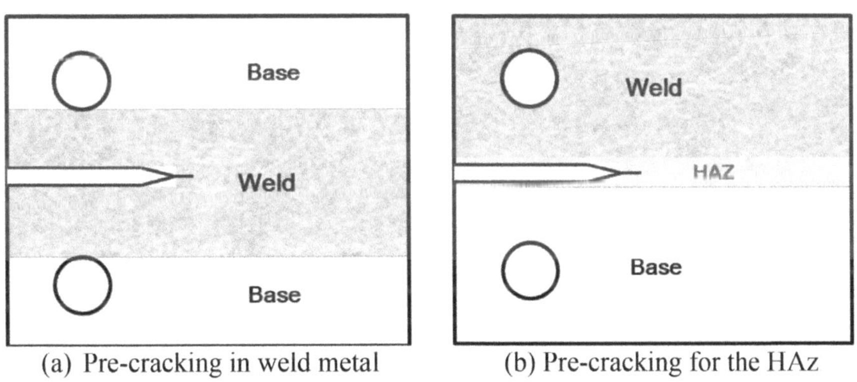


**Fig. 4.** Pre-cracking locations in front of a notch for the WM and HAZ of CT specimens.

($a_{pf}$-$a_o$) can be obtained by substrating the initial crack length, $a_o$ from the finally predicted crack length, $a_{pf}$ [12]. More details for the CCG tests and crack calculations are reported in the author's previous study [1].

According to these procedures, the validity of all tested CT specimens was checked. The CCG data for the BM were obtained for seven samples with the applied load ranges of 6200–7000 N. The CCG data for the WM were for six samples with the applied load ranges of 5700–6300 N. The CCG data for the HAZ were for five samples with the applied load ranges of 6600–7000 N. As described above, the applied load levels of the WM were lower than those of the BM and HAZ. The reason for this is that the valid 6-samples within ASTM requirements after the CCG tests of the WM corresponded to the load ranges of 5700–6300 N, although the applied stress levels for BM, WM, and HAZ were same in the CCG tests.

## 3. Results and discussion

### 3.1. Tensile and creep rupture behaviors

Tensile material constants of $D$ and $m$ were determined by the tensile tests for the BM, WM, and HAZ at 550 °C. Using the tensile curves, as shown in Fig. 6, tensile plastic constants were determined by $\varepsilon_p = D(\sigma/\sigma_y)^m$, where $\varepsilon_p$ is the true plastic strain, σ is the true stress, and $\sigma_y$ is the yield stress in 0.2% offset. The tensile strength is lowest in the HAZ as the order of HAZ < BM < WM. This results are in accordance with the results of the micro-hardness measurement, as described above. A log-log plot of plastic strain and true stress data was used for the data up to necking point. The $D$ and $m$ values were obtained by a linear regression to give the best fit to the data of the BM, WM, and HAZ, as shown in Fig. 7. In addition, their creep rupture data were obtained at 550 °C.

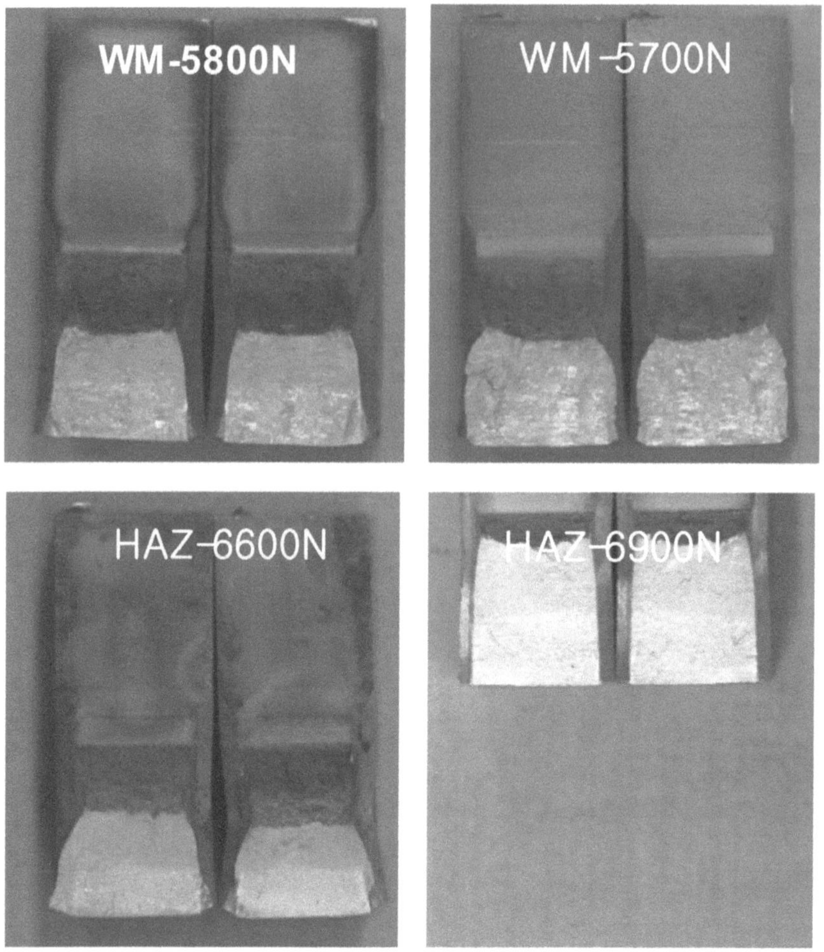


**Fig. 5.** Typical fracture photos after the CCG tests for the WM and HAZ.

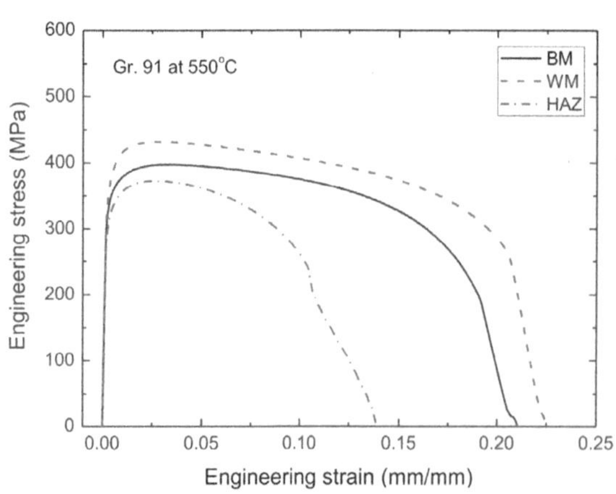


**Fig. 6.** Comparison of tensile curves for the BM, WM, and HAZ at 550 °C.

Fig. 8 shows typical creep curves obtained for the BM, WM, and HAZ at the identical stress of 170 MPa at 550 °C. A long-term rupture data of $t_r$ = 50,799 h (5.79 year) in the BM was successfully acquired. Creep curves display well-defined three creep stages such as the primary creep stage, secondary creep stage, and tertiary creep stage. It can be seen well that the secondary creep stage occupied most of total creep life by delaying the onset of the tertiary creep stage. A constant increase with times in the secondary creep stage is due to a balance of hardening and softening. The onset of tertiary creep stage starts just before failure and its period is short. The WM and HAZ are almost similar in the creep rupture time, but they are much shorter in the rupture life than the BM. Therefore, it is highly suggestive that the creep failure in Gr. 91 weld joint will occur in the weld metal region or especially a weak HAZ.

Fig. 9 shows a log-log plot of stress vs. rupture time tested at 550 °C for the BM, WM, and HAZ. The WM and HAZ reveal a significant reduction in the rupture time compared with the BM. Furthermore, the reduction gap of the WM and HAZ to the BM becomes larger with an increase in rupture time (or low stress). However, the WM and HAZ is very close in creep stress, as shown in the plot of stress vs. rupture time.

The creep constants $A$ and $n$ in Norton's power law are obtained by the creep tests for the BM, WM, and HAZ at 550 °C. Norton's equation is given by

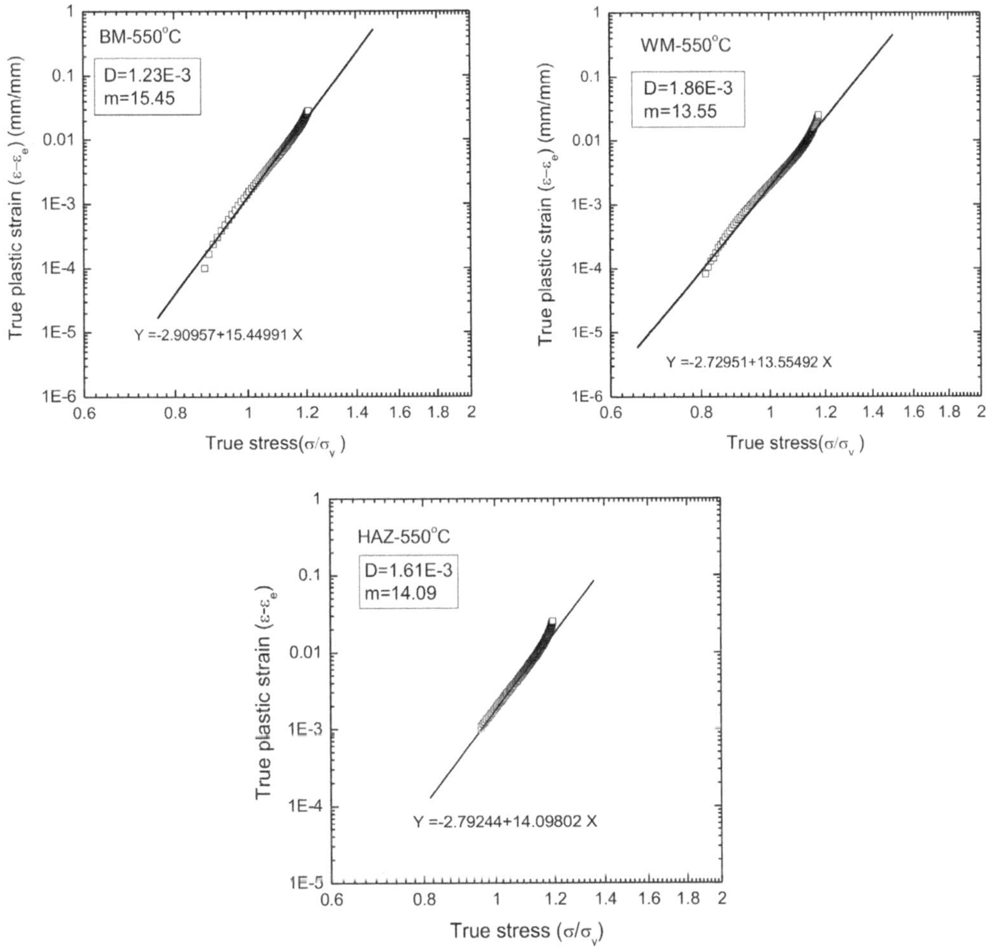


**Fig. 7.** Determination of the $D$ and $m$ values in tensile curves for the BM, WM, and HAZ at 550 °C.

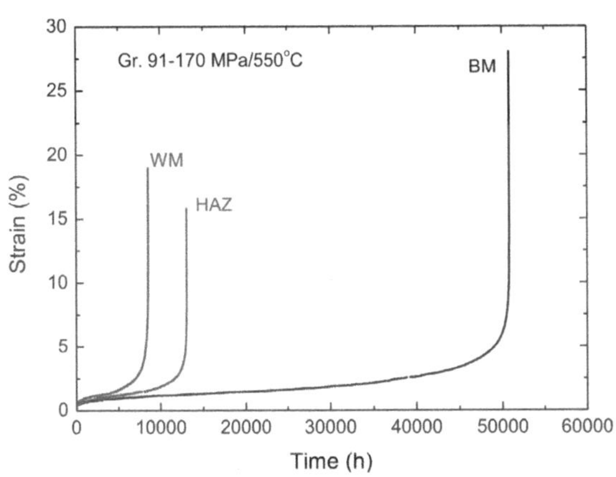


**Fig. 8.** Typical creep curves obtained for the BM, WM, and HAZ at 550 °C.

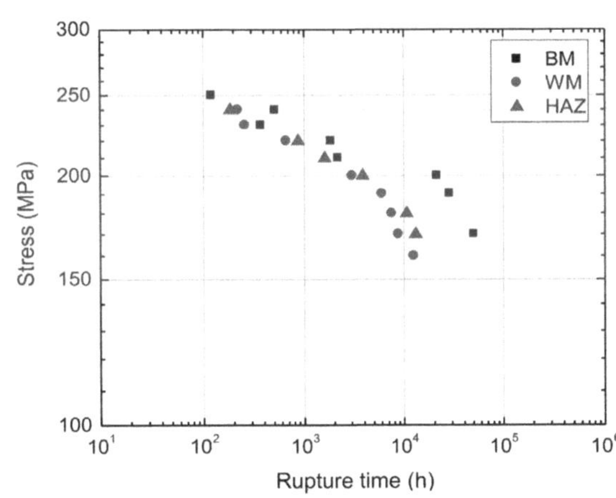


**Fig. 9.** Plot of stress vs. rupture time for the BM, WM, and HAZ at 550 °C.

$$\dot{\varepsilon}_{ss} = A\sigma^n \tag{1}$$

where, $\dot{\varepsilon}_{ss}$ is the steady state creep rate. The relationships between the steady state creep rate and applied stress for the BM, WM, and HAZ are presented in Fig. 10. Each fitting line for the BM, WM, and HAZ obeys a good linearity. The WM and HAZ are faster in the creep rate than the BM, and the WM and HAZ are very close. At the high stress levels, the three regions approach to be same, but at the low stresses, their different gaps are larger. Accordingly, from the results of the tensile and creep tests, the constants of *D*, *m*, *A*, and *n*, which are needed for calculating *C**-fracture parameter, were determined on the basis experimental data for the BM, WM, and HAZ. The constants are summarized in Table 2.

Fig. 11 shows a comparison of the rupture elongation and reduction of area (ROA) tested at 550 °C for the BM, WM, and HAZ. In the rupture elongations, the WM and HAZ are lower for about 10% than the BM, and they are also lower in the ROA. The reduction of creep life (or stress) in the WM and HAZ is due to faster creep rate and lower rupture ductility than the BM.

### 3.2. Analysis of creep rupture data

To accurately estimate a long-term creep life using short-term creep data, most researchers have used well Monkman-Grant (M-G) empirical equation. A generalized form of the M-G relation (MGR) is given as [13,14].

$$\log(t_r) + k\log(\dot{\varepsilon}_{ss}) = C \tag{2}$$

where $k$ is the slope in a log-log plot of $t_r$ vs. $\dot{\varepsilon}_{ss}$ and constant $C$ is the intercept. The validity of the MGR can be identified from the slope in the log-log plot equal to unity, and it is given as:

$$\log(t_r) + \log(\dot{\varepsilon}_{ss}) = C_{MG} \tag{3}$$

If $k = 1$, it becomes $C = C_{MG}$. The $C_{MG}$ means a real constant independent of creep testing conditions. If $k \neq 1$, it indicates that the product of the creep rate and rupture time varies with creep test conditions. Here, using the creep rupture data for the BM, WM, and HAZ, the validity of the MGR is examined, and the constant values of $k$ and $C$, which are being used to predict creep life, are determined.

Fig. 12 shows a M-G plot of the rupture time and steady state creep rate for the BM, WM, and HAZ. By the best fitting the total data of the BM, WM, and HAZ, a linear equation of $y = 0.298\ x^{-0.795}$

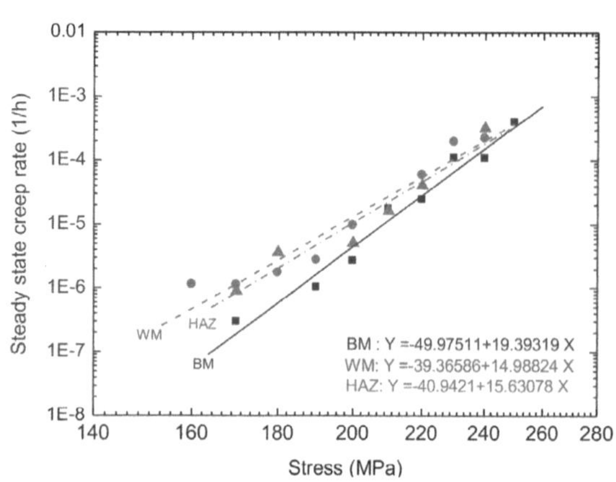


**Fig.10.** Comparison of the steady state creep rates for the BM, WM, and HAZ at 550 °C.

**Table 2**
Material constants obtained for the BM, WM, and HAZ at 550 °C.

| Materials | $\sigma_{uts}$ (MPa) | $\sigma_y$ (MPa) | Tensile constants ($\varepsilon_p = D(\sigma/\sigma_y)^m$) | | Creep constants ($\dot{\varepsilon}_{ss} = A\sigma^n$) | |
|---|---|---|---|---|---|---|
| | | | D | m | A (MPa$^{-n}$h$^{-1}$) | n |
| BM | 397 | 338 | 1.23E-03 | 15.45 | 1.05E-50 | 19.39 |
| WM | 431 | 376 | 1.86E-03 | 13.55 | 4.31E-40 | 14.98 |
| HAZ | 372 | 321 | 1.61E-03 | 14.09 | 1.14E-41 | 15.63 |

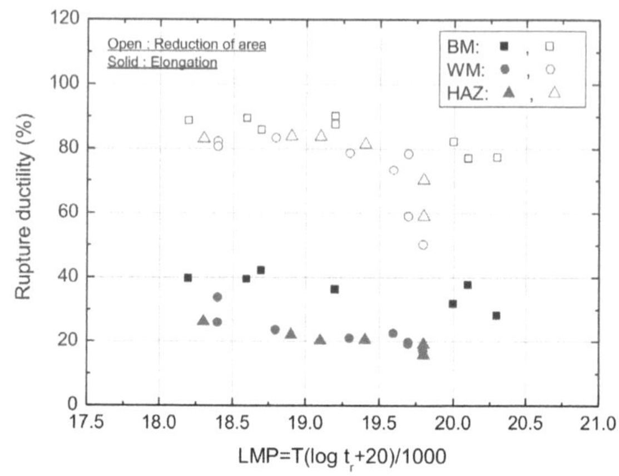


**Fig. 11.** Comparison of creep rupture ductility for the BM, WM, and HAZ at 550 °C.

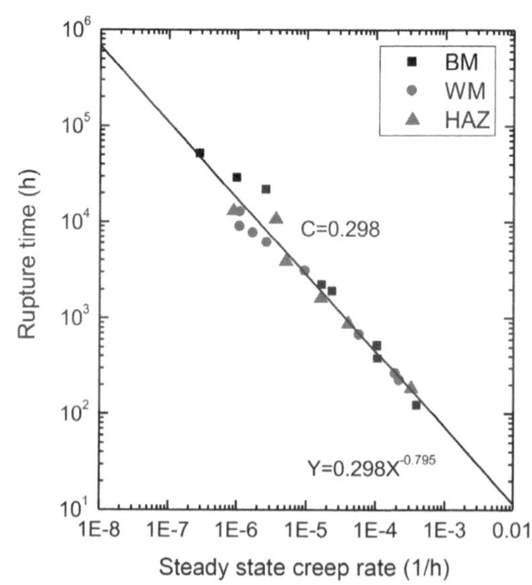


**Fig. 12.** Monkman-Grant plot showing a relationship between the rupture time and steady state creep rate for the BM, WM, and HAZ.

was obtained. Hence, it means that y-axis is the rupture time and x-axis is the steady state creep rate. Thus, it is determined as $C = 0.298$ and $k = -0.795$ in Eq. (2). Also, when $k = 1$, it is obtained as $C_{MG} = 0.030$ in Eq. (3). Therefore, the validity in the MGR is identified to be very weak because of $C \neq C_{MG}$.

In addition, creep damage tolerance factor ($\lambda$) becomes a useful parameter to assess susceptibility of material in creep deformation, and it is often used to assume creep damage mechanism. The $\lambda$ equation is defined as [13,14].

$$\lambda = \frac{\dot{\varepsilon}_{avg}}{\dot{\varepsilon}_{ss}} = \frac{\varepsilon_f}{\dot{\varepsilon}_{ss} \cdot t_r} \tag{4}$$

where, $\varepsilon_f$ is the creep rupture elongation, and $\dot{\varepsilon}_{avg}$ is the average creep rate defined as $\varepsilon_f/t_r$. Thus, the $\lambda$ is the ratio of strain to failure to M-G relation, and it means the ratio of the average creep rate to steady state creep rate. The $\lambda$ value can be determined by the plot of log $\dot{\varepsilon}_{ss}$ vs. log $\varepsilon_f/t_r$, as shown in Fig. 13. In the present study, the $\lambda$ value is approximately optimized for the $\lambda$ = 8.50 as $\dot{\varepsilon}_{avg} = 8.50(\dot{\varepsilon}_{ss})^{1.0}$ for all data of the BM, WM, and HAZ, although the best fit equation is determined for $\dot{\varepsilon}_{avg} = 1.88(\dot{\varepsilon}_{ss})^{0.863}$ for the total data. For a higher value of $\lambda$=8.50 in the present study, it can be analyzed because of higher rupture elongation to lower secondary creep rate, as shown in Fig. 8. It is known that when $\lambda$ = 1.5 to 2.5, damage mechanism in creep corresponds to "cavitation", and when $\lambda$ = 5 or higher, it occurs from "microstructural degradation" of the coarsening of precipitates and/or softening of dislocation substructures, etc. [14,15]. Thus, creep deformation in the present study can be inferred to be governed from the microstructural degradation.

Fig. 14 shows typical microstructures ($t_r$ = 50,799 h (5.8 year)) crept under 170 MPa at 550 °C for the BM. The SEM photos displays transgranular fracture resulted from necking rather than the growth of cavity or microvoids (Fig. 14a and b). A core region displays a final-fracture surface by creep damage (Fig. 14b). In the deformed region of necking area (Fig 14c), the prior austenitic grain boundary (PAGB) was elongated with grain flow along the stress direction in near fracture zone by creep stresses. In addition, in the gauge section of a rupture specimen, many precipitates in the matrix are widely distributed in the PAGB, block and packet boundaries owing to long-term creep exposure of 5.8 year (Fig. 14d). The precipitates of $M_{23}C_6$ and MX (M – V, Nb; X = C, N) formed in these boundaries coarsened significantly with an increase in creep life (or stress) [16,17]. The fine and stable carbides and carbonitrides, which are formed from the addition of Nb and V in Gr. 91 steel, improve high-temperature creep resistance by disrupting the dislocation movement and delaying the plastic deformation by inhibiting the GB sliding and delaying the onset of tertiary creep stage [14].

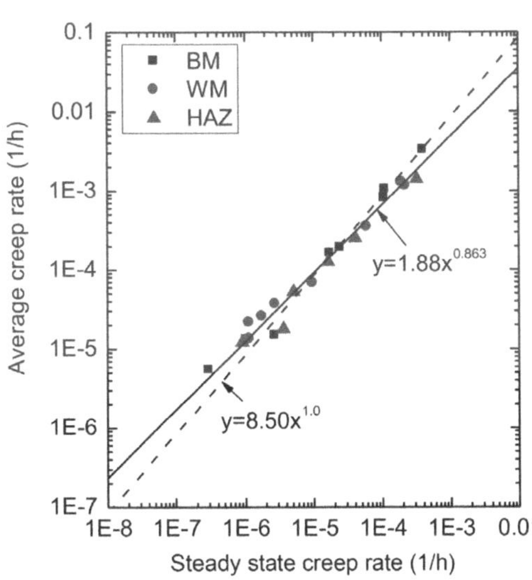


**Fig. 13.** Creep damage tolerance factor showing a relationship between the average creep rate and steady state creep rate for the BM, WM, and HAZ.

Fig. 15 displays OM structures for Type-IV failure in a single weld joint of Gr. 91 steel. The creep failure occurs in the fine-grained heat affected zone (FGHAZ) and intercritical HAZ (ICHAZ). Subzones of the HAZ part can be classified as the coarse-grained HAZ (CGHAZ), FGHAZ, ICHAZ, and over-tempered region, as shown in the figure well. The HAZ failure of Gr. 91 weldment brings about most commonly in the weak HAZ regions of the FGHAZ + ICHAZ. It is known that this HAZ failure was originated from Type-IV cracking mostly associated with the adjacent fine grain region or the intercritical zone [9].

### 3.3. CCGR laws for the BM, WM, and HAZ

The C*-fracture parameter was widely applied to evaluate the CCGR in metals under a steady state creep [3,18]. The steady state creep rate is given by Norton's power law. A general equation of the creep crack growth rate (*da/dt*) is defined as

$$\frac{da}{dt} = B(C^*)^q \tag{5}$$

where the *B* and *q* are the material constants, which are determined from a regression line of the CCGR data. The constants are the intercept and slope in the plot of *da/dt* vs. C*. To calculate the C* values, the material constants given in Table 2 are used. The plots of *da/dt* and C* are obtained for all samples of the BM, WM, and HAZ.

For the CT specimen, the C* value and load-line displacement rate ($\dot{V}_c$) due to creep strain are obtained by calculating following Eqs. (6) and (7) [11]:

$$C^* = \frac{P\dot{V}_c}{B_N(W-a)}\left(\frac{n}{n+1}\right)\cdot(2.2) \tag{6}$$
$$(0.4 > a/W > 0.7)$$

$$\dot{V}_c = \dot{V} - \frac{\dot{a}B_N}{P}\left(\frac{2K^2}{E'} + (m+1)J_p\right) \tag{7}$$

where $J_p$ is the fully-plastic contributions to J-integral, *P* is the applied load, *a* is the crack size, *W* is the width of the specimen, $\dot{V}$ is the total load-line displacement rate, $B_N$ is the net thickness of specimen, *E'* is the elastic modulus for plane strain, *K* is the stress intensity factor, $\dot{a}$ is the crack growth rate, and *m* is the stress exponent in the Ramberg-Osgood stress vs. strain relationship. In addition, the value of $J_p$ contribution for CT specimen can be obtained as

$$J_p = \frac{D_1 h_1\left(a/W,\ m\right)}{\sigma_y (W-a)^m}\left(\frac{P}{1.455B_N}\right)^{m+1} \tag{8}$$

where $h_1$ is the function of *a/W* and *m*, and it value is obtained using the table result of ASTM E1457.

Fig. 16 illustrates the plot of *da/dt* vs. C* determined at 550 °C for the BM, WM, and HAZ. The solid lines are for the regression lines determined by the least squares fit method for each CCG data for the BM, WM, and HAZ. For a given C* value, the CCGR laws for the BM, WM, and HAZ are determined respectively, as follows:

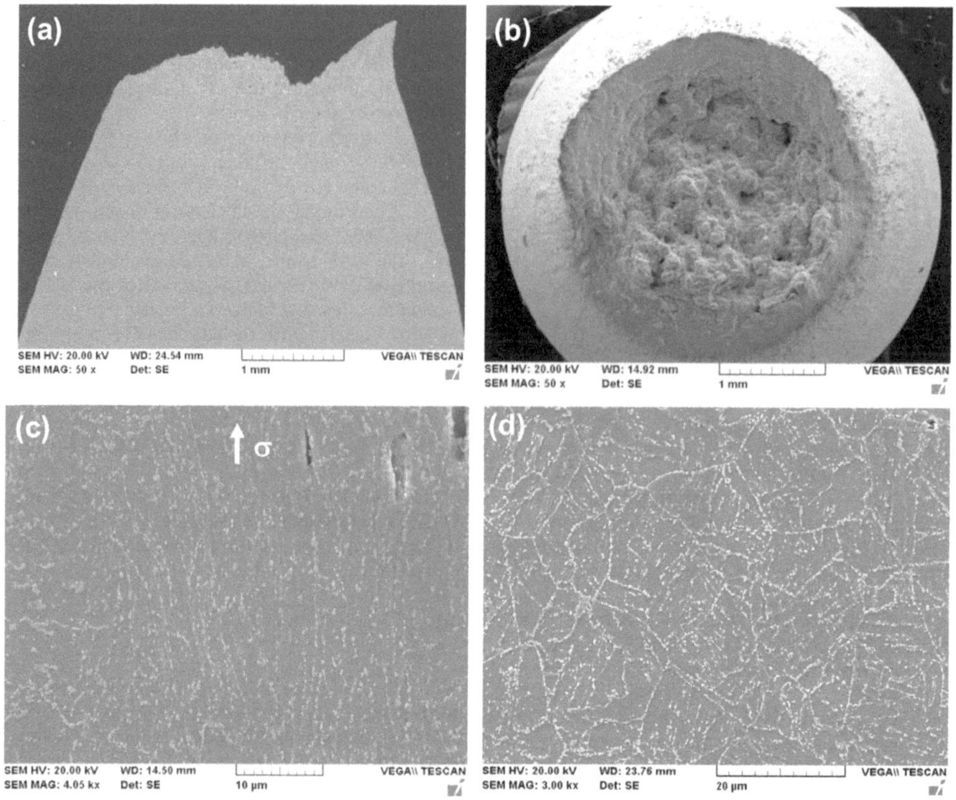


**Fig. 14.** Microstructures of a rupture specimen for the BM under 170 MPa at 550 °C ($t_r$ = 50,799 h); the photos display (a) creep failure by necking, (b) transgranular fracture, (c) deformed precipitates in near necking region, and (d) precipitates formed in the matrix.

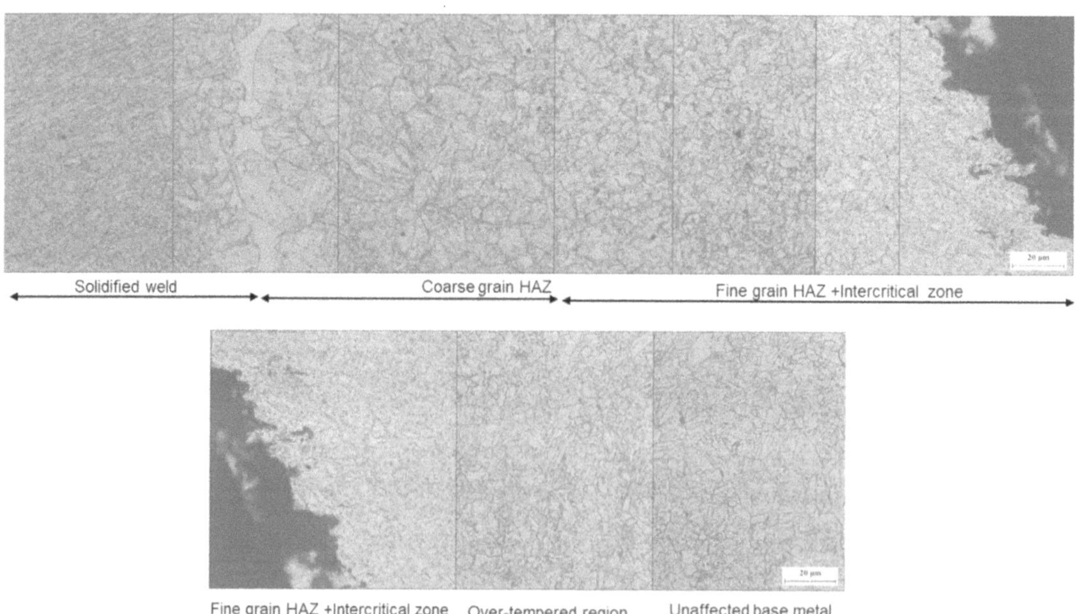


**Fig. 15.** OM structure showing typical Type-IV failure crept in the FGHAZ and intercritical zone in a single weld joint of Gr. 91 steel.

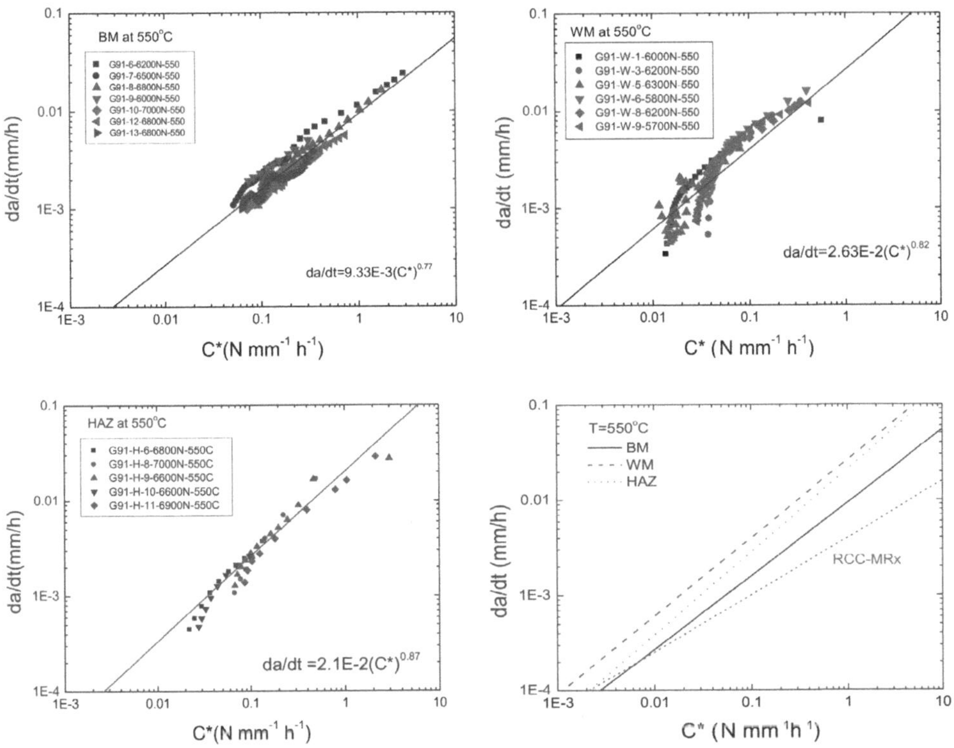


**Fig. 16.** Comparison of the *da/dt* vs. *C** obtained for the BM, WM, and HAZ at 550 °C.

$$da/dt = 9.33x10^{-3}(C^*)0.77 \quad (for\ BM) \quad (validity\ range:\ 0.05 < C^* < 3.05\ N/mm.h) \tag{9}$$

$$da/dt = 2.63x10^{-2}(C^*)0.82 \quad (for\ WM) \quad (validity\ range:\ 0.013 < C* < 0.60\ N/mm.h) \tag{10}$$

$$da/dt = 2.13x10^{-2}\ (C^*)0.87 \quad (for\ HAZ) \quad (validity\ range:\ 0.02 < C^* < 3.27\ N/mm.h) \tag{11}$$

When all curves of the BM, WM, and HAZ were overlapped together, the WM and HAZ are faster crack growth than the BM. The WM is faster for about 2.8 times than the BM. Therefore, this result is in accordance with a current research result reporting the fact that the WM and HAZ showed faster CCGRs than the BM [20]. The WM and HAZ is almost similar although the HAZ is a little lower in the CCGRs than the WM. The HAZ is assumed to be inhomogeneity as a transition region between the BM and WM among these materials affecting the state of the stress field or strain rate field near the weld fusion line and near the BM and HAZ interface. It is identified that the CCGRs in the HAZ location are conformed well to the result of the creep rate.

In addition, the CCGR law of the BM obtained in the present investigation is compared with the currently elevated temperature design (ETD) code, RCC-MRx in France [19], as shown in Fig. 16. As shown well in the figure, a comparison of the CCGR data obtained in the present tests to RCC-MRx code data exhibits a marginal difference although the CCGR value in RCC-MRx is a little low. Therefore, the CCGR results in the present tests will be reasonable in validity. However, to more clearly define the CCGR laws for Gr. 91 weldments, more CCG data should be accumulated for the design use because the CCGR behavior for its weldments will depend on different materials and experimental conditions; welding methods, heat-treatment conditions, CCG test methods, specimen sizes, and environments, etc.

Fig. 17 illustrates a comparison of the CCGR lines at 550 °C and 600 °C for the BM, WM, and HAZ. The CCGR lines plotted at 600 °C were obtained from the CCG tests in the author's previous study [1]. In the figure, the solid lines indicate the CCGRs at 550 °C, and the

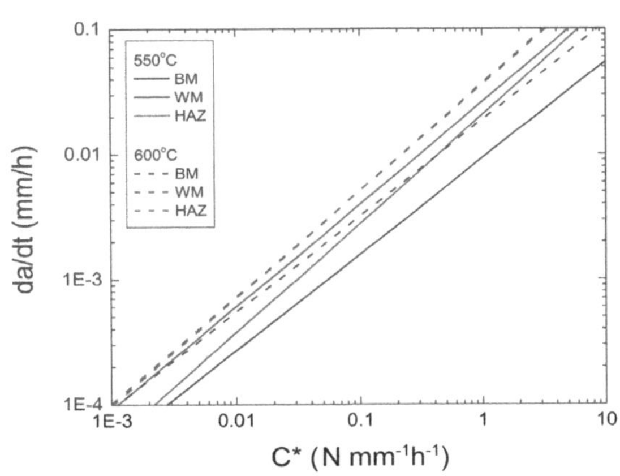


**Fig. 17.** Comparison of the CCGRs at 550 °C and 600 °C for the BM, WM, and HAZ of Gr. 91 steel.

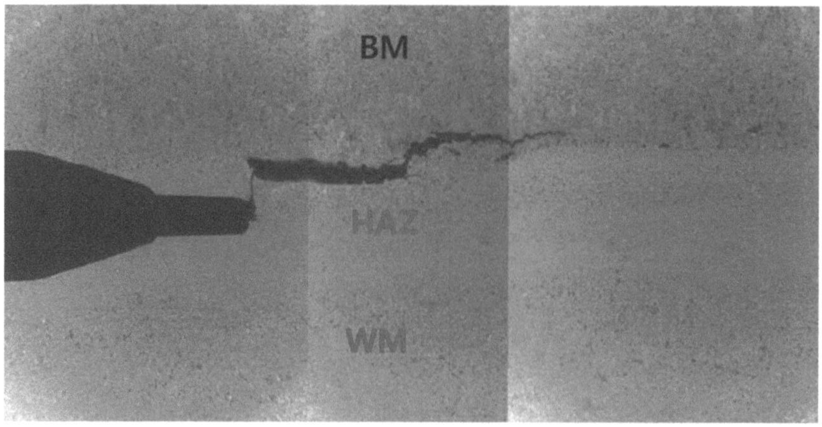


**Fig. 18.** A typical photo showing crack propagation along a weak HAZ plane between the BM and WM in the CCG tests.

dotted lines indicate the CCGRs at 600 °C. Regardless of the BM, WM, and HAZ, the CCGRs at 600 °C are higher than those at 550 °C owing to severe creep damage at a higher-temperature. The CCGRs of the WM and HAZ are almost similar, and their CCGRs are faster than those of the BM.

Fig. 18 shows an evident photo for a crack path in the HAZ plane during the CCG tests. A crack path progresses towards a weak FGHAZ adjacent to the BM. Namely, although a pre-cracking route in front of notch starts at the centre of the HAZ, the cracking route changes towards a weak FGHAZ during the CCG tests. In front of a crack path, the microvoids are formed owing to accumulated creep damage during the CCG tests. It is thought that a creep cracking is propagated by the connection and combination of these microvoids. The failure location of Type-IV cracking is most commonly reported to be parallel or offset from the fusion line in the FGHAZ adjacent to the BM [9].

## 4. Conclusions

In the creep test results, the WM and HAZ revealed a considerable reduction in the rupture life compared with the BM. A significant reduction of the creep life in the WM and HAZ was attributed to faster creep rate and lower rupture ductility than the BM, and the WM and HAZ were very close in creep behavior. The creep rates of the WM and HAZ to BM were gradually approached to be same at the high stresses, but at the low stresses, its different gap was larger. The material constants to estimate a long-term creep life in the MGR were optimized for $C = 0.298$ and $k = -0.795$. A value of creep damage tolerance factor ($\lambda$) conformed approximately to $\lambda = 8.50$ for all data of the BM, WM, and HAZ. In the CCG test results, the WM and HAZ were found to be faster crack growth than BM, and they were similar in the CCG behavior. The CCGR behavior for the three regions was conformed well to the results of the creep rates. In addition, the CCGRs obtained in the present study exhibited a marginal difference when compared with that of RCC-MRx code. A crack path in the HAZ plane progressed towards a weak FGHAZ adjacent to the BM.

## Declaration of competing interest

The authors declare that they have no known competing financial interests or personal relationships that could have appeared to influence the work reported in this paper.

## Acknowledgement

This work was supported by the National Research Foundation (NRF) grant funded by the Korea government (2020M2D4A 2068407).

## Appendix A. Supplementary data

Supplementary data to this article can be found online at https://doi.org/10.1016/j.net.2020.07.015.

## References

[1] W.G. Kim, J.Y. Park, H.Y. Lee, S.D. Hong, Y.W. Kim, S.J. Kim, Time-dependent crack growth behavior for a SMAW weldment of Gr. 91 steel, Int. J. Pres. Ves. Pip. 110 (2013) 66–71.
[2] B.K. Choudhary, E. Issac Samual, Creep behavior of modified 9Cr-1Mo ferritic steel, J. Nucl. Mater. 412 (2011) 82–89.
[3] K.M. Nikbin, D.J. Smith, G.A. Webster, An engineering approach to the prediction of creep crack growth, Trans. ASME. 108 (1986) 186–191.
[4] W.G. Kim, J.Y. Park, S.D. Hong, S.J. Kim, Probabilistic assessment of creep crack growth rate for Gr. 91 steel, Nucl. Eng. Des. 241 (2011) 3580–3586.
[5] Y. Takahasi, T. Igari, F. Kawashima, S. Date, N. Isobe, T. Itoh, Y. Noguchi, K. Kobayashi, High-temperature crack growth behavior of high-chromium steels, SMiRT 18 (7) (2005) 1904–1915. SMiRT18-G11.
[6] E.I. Samuel, B.K. Choudhary, W.G. Kim, Reliability Analysis on Long-Term Creep Life Prediction of Grade 91 Steel Weld Joints, 7th International Conference on Creep, Fatigue and Creep-Fatigue Interaction (CF-7), Tamil Nadu, India, 2016, pp. 923–928.
[7] Y. Wang, L. Li, R. Kannan, Transition from Type IV to Type I cracking in heat-treated grade 91 steel weldments, Mater. Sci. Eng. 714 (2018) 1–13.
[8] M. Yamamoto, N. Miura, T. Ogata, Applicability of C* parameter in assessing Type-IV creep cracking in mod. 9Cr-1Mo steel welded joint, Eng. Fract. Mech. 77 (2010) 3022–3034.
[9] D.J. Abson, J.S. Rothwell, Review of Type IV cracking of weldments in 9-12%Cr creep strength enhanced ferritic steels, Int. Mater. Rev. 58 (8) (2013) 437–473.
[10] ASTM E139-17, Standard Test Method for Conducting Creep, Creep-Rupture and Stress Rupture Tests of Metallic Materials, ASTM standard, 2017.
[11] ASTM E1457-17, Standard Test Method for Measurement of Creep Crack Growth Rates in Metals, ASTM standard, 2017.
[12] B. Dogan, B. Petrovski, Creep crack growth of high temperature weldments, Int. J. Pres. Ves. Pip. 78 (2001) 795–805.
[13] W.G. Kim, J.Y. Park, I.M.W. Ekaputra, S.J. Kim, M.H. Kim, Y.W. Kim, Creep deformation and rupture behavior of Alloy 617, Eng. Fail. Anal. 58 (2015) 441–451.
[14] K. Guguloth, N. Roy, Creep deformation behavior of 9CrMoVNb (ASME Grade 91) steel, Mater. Sci. Eng. 680 (2017) 386–404.
[15] B.K. Choudhary, Tertiary creep behaviour of 9Cr-1Mo ferritic steel, Mater. Sci. Eng. 585 (2013) 1–9.
[16] W.G. Kim, J.Y. Park, I.M.W. Ekaputra, S.J. Kim, J. Jang, Influence of hold time and stress ratio on cyclic creep properties under controlled tension loading cycles of Grade 91 steel, Nucl. Eng. Tech. 49 (2017) 581–591.
[17] G. Qin, S.V. Hainsworth, A. Strang, P.F. Morris, P.D. Clarke, A.P. Backhouse, TEM studies of microstructural evolution in creep exposed E911, in: I.A. Shibli,

S.R. Holdsworth (Eds.), Proceedings of the 2nd ECCC Creep Conf. On Creep & Fracture in High Temperature Components, Zurich ETD, 2009, pp. 889–899.
[18] A. Saxena, Nonlinear Fracture Mechanics for Engineers, CRC Press, New York, USA, 1997.
[19] Afcen RCC-MRx, Code Section III – Tome 6, Probationary phase rules, France, 2015.
[20] M. Swei, A.S. Sedmak, B. Petrovski, Z.Z. Golubovi, S.A. Sedmak, M. Katini, K.I. Azzabi, Creep crack growth behavior of P91 steel weldments, Therm. Sci. 23 (2B) (2019) 1203–1209.

KSME

Springer

Journal of Mechanical Science and Technology 34 (7) 2020

Original Article
DOI 10.1007/s12206-020-0614-5

Keywords:
· Alloy 617
· Creep
· Creep rupture
· Tension
· Thermal ageing

Correspondence to:
Woo-Gon Kim
wgkim@kaeri.re.kr

Citation:
Kim, W.-G., Sah, I., Kim, E.-S., Kim, S.-J., Kim, M.-H. (2020). Tension and creep rupture behaviors of Alloy 617 thermally aged for a year at 900 °C. Journal of Mechanical Science and Technology 34 (7) (2020) 2813~2819.
http://doi.org/10.1007/s12206-020-0614-5

Received March 3rd, 2020
Revised March 31st, 2020
Accepted April 20th, 2020

† Recommended by Editor Chongdu Cho



# Tension and creep rupture behaviors of Alloy 617 thermally aged for a year at 900 °C

**Woo-Gon Kim[1], Injin Sah[1], Eung-Seon Kim[1], Seon-Jin Kim[2] and Min-Hwan Kim[1]**

[1]Korea Atomic Energy Research Institute, 989-111 Daedeok-daero, Yuseong-gu, Daejeon 34057, Korea,
[2]Dept. of Mechanical Design Engineering, Pukyong National University, Busan 48513, Korea

**Abstract** Tension and creep rupture behaviors of Alloy 617 after a year of thermal ageing at 900 °C were investigated in comparison with those of unaged (virgin) material. Tension tests were carried out under a controlled strain rate of 5.85 x $10^{-4}$ (1/s) from room temperature to 950 °C, and creep tests were conducted with different stresses at 900 °C. Tensile and creep rupture properties were obtained, then creep microstructures and the oxidation layer were measured and analyzed. The mechanical properties of Alloy 617 aged for a year showed effects of thermal damage. The mechanical properties were reduced in terms of tensile and creep strengths compared with the unaged. However, the tensile and creep rupture elongation increased. The reduction of creep life (or strength) in the aged material was caused by the rapid creep strain rate, which resulted from material softening due to thermal aging. In the micro-hardness values, the aged material decreased by approximately 26 % compared with the unaged one. The aged material formed a $Cr_2O_3$ layer of approximately 20 μm thickness on its outer surface. The outer and internal oxide-layer thickness revealed a typical trend of a parabolic increase.

## 1. Introduction

A very high temperature reactor (VHTR) system is a gas-cooled reactor designed to produce hydrogen at a temperature of up to 900-1000 °C, generate pressure of up to 7 MPa, and complete a design life of 60 years. Nickel-based Alloy 617 is verified as one of the candidate materials for the components in Generation-IV reactor systems because of its excellent mechanical properties and corrosion resistance within a temperature range of 760-1000 °C [1-8].

Long-term and high-temperature exposures during service cause metallic materials to inevitably suffer from thermal-aging processes, which result in a microstructure evolution and changes in mechanical properties. A mechanistic understanding on thermal aging due to long-term exposure at elevated temperatures is highly important [9]. However, the design guidelines of mechanical properties on long-term aging have not been provided in the current elevated temperature design (ETD) codes ASME, RCC-MRx, or elsewhere. In these circumstances, next generation nuclear plant (NGNP) research and development plans include material works to develop the ASME code for Alloy 617 in high temperature reactors. A wide variety of material data for Alloy 617 for design application are gathered and uploaded to the Generation-IV Forum (GIF) materials handbook database (DB) website organized by GIF constitutive countries for material works. Therefore, a design guideline on thermal aging effects of Alloy 617 is necessary, and relevant experimental data should be sufficiently produced for design use. In addition, the change in oxidation microstructure on thermal aging requires better understanding.

In the present study, the tension and creep rupture properties of Alloy 617 after a year of thermal ageing at 900 °C were investigated in comparison with the unaged material. The microhardness and oxidation layers for the aged materials were measured and analyzed. In addition, the evolution of crept microstructures was observed and analyzed.

## 2. Experimental procedures

Nickel-based superalloy, Alloy 617 (brand name: Haynes 617 of commercial grade) of a hot-rolled plate with a thickness of 25.9 mm is used for this study. Chemical compositions are given as (wt.,%), Al: 1.06, B: < 0.002, C: 0.08, Co: 12.3, Cr: 22.2, Cu: 0.0268, Fe: 0.9496, Mn: 0.0295, Mo: 9.5, Ni: 53.11, P: 0.003, S: < 0.002, Si: 0.0841, and Ti: 0.41. The blocks of thermally aged materials for tensile and creep tests were prepared with rectangular shape of 26 mm H, 42 mm W, and 90 mm L. Each block is constantly maintained for a year (8760 h) at 900 °C in an air environment within the box furnace.

After a year of thermal aging, the material blocks were taken out from the box furnace, and the tensile and creep materials were obtained using electric discharge machine processing. The tensile materials were plate type with 2.0 mm thickness, 25 mm gauge length, and 6.25 mm width within gage length while the creep materials were round type with 30 mm gauge length and 6 mm diameter. Tensile tests were carried out using the same strain rate of $5.85 \times 10^{-4}$ (1/s) at room temperature, 400, 600, 700, 800, 850, 900, and 950 °C. Creep tests were performed under different stress levels at 900 °C. The creep strain-time data were obtained automatically using a personal computer through a precision extensometer fit to the creep materials. Creep curves were obtained, and the minimum creep rate was determined by fitting the secondary creep stage in the creep strain-time curves. In addition, the materials for measuring micro-hardness and oxidation were coin type with 15 mm diameter and 3 mm thickness. After the creep tests of aged materials, all rupture materials were observed using SEM fracture micrographs. The precipitates formed in the creep materials were investigated using EDS analysis.

## 3. Results and discussion

### 3.1 Tension and creep rupture properties

Hot-tensile properties for the aged Alloy 617 were obtained and compared with the unaged material. The aged material showed slight declines in the ultimate tensile strength (UTS) and yield strength (YS) values for approximately 6-16 % in YS and 8-18 % in UTS at all temperature ranges, as shown in Fig. 1. However, in its tensile elongation, the trend increases reversely, as shown in Fig. 2. Notably, the elongation of the aged material is reversely reduced at the temperature range of 800-950 °C, and the possible reason is that the aged material became brittle owing to thermal damage at high temperatures.

Fig. 3 shows the change in micro-hardness values measures for the unaged and aged materials. The aged material declined by approximately 26 % with Hv = 74 from Hv = 214 in the aged material from Hv = 288 in the unaged one. Apparently, the softening of the aged material is caused by thermal-ageing damage.

Fig. 4 shows microstructural changes before and after the thermal ageing of Alloy 617. No difference in grain sizes are observed between the aged and unaged materials, but the aged material shows precipitates of fine carbides in grain boundaries due to the thermal ageing effect. The carbides are clearly shown in a high magnification of Fig. 4(c). The unaged material reveals carbide stringers parallel to the rolling direction. However, in the aged material, the carbide stringers are grown by combination during ageing. Thus, the differences of tensile

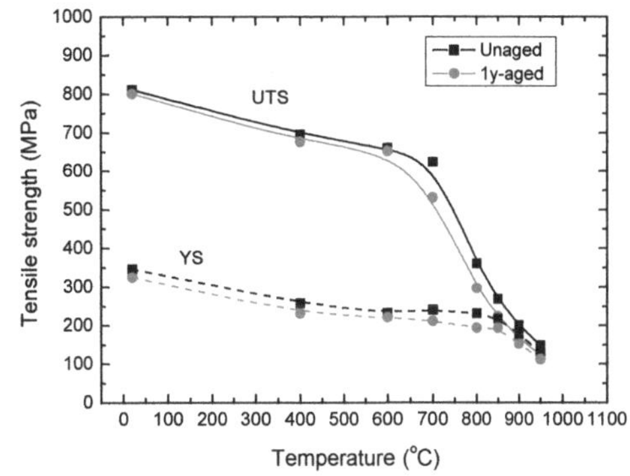


Fig. 1. Comparison of tensile strengths for the aged and unaged materials with temperatures.

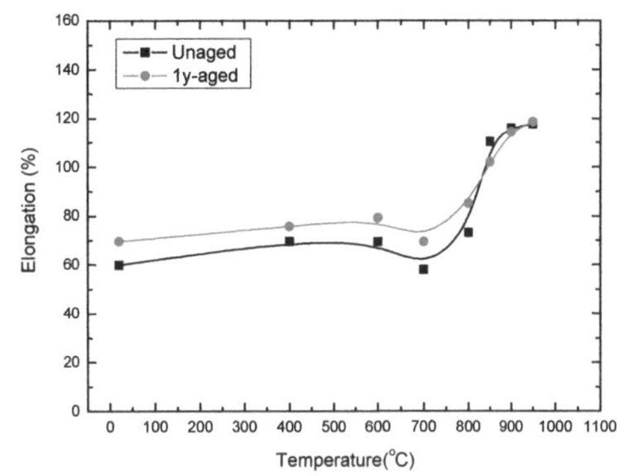


Fig. 2. Comparison of tensile elongation for the aged and unaged materials with temperatures.

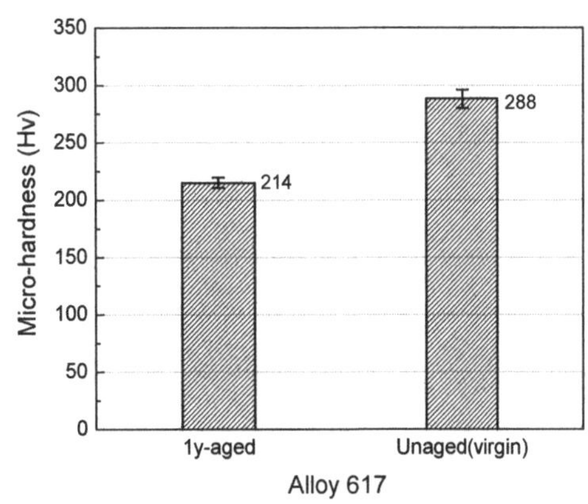


Fig. 3. Comparison of micro-hardness values measured for the aged and unaged materials.

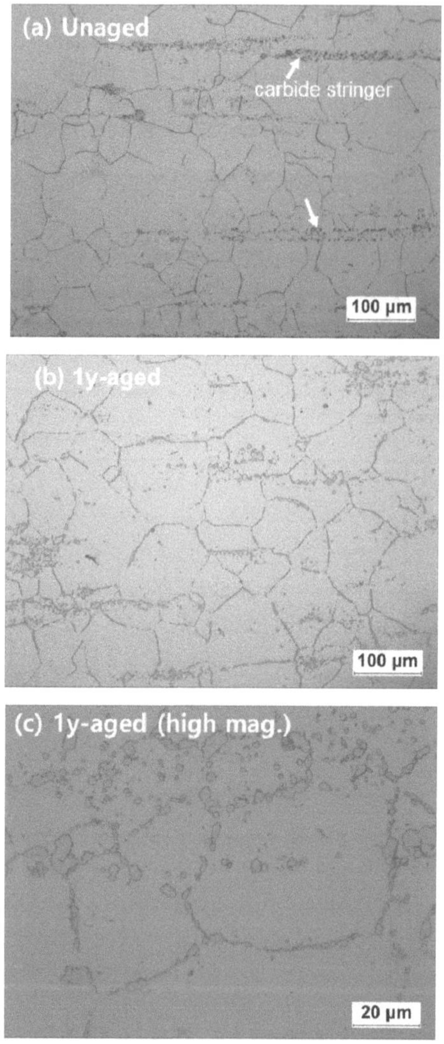


Fig. 4. Comparison of microstructures of before and after the thermal ageing of Alloy 617: (a) Unaged; (b) 1y-aged; (c) high magnification of 1y-aged material.

properties between the unaged and aged materials are caused by the slight microstructural changes.

In addition, creep rupture properties, such as the rupture time, minimum creep rate, rupture elongation, and reduction of area for the thermally aged Alloy 617 are obtained using a series of creep tests. Long-term creep rupture data reaching 12731 h (1.45 y) under 25 MPa and 7363 h (0.84 y) under 22 MPa are achieved for the unaged and aged materials, respectively.

Fig. 5 shows a comparison plot of the log stress vs. Larson-Miller parameter (LMP) obtained using the creep tests for the unaged and aged materials at 900 °C. Hence, it was used as $C$ = 20 in the equation of LMP = $T$ (log $t_r$ + $C$). The aged material shows approximately 5 % decline in creep stress compared with the unaged one. The aged material softened with the re-

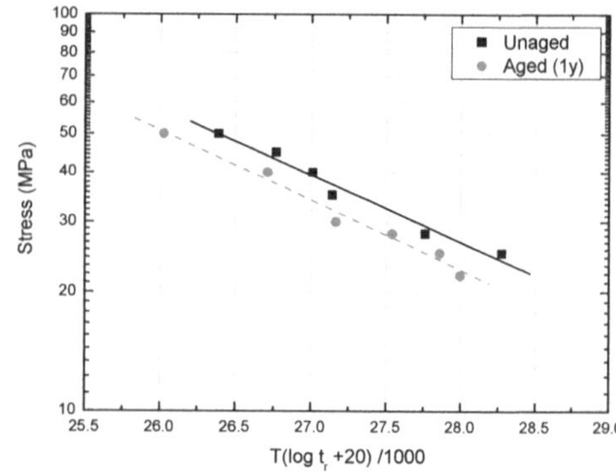


Fig. 5. Comparison of log stress vs. LMP obtained for the unaged and aged material.

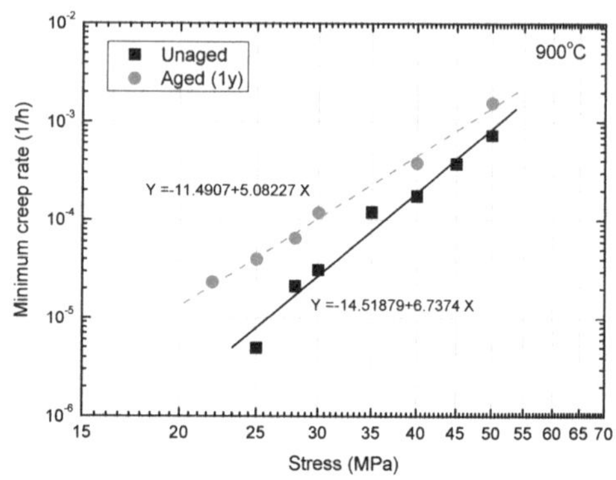


Fig. 6. Comparison of creep rate vs. stress obtained for the unaged and aged materials.

duction in micro-hardness value (Hv) and tensile strength because of the aging effect. Under this test, the creep deformation is assumed to correspond to the power-law creep region because creep deformation is mainly controlled by the dislocation climb [1, 8]. Thus, the Norton's power-law relationship of the minimum creep rate and stress obeys a linear line, as shown in Fig. 6. The equation is

$$\dot{\epsilon}_m = A\sigma^n. \tag{1}$$

From the power-law equation, the material constants can be reasonably obtained. $A$ values were obtained with approximately 3.09 x $10^{-15}$ ($MPa^{-n}$ $h^{-1}$) for the unaged material, and 3.23 x $10^{-12}$ ($MPa^{-n}$ $h^{-1}$) for the aged material. The $n$ values of creep exponent were 6.73 and 5.08 for the unaged and aged materials. A slight difference is observed for the two materials in the $A$ and $n$ values. The differences in the creep strain rates between the two materials gradually increase in the lower stress ranges. In addition, a Monkman-Grant (M-G) plot of the minimum creep rate and rupture time reveals a linear relation, as shown in Fig. 7. The M-G equation is

$$\log t_r + m \log \dot{\epsilon}_m = C_{MG}. \tag{2}$$

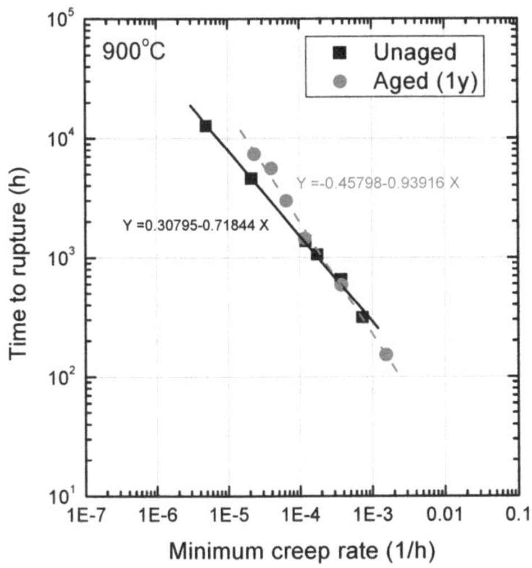


Fig. 7. Comparison of Monkman-Grant plot obtained for the unaged and aged materials.

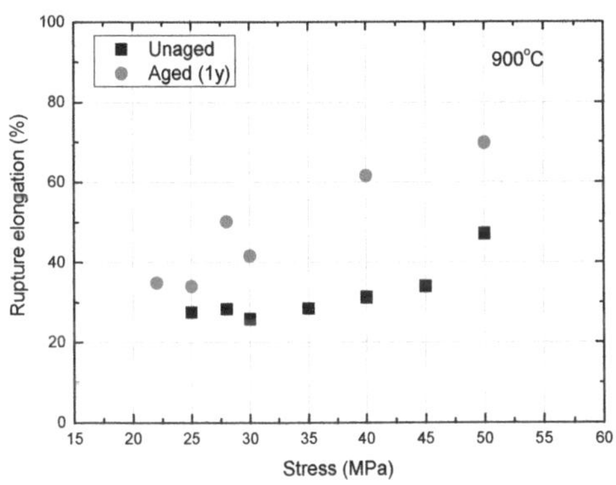


Fig. 8. Comparison of rupture elongation vs. stress for the unaged and aged materials.

The M-G plot shows the similarity of the two materials. Given the 0.71 and 0.93 m values for the unaged and aged materials, respectively, the slopes have a marginal difference that gradually increases in the lower creep stress ranges.

Fig. 8 compares the creep rupture elongation vs. stress tested for the unaged and aged materials at 900 °C. The aged material is higher in rupture elongation than the unaged one, and the rupture elongation of the two materials increases as stress increases. However, in the low stress region of 25 MPa, the difference between the two materials is small. The reason for this is that in the lower stress range of a longer time, the creep rupture mechanism of Alloy 617 may occur due to cavity formation from intergranular fractures rather than failure by necking that results in ductile fracture [8].

Fig. 9 shows a comparison of the creep rupture elongation vs. rupture time obtained for the unaged and aged materials at 900 °C. The aged material is higher in terms rupture elongation

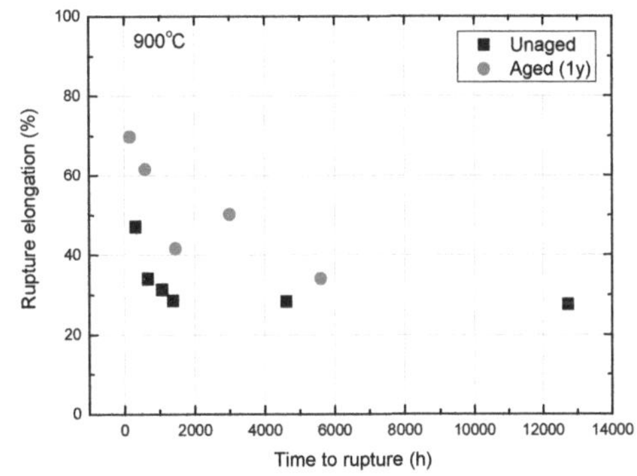


Fig. 9. Comparison of rupture elongation vs. rupture time for the unaged and aged materials.

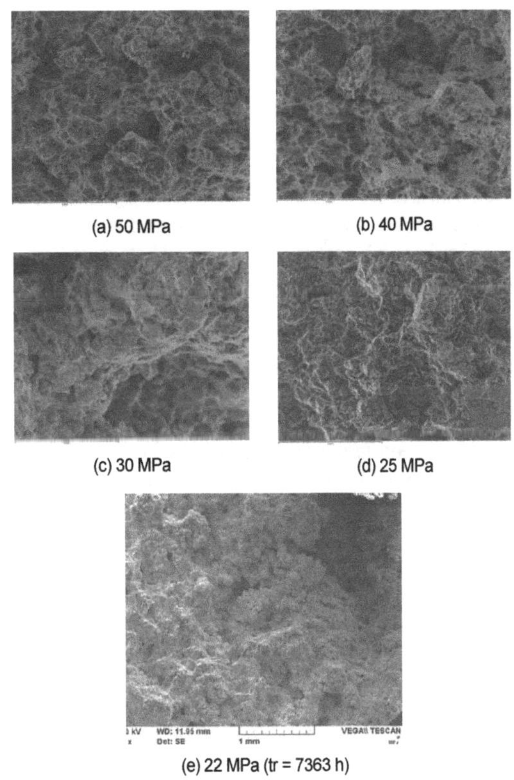


Fig. 10. SEM fracture photos with creep stresses of aged Alloy 617 materials crept at 900 °C.

than the unaged one. However, rupture elongation slightly decreases with an increase of rupture time regardless of material age. As described above, the aged material softened because of the thermal damage at high temperatures.

## 3.2 Creep fracture microstructures

Fig. 10 shows typical SEM fracture surfaces with applied

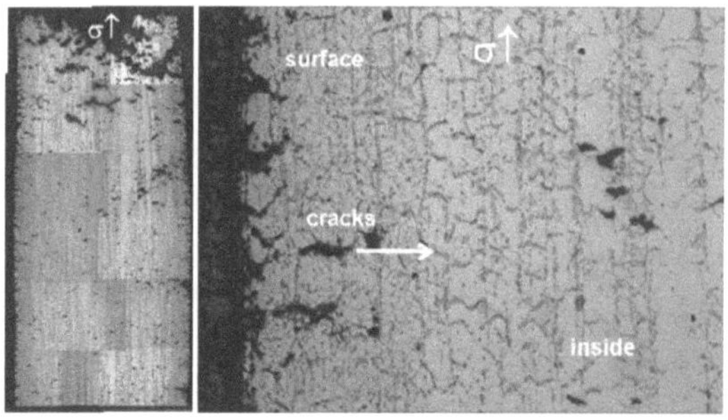


(a) Cross-section view (b) Enlarged view of left

Fig. 11. Typical OM fracture photo crept under 25 MPa Alloy 617 aged for a year at 900 °C.

stresses for materials aged for a year at 900 °C. The fracture surface exhibits a combination of dimple rupture in the high stress ranges of 50, 40, and 30 MPa. In a low stress of tr = 7363 h under 22 MPa, the fracture surface reveals a dominant intergranular fracture.

Fig. 11 shows typical OM structures on creep cracks along grain boundaries (GB) for a fractured material (tr = 5600 h) under 25 MPa at 900 °C after a year of thermal aging at 900 °C. The creep cavities and cracks were generated along the GB perpendicular to the applied stress. This OM structure kept at 900 °C was identified to be similar to the unaged structures in the author's previous study investigated at 800, 850, 900, and 950 °C [8]. Typical intergranular cracking appears to be propagated by connecting of minor cavities formed on the GB for prolonged creep duration. The cavities are widely distributed over the whole gage length of fracture material, as shown in Fig. 11(a). Creep failure finally occurs by incorporating and connecting these cavities (or voids) caused from creep damage. These cavities seem to be initiated from precipitates formed along the GB. The cracks start from the surface and develop inside the matrix, as shown in Fig. 11(b).

The SEM micrographs show no evidence on changes in microstructures between the unaged and aged materials given the close similarities of the SEM fractures. However, the two materials exhibit several differences for oxidation microstructures in the crept materials. This result is described in detail for oxidation microstructures in the following section.

### 3.3 Oxidation behavior

Fig. 12 shows the precipitates formed for (a) a non-creep test material (only aged for a year state at 900 °C) and (b) a creep tested material: Crept under 25 MPa after a year of aging at 900 °C. Only the year-aged material presents a Cr-oxide layer of $Cr_2O_3$ on the outer oxide layer. The Cr-oxide layer reveals a smooth surface and dense structure. Voids of black color are observed below the internal oxide layer, as shown in Fig. 11(a). The outer-oxide layer thickness of $Cr_2O_3$ was measured to be approximately 20 μm.

The creep tested material after a year of aging exhibits

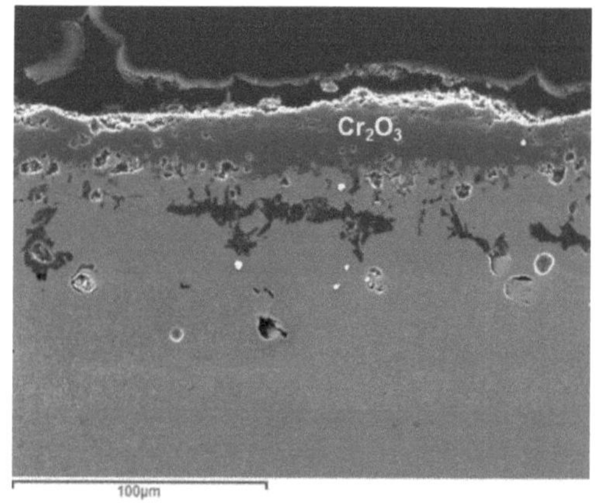


(a) Oxidation of a year-aged material at 900 °C (non-creep test)

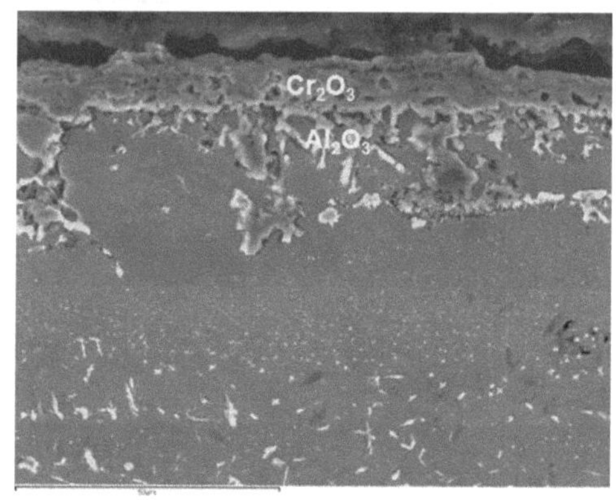


(b) Precipitates formed in a crept material under 25 MPa after a year of aging at 900 °C

Fig. 12. Precipitates formed for two materials: (a) Year-aged material of non-creep test; (b) a creep tested material: Crept under 25 MPa after a year of aging at 900 °C.

coarse and blocky carbides caused by creep stress, as shown in Fig. 12(b). Just below the outer oxide layer, the internal-oxide layer is developed between the surface and matrix, and are mainly composed of Al-oxide ($Al_2O_3$). The Al-oxide intergrowth is easy in the crept material because creep voids or cracks are generated by the applied stresses. The internal oxide acts like a chain between the outer oxide layer and matrix. Below the Al-oxide layer, a Cr-depleted zone exists between the Al-oxide layer and matrix owing to diffusion of Cr during creep at high temperature.

In the author's previous study, cylindrical coarse carbides and Cr-rich $M_{23}C_6$ mainly developed on the GB in the fractured materials [8]. In the precipitate-denuded regions, the carbides developed near the GB. According to other studies on Alloy 617, $M_{23}C_6$ carbide were abundant at temperatures of 649-1093 °C. Other carbides ($Cr_{23}C_6$), carbon nitrides [Cr Mo (C, N)], and nitride (TiN) were observed but were rare [10-12].

Fig. 13 shows the change in outer oxide-layer thickness with the creep rupture time measured for the unaged and aged

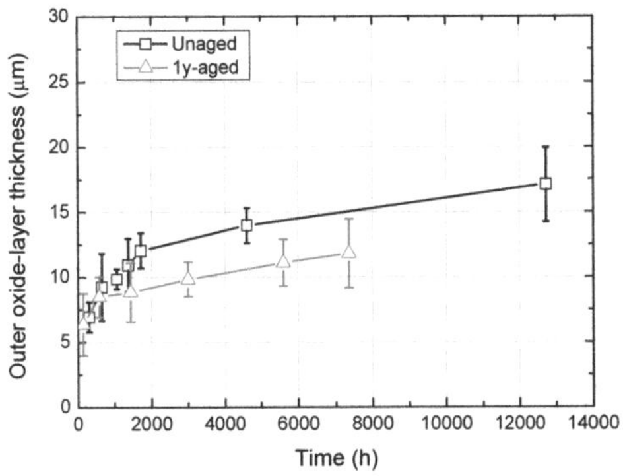


Fig. 13. Change of outer oxide-layer thickness with creep rupture time measured for the unaged and aged materials after creep tests at 900 °C.

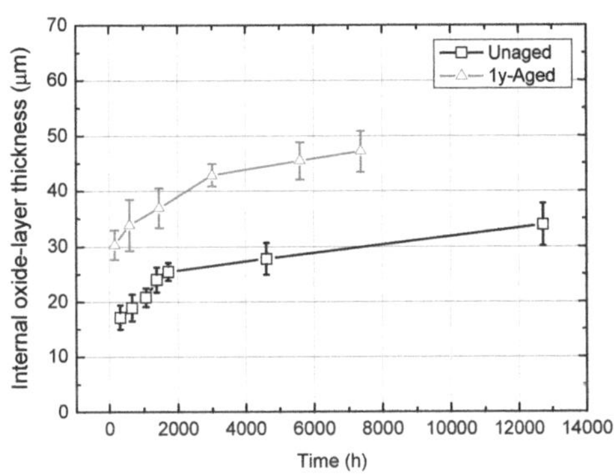


Fig. 14. Change of internal oxide-layer thickness with creep rupture time measured for the unaged and aged materials after creep tests at 900 °C.

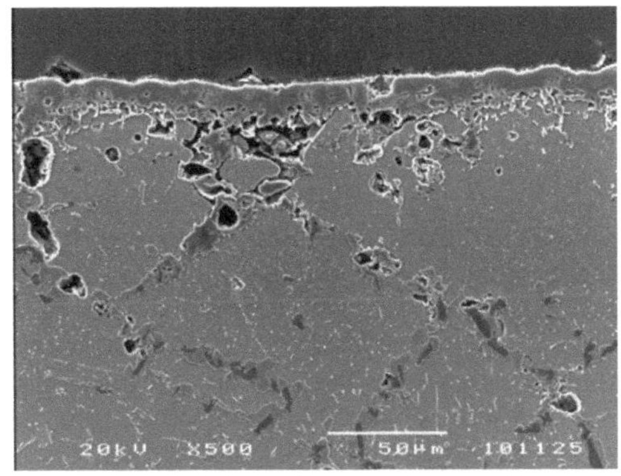

(a) Precipitates formed in the unaged material crept under 28 MPa / 900 °C (tr = 4616 h)

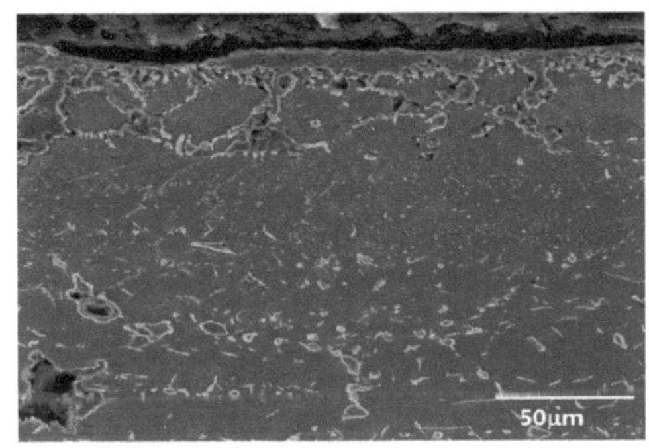


(b) Precipitates formed in the aged material crept under 25 MPa / 900 °C (tr = 5600 h)

Fig. 15. Precipitates for the unaged and aged materials after creep tests at 900 °C.

materials after creep tests at 900 °C. For each material, the average thickness was measured from over 20 points from the surface images. As the creep rupture time increases, the thickness reveals a parabolic increase. The increase in the outer-oxide layer thickness assumes a smooth slope. Under approximately 1500 h in creep rupture time, the thickness for the two materials is almost the same. At approximately over 2000 h in creep rupture time, the thickness of the unaged material is higher than that of the aged one. However, the difference is slight at 5-6 μm regardless of an increase in rupture time.

Fig. 14 shows the change in internal oxide-layer thickness with the creep rupture time measured for the unaged and aged materials after creep tests at 900 °C. The aged material reaches approximately 15 μm higher in thickness than the unaged one. In addition, the thickness increases with an almost parabolic trend with an increase in creep rupture time.

Measuring the depth of the internal oxide layer is difficult, but the depth roughly increases as creep time lengthens. Excessive internal oxide layer is detrimental to creep resistance. For the measurement results of the outer- and internal-oxide layers, the analysis can be given as follows. The decrease in the outer-oxide layer in the aged material in comparison with the unaged one is caused by the surface Cr oxides that already formed from a year of thermal aging dissipated for intergrowth of the internal Al oxides owing to stress and time in the creep tests. As a result, the outer-oxide layer of the unaged material is thicker than the aged one. By contrast, the internal oxide layer of the aged material is deeper than that of the unaged one. Figs. 15(a) and (b) show the evidence of crept microstructures of the unaged and aged materials. Moreoever, in the carbide depleted zone, the aged material is thought to be wider than the unaged one, because the former had a longer thermal exposure at a high temperature. The precipitates formed in the matrix also differ for the two materials.

The unaged material forms precipitation mainly along the GB and fine precipitation in the matrix. However, the aged material forms coarse precipitates in the matrix and in the GBs, as shown in Figs. 15(a) and (b). The outer-and internal-oxide layers thickened with the increase in time. The deterioration of creep resistance for the aged material is allegedly caused by the changes in microstructures of the surface oxide layer, internal oxide layer, and the depth of the Cr-depleted zone, which significantly affect the creep properties. They will have detrimental effects when the Cr-depleted zone is enhanced depending on various microstructural characteristics of surface cracking along the GB, such as voids below the surface and in the matrix [10].

## 4. Conclusions

The tensile and creep rupture properties of Alloy 617 thermally aged for a year at 900 °C were investigated and comparison with those of an unaged material. The aged material exhibited changed in mechanical properties owing to aging damage. For the tensile and creep strengths, the aged material showed lower than the unaged one. In the tensile elongation, creep rupture elongation occurred. For the micro-hardness values, the aged material decreased by approximately 26 % as a reduction of Hv = 74 compared with the unaged one. The reduction of creep life of the aged material is assumed to be caused by the faster creep strain rate according to the change of oxidation microstructures. The creep rupture ductility decreased with an increase in rupture time. The aged material forms an outer oxide layer of $Cr_2O_3$ with approximately 20 μm thickness. The outer and internal oxide-layer thickness reveals a parabolic increase with increasing creep rupture time for both materials. In the outer-oxide thickness, the unaged material was higher than the aged one, but the difference was slight at 5-6 μm thickness. In the internal oxide, the aged material was thicker by approximately 15 μm thickness than the unaged one.

## Acknowledgments

This work was supported by the National Research Foundation of Korea (NRF) grant funded by the Korean government (2020M2D4A2068407).

## References

[1] W. G. Kim, S. N. Yin, G. G. Lee, Y. W. Kim and S. J. Kim, Creep oxidation behavior and creep strength prediction for Alloy 617, *Int. J. Press. Vessel Pip.*, 87 (2010) 289-295.
[2] W. G. Kim, S. N. Yin, J. Y. Park, S. D. Hong and Y. W. Kim, An improved methodology for determining tensile design strengths of Alloy 617, *J. Mech. Sci. Technol.*, 26 (2) (2012) 379-387.
[3] J. H. Chang et al., A study of a nuclear hydrogen production demonstration plant, *Nucl. Eng. and Tech.*, 39 (2) (2010) 111-122.
[4] W. G. Kim, J. Y. Park, I. M. W. Ekaputra, S. D. Hong, S. J. Kim and Y. W. Kim, Comparative study on the high-temperature tensile and creep properties of Alloy 617 base and weld metals, *J. Mech. Sci. Technol.*, 27 (8) (2013) 2331-2340.
[5] W. G. Kim, J. Y. Park, G. G. Lee, S. D. Hong and Y. W. Kim, Temperature effect on the creep behavior of Alloy 617 in air and helium environments, *Nucl. Eng. Des.*, 271 (2014) 291-300.
[6] W. G. Kim, S. N. Yin, W. S. Ryu and J. H. Chang, Analysis of the creep rupture data of Alloy 617 for a high temperature gas cooled reactor, *Proceedings of the ASME 2007 Pressure Vessels and Piping Conference. Volume 9: Eighth International Conference on Creep and Fatigue at Elevated Temperatures*, Texas, USA (2007).
[7] S. Dewson and X. Li, Selection criteria for the high temperature reactor intermediate heat exchanger, *Proc. of ICAPP 05*, Seoul, Korea (2005).
[8] W. G. Kim, J. Y. Park, I. M. W. Ekaputra, S. J. Kim, M. H. Kim and Y. W. Kim, Creep deformation and rupture behavior of Alloy 617, *Eng. Fail. Anal.*, 58 (2015) 441-451.
[9] W. Ren and R. Swimdeman, A review of aging effects in Alloy 617 for Gen-IV nuclear reactor applications, *Proc. of ASME PVP 2006*, Vancouver, Canada (2006).
[10] C. Jang, D. Lee and D. Kim, Oxidation behavior of an Alloy 617 in very high temperature air and helium environments, *Int. J. Press. Vessel. Pip*, 85 (2008) 368-377.
[11] R. Cook, Creep properties of Inconel-617 in air and helium at 800 to 1000 °C, *Nucl. Technol.*, 66 (1984) 283-288.
[12] W. G. Kim, I. M. W. Ekaputra, J. Y. Park, M. H. Kim and Y. W. Kim, Investigation of creep rupture properties in air and He environments of Alloy 617 at 800°C, *Nucl. Eng. Des.*, 306 (2016) 177-185.

**Woo-Gon Kim** received his Ph.D. at Chungbuk National University at Dept. of Mechanical Engineering in 1998. He is a Principal Researcher at Korea Atomic Energy Research Institute. His specialty is mechanical assessment, analysis, modeling, and database establishment of high temperature nuclear materials for Gen-IV reactor systems, and especially, time-dependent creep and crack growth behaviors at elevated temperatures.

Engineering Failure Analysis 116 (2020) 104736

Contents lists available at ScienceDirect

Engineering Failure Analysis

journal homepage: www.elsevier.com/locate/engfailanal

# Evaluation of tension and creep rupture behaviors of long-term exposed P91 steel in a supercritical plant

Woo-Gon Kim[a,*], Hyeong-Yeon Lee[a], Hyun-Uk Hong[b]

[a] Korea Atomic Energy Research Institute (KAERI), 989-111 Daedeokdaero, Yuseong-gu, Daejeon 305-353, Republic of Korea
[b] Changwon National University, 20 Changwondaehakro, Uichang-gu, Changwon, Gyeongnam 51140, Republic of Korea

ARTICLE INFO

*Keywords:*
Creep
Creep rupture
P91 steel
Exposure
Damage tolerance factor

ABSTRACT

Tension and creep rupture behaviors of long-term exposed P91 steel, sampled from of a T-fitting part of reheat steam piping serviced for 73,716 h at 569 °C and 46.7 bars in a supercritical plant, were evaluated in comparison with those of virgin Gr. 91 steel. Tensile strength reduction factors were proposed for the exposed P91, and creep rupture behaviors were analyzed using various creep laws: Norton's power law, the Monkman-Grant relations, and a damage tolerance factor (λ). The exposed P91 revealed a significant reduction in tensile strengths, but an increase in elongation. It was found to have a lower creep strength, faster creep strain rate, and higher rupture ductility than the virgin steel. The reduction of mechanical strengths in the exposed P91 was due to the coarsening of precipitates, dissolution of useful fine MX precipitates, and recovery of dislocation substructure during the long-term thermal exposure.

## 1. Introduction

Due to its high-temperature strength, high thermal conductivity, low thermal expansion, and good corrosion resistance, modified 9Cr-1Mo steel (ASME Grade 91, hereafter "Gr. 91") is a prime candidate material for structural components such as steam generators, intermediate heat exchangers, and hot pipes in sodium-cooled fast reactors (SFRs) [1-4]. The components are designed to last for up to 60 years, and they are exposed in the plant during the long-term design life at elevated temperatures.

Long-term exposure of Gr. 91 steel at elevated temperatures leads to drops in various mechanical properties due to degradation of the microstructure as a result of the coarsening of carbide precipitates and precipitation of new phases such as Laves phases and modified Z-phases [5-9]. It was reported that Gr. 91 steel was more steeply degraded due to thermal aging than expected in design codes regarding mechanical strengths [10]. Also, compared with austenitic stainless steel, Gr. 91 steel can be more significantly affected by thermal aging when it is exposed over the long term in power plant environments.

Currently elevated temperature design (ETD) codes such as ASME Section III Division 5 (ASME Sec. III, Div. 5) in the United States and RCC-MRx in France do not take thermal aging effects into account as much as is required [11,12]. The ASME code provides some guidelines on thermal aging effects in limited cases, and in Gr. 91 steel (written as 9Cr-1Mo-V in the ASME code), reduction factor (RF) values of yield strength (YS) and tensile strength (TS) are manifested at temperatures above 480 °C. The code provides only TS values with different RFs as a function of time at a given temperature above 480 °C. In the case of RCC-MRx [12-14], the thermal aging effect is considered only for austenitic stainless steel. In addition, RF data on creep strength owing to thermal aging are not supplied elsewhere, and the effects of thermal aging are still not well understood, especially the effects on microstructural

* Corresponding author.
*E-mail address:* wgkim@kaeri.re.kr (W.-G. Kim).

https://doi.org/10.1016/j.engfailanal.2020.104736
Received 28 August 2019; Received in revised form 4 June 2020; Accepted 7 July 2020
Available online 18 July 2020
1350-6307/

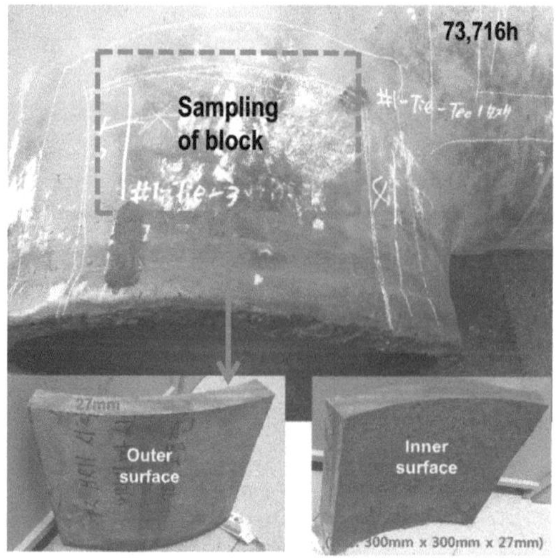


**Fig. 1.** Shape of service-exposed P91 steel sampled from T-fitting part of reheat steam piping serviced for 73,716 h at 569 °C in a SC plant.

degradation. Therefore, the loss level of creep strength after long-term thermal exposure for Gr. 91 steel should be investigated to establish a material database for design use and/or to incorporate situations of allowable stress in service.

In this study, tensile and creep rupture behaviors of service-exposed P91 steel (hereafter "exposed P91"), which was sampled from the T-fitting part of the reheat steam piping serviced for 73,716 h at 569 °C and 46.7 bar in a supercritical (SC) plant in Korea, were comparatively investigated with those of unexposed (hereafter "virgin") steel. The TS RFs of the exposed steels were obtained, and creep rupture behaviors were analyzed using various creep laws of Norton's power law, the Monkman-Grant (M-G) relations, and damage tolerance factor (λ). Microstructure changes before and after exposures were examined and discussed.

## 2. Experimental procedures

Virgin Gr. 91 steel was used for commercial grade hot-rolled Gr. 91 steel plate of 16 mm thickness. The heat treatment condition of the steel was normalized and tempered (N + T) with 1 min/mm at 1050 °C and 3 min/mm at 770 °C. The exposed P91 steel was sampled from the T-fitting part of the reheat steam piping used for 73,716 h (≈ 8.4 years) under operating conditions of high temperature (569 °C) and high pressure (46.7 bars) in the SC plant in Korea. The pipe dimensions of the T-fitting part were 819.4 mm

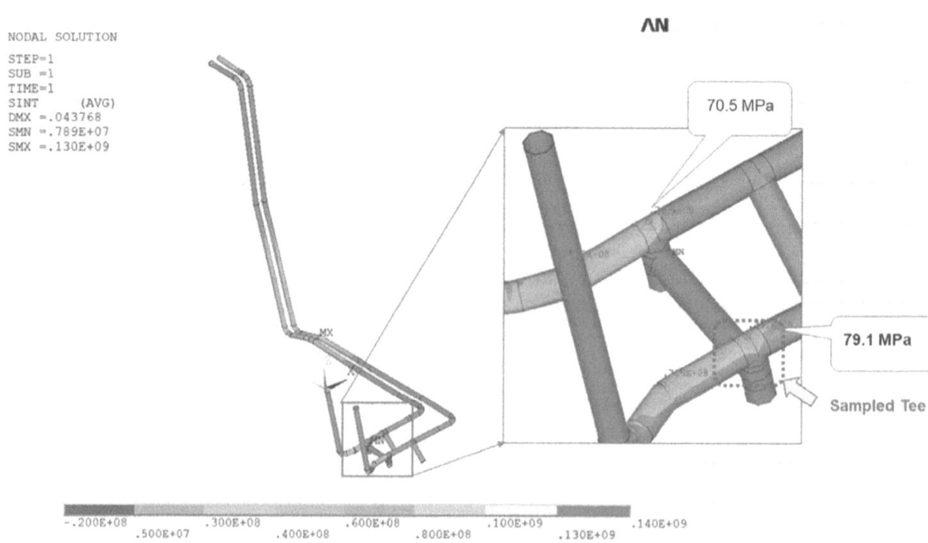


**Fig. 2.** Stress analysis results carried out by finite element analysis in the reheat steam piping system.

**Table 1**
Chemical compositions of virgin Gr. 91, exposed P91, and ASME Gr. 91 (wt.%).

| | C | Si | Mn | P | S | Ni | Cr | Mo | V | Others |
|---|---|---|---|---|---|---|---|---|---|---|
| Virgin Gr. 91 (Unexposed) | 0.115 | 0.23 | 0.415 | 0.012 | 0.0014 | 0.22 | 8.9 | 0.869 | 0.194 | Al 0.020<br>N 0.0513<br>Nb 0.073 |
| Exposed P91 (73.716 h) | 0.08 | 0.34 | 0.52 | 0.018 | 0.003 | 0.29 | 9.28 | 2.43 | 0.32 | |
| ASME II-A (Gr. 91) | 0.08<br>-<br>0.12 | 0.2<br>-<br>0.5 | 0.3<br>-<br>0.6 | 0.020 max. | 0.010 max | 0.4 max | 8.0<br>-<br>9.5 | 0.85<br>-<br>1.05 | 0.18<br>-<br>0.25 | Nb 0.03–0.07<br>Al 0.020<br>Ti 0.01<br>Zr 0.01 |

outer diameter and 29.4 mm thickness. The size of the sampled block was 300 mm width and 300 mm height, as shown in Fig. 1.

The geometry of the reheat steam piping system in the SC plant in Korea is shown in Fig. 2. Pipe stress analysis was conducted, and the stress intensity under the operating steam condition (569 °C and 46.7 bar) at the sampled location of the Tee was assumed for 79.1 MPa, as noted in the figure. The stress level is not high, but effects of thermal aging seem to degrade creep and material strength of the P91 steel.

Chemical compositions of the virgin Gr. 91, service-exposed P91, and T-fitting (forged Gr. 91: F91) given in ASME Section II Part A, are given in Table 1. Overall, the chemical compositions of virgin Gr. 91 are within the limit of ASME II-A specification [15]. The chemical compositions of the exposed P91 were analyzed by energy dispersion X-ray spectroscopy (EDS). Although the exposed P91 saw slight increases in Mn, Cr, Mo, and V elements compared with virgin Gr. 91 during the service duration of 73,716 h, the analysis results seem to show only marginal differences.

High-temperature tension tests were performed under a constant strain rate of $6.67 \times 10^{-4}$ (1/s) at intervals of 50 °C from room temperature (RT) to 650 °C. Tensile test specimens were cylindrical form with 45 mm gauge length and 9 mm diameter, which match ASTM standard specimen size. Creep tests were performed using constant-load creep testers under different applied loads at an identical temperature of 600 °C. Creep test specimens were cylindrical form with 30 mm gauge length and 6 mm diameter. In the creep tests, real strain-time data were collected on a PC by using a high-precision extensometer attached to the specimen. The steady state creep rate was determined from the slope value at the secondary creep stage. In addition, microstructure evolution of the exposed P91 was examined using an optical microscope (OM), scanning electron microscope (SEM), and transmission electron microscope (TEM).

## 3. Results and discussion

### 3.1. Tensile reduction factors

To investigate the reduction factor (RF) of the exposed P91 steel, tensile tests were performed for virgin Gr. 91 and exposed P91 at temperatures from RT to 650 °C. The RF values of the yield and tensile strengths (YS and TS) were obtained as a function of temperature. In addition, to clarify the change of hardness before and after exposures, micro-hardness measurements were conducted at RT. Table 2 provides a summary of the micro-hardness measurements for the two steels. The exposed P91 shows a hardness reduction of about 15%, with a value of Hv = 194 for exposed steel, down from the initial value of Hv = 229 for the virgin specimen. This means that the exposed P91 was subjected to softening of material during the 73,716 h of accumulated thermal exposure. In Section 3.3, this softening will be described in detail from examination of microstructural degradation of exposed P91.

Fig. 3 shows a comparison of typical tensile stress-strain curves for the virgin and exposed steels tested at 600 °C. The exposed P91 saw decreases in strengths of YS and TS, but a reverse trend was evident for the elongation, which increased. From the results of the tensile curves, the RF values for YS and TS were obtained at each temperature. Fig. 4 shows comparisons of (a) YS and (b) TS, and (c) elongation tested for the virgin and exposed P91. Also, experimental test results for the two steels were compared with the values of ETD design codes. In the changes of the YS values, virgin steel had reductions of 27% at RT and 32% at 550 °C after exposure for 73,716 h at operation condition of 569 °C and 46.7 bars. It is also clearly shown that the exposed steel dropped partly below the minimum strengths in RCC-MRx and ASME Sec. III, Div. 5 [11,13,16]. Therefore, the YS values obtained for the exposed P91 steel will drop far below the code strengths when the plant operation time accumulates more than 60 years of the target design lifetime of Gen-IV reactor systems [17].

In addition, in terms of changes of TS values, virgin steel saw reductions of about 12% at RT and 26% at 550 °C after exposure.

**Table 2**
Micro-hardness reduction of exposed P91.

| | Virgin Gr. 91 | Exposed P91 |
|---|---|---|
| Hv (avg. of 20 times) | 229 | 194 |
| Reduction (%) | About 15% reduction in exposed P91 | |

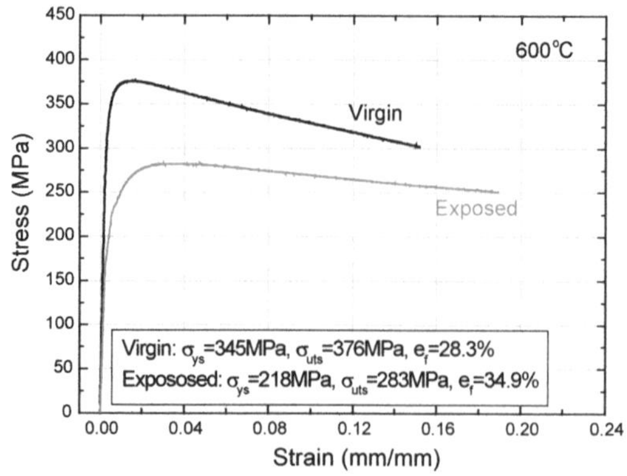


**Fig. 3.** Typical tensile stress-strain curves for virgin and exposed steels at 600 °C.

The exposed steel dropped almost below the average strengths in RCC-MRx and ASME Sec. III, Div. 5, although its TS value was a little higher than the minimum value in $RCC-MR_X$. In the case of the long-term design life of 60 years in Gen-IV reactor systems, it is assumed that the exposed steel will drop more rapidly than those of the design codes. In addition, the exposed steel presents a contrasting increase of about 18–2 3% in tensile elongation regardless of temperature. It is clear that the increase in tensile elongation arose from softening of the material owing to the long-term thermal exposure of 73,716 h.

Using the results shown in Fig. 4, the RF values for YS and TS in the exposed steel were obtained. Here, the RF is defined as RF = $\sigma_{exposed} / \sigma_{virgin}$. As can be seen in Fig. 5, the RF values for YS and TS largely dropped with increases in temperature; at the studied temperature conditions of RT to 650 °C, the RF ranges were within 0.6–0.7 for YS and 07–0.9 for TS. Thus, the reduction amount for YS is larger than that for TS. Best fit equations for YS and TS were optimized for the second-order polynomials as a function of temperature, as:

$$\text{RF values for TS} = 0.88389 - 1.09126E6 \cdot T - 3.87777E - 7\ T^2 \quad (1)$$

$$\text{RF values for YS} = 0.72162 - 4.106\ E-5\ T - 1.56471\ E-7\ T^2 \quad (2)$$

where, T is the temperature (°C). Using Eqs. (1) and (2), the RF values for YS and TS can be estimated at a specific temperature of the exposed P91 steel serviced for 73,716 h.

In addition, the TS RF of P91 steel exposed to service for 73,716 h at 569 °C is determined using the RFs given in ASME BPV Sec. III, Div. 5, as shown in Fig. 6, which is plotted from Table HBB-3225-4 in ASME and shows TS RFs for 9Cr-1Mo-V steel [11]. From Fig. 7, the TS RF with temperatures for the given time of 73,716 h can be estimated. The TS RF of the exposed P91 was obtained as RF = 0.908 at 569 °C, as shown in Fig. 7. The value of RF = 0.908 obtained from the ASME code data is almost identical to that of RF = 0.882 obtained from the tensile tests in the present investigation, as noted in Fig. 5. The reason for the greater reduction of RF = 0.882 in the experimental data is assumed to be that the exposed P91 was damaged by the pressure of 46.7 bar in the USC plant. From the tensile test results, the service-exposed steel was significantly reduced in TS, but the reverse was found for the elongation, which increased. It is suggested that the reason for this is that the exposed P91 is softened by degradation of the microstructure owing to long-term thermal exposure.

### 3.2. Creep rupture behavior

Creep properties such as rupture time, creep strain rate, and rupture elongation were obtained through a series of creep tests at 600 °C for the virgin and exposed P91 steels. Creep rupture data for the two steels were comparatively analyzed using various creep laws.

Fig. 8 shows comparisons of typical creep curves of virgin and exposed P91 under 150 MPa and 130 MPa at 600 °C. As can be clearly seen in the creep curves, the rupture time of the exposed steel decreases greatly; however, contrarily, there is an increase in rupture elongation compared with that of virgin steel. In the comparison results of creep rupture properties, the exposed P91 reveals lower creep stress, faster creep rate, and higher creep rupture ductility than those of the virgin sample, as shown in Figs. 9-11. The exposed sample shows a drop of creep strength of about 100 MPa (50%) compared with that of the virgin sample. Furthermore, the different gaps between the two steels become larger with increases in rupture time. The exposed P91 ruptures much more quickly with an approximately 3-orders faster creep strain rate than that of the virgin sample. The values of creep constants $A$ and $n$ in Norton's power law, $\dot{\varepsilon} = A\sigma^n$, are obtained for the two steels as $A$ = 7.76E−33 ($MPa^{-n}\ h^{-1}$) and $n$ = 12.4 for the virgin steel, and $A$ = 3.01E−20 ($MPa^{-n}\ h^{-1}$) and $n$ = 7.7 for the exposed steel. The faster creep rate of the exposed steel is attributed to softening of the strength and hardness as described above. In addition, creep rupture elongation increased by about 10% for the exposed steel, as shown in Fig. 11-(a). It is thus obvious there was an increase in creep rupture ductility owing to softening during long-term thermal exposure. Also, rupture elongation is slightly lower (or almost identical) at the same rupture time compared with that of virgin steel,

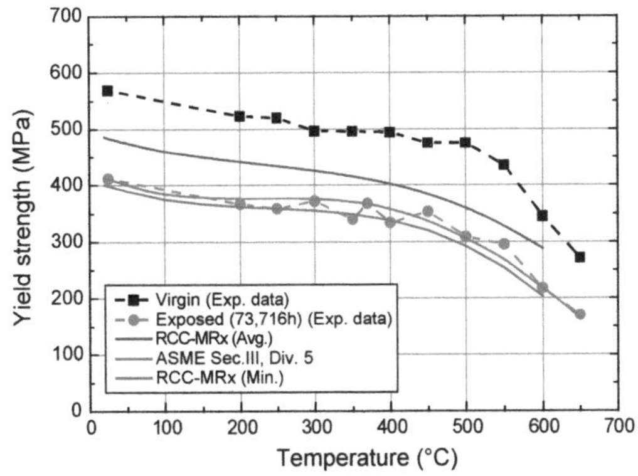


(a) yield strength (YS)

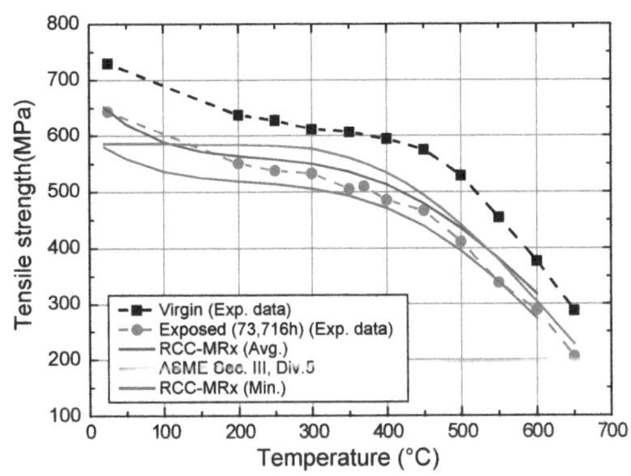


(b) tensile strength (TS)

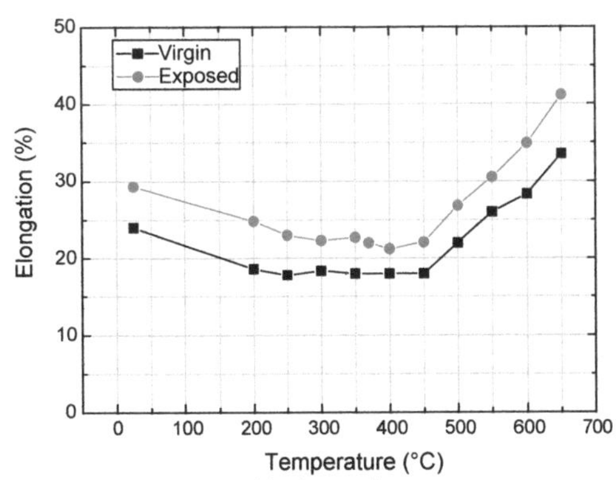


(c) elongation

**Fig. 4.** Comparison of (a) YS and (b) TS, and (c) elongation with temperatures for virgin and exposed steels.

although the rupture data are somewhat scattered, as shown in Fig. 11-(b).

Fig. 12 shows a comparison of the Monkman-Grant (M-G) plot of the virgin and exposed steels. The M-G relation is expressed as [18-20]:

$$\log(t_r) + m\log(\dot{\epsilon}_{ss}) = C_{MG} \quad (3)$$

where $t_r$ is the rupture time and $\dot{\epsilon}_{ss}$ is the steady state creep rate. When $m = 1$, the $C_{MG}$ value is different between the two steels. The exposed steel is higher for 2.9 times in the $C_{MG}$ value than the virgin sample, as $C_{MG} = 0.086$ in the exposed steel and $C_{MG} = 0.030$ in the virgin sample. The reason for this is that the creep rate of the exposed steel was faster than that of the virgin sample at the same

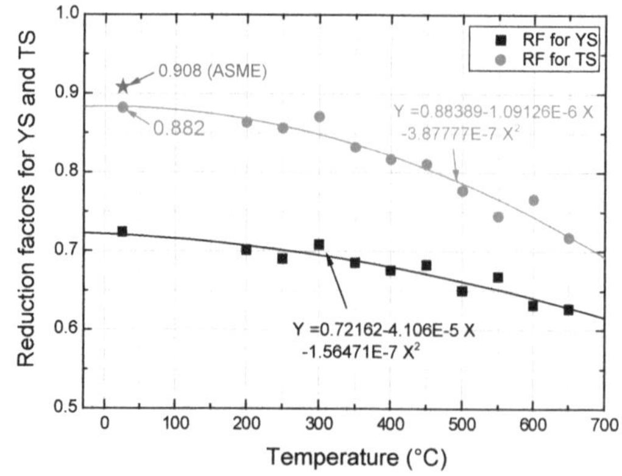


**Fig. 5.** Variations of YS and TS reduction factors with temperatures for virgin and exposed steels.

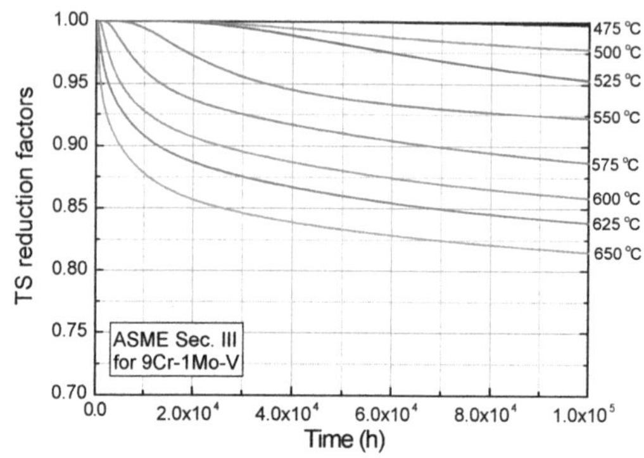


**Fig. 6.** TS reduction factors for 9Cr-1Mo-V steel with time variations at a specific temperature in ASME code.

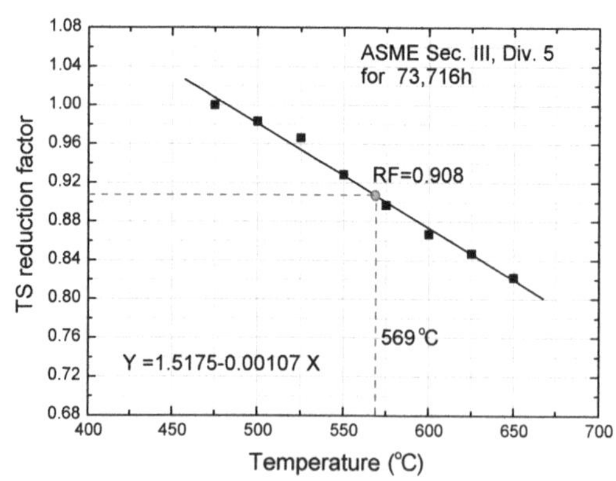


**Fig. 7.** A TS reduction factor determined for service-exposed P91 for 73,716 h at 569 °C.

rupture time. The faster creep rate arises from the higher rupture ductility owing to softening of the exposed steel.

In addition, the creep damage tolerance factor (λ) is often used as an important parameter to assess susceptibility of a material to localized cracking at strain concentrations and to assume damage mechanism [21]. The λ equation is defined as the ratio of the average creep rate to the secondary creep rate:

$$\lambda = \frac{\dot{\epsilon}_{avg}}{\dot{\epsilon}_{ss}} = \frac{\epsilon_f}{\dot{\epsilon}_{ss} \bullet t_r} \tag{4}$$

where $\dot{\epsilon}_{ss}$ is the steady state creep rate, $\epsilon_f$ is the rupture elongation, and $\dot{\epsilon}_{avg}$ is the average creep rate defined by $\frac{\epsilon_f}{t_r}$. The values of the

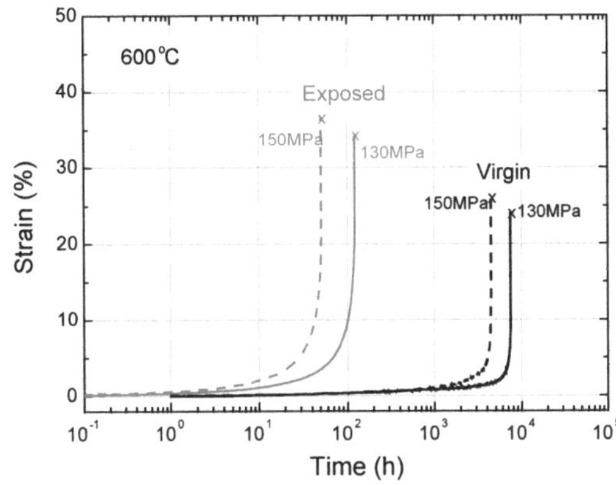


**Fig. 8.** Comparisons of typical creep curves under 150 MPa and 130 MPa at 600 °C for virgin and exposed steels.

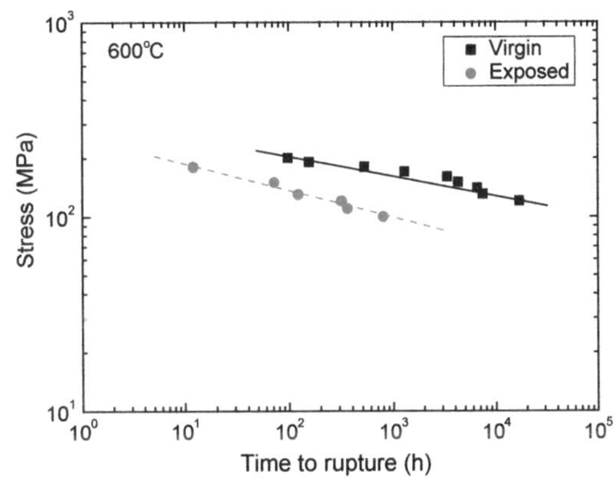


**Fig. 9.** Comparison of stress vs. time to rupture obtained for virgin and exposed steels.

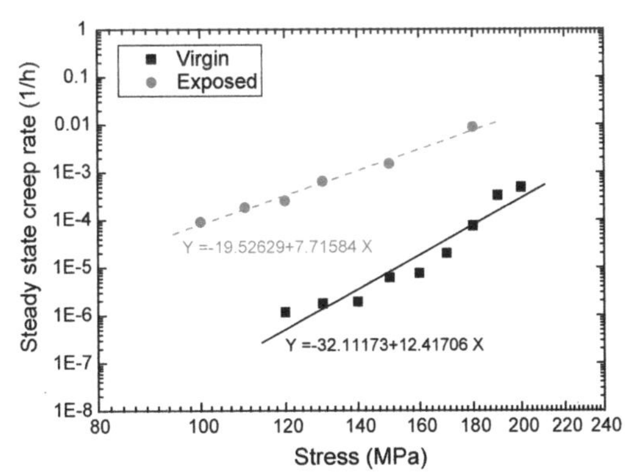


**Fig. 10.** Comparison of steady state creep rate vs. stress for virgin and exposed steels.

average creep rate is determined from the data of the rupture elongation and rupture time obtained for a creep tested specimen. In the present investigation, it was obtained as $\lambda = 3.68$ for the exposed steel and $\lambda = 9.45$ for the virgin steel, respectively, as shown in Fig. 13. The exposed steel is lower for 2.56 times in $\lambda$ value than the virgin steel. This is because the steady state creep rate of the exposed steel was significantly higher than that of the virgin steel. It is also assumed that the rupture elongation was a little lower than that of the virgin steel at the same rupture time, as shown in Fig. 11 (b). It is known that when $\lambda = 1.5$–$2.5$, the creep damage mechanism is "cavitation", and when $\lambda = 5$ or more, it is "microstructural degradation" with coarsening of precipitates and/or, for

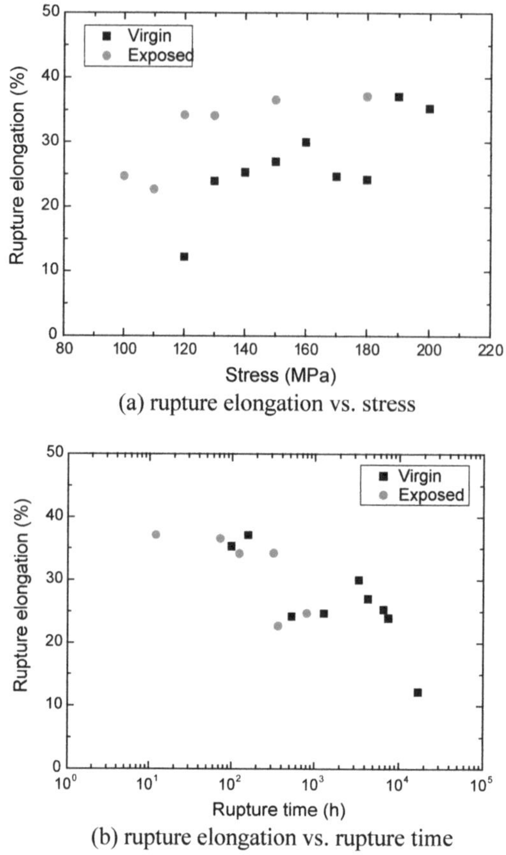


**Fig. 11.** Comparison of rupture elongation vs. (a) stress and (b) rupture time for virgin and exposed steels.

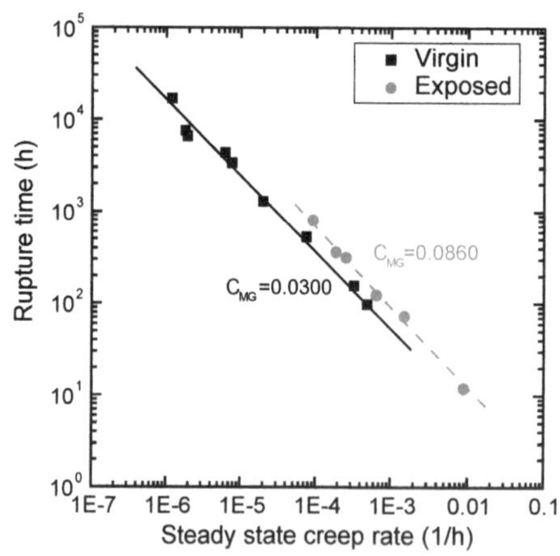


**Fig. 12.** Comparison of Monkman-Grant plot for virgin and exposed steels.

example, dislocation substructural softening [3,21]. Thus, it can be inferred that the creep deformation mechanism of the exposed P91 is dominantly governed by the combination of cavitation and microstructural degradation.

Fig. 14 shows a comparison of the rupture stresses given in RCC-MRx for virgin Gr. 91 steel and experimental creep strengths obtained at 600 °C for the exposed P91. The figure indicates that $(S_r)_{moy}$ is the average rupture stress (MPa), $S_r$ is the minimum rupture stress (MPa), and $S_t$ is the allowable stress (MPa). Experimental rupture data for the exposed P91 are higher than the $S_t$ of

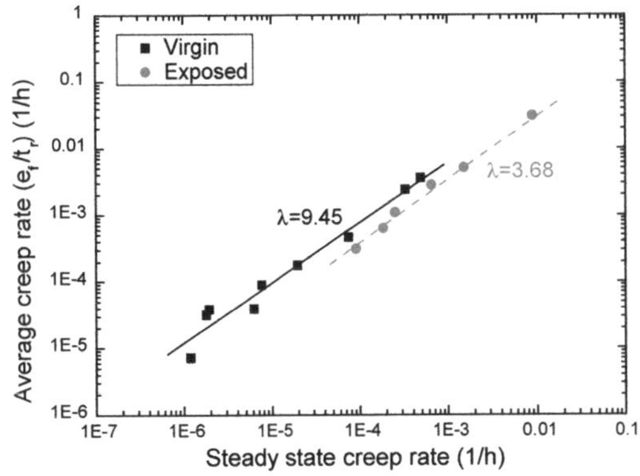


**Fig. 13.** Comparison of average creep rate vs. state creep rate for virgin and exposed steels.

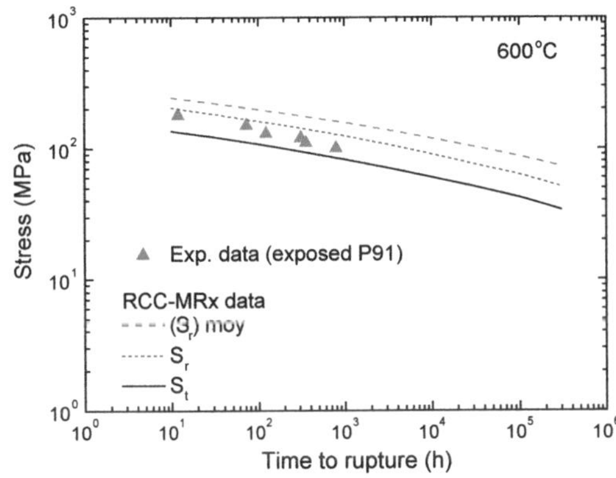


**Fig. 14.** Comparison of creep rupture stress obtained for exposed P91 steel and given in RCC-MRx code.

RCC-MRx, but lower than $S_r$ and $(S_r)_{moy}$ of RCC-MRx [13]. Thus, in the present test conditions, the exposed P91 appears to be sound in the stress status because it is higher than the allowable stress. However, when rupture time is above $10^4$h, it is questionable if the soundness of the exposed steel would be insured because it will suffer severe degradation during the longer-term creep.

Fig. 15 shows a comparison of the creep strain rate as a function of stress for the virgin Gr. 91 and exposed P91. Data on creep strain rate for virgin steels were collected from different sources, and the exposed P91 data line is for experimental data tested at 600 °C. Data for virgin steels are almost identical for creep rate as a whole, although the three lines show some differences, namely, the line slope of RCC-MRx [13] is a little different from the experimental data at KAERI and the data from and Isaac Samuel [3]. For

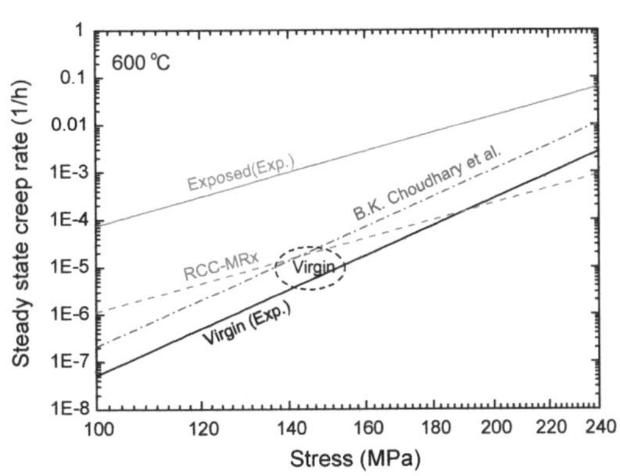


**Fig. 15.** Comparison of creep strain rates from different sources for virgin and exposed steels.

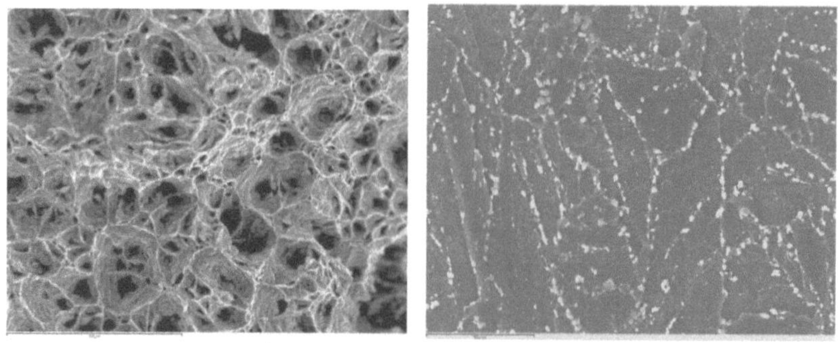

(a) virgin ($t_r$ = 7,475 h, 130MPa / 600 °C)

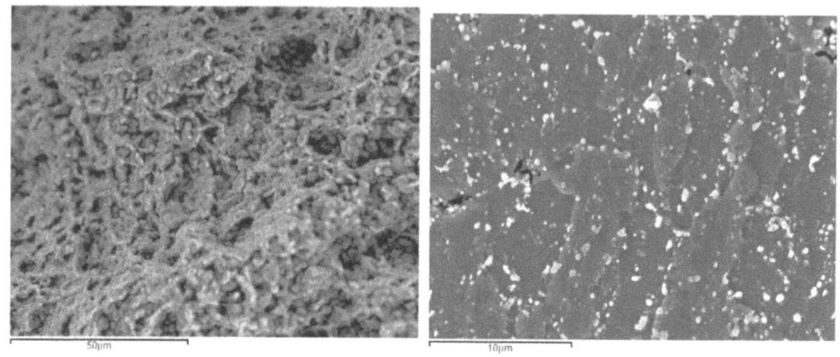


(b) exposed ($t_r$ = 803 h, 100 MPa / 600 °C)

**Fig. 16.** SEM fracture surface and precipitates observed for crept specimens for virgin and exposed steels.

creep rate, the exposed P91 is found to be accelerated by higher values of about two orders compared with RCC-MRx.

Fig. 16 shows SEM fracture surfaces and precipitates observed after creep tests for the virgin Gr. 91 and exposed P91 steels. The two steels present transgranular fractures on the fracture surfaces, as shown in the two images on the left hand side, but the exposed steel reveals severe brittle fractures owing to service exposure. The exposed steel is coarsened in the precipitates, but the precipitates along the prior austenite grain boundary (PAGB) and lath boundary are not clear, as observed for crept specimens in the right-hand side photos.

The precipitates are extensive and relatively coarse $M_{23}C_6$ carbides along the grain boundaries (GBs), and the presence of carbide precipitates plays a significant role in maintaining proper mechanical properties of heat-resistant steels at elevated temperatures. Virgin Gr. 91 steel forms inside the matrix as well as along the GBs. The precipitates are of coarsely distributed $M_{23}C_6$ at the GBs and finely distributed MX (M = V, and X = C, N) carbonitride precipitates within the matrix. It is known that $M_{23}C_6$ carbides work as a barrier for migrating boundaries, and MX particles act as obstacles to the movement of dislocations in the matrix to sustain the higher creep strength at elevated temperature [9,21-23].

From the creep results for the exposed P91 steel, it is considered that the reduction of the creep strength can be largely attributed to the coarsening of $M_{23}C_6$ precipitates, which coarsen easily during creep, because the precipitates are mainly composed of iron, chromium, and molybdenum (M = Fe, Cr, Mo), whose solubility levels are rather high [5,24].

### 3.3. Microstructure evolution

As described above, since the exposed P91 steel was identified as undergoing a significant reduction in tension and creep rupture properties, the microstructural changes before and after exposures are examined using OM, SEM, TEM, and the replica technique.

Fig. 17 shows a comparison of OM structures observed for the virgin Gr. 91 and exposed P91 steels. The virgin sample shows typical tempered-martensite, well-strengthened by carbides. Its grain size is increased owing to long-term exposure. As is known from previous studies [5,24-26], the initial microstructure of ASME Gr. 91 steel consists of a tempered martensite matrix with dense dislocations and precipitates. There are two kinds of precipitates of $M_{23}C_6$ carbides located at prior austenite grain boundaries (PAGBs) and at other boundaries of packets, blocks, and martensite lath, and finely dispersed MX carbonitrides within laths. Most of the precipitates formed along the boundaries such as PAGBs, laths, blocks, and packets are present for $M_{23}C_6$ carbides.

Fig. 18 shows typical SEM structures observed for fractions of lath boundaries and lath width for virgin and exposed samples. In the exposed P91, the grain boundary carbides coarsened and lath width increased compared with those characteristics of virgin Gr.

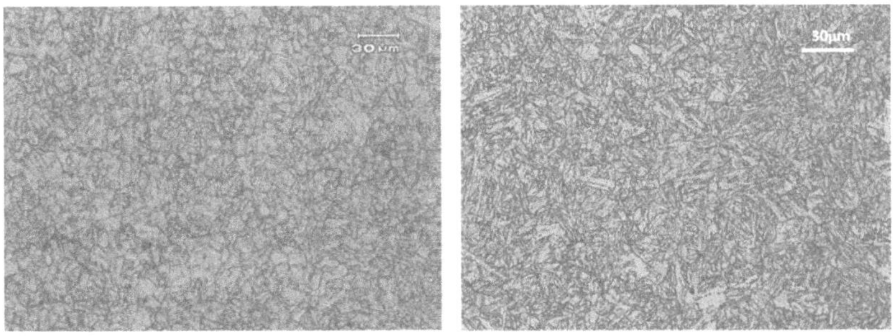


(a) virgin (b) exposed

**Fig. 17.** Initial OM structures observed for virgin and exposed samples.

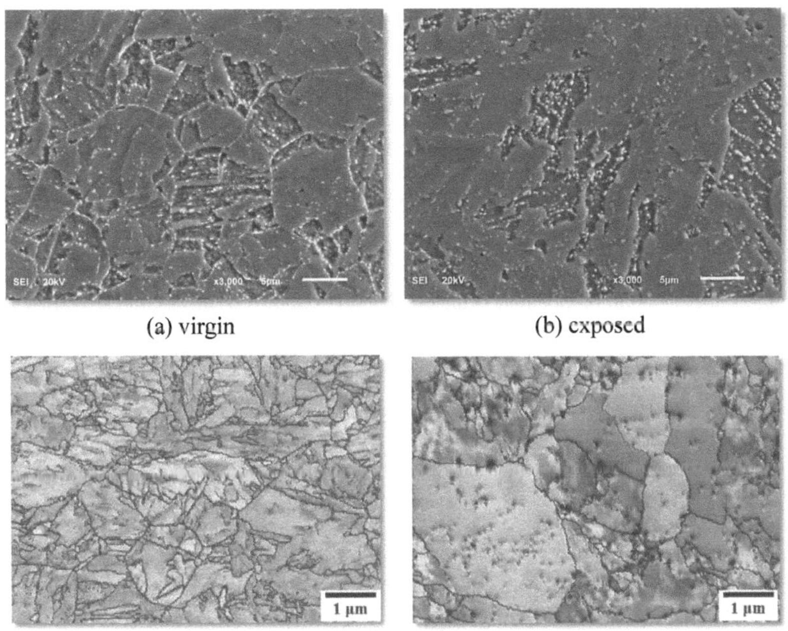


(a) virgin (b) exposed

(c) virgin (d) exposed

**Fig. 18.** Typical SEM structures observed for virgin and exposed samples.

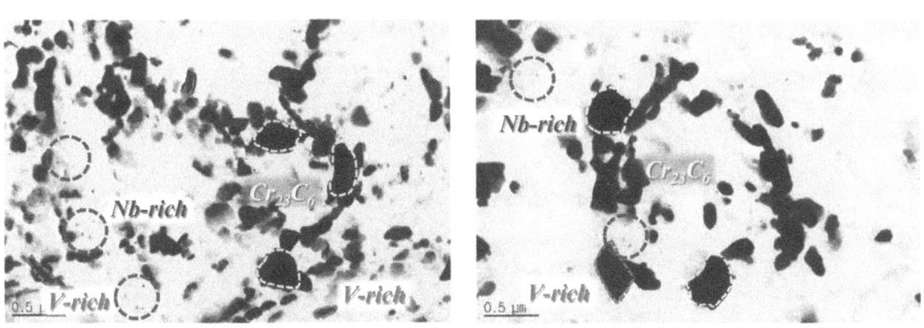


(a) virgin (b) exposed

**Fig. 19.** Precipitates formed for virgin and exposed samples observed by replica technique.

**Table 3**
Fraction and size of precipitates formed from virgin and exposed samples.

| Materials | Virgin Gr. 91 | | Exposed P91 | |
|---|---|---|---|---|
| | Fraction (%) | Size (μm) | Fraction (%) | Size(μm) |
| $M_{23}C_6$ | 7.0 | 0.094 | 5.6 | 0.130 |
| Nb-rich | 2.6 | 0.020 | 0.1 | 0.028 |
| V-rich | 1.6 | 0.088 | 3.1 | 0.159 |

91. The fraction of lath boundaries was found to be 38.1% for the virgin sample and 34.2% for the exposed sample, and lath width (μm) was 1.34 for the virgin sample and 2.07 for the exposed one. The Laves phase was sparsely observed by SEM analysis.

Fig. 19 provides a comparison of the precipitates observed for the virgin and exposed samples, obtained using the replica technique. Grain boundary carbides consisted almost entirely of $M_{23}C_6$. Nb-rich MX precipitates were uniformly precipitated with sizes under 20 nm. In the exposed sample, the precipitates of $M_{23}C_6$, Nb-rich MX precipitates, and V-rich MX precipitates had coarsened in size. However, there was no significant increase in the size of the MX precipitates, as previously reported [27]. This is because MX-type precipitates do not grow easily during high-temperature exposure owing to the solubility levels of Nb and V being quite low in the ferritic matrix [5]. The fraction of fine Nb-rich and V-rich MX precipitates significantly decreased. The reason is due to the dissolution of the MX precipitates during thermal exposure for 73,716 h at an elevated temperature of 569 °C. The Laves phase was intermittently observed by SEM. The Z-phase was not observed by either SEM or TEM. The fraction and size of precipitates measured for the virgin and exposed samples are summarized in Table 3.

Fig. 20 shows a comparison of dislocation density observed by TEM for the virgin and exposed samples. In the virgin sample, a high density of dislocations was observed in the narrow laths of about 1.3 μm size, which were well distinguished by TEM. In contrast, in the exposed sample, the density of dislocations decreased due to matrix recovery. The size of the laths and precipitates increased owing to long-term thermal exposure.

From microstructure examination, it is suggested that reductions of the tension and creep strengths in the exposed P91 mainly depend on the coarsening of precipitates, dissolution of useful fine MX precipitates, and recovery of dislocation substructure during thermal exposure. It is considered that the precipitation of undesirable Laves phase (rich in Mo) with associated loss of solid solution

**Fig. 20.** Dislocation density structures of virgin and exposed samples observed by TEM.

strengthening can accelerate subgrain coarsening [28].

## 4. Conclusions

Tension and creep rupture behaviors of exposed P91 steel, which was sampled from the T-fitting part of the reheat steam piping after service for 73,716 h at 569 °C and 46.7 bars in a supercritical (SC) plant, were investigated in comparison with those of virgin Gr. 91 steel. The exposed P91 showed a significant reduction in tensile strength, but there was an increase of about 18–23% in tensile elongation regardless of temperature. In the exposed P91 that had been in service for 73,716 h at 569 °C, the value of the reduction factor, RF = 0.908, obtained from ASME Sec. III, Div. 5 was found to be similar to the RF = 0.882 obtained from the experimental tensile tests. The RF values for yield strength (YS) and tensile strength (TS) were significantly reduced with increases in temperature owing to microstructure degradation in the exposed steel. For the present temperature conditions, the RF ranges were within 0.6–0.7 for YS and 0.7–0.9 for TS. The exposed P91 steel was found to have lower creep strength, faster creep strain rate, and higher rupture ductility than virgin steel. Damage tolerance factors ($\lambda$) were found to be $\lambda$ = 3.68 for the exposed steel and $\lambda$ = 9.45 for the virgin steel. The differences for the two steels resulted from the steady state creep rate in the exposed steel which was significantly higher than that of the virgin steel. It is evident from microstructural examination that the reductions of tension and creep strengths in the exposed P91 arose from the coarsening of precipitates, dissolution of useful fine MX precipitates, and recovery of dislocation substructure during long-term thermal exposure.

## Declaration of Competing Interest

The authors declared that there is no conflict of interest.

## Acknowledgment

This research was supported by the International Research & Development Program of the National Research Foundation (NRF) of South Korea funded by the Ministry of Science and ICT (NRF-2017K1A3A7A03093782).

## Appendix A. Supplementary material

Supplementary data to this article can be found online at https://doi.org/10.1016/j.engfailanal.2020.104736.

## References

[1] W.G. Kim, J.Y. Park, S.J. Kim, J. Jang, Reliability assessment of creep rupture life for Gr. 91 steel, Mater. Des. 51 (2013) 1045–1051.
[2] S.L. Mannan, S.C. Chetal, B. Raj, S.B. Bhoje, Selection of materials for prototype fast breeder reactor, Trans. Ind. Inst. Met. 56 (2003) 155–178.
[3] B.K. Choudhary, E. Isaac Samuel, Creep behavior of modified 9Cr-1Mo ferritic steel, J. Nucl. Mater. 412 (2011) 82–89.
[4] W.G. Kim, J.Y. Park, S.J. Kim, E.S. Kim, J. Jang, Improvement of long-term creep life extrapolation using a new master curve for Grade 91 steel, J. Mech. Sci. Technol. 32 (9) (2018) 4165–4172.
[5] M. Taneike, K. Sawada, F. Abe, Effect of carbon concentration on precipitation behavior of $M_{23}C_6$ carbides and MX carbonitrides in martensitic 9Cr steel during heat treatment, Metall. Mater. Trans. A. 35 (4) (2004) 1255–1262.
[6] C.G. Panait, W. Bendick, A. Fuchsmann, A.F. Gourgues-Lorenzon, J. Besson, Study of the microstructure of the Grade 91 steel after more than 100,000 h of creep exposure at 600°C, Int. J. Press. Vessels Piping. 87 (6) (2010) 326–335.
[7] B.S. Srinivas Prasad, V.B. Rajkumar, K.C. Hari Kumar, Numerical simulation of precipitate evolution in ferritic-martensitic power plant steels, CALPHAD, 36(2012) 1–7.
[8] L. Korcakova, J. Hald, M.A.J. Somers, Quantification of laves phase particle size in 9CrW steel, Mater. Charact. 47 (2) (2001) 111–117.
[9] A. Baltusnikas, I. Lukosiute, V. Makarevicius, R. Kriukiene, A. Grybenas, Influence of thermal exposure on structural changes of $M_{23}C_6$ carbide in P91 steel, J. Mater. Eng. Perform. 25 (5) (2016) 1945–1951.
[10] H.Y. Lee, W.G. Kim, S.K. Son, S.W. Hong, C.S. Seok, Effect of thermal aging on material strength and fracture behaviour in mod.9Cr-1Mo steel, KSME 4(2) (2016) 1–7 (Korean).
[11] ASME Boiler and Pressure Vessel Code, Section III Division 5, Rules for construction of nuclear facility components-high temperature reactors (2015).
[12] RCC-MRx, Section III Tome 1, Subsection NB, Class N1Rx Reactor Components, 2012 Edition Ad-dendum, AFCEN, (2013).
[13] RCC-MRx, Section III Tome 1, A3, 2012 Edition, AFCEN, (2013).
[14] RCC-MRx, Section III Tome 1, Appendix A9, Property Groups for Welded Joints, 2012 Edition, AFCEN, (2012).
[15] ASME Boiler and pressure vessel code Section II Part A, ASME, (2015).
[16] ASME Boiler and pressure vessel code Section II Part D, ASME, (2015).
[17] Y.I. Kim, J.W. Jang, J.H. Lee, S.J. Kim, S.O. Kim, J.B. Kim, H.Y. Jung, Conceptual design report of SFR Demonstration Reactor of 600MWe capacity, KAERI/TR-4598/2012, (2012).
[18] W.G. Kim, J.Y. Park, I.M.W. Ekaputra, S.J. Kim, M.H. Kim, Y.W. Kim, Creep deformation and rupture behavior of Alloy 617, Eng. Fail. Anal. 58 (2015) 441–451.
[19] F.C. Monkman, N.J. Grant, An empirical relationship between rupture life and minimum creep rate in creep rupture tests, ASTM Proc. 56 (1956) 593–620.
[20] F. Dobes, K. Milicka, The relation between minimum creep rate and time to fracture, Met. Sci. 10 (1976) 382–384.
[21] B.K. Choudhary, Tertiary creep behaviour of 9Cr-1Mo ferritic steel, Mater. Sci. Eng. A. 585 (2013) 1–9.
[22] P.K. Parida, A. Dasgupta, S. Saibaba, Study of microstructure and microtexture of modified 9Cr-1Mo steel subjected to high deformation, J. Nucl. Mater. 432 (1–3) (2013) 450–454.
[23] G. Golanski, J. Kepa, The Effect of ageing temperatures on microstructure and mechanical properties of GX12CrMoVNbN9-1 (Gp91) cast steel, Arch. Metall. Mater. 57 (2) (2012) 575–582.
[24] W.G. Kim, J.Y. Park, I.M.W. Ekaputra, S.J. Kim, J. Jang, Cyclic creep behavior under ten-tension loading cycles with hold time of modified 9Cr-1Mo steel, Mater. High Temp. 31 (2014) 249–257.
[25] L. Cipola, Conversion of MX nitrides to modified Z-phase in 9-12%Cr ferritic steels, Ph.D. Thesis, Technical University of Denmark, (2010).
[26] V. Thomas Paul, Microstructural stability of modified 9Cr-1Mo steel during long term exposures at elevated temperatures, J. Nucl. Mater. 378 (2008) 273–281.
[27] C.G. Panait, Evolution of dislocation density, size of subgrains and MX-type precipitates in a P91 steel during creep and during thermal ageing at 600°C for more than 100,000h, Mater. Sci. Eng. A. 527 (2010) 4062–4069.
[28] X. Wang, Laves-phase evolution during aging in 9Cr-1.8W-0.5Mo-VNb steel for USC power plants, Mater. Chem. Phys. 163 (2015) 219–228.

Springer

KSME

Journal of Mechanical Science and Technology 32 (9) (2018) 4165~4172
www.springerlink.com/content/1738-494x(Print)/1976-3824(Online)
DOI 10.1007/s12206-018-0814-4

# Improvement of long-term creep life extrapolation using a new master curve for Grade 91 steel †

Woo-Gon Kim[1,*], Jae-Young Park[2], Seon-Jin Kim[2], Eung-Seon Kim[1] and Jinsung Jang[1]

[1]*Korea Atomic Energy Research Institute, 989-111 Daedeok-daero, Yuseong-gu, Daejeon 305-353, Korea*
[2]*Pukyong National University, 365 Shinsunro, Nam-gu, Busan 608-739, Korea*

(Manuscript Received October 29, 2017; Revised May 14, 2018; Accepted May 14, 2018)

**Abstract**

This study aims to improve the long-term creep life extrapolation of Grade 91 (Gr. 91) steel using a new master curve of a hyperbolic sine ("sinh") form. In the master curve for extrapolating creep rupture life, a controversial point of a polynomial form, which has usually been used in time–temperature parametric methods, was preliminarily indicated. The validity of the new master curve was verified. A large body of creep rupture data for Gr. 91 steel was prepared from different sources of available worldwide data at 500 °C to 700 °C for Gr. 91 steel. The material constants of the Larson–Miller, Orr–Sherby–Dorn, and Manson–Harferd parameters were obtained using the polynomial and sinh forms on the basis of the rupture data. Long-term creep rupture life was extrapolated for up to 60 years using the two forms that defined the master curves. Results of long-term creep life extrapolation showed that the master curve of the polynomial form exhibited overestimation due to the divergent curves in the low stress ranges, whereas the master curve of the sinh form revealed goodness in the low stress ranges beyond the experimental data. The proposed sinh form was superior to the polynomial form. The sinh form could be applied to extrapolate the long-term creep rupture life of other heat-resistant steels. The creep rupture stresses predicted by the sinh function were comparable to those of RCC-MRx code.

*Keywords*: Creep; Creep life extrapolation; Master curve; RCC-MRx; Time-temperature parameter

## 1. Introduction

Grade 91 (Gr. 91) steel is a major candidate material to apply to steam generators, intermediate heat exchangers, and hot pipes in sodium-cooled fast reactors (SFRs) [1-6]. Extrapolation methods of long-term creep rupture life (stress) are important because the SFR components will be designed for up to 60 years of target life of at high temperatures. Designing SFRs for a 60-year operation requires the experimental creep rupture data for Gr. 91 steel close to 20 years because creep life extrapolation is known to be proper for three-time factor rule. However, in practice, producing experimental data close to 20 years and/or at least $10^5$ h is an expensive and time-consuming work.

The french nuclear design code, RCC-MRx [7], which is useful in designing fast reactors, provides creep design curves that reach $3 \times 10^5$ h (34.2 years). However, the design curves are recognition of overestimation for Gr. 91 steel. This phenomenon for Gr. 91 steel is that the practical rupture life at low stresses is lower than the expected life (stress) because of the nonlinear characteristics in the log $\sigma$ versus log $t_r$ relation due to microstructural degradation during long-term creep. Thus, reliable techniques for creep life extrapolation should be developed to meet a safe design life of 60 years.

Several different time–temperature parameters (TTPs), such as Larson–Miller (LM), Orr–Sherby–Dorn (OSD), Manson–Harferd (MH), Manson–Succop, and Goldhoff–Sherby parametric methods, have been typically used for extrapolating creep data to longer times than those for which experimental data are available [8-10]. In TTPs, a single curve called the "master rupture curve" (hereafter, "master curve") is important, and the master curve that depicts the relationships between parameter and stress has been commonly used. However, using the master curve to extrapolate long-term creep life is often improper because polynomial forms are sharply bent with convex or concave curves in the low stress ranges beyond the experimental test durations. Therefore, a new master curve is required to solve this problem in the master curve in TTP methods.

In this study, a new master curve of a hyperbolic sine ("sinh") form in the LM, OSD and MH methods was proposed to extrapolate accurately the long-term creep rupture life (stress) of Gr. 91 steel. Constant values used in the parametric equations were optimized using the creep rupture data.

*Corresponding author. Tel.: +82 42 868 2493, Fax.: +82 42 868 8549
E-mail address: wgkim@kaeri.re.kr
†Recommended by Associate Editor Nam-Su Huh

Average stress values were extrapolated by the sinh function in the three methods, and their results were compared with the RCC-MRx for up to 60 years.

## 2. Creep life extrapolation approaches

### 2.1 TTP methods

The TTP methods define the interrelation between time and temperature that contributes to creep damage and results in a definite rupture life at given stress. Time and temperature variables are combined into a single parameter, which is expressed as a function of stress. A semi–log plot of log stress versus parameter results in a single master curve for creep rupture data that corresponds to various temperatures and stresses. These TTP parameters are useful in extrapolating long-term creep life (stress) that corresponds to longer rupture lives beyond the experimental test durations.

In the TTPs, the LM, OSD and MH methods were applied herein. The LM [8], OSD [9] and MH [10] equations are as follows:

$$LM(P) = T(\log t_r + C)\,, \tag{1}$$

$$OSD(P) = \log t_r - \frac{Q}{2.3RT}\,, \tag{2}$$

$$MH(P) = \frac{\log t_r - \log t_a}{T - T_a} \times 10^3\,, \tag{3}$$

where $T$ is the absolute temperature, $t_r$ is the time to rupture, and $C$ is the constant in the LM equation; $Q$ is the activation energy, and $R$ is the universal gas constant ($R$ = 8.314 J mol$^{-1}$ K$^{-1}$) in the OSD equation; and $T_a$ and log $t_a$ are the constants in the MH equation. The parameter ($P$) in each equation can be utilized as a function of stress.

### 2.2 Proposition of a new master curve of sinh form

The master curve that depicts the relationships between log (stress) and parameter ($P$) should be suitably determined to achieve long-term life extrapolation. As a typical master curve in the TTP method, a polynomial form has been conventionally used. However, using polynomial forms to predict long life accurately is improper because it is convex or concave in the low stress ranges beyond experimental creep durations. The reason for this limitation is the intrinsic characteristics in the polynomial equations. Consequently, the extrapolated lives (stresses) do not meet the experimental rupture data, and they cause overestimation or underestimation according to the degree of orders of polynomial forms. To overcome this problem, a new master curve of sinh function instead of the polynomial form was proposed herein, as shown as follows:

$$P(t_r, T) \Rightarrow f(\sigma)\,, \tag{4}$$

$$P = b_1 \log(\sigma) - b_2 \log(\sigma)^2 - b_3 \log(\sigma)^3 + ......b_n \log(\sigma)^n\,, \tag{5}$$

$$P = b_1 + b_2 \sinh(b_3 \log\sigma) - b_4(\log\sigma)\,, \tag{6}$$

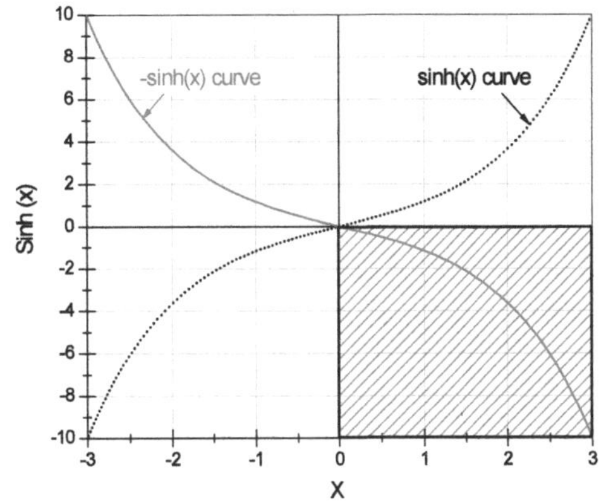


Fig. 1. Definition of sinh function.

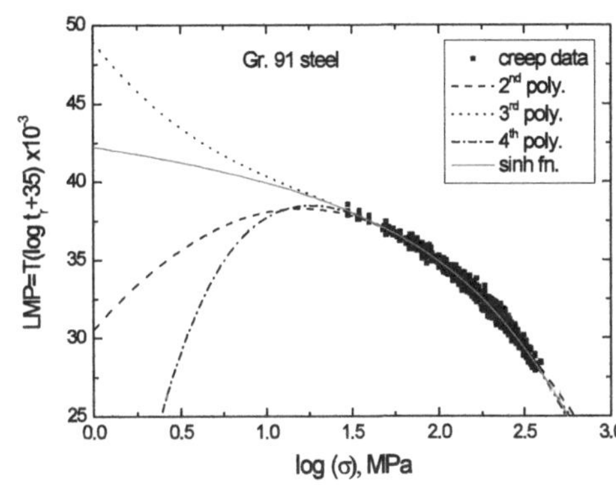


Fig. 2. Comparison of the master curves fitted for the sinh function and different polynomials for creep rupture data of Gr. 91 steel.

where $b_1$, $b_2$, $b_3$ and $b_4$ are the coefficients, and $\sigma$ is the stress. The TTP ($P$) can be expressed as a function of stress as Eq. (4). The parameter was used as the polynomial form of Eq. (5). Herein, the sinh function of Eq. (6) can be used instead of Eq. (5). Eq. (6) is stable without a sharp bent in the extrapolation of the low stress ranges beyond the experimental test durations.

A single master curve defined as the sinh(x) = ($e^x - e^{-x}$)/2 can be plotted, as shown in Fig. 1. A slashed region is used for long-term life extrapolation. The sinh(x) function has only an inflection point at x (stress) = 0; therefore, the master curve can generate a stable extrapolation for up to extremely low stress level of x = 0. That is, the master curve of the sinh form in LM parameter fitted for the creep rupture data of Gr. 91 steel is steadily extended up to x = 0 without convex or concave in the low stress level, as shown in Fig. 2.

The polynomials of the 2nd, 3rd and 4th forms are shown to be convex or concave in the extremely low stress level. Thus, Eq. (6) is verified to be reasonable as a master curve in the extrapolation of the low stress range beyond the experimental test durations of Gr. 91 steel.

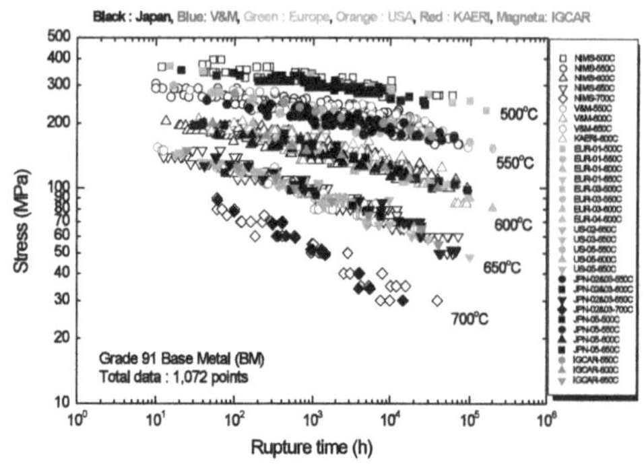

Fig. 3. Large body of creep rupture data collected for Gr. 91 steel.

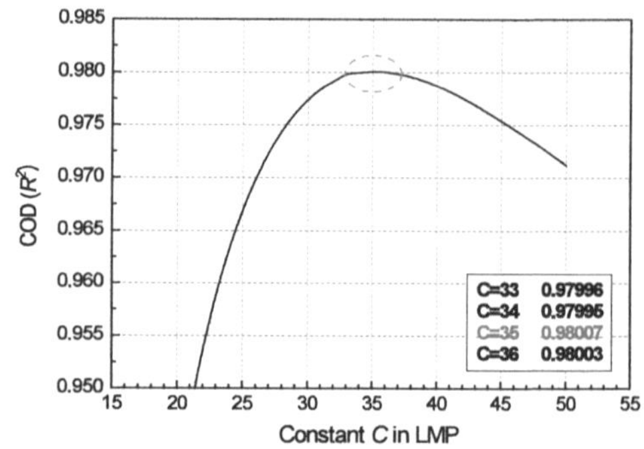

Fig. 4. Plot of the $R^2$ values for an optimum value of $C = 35$ in the LM.

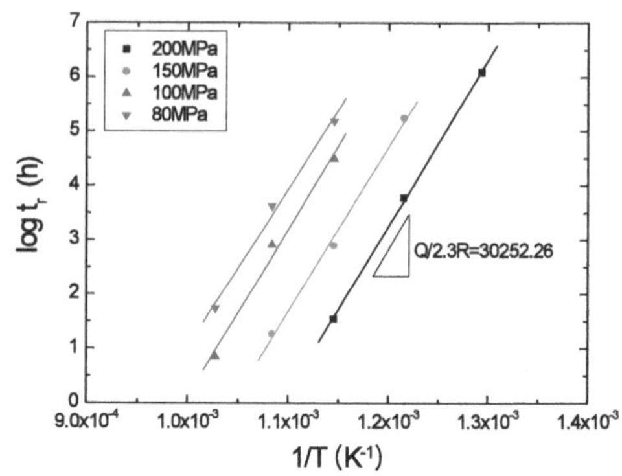

Fig. 5. Plot of log $t_r$ versus $1/T$ for a $Q$ value in the OSD parameter.

## 3. Data preparation and error evaluation method

A large body of creep rupture data for Gr. 91 steel was collected through worldwide literature surveys and the experimental data produced in Korea Atomic Energy Research Institute. The total rupture data were 1072 points for the plate, pipe, and forging types. The chemical compositions for all collected data were identified to be within ASTM and ASME Gr. 91 specifications. Detailed material information for the collected data was reported in the author's previous study [11]. Creep life extrapolation was performed using the master curves of the polynomials and sinh form in the LM, OSD and MH methods through the data bank.

The goodness-of-fitted master curves and the predicted life were evaluated using statistical parameters, such as coefficient of determination (COD, $R^2$) and standard error of estimate (SEE). The $R^2$ values are obtained as follows [12]:

$$R^2 = 1 - \sum_{i=1}^{n} \frac{(y_i - \hat{y}_i)^2}{(y_i - \bar{y})^2}, \tag{7}$$

where $y_i$ is the log $t_r$ at $i$ data, $\hat{y}_i$ is the average value estimated at $i$ data, and $\bar{y}$ is the average value of log $t_r$. The $R^2$ varies between 0 and 1. A value close to 1 indicates that the fit is a good. To judge the goodness of life prediction, the SEE values are obtained as [13]

$$SEE = \sqrt{\sum_{i=1}^{n} \frac{(y_i - \hat{y}_i)^2}{(n - n_p)}}, \tag{8}$$

where $n$ is the total creep data, and $n_p$ is the number of undetermined coefficients involved in the regression equation. Here, the LM and OSD are $n_p = k + 2$, and the MH is $n_p = k + 3$. $k$ is the degree of the regression equation.

## 4. Results and discussion

### 4.1 Extrapolation for polynomial form

An optimum value for the parametric constant of $C$, $Q$, $T_a$ and $t_a$ used in the parametric equations should be reasonably determined on the basis of experimental data to extrapolate the long-term creep life accurately. In the LM parameter of Eq. (1), the 3rd-order polynomial showed the best fit to the experimental data compared with the 2nd- and 4th-order polynomials. The $C$ value was optimized using the $R^2$ to obtain the best fit with various $C$ values from the plot of $P$ versus log stress using the 3rd-order polynomial. As described in Eq. (7), an optimum $C$ value was obtained for $C = 35$ of the highest $R^2$ value, and it was nearly close to 1 as $R^2 = 0.98$, as shown in Fig. 4. Therefore, the master curve with $C = 35$ was a good match with the tested creep data. In the OSD parameter of Eq. (2), an optimum $Q$ value was obtained for $Q = 578.4$ kJ/mol when taking an average slope ($Q/2.3RT = 30252.26$) of the families of the paralleled lines on the coordinate axes of log $t_r$ versus $1/T$, as shown in Fig. 5. In the MH parameter of Eq. (3), the $T_a$ and $t_a$ values were obtained at the intersection point of the families of the isostress lines plotted on the coordinate axes of log $t_r$ versus $T$, as shown in Fig. 6. The constant values determined for the LM, OSD and MH methods are summarized in Table 1.

The prediction of the long-term creep life was performed using the master curves of the 2nd-, 3rd- and 4th-order polynomials in the LM, OSD and MH methods using the obtained constant values. The master curves fitted for the 2nd- and 4th-order polynomials showed a sharp bent with convex curves in

Table 1. Constant values determined for the LM, OSD and MH methods.

| LM | OSD | MH |
|---|---|---|
| $C = 35$ | $Q = 578.4$ (kJ/mol) ($Q/2.3R = 30252.6$) | $T_a = 526.34$ (K) $t_a = 1.185 \times 10^{17}$ (h) |

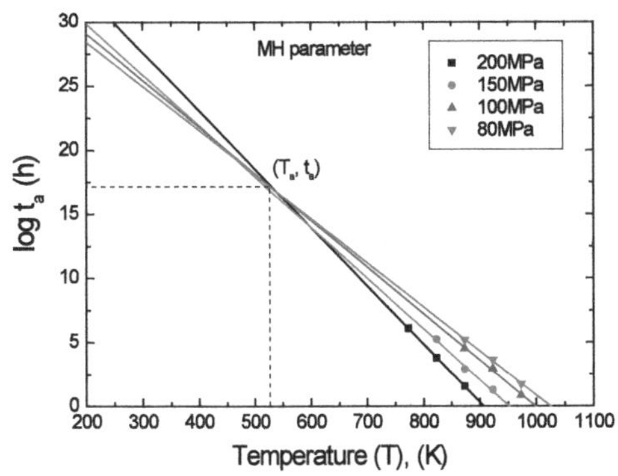


Fig. 6. Determining $t_a$ and $T_a$ values in the MH parameter.

Table 2. Coefficients optimized for the LM, OSD and MH methods for the master curves of the sinh function.

$$P = b_1 + b_2 \sinh(b_3 \log\sigma) - b_4(\log\sigma)$$

| Coefficients | LM | OSD | MH |
|---|---|---|---|
| $b_1$ | 42.26582 | −22.2307 | −21.1700 |
| $b_2$ | −0.77136 | −0.04410 | −0.01601 |
| $b_3$ | 1.31901 | 2.31323 | 3.19588 |
| $b_4$ | −1.00485 | −2.71087 | −4.84569 |
| $R^2$ (COD) | 0.98008 | 0.97789 | 0.97400 |

Table 3. Comparisons of SEE values for the experimental data and predicted values using the sinh function in the LM, OSD and MH methods.

| Methods | Total data | < 5000 h | > 5000 h |
|---|---|---|---|
| LM | 0.36797 | 0.36964 | 0.36096 |
| OSD | 0.35481 | 0.35510 | 0.32303 |
| MH | 0.37062 | 0.37094 | 0.31922 |

the low stress ranges beyond the creep data due to the intrinsic characteristics of the polynomials. The predicted stresses were underestimated in the low stress ranges beyond the creep data at 650 °C and 700 °C in the OSD and MH methods due to the convex master curves in the 2nd- and 4th-order polynomials. Therefore, the master curve of the 3rd-order polynomial was proper in long-term life extrapolation compared with those of the 2nd- and 4th-order polynomials for all three methods. The master curve of the 3rd-order polynomial in the LM method exhibited a stable increase without the inflection curve, as shown in Fig. 7. However, the master curves of the 3rd-order polynomials in the OSD and MH methods revealed a sharp increase as a divergent master curve in a low stress range beyond the creep data, as shown in Figs. 8 and 9. That is, the rupture stress curves predicted for the 3rd-order polynomial were reversely overestimated in the low stress ranges beyond the test data at 700 °C in the OSD and at 650 °C and 700 °C in the MH, as marked by the dotted lines in Figs. 8 and 9. Thus, although the 3rd-order polynomial was applicable at the temperatures of 500 °C, 550 °C and 600 °C, it was improper to be used in the low stress ranges at 650 °C and 700 °C for the accurate extrapolation of the lower stress ranges beyond 60 years. Accordingly, a new function was required instead of the polynomial forms, which have been commonly used, to extrapolate the longer life over the design life (60 years) of the Prototype Generation-IV SFR system.

### 4.2 Extrapolation for sinh form

Figs. 10-12 show the results of the predicted curves for up to $10^6$ h at temperatures of 500 °C, 550 °C, 600 °C, 650 °C and 700 °C using the master curves, which were optimized using the sinh function in the LM, OSD and MH methods. As shown in each figure, the fitted master curves exhibited a stable increase in the extrapolation beyond the creep data. All master curves of the three methods were a good match with the creep test data as a high value of $R^2 \cong 0.98$, as shown in Table 2. The coefficients of $b_1$, $b_2$, $b_3$ and $b_4$ were optimized for the master curves with the sinh function for the LM, OSD and MH methods, as listed in Table 2. The $R^2$ values defined in Eq. (7) were obtained to verify the goodness of the fitted master curves. The plots of log (stress) versus log (time to rupture) showed that the predicted curves for all temperatures were reasonably extrapolated in the low stress ranges. The LM and OSD methods were similar in the predicted curves; however, the MH method indicated a greater reduction in the long-term rupture life at 650 °C and 700 °C than the LM and OSD methods.

The goodness of the creep life extrapolation obtained using the sinh function for the three methods was assessed using the SEE values defined in Eq. (8). The SEE values were obtained by taking the total data under 5000 h and beyond 5000 h in all rupture data at 500 °C, 550 °C, 600 °C, 650 °C and 700 °C, as listed in Table 3. For the total data and under 5000 h data, the predicted results for the three methods were almost similar to the range of SEE = 0.35–0.37. However, beyond the 5000 h data, the OSD and MH methods were obtained for SEE ≅ 0.32, and the LM method was obtained for SEE = 0.36. The MH method was the lowest in SEE value in the order of MH > OSD > LM. Therefore, the MH method could be applied to extrapolate the long-term rupture stress (or life) of Gr. 91 steel accurately.

However, the extrapolated curves of the MH were slightly underestimated compared with the tested creep rupture data in the longer-term life beyond $10^5$ h at 650 °C and $10^5$ h at 700 °C, as shown in Fig. 12(b).

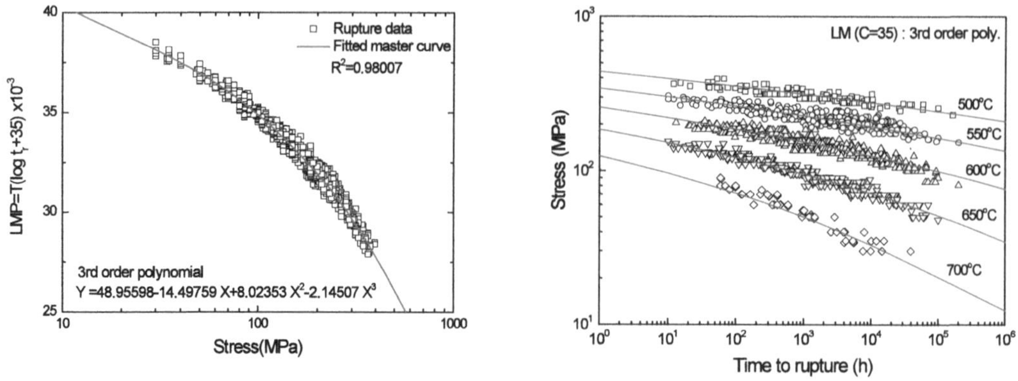


(a) Master curve fitted by the 3rd-order polynomial (b) Life prediction curves

Fig. 7. Plots of (a) a master curve; (b) the rupture stress curves predicted for the 3rd-order polynomial in $C = 35$ of the LM parameter.

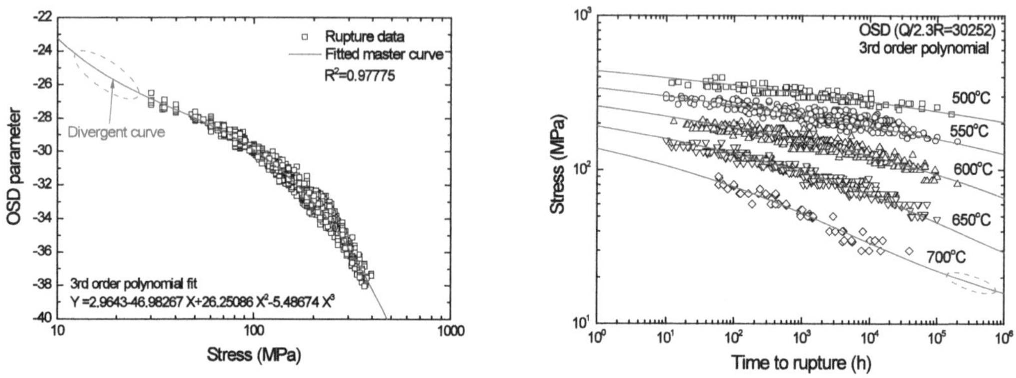


(a) Master curve fitted by the 3rd-order polynomial (b) Life prediction curves

Fig. 8. Plots of (a) a master curve; (b) the rupture stress curves predicted for the 3-order polynomial in the OSD parameter.

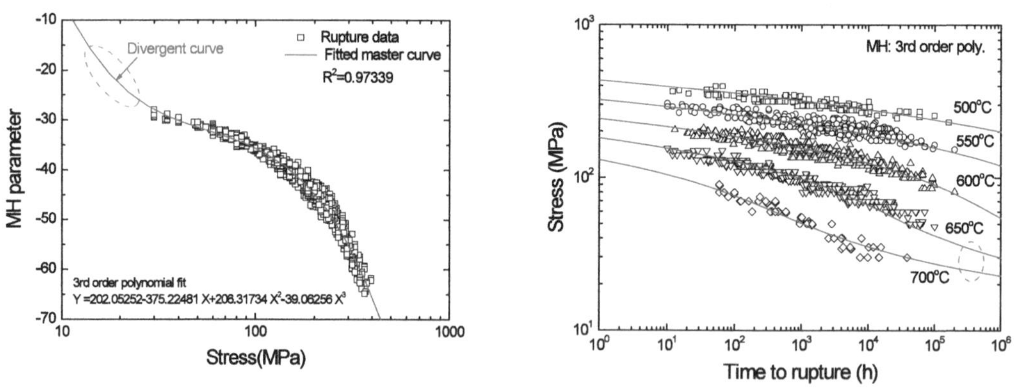


(a) Master curve fitted by the 3-order polynomial (b) Life prediction curves

Fig. 9. Plots of (a) a master curve; (b) the rupture stress curves predicted for the 3-order polynomial in the MH parameter.

### *4.3 Comparison of average rupture stress to RCC-MRx*

Average rupture stress was predicted for up to 60 years at 500 °C, 550 °C, 600 °C, 650 °C and 700 °C of Gr. 91 steel using the master curves of the sinh function. The predicted stress values were compared with RCC-MRx data. RCC-MRx data are only available for up to $3 \times 10^5$ h (34.2 years) at different temperatures. A linear extrapolation based on RCC-MRx recommendation was conducted by fitting a straight line that passed through the last two data points in the log (stress)

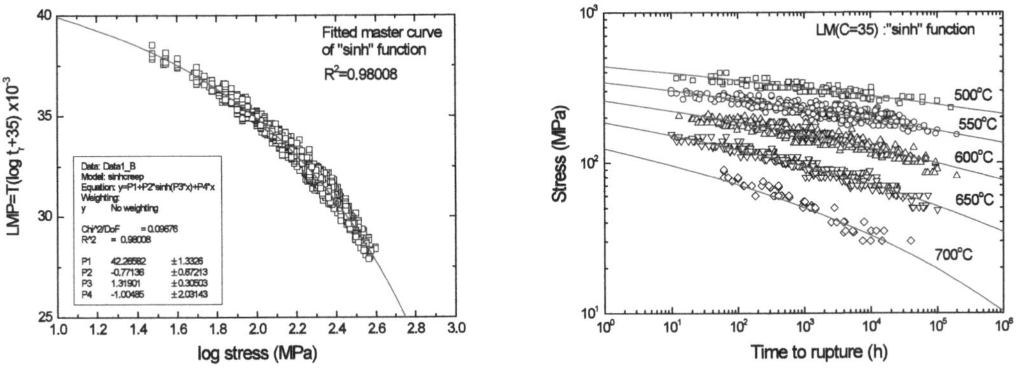


(a) Master curve fitted by the sinh function (b) Life prediction curves

Fig. 10. Plots of (a) a master curve; (b) the rupture stress curves predicted for the sinh function in $C = 35$ of the LM parameter.

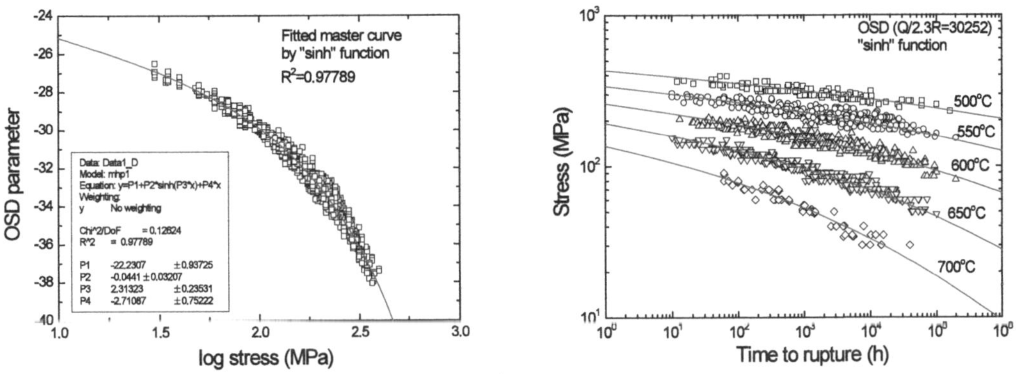


(a) Master curve fitted by the sinh function (b) Life prediction curves

Fig. 11. Plots of (a) a master curve; (b) the rupture stress curves predicted for the sinh function in the OSD parameter.

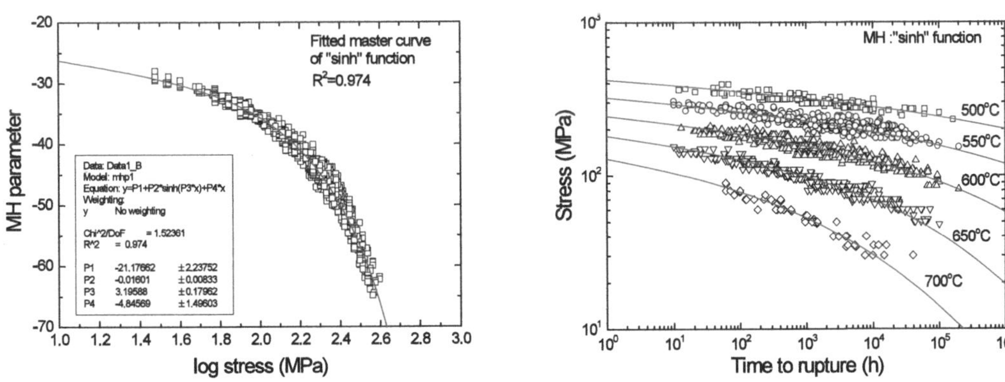


(a) Master curve fitted by the sinh function (b) Life prediction curves

Fig. 12. Plots of (a) a master curve; (b) the rupture stress curves predicted for the sinh function in the MH parameter.

versus log (time) plot to obtain the rupture stress values of up to $5.256 \times 10^5$ h (60 years). In RCC-MRx, the average rupture stresses for Gr. 91 steel are obtained at 425 °C–675 °C in Table A3.18AS.53b [7].

A comparison at 700 °C was not considered because RCC-MRx data at 700 °C are unavailable. Fig. 13 shows the comparisons of the average rupture stresses predicted using the sinh function for the three methods and RCC-MRx data at each temperature. In each figure, the black lines indicate the average rupture stress curve given by RCC-MRx, and the symbols indicate the stress values extrapolated using the sinh function.

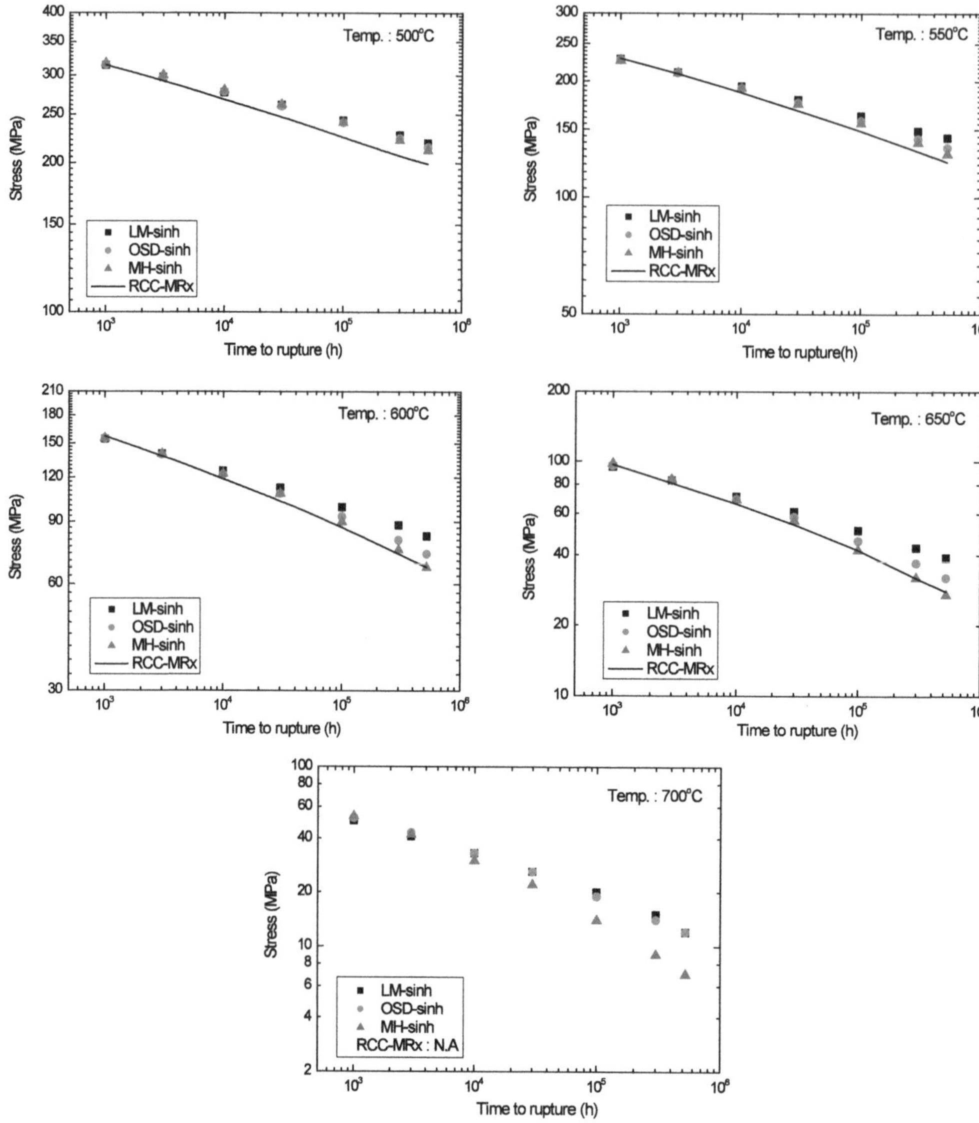


Fig. 13. Comparison of average rupture stresses predicted for up to 60 years using the sinh function of the three methods to RCC-MRx at 500 °C, 550 °C, 600 °C, 650 °C and 700 °C of Gr. 91 steel.

The LM method for the predicted stress was the highest in the order of LM > OSD > MH at 500 °C–700 °C, whereas the MH method was the lowest. Although the gap among the three methods was small, it increased with the rupture time. The predicted stress values evaluated from the sinh function in the LM, OSD and MH methods were comparable and consistently higher than those provided by RCC-MRx. The MH method was considerably closer to RCC-MRx data than the LM and OSD methods. However, marginal variations in the predicted stresses deduced from the three methods were observed at 500 °C–650 °C. This phenomenon was not of serious concern because of the closeness of the predicted stress values, particularly for up to 60 years. Thus, the master curve of the sinh function was applicable for other heat-resistant steels instead of that of the polynomial equation, which has been commonly used in the TTP method. RCC-MRx data could be safely used for the design life of 60 years for Gr. 91 steel.

## 5. Conclusions

A master curve of a hyperbolic sine (sinh) form was proposed instead of a polynomial form that is commonly used in TTP methods to improve the long-term creep life extrapolation of Gr. 91 steel. Various constant values used in the TTP equations were obtained for $C$ = 35 in the LM, $Q$ = 578.4 kJ/mol in the OSD, and $T_a$ = 526.34 and log $t_a$ = 17.074 in the MH. In the master curves, the best-fitted 3rd-order polynomials exhibited overestimation in the predicted stress values due to divergent curves. The sinh function proposed in the present investigation indicated more goodness than the polynomial equation in the extrapolation of extremely low stress ranges beyond the experimental creep durations. The MH method demonstrated a better agreement with experimental data than the LM and OSD methods. In the comparison of the predicted stress values to RCC-MRx for up to 60 years, the three methods were comparable to RCC-MRx. Particularly, the MH method was in accordance with RCC-MRx data. The master curve of the sinh function was applicable for other heat-resistant steels instead of that of the polynomial equation. RCC-MRx data could be safely used for the design life of 60 years of Gr. 91 steel.

## Acknowledgments

This work was supported by the Innovative R&D Project of the Korea Atomic Energy Research Institute and the National Research Foundation grant (Project No.: 2017M2A8A10 14758) funded by the Korea government (MSIP).

## References

[1] W. G. Kim, J. Y. Park, M. G. Won, H. Y. Lee and N. Su. Huh, Non-linear modeling of stress relaxation curves for Grade 91 steel, *J. Mech. Sci. Technol.*, 32 (2018) 1143-1151.

[2] B. K. Choudhary and E. I. Samual, Creep behavior of modified 9Cr-1Mo ferritic steel, *J. Nucl. Mater.*, 412 (2011) 82-89.

[3] S. L. Mannan, S. C. Chetal, B. Raj and S. B. Bhoje, Selection of materials for prototype fast breeder reactor, *Trans Indian Inst Met*, 56 (2013)155-178.

[4] W. G. Kim, S. H. Kim and W. S. Ryu, Evaluation of Monkman-Grant parameters for type 316LN and modified 9Cr-Mo stainless steels, *J. Mech. Sci. Technol.*, 16 (2002) 1420-1427.

[5] W. G. Kim, J. Y. Park, S. J. Kim and J. Jang, Reliability assessment of creep rupture life for Gr. 91 steel, *Mater Design*, 51 (2013) 1045-1051.

[6] V. S. Srinivasan, B. K. Choudhary, M. D. Mathew and T. Jayakumar, Long-term creep-rupture strength prediction for modified 9Cr–1Mo ferritic steel and type 316L(N) austenitic stainless steel, *Mater High Temper*, 29 (2012) 41-48.

[7] French Nuclear Design Code, RCC-MRx, Section III-Tome1-Subsection Z, *Technical Appendices A3-18AS* (2010) 14-15.

[8] F. R. Larson and J. Miller, A time-temperature relationship for rupture and creep stresses, *Trans. ASME*, 74 (1952) 765-771.

[9] R. L. Orr, O. D. Sherby and J. E. Dorn, Correlations of rupture data for metals at elevated temperature, *Trans. ASME*, 46 (1954) 113-118.

[10] S. S. Manson and A. M. Haferd, A linear time-temperature relation for extrapolation of creep and stress rupture data, *NASA Technical Note 2890*, Mar (1953).

[11] W. G. Kim, J. Y. Park, B. K. Choudhary, S. J. Kim, M. H. Kim and J. Jang, Influence of data size on the reliability assessment of creep life of Grade 91 steel, *J. Mech. Sci. Technol.*, 28 (2014) 4493-4501.

[12] W. G. Kim, S. N. Yoon and W. S. Ryu, Application and standard error analysis of the parametric methods for predicting the creep life of type 316LN SS, *Key Engineering Materials*, 297-300 (2005) 2272-2277.

[13] *Origin 8 user guide*, OriginLab Corporation (2007) 426-427.

**Woo-Gon Kim** received his Ph.D. from Chungbuk National University at the Department of Mechanical Engineering in 1998. He is a Principal Researcher at the Korea Atomic Energy Research Institute. His specialty includes mechanical assessment, analysis, modeling, and database establishment of high-temperature nuclear materials for Generation-IV reactor systems, particularly time-dependent creep and crack growth behaviors at elevated temperature.

Springer

KSME

Journal of Mechanical Science and Technology 32 (3) (2018) 1143~1151
www.springerlink.com/content/1738-494x(Print)/1976-3824(Online)
DOI 10.1007/s12206-018-0217-6

# Non-linear modeling of stress relaxation curves for Grade 91 steel †

Woo-Gon Kim[1,*], Jae-Young Park[1], Min-Gu Won[2], Hyeong-Yeon Lee[1] and Nam-Su Huh[3]

[1]*Korea Atomic Energy Research Institute (KAERI), 989-111, Daedeok-daero, Yuseong-gu, Daejeon 34057, Korea*
[2]*School of Mechanical Engineering, Sungkyunkwan Univ., Seobu-ro, Jangan-gu, Suwon, Gyeonggi-do, Korea*
[3]*Dept. of Mechanical System Design Engineering, Seoul National University of Science and Technology, 232 Gongreung-ro, Nowon-gu, Seoul 01811, Korea*

(Manuscript Received August 28, 2017; Revised November 28, 2017; Accepted November 29, 2017)

**Abstract**

We focused on the modeling of the high-temperature Stress relaxation (SR) curves for Gr. 91 steel using two nonlinear models of a spring-dashpot and Peleg model to get a better fit to experimental relaxation data. A series of SR tests were performed under a constant strain of 0.6 % at 500, 550, 600 and 650 °C. The SR behavior of Gr. 91 steel was clarified based on the experimental data. In the modeling of the SR curves, each parameter value for the two models was optimized to give the best fit to the experiment SR data. The spring-dashpot model did not match the experimental data at 500 and 550 °C, although the model showed good agreement with the experimental data at 600 and 650 °C. However, the Peleg model was superior to the spring-dashpot model for all temperature conditions, and its parameter values followed temperature dependence well. Using the Peleg model, the SR curves for Gr. 91 steel could be suitably modeled at different temperature ranges. The confidence of the Peleg model was validated from elastic-creep finite element analysis. It is suggested that the Peleg model can be a useful tool to characterize the relaxation behavior of Gr. 91 steel.

*Keywords*: Relaxation; Stress relaxation; Gr. 91 steel; Spring-dashpot model; Peleg model

## 1. Introduction

Grade 91 steel (Called modified 9Cr-1Mo steel, hereafter Gr. 91 steel) is being considered as a potential candidate for major structural materials in Sodium-cooled fast reactors (SFR) and for Reactor pressure vessel (RPV) in Gen-IV Very high temperature reactor (VHTR). The Gr. 91 steel, known as P91 or T91, was modified by addition of strong carbide forming elements such as Nb and V to enhance creep properties in earlier version of plain 9Cr-1Mo steel, known as P9 or T9. The high creep strength and low cycle fatigue strength of the Gr. 91 steel allow the application of thinner pipes or tubes to improve the heat transfer efficiency in the SFR system [1-7].

Presently, an R&D program to develop a Prototype Gen-IV SFR (PGSFR) system by 2028 is ongoing in the Korea Atomic Energy Research Institute (KAERI), and important key technologies for the PGSFR are under development through the R&D program. The Gr. 91 and 316 stainless steels are used as the two main materials in PGSFR system, as shown in Fig. 1. The structural components are designed to last for up to 60 years at elevated temperatures [8, 9]. To design the components, the Stress relaxation (SR) data decipting a material response from stress under constraint conditions relaxed by creep strain are sufficiently required. In a general experiment test, the total strain of a uniaxial test specimen is fixed and load (or stress) is monitored as it is gradually decreased by conversion of elastic to anelastic strain at a constant temperature. The engineering applications of relaxation data cover a wide range of displacement or strain controlled loading such as bolted joint, thermal stresses giving rise to secondary loads or to a creep-fatigue cycle and welding residual stresses. In addition, SR offers a means of evaluating the deformation dynamics of materials. However, unlike creep rupture data, which are used for design under primary stresses, most SR models are limited in applicability by either strain range, temperature, or both [10]. Sufficient studies for the SR behavior of Gr. 91 steel were not done, and it was not understood well. Especially, systematic studies for modelling the SR curve based on relaxation data are needed.

In the present study, a series of SR tests were conducted under a constant strain of 0.6 % at 500, 550, 600 and 650 °C for Gr. 91 steel. Based on the experimental data, the SR curves were characterized using two non-linear models in terms of the spring-dashpot model and the Peleg model, which were proposed as the appropriate models among various SR models. Each parameter value used in the two models was optimized using a Non-linear least square fit (NLSF) method to give the

*Corresponding author. Tel.: +82 42 868 2493, Fax.: +82 42 868 8549
E-mail address: wgkim@kaeri.re.kr
† Recommended by Editor Chongdu Cho

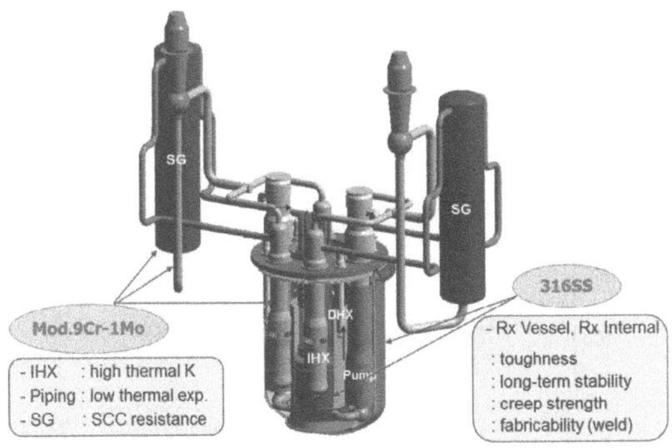


Fig. 1. Two main materials for structural components in a PGSFR system.

Table 1. Chemical compositions of Gr. 91 steel (wt. %).

| C | Si | Mn | P | S | Ni | Cr |
|---|---|---|---|---|---|---|
| 0.115 | 0.23 | 0.415 | 0.01 | 0.0014 | 0.22 | 8.9 |
| Mo | Cu | V | Al | N | Nb | - |
| 0.869 | 0.038 | 0.194 | 0.020 | 0.0513 | 0.073 | |

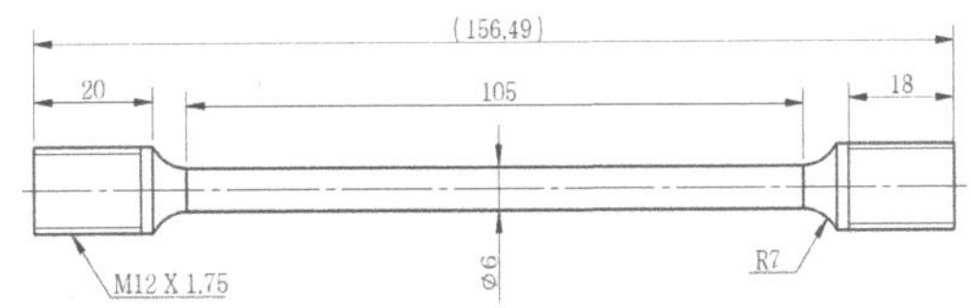


Fig. 2. Specimen dimensions for stress relaxation tests.

best fit to the experimental SR data. The temperature dependence for each parameter was investigated, and using the temperature dependence, the SR curves were numerically modeled at any given temperature. In addition, a Finite element analysis (FEA) was performed using the Peleg model identified as a suitable model.

## 2. Experimental procedure

A commercial grade hot-rolled Gr. 91 steel plate with a thickness of 16 mm was used for the SR tests. The heat treatment was normalized and tempered (N+T) at 1050 °C/1 mim per mm and 770 °C/3 mim per mm. The chemical composition is listed in Table 1. The test specimens were prepared with a round type of 6.0 mm in diameter and 105 mm in gauge length, as shown in Fig. 2. The gauge length was very long so as to precisely measure the SR data with time accumulation, and the specimen' dimensions were designed by referring to the ASTM E328 standard [11].

The SR tests were performed by slowly applying a finite amount of initial strain, $\varepsilon_o$ (Total or global strain), to a specimen and then maintaining its constant strain. After loading up to the initial constraint, the stress σ, dropped with time variations. In this study, the SR tests were carried out under a controlled constant tensile strain of 0.6 % at the four temperatures of 500, 550, 600 and 650 °C. The initial strain value of 0.6 % was determined from a stress value corresponding to a 0.6 % strain using the tensile curves obtained by tensile tests at 500, 550, 600 and 650 °C. An initial loading rate reaching 0.6 % was applied with the strain rate of $3.33\text{x}10^{-4}$ (1/s). After reaching the target strain level of 0.6 %, the strain was constantly maintained with a stop, and then the relaxation data with elapsed time were automatically acquainted using a personal computer. A three-zone furnace was used to provide the constant temperature environment. A K-type thermocouple was attached to measure the temperature of the specimens during the SR tests. The jigs and pull rods attached to each specimen used an Ni-based superalloy to prevent high-temperature deformation during the test period at high temperature. All SR tests were referred according to the ASTM E328 procedures [11].

## 3. Nonlinear stress relaxation models

To characterize the SR behavior of Gr. 91 steel, two nonlinear models in terms of a spring-dashpot and Peleg model were newly considered excepting classical relaxation models such as Kohlrausch, Kuhlmann, and Felthan, etc. [10, 12]. The equations of the two models are analyzed as follows.

First, the spring-dashpot model can be derived from a single paralleled spring-dashpot combination, as shown in Fig. 3(a). Under a constant strain condition, the equation of the spring-dashpot model can be given by [13]

$$\sigma_e + \sigma_a = eE + \left[\frac{e_a}{a(1-e^{-t/\tau})}\right] \tag{1}$$

where $\sigma_e$ is the elastic part of the stress and $\sigma_a$ is the anelastic part, and also, $E$ is Young's modulus, $e$ is the elastic strain, $e_a$ is the anelastic strain, $a$ is the spring constant, $t$ is the time, and τ is the viscous constant in dashpot. This equation was derived by converting strain into stress. In addition, Eq. (1) is modified into multi terms to give a better fit with the experimental data in an NLSF method. The modified equation can be proposed by

$$\sigma_e + \sigma_a = eE + \left[\frac{e_a}{a(1-e^{-t/\tau})} + \frac{e_a}{a(1-e^{-t/\tau})}\right], \tag{2}$$

$$y = p_1 + \left[\frac{p_2}{p_3(1-e^{-x/p_4})} + \frac{p_5}{p_6(1-e^{-x/p_7})}\right]. \tag{3}$$

Eq. (2) means that metal behaves in the same way as a spring in series with two parallel spring-dashpot combinations, as shown in Fig. 3(b). The SR values of the y-axis can be obtained with the time variations of the x-axis by determining the values for seven parameters: $P_1$, $P_2$, $P_3$, $P_4$, $P_5$, $P_6$ and $P_7$, as

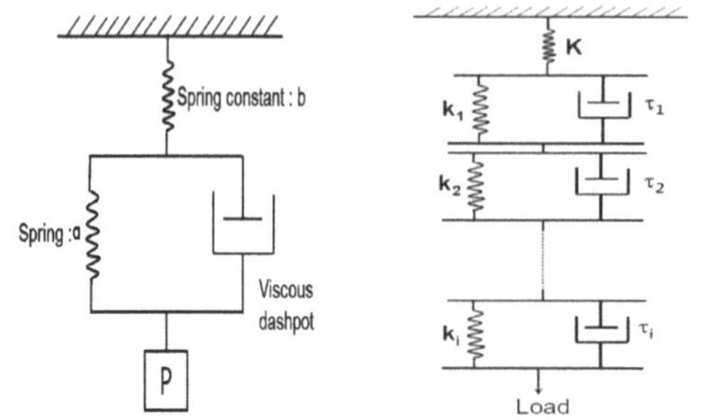


(a) Single spring-dashpot system (b) Multi spring-dashpot system

Fig. 3. Schematic drawing of simplified spring-dashpot analogue to anelastic strain.

given in Eq. (3).

Second, the Peleg model, which is known as a rheological model of non-linear SR, is proposed herein [14]. Myhan et al. used this model as a useful tool for evaluating the viscoelastic properties for food materials [15]. It was reported that the Peleg model was logically and empirically validated to accurately describe various SR behaviors in the compression and expansion phases of biological material. Therefore, we propose the temperature-dependent Peleg law to verify its applicability to Gr. 91 steel, because Gr. 91 steel has viscoelastic properties at higher temperature. The equation is given by

$$\sigma(t)=\sigma_o\left(1-\frac{t^n}{k_1+k_2t^n}\right), \tag{4}$$

where $\sigma_o$ is the initial stress (or maximum stress), $t$ is the time, $k_1$ and $k_2$ are the rheological parameters (or coefficients), and $n$ is the time exponent. The $\sigma_o$ value in Eq. (4) was determined from a stress corresponding to a 0.6 % strain using the tensile curves, which were obtained from uniaxial tensile tests at each temperature of 500, 550, 600 and 650 °C.

## 4. Modeling of the SR curves

### 4.1 Feature of the SR curves

The SR curves were obtained from a series of SR tests at 500, 550, 600 and 650 °C under a 0.6 % constraint of Gr. 91 steel, as shown in Fig. 4. The four SR curves at each temperature follow temperature dependence well: The SR curves decrease in order with an increase in temperature. The stress drops of Gr. 91 steel are very large at the beginning of the initial stage. The drops of elastic stress are instantaneous due to elastic recovery at the high temperatures, but soon diminish to a very small value, as shown in Fig. 5. The initial stress ($\sigma_o$) at zero time ($t_o$) is divided as the three components of $\sigma_o=\sigma_e+\sigma_a+\sigma_p$ during relaxation tests. Hence, $\sigma_e$ is the elastic stress, $\sigma_a$ is the anelastic stress due to creep recovery, and $\sigma_p$ is the plastic stress, which is a permanent stress remained finally by stress relaxation tests. Based on the SR

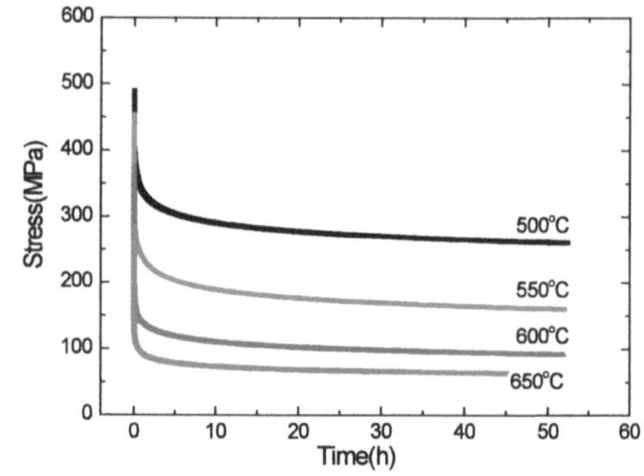


Fig. 4. Stress relaxation curves obtained at 500, 550, 600 and 650 °C under a 0.6 % constraint of Gr. 91 steel.

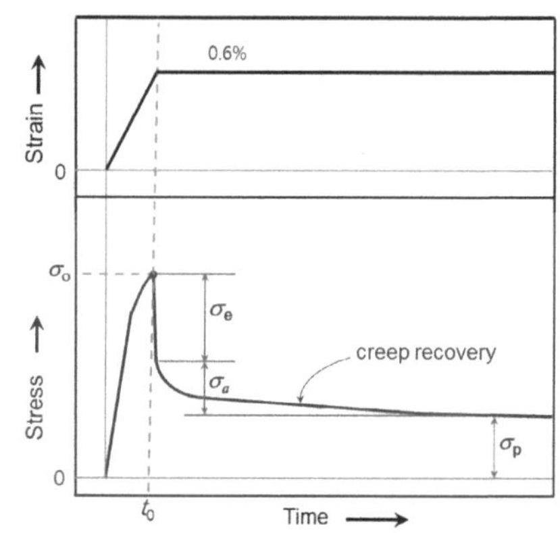


Fig. 5. Schematic drawing for three stress components under a 0.6 % constraint of Gr. 91 steel.

curves, the $\sigma_e$ value was determined for the four temperatures, as shown in Fig. 6. The $\sigma_e$ value shows an increase with an increase in temperature. It means that the stress drops of elastic stress (Elastic recovery) were larger at the high temperature than at the low temperature.

In addition, the values of stress relaxation rate ($\dot{\sigma}=d\sigma/dt$) at each temperature were obtained by differentiating between stress and time based on the SR tested data, as shown in Fig. 7. The relaxation rate in the elastic range at the initial stage increases very steeply, and then the relaxation rate in the anelastic range increases smoothly due to creep recovery. In the anelastic range (above 40000 seconds), as the relaxation time increases, the gap of each curve is gradually reduced, and the relaxation rates are saturated in the longer time regardless of the temperatures. From the results of Fig. 7, when the relaxation rate at a given time of 120000 seconds (33.3 h) is compared with the four temperatures, the relaxation rate decreases with an increase in temperature, as shown in Fig. 8. The difference between 500 °C and 550 °C is not much, but after 600 °C, it decreases significantly. Accordingly, it can be clarified that Gr. 91 steel is so large in the amount of stress relaxation at first, but after relaxing at first, its amount is not

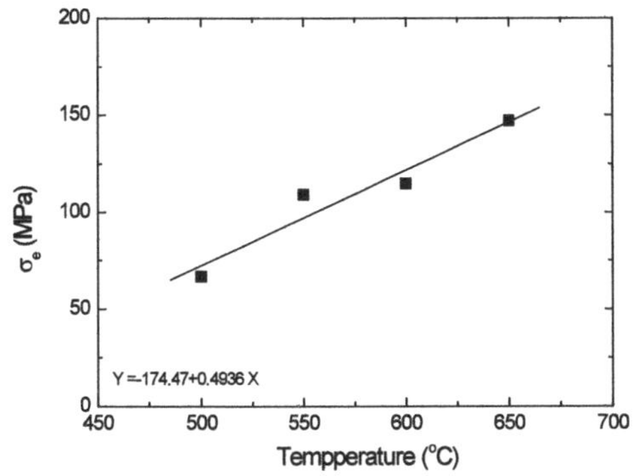


Fig. 6. Temperature dependence of elastic stress component obtained from the SR curve.

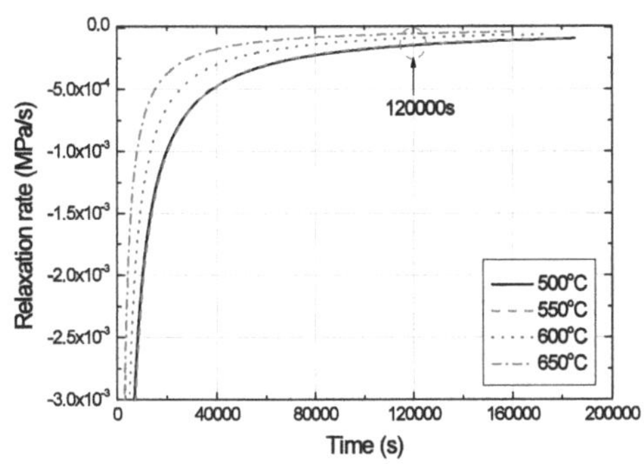


Fig. 7. Relaxation rate curves obtained at each temperature of Gr. 91 steel.

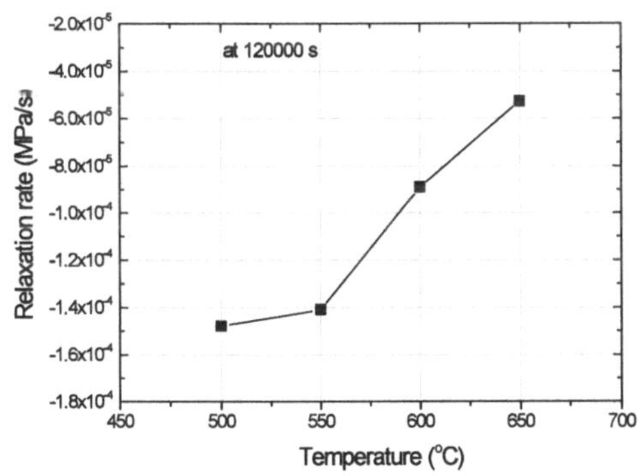


Fig. 8. Variation of relaxation rates with temperatures at the given time of 120000 s.

much due to anelastic creep recovery. In the present test condition, the relaxation rate of the anelastic creep recovery is slow as the order of about $10^{-4}$ to $10^{-5}$ (MPa/s) at 500 °C to 650 °C.

### 4.2 Application of the two nonlinear models

Figs. 9 and 10 illustrate the NLSF results of the SR curves

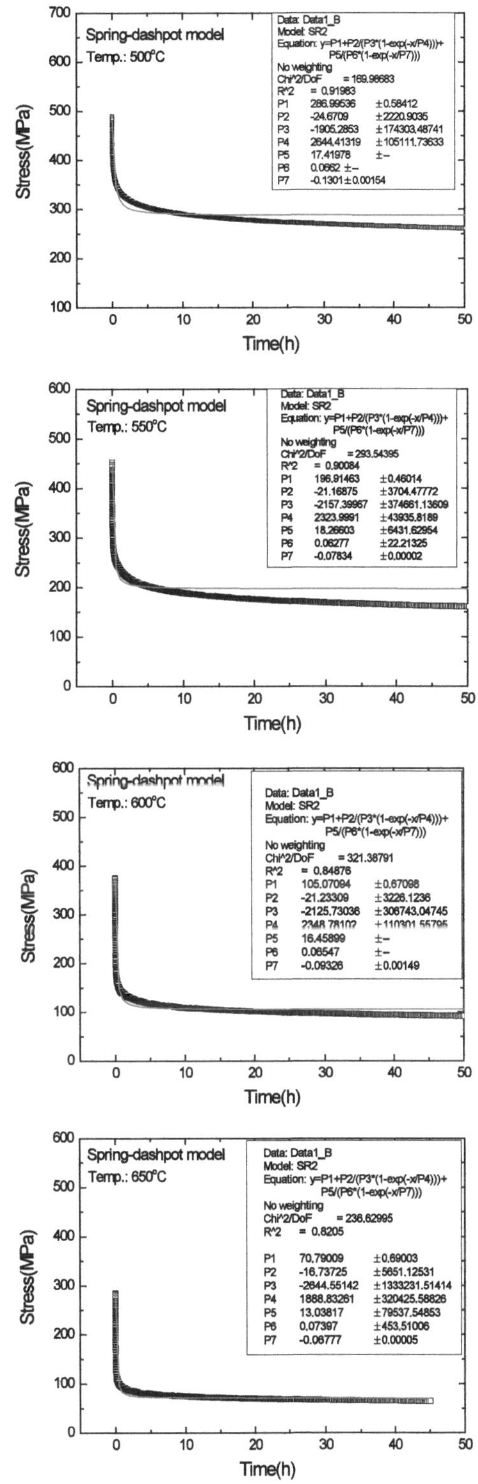


Fig. 9. NLSF results of the spring-dashpot model to the experimental data at 500, 550, 600 and 650 °C.

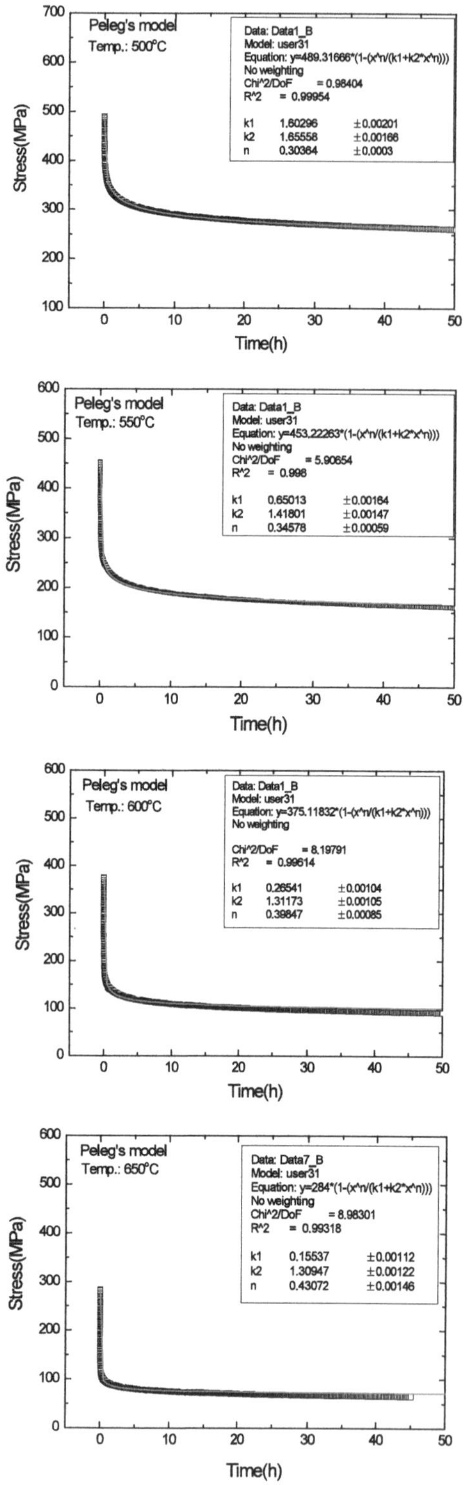


Fig. 10. NLSF results of the Peleg model to the experimental data at 500, 550, 600 and 650 °C.

Table 2. Summary of parameter values optimized at each temperature in the Peleg model.

| Temp.(°C) | $\sigma_o$ | $k_1$ | $k_2$ | $n$ | $R^2$(COD) |
|---|---|---|---|---|---|
| 500 | 489.316 | 1.602 | 1.655 | 0.303 | 0.999 |
| 550 | 453.222 | 0.650 | 1.418 | 0.345 | 0.998 |
| 600 | 375.118 | 0.265 | 1.311 | 0.398 | 0.996 |
| 650 | 284.000 | 0.155 | 1.309 | 0.431 | 0.993 |

in terms of the spring-dashpot model and the Peleg model to experimental data at each temperature of 500, 550, 600 and 650 °C of Gr. 91 steel. For the spring-dashpot model, at 600 and 650 °C, the fitted curves show good agreement with the experimental data, but at 500 and 550 °C, the fitted curves do not match with the experimental data because the spring-dashpot model is very small in the amount of SR. Therefore, to give better fit to the experimental data, more spring-dashpot terms should be added. As the number of the spring-dashpots increases, the model will be more and more accurately fitted with the experimental data. However, as the number of parameters increases, it will be difficult to determine the optimum values of the parameters and/or to define the temperature dependence of each parameter.

On the other hand, for the Peleg model, the NLSF results of the SR curves reveal a good agreement with the experimental data regardless of four temperatures, as shown in Fig. 10. Thus, the Peleg model was superior in modeling the SR curves of Gr. 91 steel compared with the spring-dashpot model. Thus, the parameter values of $k_1$, $k_2$ and $n$ in the Peleg model were optimized through many iteration processes to give the best regression to the experimental data. A summary of $\sigma_o$, $k_1$, $k_2$ and $n$ is given in Table 2. Hence, as described above, the $\sigma_o$ value is for a stress corresponding to a 0.6 % strain using the experimental tensile curves at each temperature. As listed in Table 2, the $R^2$ value (Coefficient of determination, COD) is closer to unity. It means that the Peleg model can describe well the SR curves of Gr. 91 steel. Using the results of Table 2, to model the SR curves in other temperature ranges, temperature dependence for four parameters was investigated. As shown in Figs. 11(a) and (b), the $\sigma_o$ (a) follows the second polynomial, and $k_1$, $k_2$ and $n$ (b) follow a linear relation well when given by semi-log coordinate of parameters vs. temperatures. Therefore, it is suggested that this result is reasonable because the parameters of the Peleg model follow the temperature dependence well. Each equation on temperature dependence of four parameters can be obtained as follows

$$\sigma_o = -603.4622 + 4.93966T - 0.0055T^2 \quad (5)$$

$$\mathrm{Log}(k_1) = 3.60155 - 0.00686T \quad (6)$$

$$\mathrm{Log}(k_2) = 0.54185 - 6.79297E\text{-}4T \quad (7)$$

$$\mathrm{Log}(n) = -1.03594 + 0.00104T. \quad (8)$$

Using the four equations obtained on temperature dependence, the SR curves at different temperature ranges for Gr. 91

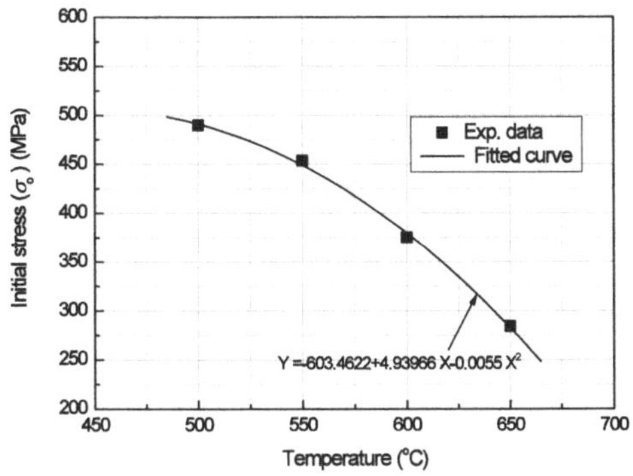


(a) $\sigma_o$

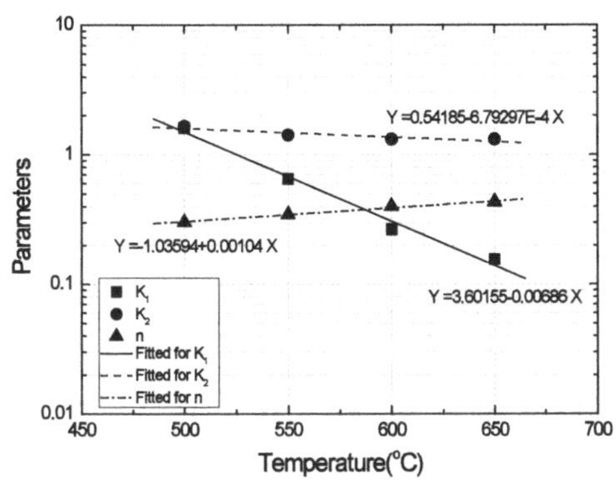


(b) $k_1$, $k_2$ and n

Fig. 11. Temperature dependence of (a) $\sigma_o$; (b) $k_1$, $k_2$ and n in the Peleg model.

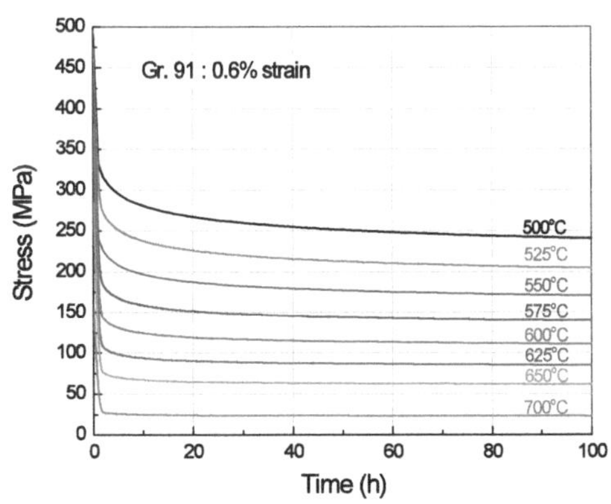


Fig. 12. Modeling of SR curves by the Peleg model at high temperatures.

steel can be suitably modeled, as shown in Fig. 12. It is drawn for up to 100 h in the temperature range of 500–700 °C using the Peleg model. The SR curves are low with an increase in temperature, and the stress recovery rate is slow with an increase in temperature. Especially at 700 °C, stress recovery

Table 3. Each parameter value predicted in the Peleg model.

| Temp.(°C) | $\sigma_o$ | $k_1$ | $k_2$ | n |
|---|---|---|---|---|
| 500 | 491.830 | 1.484 | 1.592 | 0.304 |
| 525 | 474.384 | 1.000 | 1.531 | 0.323 |
| 550 | 450.063 | 0.673 | 1.473 | 0.343 |
| 575 | 418.867 | 0.453 | 1.416 | 0.364 |
| 600 | 380.796 | 0.305 | 1.362 | 0.387 |
| 625 | 335.850 | 0.206 | 1.310 | 0.411 |
| 650 | 284.029 | 0.138 | 1.259 | 0.436 |
| 700 | 159.762 | 0.063 | 1.165 | 0.492 |

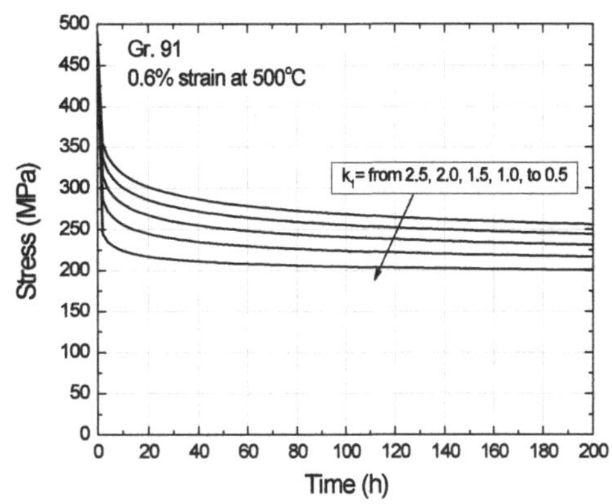


Fig. 13. Effect of $k_1$ on the SR curves in the Peleg model.

does not occur with almost constant value regardless of time accumulation. Table 3 provides a summary for the four parameters determined from temperature dependence to give a numerical calculation of the SR curves. Using the provided parameters, the SR curves for Gr. 91 steel can be conveniently obtained for various temperatures between 500–700 °C.

### *4.3 Effect of each parameter in the Peleg model*

As discussed, because the Peleg model was better for modeling SR curves than the spring-dashpot model, the Peleg model was considered here. The effect of $k_1$, $k_2$ and $n$ influencing the SR curves in the Peleg model was analyzed in detail. As given in Table 3, the $k_1$ value decreases with an increase in temperature from 1.484 to 0.063 at 500-700 °C, and the $k_2$ value also decreases from 1.592 to 1.165. However, the difference of the $k_2$ values is not large with an increase in temperature. In addition, the $n$ value is almost similar at 0.304–0.492, although its value increases slightly with an increase in temperature.

Figs. 13-15 present typical plots for the effect of variables $k_1$, $k_2$ and $n$ influencing the SR curves in the Peleg model. The figures are illustrated for 0.6 % at 500 °C of Gr. 91 steel. For the effect of the SR curves, $k_2$ is more significant than $k_1$, as shown in Figs. 13 and 14. The SR curve decreases largely

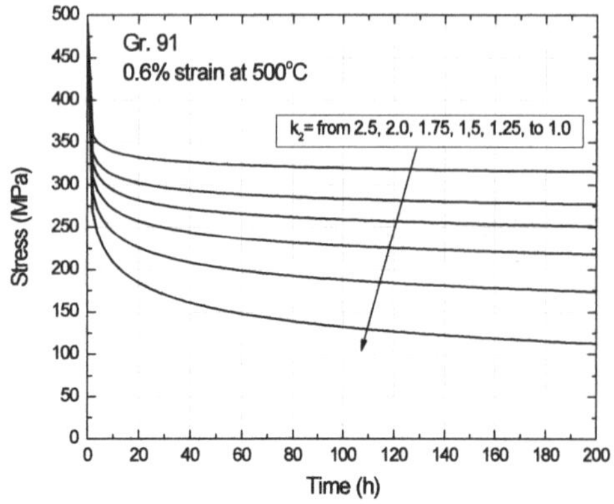


Fig. 14. Effect of $k_2$ on the SR curves in the Peleg model.

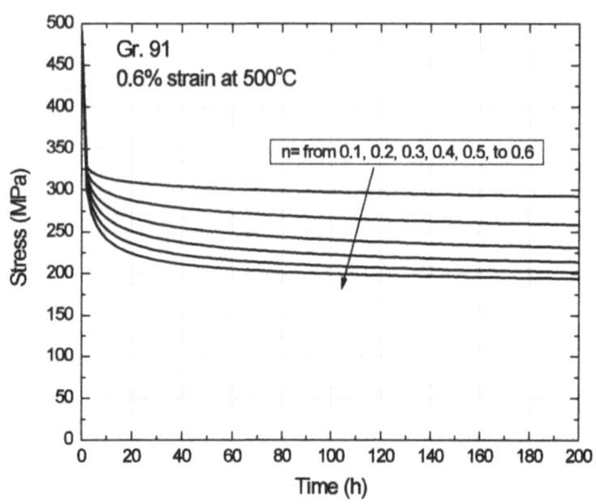


Fig. 15. Effect of $n$ on the SR curves in the Peleg model.

with a decrease in the $k_2$ value compared with the $k_1$ value. The reason for this is that the time values of the x-axis are large in the SR tests. It means that the SR curves of the Peleg model can be very sensitively affected in spite of the small change of the $k_2$ value. For the effect of the $n$ value, the SR curves decrease gradually with an increase in the $n$ value, as shown in Fig. 15. In the above value of n = 0.5, the difference between the SR curves is reduced. From the results, it can be analyzed that the variable $k_2$ is found to be more sensitive in determining a relaxation-curve shape than the variable $k_1$. The exponent $n$ is clarified to be involved in the range of $n \cong 0.3$-$0.5$ at the high temperatures of Gr. 91 steel.

### 4.4 FE analysis

The SR curves were obtained from a series of SR tests at 500, 550, 600 and 650 °C under a 0.6 % constraint of Gr. 91 steel, as shown in Fig. 4. To numerically validate the proposed temperature-dependent Peleg model for predicting the SR behavior, i.e., Eqs. (5)-(8), and to develop numerical implementation system to evaluate the SR of components at elevated temperature, elastic-creep Finite element (FE) analyses of SR were conducted using the general purpose FE software, ABAQUS. To implement the proposed temperature-dependent Peleg model for numerical prediction of SR at elevated temperature, a user defined subroutine needed to define the proposed temperature-dependent Peleg creep law (CREEP within ABAQUS user subroutines) was developed in the present study.

Fig. 16. Typical FE mesh of half-axisymmetric FE model for the gauge section of the SR specimen.

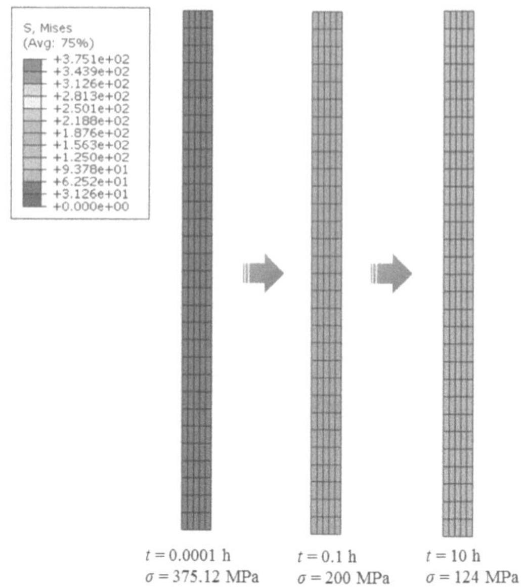


Fig. 17. FE contour results due to stress relaxation according to the time at 600 °C of Gr. 91 steel.

Fig. 16 depicts the typical FE mesh employed in the present study where only the gauge length section of SR test specimen was modeled. To reduce the computing time, 1/2 axisymmetric model of SR test specimen was used, and reduced integration 4-nodes axisymmetric elements (CAX4R in ABAQUS element library) were used. The number of elements and nodes in the FE model are 150 elements/185 nodes, respectively. In terms of boundary condition, axisymmetric condition was applied to the top section and the left section of the specimen was fixed to consider symmetric condition. As for loading condition, constant tensile strain of 0.6 % was applied and the corresponding initial stress was calculated elastically. Then, the strain was maintained with constant value during the SR simulation, and the temperature-dependent creep calculation was conducted to predict SR curves based on the Peleg law. Four FE analyses were performed for four different tem-

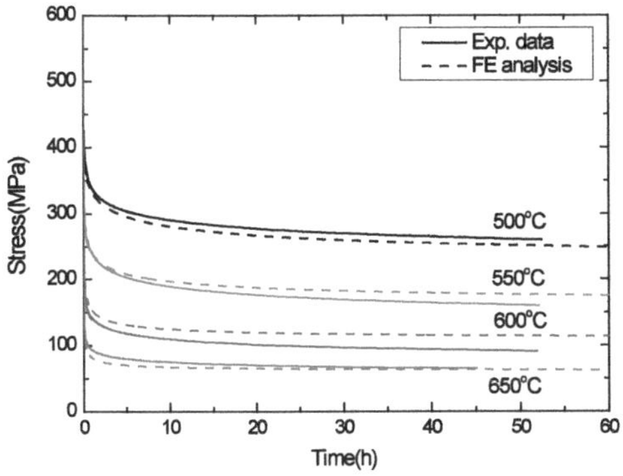


Fig. 18. Comparison of the SR curves between the FE analysis and the experiment data at 500, 550, 600 and 650 °C.

perature conditions, and the resulting SR curves from the present FE analyses were compared with those from the experiments.

Fig. 17 shows the SR behavior at 600 °C during the SR simulation according to the relaxation time. As shown in these stress contours during the analysis, the SR during the relaxation testing was well simulated by using the present user defined subroutine. Fig. 18 compares the overall FE SR results incorporating the present user defined creep based on the proposed Peleg model with the experimental results. As shown, the resulting SR curves from the present FE analyses are in good agreement with experiment results at all temperature conditions. Thus, the confidence of the proposed temperature-dependent Peleg model to evaluate the SR behavior of Gr. 91 was numerically gained.

## 5. Conclusions

The SR behavior of Gr. 91 steel showed that the stress drops were very large at the beginning of the initial stage, but soon diminished to a very small value. The relaxation rate at the initial stage increased steeply, and then the relaxation rate in the anelastic range increased smoothly due to creep recovery. As relaxation time in the anelastic range increased, the gab of each relaxation curve was gradually reduced, and saturated in the longer time regardless of the temperatures. In the modeling of the SR curves, the spring-dashpot model did not match the experimental data at 500 and 550 °C, although the model was good agreement at 600 and 650 °C. On the other hand, the Peleg model was superior to the spring-dashpot model for all temperature conditions, and its parameter values followed temperature dependence well. The $k_2$ in the Peleg model was more sensitive in determining an SR-curve shape than the $k_1$. The reason for this was that the time values of the x-axis were large in the SR tests. The exponent $n$ was clarified to be in the range of n ≅ 0.3–0.5 at high temperatures of Gr. 91 steel. The FE analysis results using the Peleg model showed a good agreement with the experimental data. The Peleg model of four parameters was more convenient compared with the spring-dashpot model of seven parameters. It is suggested that the Peleg model can be a useful tool to characterize the relaxation behavior of Gr. 91 steel.

## Acknowledgments

This work was supported by an International Research & Development Program Foundation NRF grant (NRF-2017K1A3A7 A03086464) funded by the Korea government (MSIP). The authors would like to acknowledge with gratitude Mr. Seok-Woo Hong and Professor Chang-Sung Seok of Sungkyunkwan University for experimental contributions to this work.

## References

[1] S. Spigarelli and E. Quadrini, Analysis of the creep behavior of modified P91 (9Cr-1Mo-NbV) welds, *Mater. Des*, 23 (2002) 547-552.

[2] B. K. Choudhary and E. I. Samuel, Creep behavior of modified 9Cr-1Mo steel, *J. Nucl. Mater*, 412 (2011) 82-89.

[3] W. G. Kim, J. Y. Park, I. M. W. Ekaputra, S. J. Kim and J. Jang, Influence of hold time and stress ratio on cyclic creep properties under controlled tension loading cycles of Grade 91 steel, *Nucl. Eng. Tech.*, 49 (2017) 581-591.

[4] W. G. Kim, J. Y. Park, S. J. Kim and J. Jang, Reliability assessment of creep rupture life for Gr. 91 steel, *Mater. Des*. 51 (2013)1045-1051.

[5] W. G. Kim, J. Y. Park, I. M. W. Ekaputra, S. J. Kim and J. Jang, Cyclic creep behavior under tension-tension loading cycles with hold time of modified 9Cr-1Mo steel, *Mater. High Temp.*, 31 (2014) 249-257.

[6] T. Shrestha, M. Basirat, I. Charit, G. P. Potirniche and K. K. Rink, Creep rupture behavior of Grade 91 steel, *Mater. Sci. Eng. A*, 565 (2013) 382-391.

[7] I. M. W. Ekaputral, W. G. Kim, J. Y. Park, S. J. Kim and E. S. Kim, Characterization of the $Q^*$ parameter for evaluating creep crack growth rate for type 316LN stainless steel, *J. of Mechanical Science and Technology*, 30 (2016) 3151-3158.

[8] H. Y. Lee, Comparison of elevated temperature design codes of ASME subsection NH and RCC-MRx, *Nucl. Eng. Des.*, 308 (2016) 142-153.

[9] W. G. Kim, J. Y. Park, B. K. Choudhary, S. J. Kim, M. H. Kim and J. Jang, Influence of data size on the reliability assessment of creep life of grade 91 steel, *J. of Mechanical Science and Technology*, 28 (2014) 4493-4501.

[10] S. Holmstrom, R. Pohja, P. Auerkari, V. Friedmann, A. Klenk, B. Leibing, P. Buhl, M. Spindler and A. Riva, Long term stress relaxation modeling, *ECCC Conference Creep & Fracture in High Temperature Component Design and Assessment*, Rome, Italy (2014).

[11] ASTM, Standard methods for stress relaxation tests for materials and structures, *ASTM E328-13* (2013).

[12] P. Felthan, Stress relaxation in copper and alpha-brasses

at low temperatures, *J. Inst. Metals*, 89 (1961) 210-214.
[13] J. D. Lubahn and R. P. Felgar, *Plasticity and creep of metals*, John Willy & Sons Inc, New York (1961) 363-404.
[14] M. Peleg, Considerations of a general rheological model for the mechanical behavior of viscoelastic solid food materials, *J. Text Stud*, 2 (1976) 243-255.
[15] R. Myhan, M. Markowski, T. Daszkiewicz, P. Zapotoczny and P. Sadowski, Non-linear stress relaxation model as a tool for evaluating the viscoelastic properties of meat products, *J. of Food Engineering*, 146 (2015) 107-115.

**Woo-Gon Kim** received his Ph.D. from Chungbuk National University in Mechanical Engineering in 1998. He is Principal Researcher at Korea Atomic Energy Research Institute. His specialty is mechanical assessment, analysis, modeling and database establishment of high temperature nuclear materials for Gen-IV reactor systems, and especially, time-dependent creep and crack growth behaviors at elevated temperature.

Nuclear Engineering and Technology 49 (2017) 581–591

Available online at ScienceDirect

Nuclear Engineering and Technology

journal homepage: www.elsevier.com/locate/net

Original Article

# Influence of Hold Time and Stress Ratio on Cyclic Creep Properties Under Controlled Tension Loading Cycles of Grade 91 Steel

Woo-Gon Kim [a,*], Jae-Young Park [a], I Made Wicaksana Ekaputra [b], Seon-Jin Kim [b], and Jinsung Jang [a]

[a] Korea Atomic Energy Research Institute, 989-111 Daedeokdaero, Yuseong-gu, Daejeon, 305-353, Republic of Korea
[b] Pukyong National University, 365 Shinsunro, Nam-gu, Busan, 608-739, Republic of Korea

ARTICLE INFO

*Article history:*
Received 26 July 2016
Received in revised form
21 November 2016
Accepted 28 November 2016
Available online 21 December 2016

*Keywords:*
Cyclic creep
Grade 91 steel
Hold time
Sodium-cooled fast reactors
Stress ratio

ABSTRACT

Influences of hold time and stress ratio on cyclic creep properties of Grade 91 steel were systemically investigated using a wide range of cyclic creep tests, which were performed with hold times (HTs) of 1 minute, 3 minutes, 5 minutes, 10 minutes, 20 minutes, and 30 minutes and stress ratios (R) of 0.5, 0.8, 0.85, 0.90, and 0.95 under tension loading cycles at 600°C. Under the influence of HT, the rupture time increased to HT = 5 minutes at R = 0.90 and R = 0.95, but there was no influence at R = 0.50, 0.80, and 0.85. The creep rate was constant regardless of an increase in the HT, except for the case of HT = 5 minutes at R = 0.90 and R = 0.95. Under the influence of stress ratio, the rupture time increased with an increase in the stress ratio, but the creep rate decreased. The cyclic creep led to a reduction in the rupture time and an acceleration in the creep rate compared with the case of monotonic creep. Cyclic creep was found to depend dominantly on the stress ratio rather than on the HT. Fracture surfaces displayed transgranular fractures resulting from microvoid coalescence, and the amount of microvoids increased with an increase in the stress ratio. Enhanced coarsening of the precipitates in the cyclic creep test specimens was found under all conditions.



## 1. Introduction

Grade 91 steel (hereafter referred to as Gr. 91) is regarded as a prime candidate material for structural components such as steam generators, intermediate heat exchangers, and hot pipes in sodium-cooled fast reactors. The selection of Gr. 91 steel is mainly based on its high creep and low cycle fatigue resistance compared with those properties exhibited by its counterparts such as 9Cr–1Mo and 2.25–1Mo steels [1–9].

Cyclic creep (CC) behavior is very important in practice because high-temperature structural components are exposed under the cyclic conditions of repeated loading during design life reaching 60 years in a sodium-cooled fast reactor plant. The concept of CC has evolved to describe

* Corresponding author.
E-mail address: wgkim@kaeri.re.kr (W.-G. Kim).
http://dx.doi.org/10.1016/j.net.2016.11.007

material behavior under various loading conditions, in which testing under the load control mode is involved and the applied cycling frequency simulates the creep/fatigue interaction. The parameters of cyclic loading can vary in broad regions of stress and time. There can be considerable interaction between creep and cyclic loading. The effect of cyclic loading can either accelerate or retard creep deformation depending on the material structure, temperature, and stress conditions [10–15]. Although in static loading, the response of the material is simply a static state of monotonic loading, cyclic loading is a kind of complex, dynamic loading. However, so far, data on CC response have rarely been reported, and CC behavior has not been understood well, whether accelerating or retarding, in terms of experimental variables that influence creep deformation and fracture behavior. In view of this, it is necessary to systematically clarify various test conditions, such as hold time (HT), cycling frequency, stress ratio, stress range, etc., influencing the creep deformation and fracture process.

In this study, CC tests of wide ranges were performed with HTs and stress ratios under tension–tension loading cycles of Gr. 91 steel. Influence of the HTs and stress ratios on the CC response was systemically investigated, and fracture surfaces in the CC tested specimens were observed and analyzed.

## 2. Experimental procedures

A commercial-grade hot-rolled Gr. 91 steel plate with a thickness of 16 mm was used for the CC tests. The chemical composition is listed in Table 1. The heat treatment condition of the steel was normalized at 1,050°C/1 min/mm (this means an HT of 1 min/mm at 1,050°C) followed by tempering at 770°C/3 min/mm (this means an HT of 3 min/mm at 770°C). Identical round specimens with a 30 mm gauge length and 6 mm diameter were used for the CC specimens, as shown in Fig. 1.

To clarify the cyclic loading behavior, the five-case CC test conditions were widely employed from 1 minute to 30 minutes in HT and from 0.5 to 0.95 in stress ratio (R), such that HT = 1 minute, 3 minutes, 5 minutes, 10 minutes, 20 minutes, and 30 minutes, and R = 0.5, 0.8, 0.85, 0.90, and 0.95 at 600°C, as listed in Table 2. The load schedules in the tension–tension CC tests are shown in Fig. 2. In the figure, $t_p$ is the time of a period cycle, and $t_h$ is the HT. HT is identically taken at the maximum stress ($\sigma_{max}$) and minimum stress ($\sigma_{min}$) in the loading cycles. In all CC tests, the crosshead speed of loading and unloading was controlled at 200 mm/min. The loading pattern was almost rectangular. The value of mean stress in the loading cycles was fixed at 160 MPa. Terminologies in cyclic loading are defined as follows: the mean stress ($\sigma_{mean} = 1/2(\sigma_{max} + \sigma_{min})$, stress range ($\sigma_R = \sigma_{max} - \sigma_{min}$), and stress ratio ($R = \sigma_{min}/\sigma_{max}$). It should be noted that the values of the maximum stress ($\sigma_{max}$) and minimum stress ($\sigma_{min}$) at each stress ratio were obtained by calculating the relationships given by $\sigma_{max} = 2\sigma_{mean}/(1+R)$ and $\sigma_{min} = 2\sigma_{mean}/(1+1/R)$, respectively.

Load-controlled tension–tension cyclic tests were performed using a universal testing machine with 100 kN capacity (Model No.: RB Unitech-M), manufactured by R&B Company in Daejeon, Korea. The cyclic loading of the maximum and minimum stresses was applied to a specimen using an AC servomotor; this process was periodically repeated in the clockwise and counterclockwise directions. The main components were a three-zone heating furnace, a temperature controller, an extensometer, strain gage, a data acquisition system, and a program controller. The cyclic tests were automatically conducted by a scheduled program. The real-time data of the strain and stress at the elapsed times were monitored and collected by a PC through a high-precision extensometer attached to the specimen. The steady-state creep rate was as taken as a mean value of the secondary creep strain data. The CC testing apparatus is given in detail in the authors' previous study [16]. All experimental procedures referred to the recommendations of ASTM E139 [17]. Fracture surfaces in the CC tested specimens were observed using an optical microscope and a scanning electron microscope (SEM).

**Table 1 – Chemical compositions of Gr. 91 steel (wt%).**

| C | Si | Mn | P | S | Ni | Cr | Mo | Cu | V | Al | N | Nb |
|---|---|---|---|---|---|---|---|---|---|---|---|---|
| 0.115 | 0.23 | 0.415 | 0.012 | 0.0014 | 0.22 | 8.9 | 0.869 | 0.038 | 0.194 | 0.020 | 0.0513 | 0.073 |

Gr., grade.

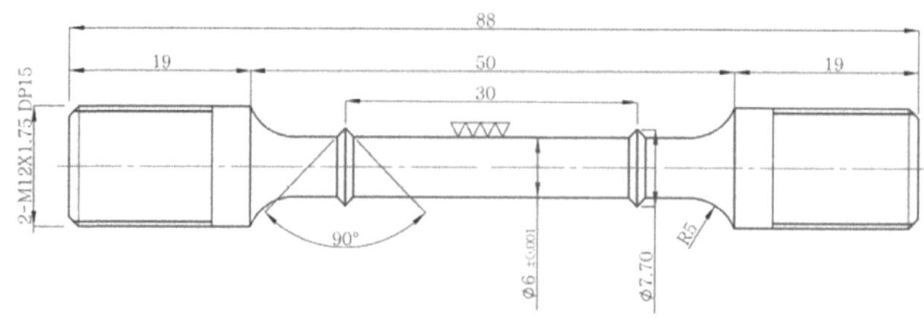


**Fig. 1 – Dimensions of the CC test specimen (unit: mm). CC, cyclic creep.**

| Table 2 – Employed CC test conditions of Gr. 91 steel at 600 °C. | | | | |
|---|---|---|---|---|
| Stress ratio (R) | Applied stress (MPa) | Mean stress (MPa) | Stress range ($\Delta\sigma$, $\sigma_R$) (MPa) | Hold time (min) |
| 0.50 | $\sigma_{max}$ = 213, $\sigma_{min}$ = 107 | 160 | 106 | 1, 3, 5, 10, 20, 30 |
| 0.80 | $\sigma_{max}$ = 178, $\sigma_{min}$ = 142 | 160 | 36 | 1, 3, 5, 10, 20, 30 |
| 0.85 | $\sigma_{max}$ = 173, $\sigma_{min}$ = 147 | 160 | 26 | 1, 3, 5, 10, 20, 30 |
| 0.90 | $\sigma_{max}$ = 168, $\sigma_{min}$ = 152 | 160 | 16 | 1, 3, 5, 10, 20, 30 |
| 0.95 | $\sigma_{max}$ = 164, $\sigma_{min}$ = 156 | 160 | 8 | 1, 3, 5, 10, 20, 30 |

CC, cyclic creep; Gr., grade.

## 3. Results and discussion

### 3.1. *Influence of HT*

Fig. 3 shows typical CC curves obtained for a 20-minute hold under a stress ratio of R = 0.85 at 600°C. The CC curves were similar in their shapes compared with the creep curves in monotonic creep tests. The black part in Fig. 3A indicates low magnification of stress profiles with time variations. Enlarged magnification of the stress and strain profiles with time for the secondary creep stages is clearly shown in Fig. 3B. It can also be clearly seen that the strain at the secondary creep stage constantly increased with time. This constancy of the secondary creep stage resulted from a balance between hardening and softening. In the present investigation, the value of the steady-state creep rate in the CC curves was determined from the mean value of the strain profile range. From the curves, it is justified to say that the stress and strain profiles with time variations were accurately monitored in the tests. For the CC test conditions of Gr. 91 steel at 600°C, data of CC properties such as the rupture time, steady-state creep rate, rupture elongation, and reduction of area were obtained. Influences of HT and stress ratio on the CC properties were systematically investigated.

Fig. 4 shows the variations of CC rupture time with an increase in HT at each stress ratio. The rupture time was influenced by the HT at R = 0.90 and R = 0.95; however, at the stress ratios of R = 0.50, 0.80, and 0.85, the rupture time was not influenced by the HT. At R = 0.90 and R = 0.95, the rupture time slightly decreases with an increase in the HT. The rupture time shows a decrease to 10-minute hold; after the 10-minute hold, the rupture time is almost similar regardless of any increase in the HT. Especially, at a low stress ratio of R = 0.50, the rupture time reaches its lowest value and there is no influence of the HT. The reason for this is that there was a much higher $\sigma_R$ = 106 MPa at R = 0.5 in the amplitude of the stress range ($\sigma_R$) than was the case for the values of R = 0.80, 0.85, 0.90, and 0.95. By contrast, the creep rate (or the steady-state creep rate) is almost constant regardless of any increase in the HT excepting the data after the 10-minute hold at R = 0.90 and R = 0.95. It is suggested that, compared with the other tested conditions, the longer rupture and decelerated creep rate after the 10-minute hold at R = 0.90 and R = 0.95 can be ascribed to the lower values in the applied load and stress range $\sigma_R$.

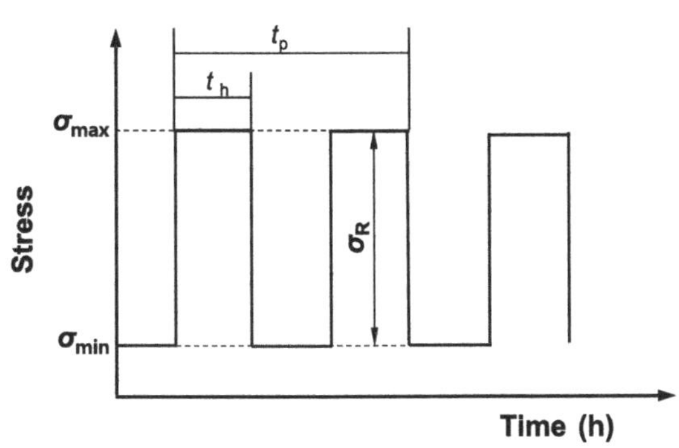


**Fig. 2 – Loading schedules of controlled CC tests. CC, cyclic creep.**

In addition, data of the rupture elongation with variation in the HT at each stress ratio display no significant difference of the HT, although they show some scatter for each test condition, as shown in Fig. 5. However, the longer-life data (HT = 1 minute and HT = 3 minutes at R = 0.90 and 0.95) were slightly lower for the rupture elongation data than for the other data. This result exactly supports that Gr. 91 steel shows a decrease for creep rupture ductility with an increase in rupture life.

### 3.2. *Influence of stress ratio*

Similarly, the influence of stress ratio on cyclic rupture properties was examined. As shown in Fig. 6, which provides the log–log plots of the rupture time versus the stress ratio at each HT, the rupture time increases with an increase in the stress ratio. However, it was found that the creep rate decreases with an increase in stress ratio (Fig. 7). As found for the influence of HT, the plot of the rupture time to stress ratio is almost linear, except for the four sets of data of HT = 1 minute and HT = 3 minutes at R = 0.9 and R = 0.95. In addition, data of the rupture elongation exhibit no significant difference from the stress ratios, although the data display some scatters for each test condition, as shown in Fig. 8.

Fig. 9 provides a plot of the number of cycles ($N_f$) as a function of the stress ratio at each HT. The value of $N_f$ increases with an increase in stress ratio. Especially, at HT = 1 minute and HT = 3 minutes, this value increases exponentially. This increase can be ascribed to the longer rupture time at this test condition (Fig. 6). It can be analyzed that, when compared with the other test conditions, the longer rupture time at this condition was due to the lower values in the load applied to the specimen during the CC tests and in the stress range $\sigma_R$.

In addition, since the peak applied stress (or $\sigma_{max}$) is not constant with each stress ratio, the measured creep rate and rupture life will vary significantly with $\sigma_{max}$, apart from the influence of the stress ratio, as shown by the data of Table 2.

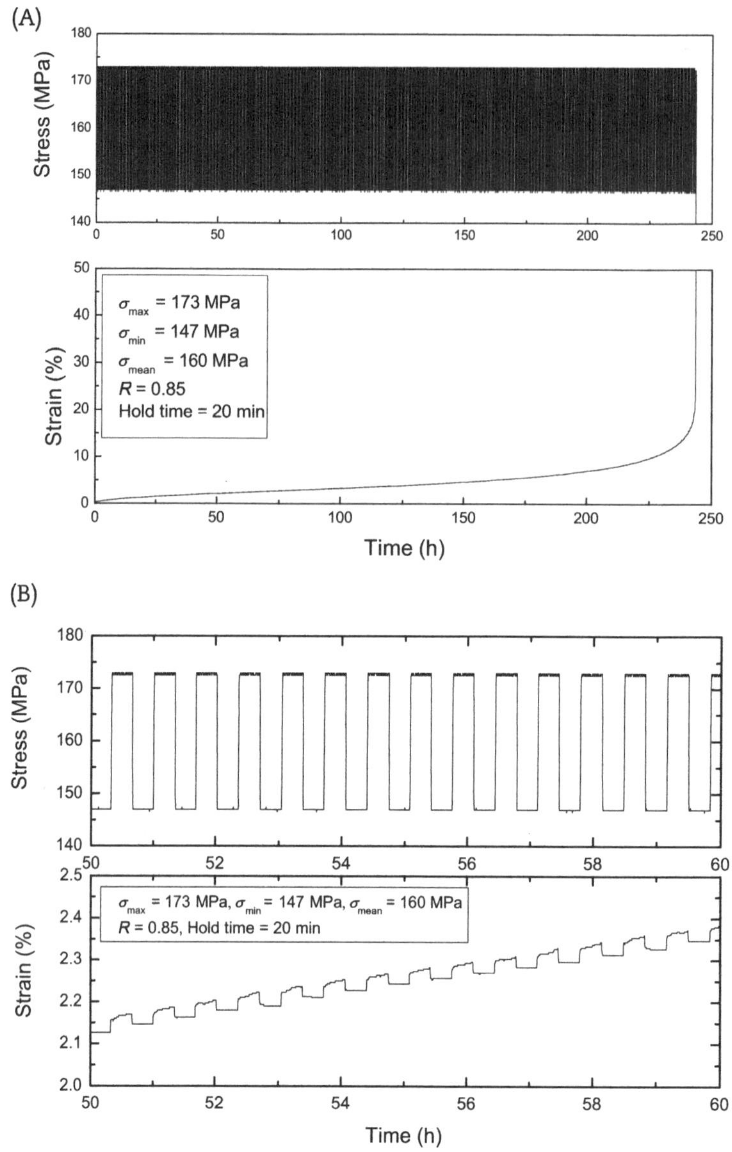


**Fig. 3 – Typical CC curves obtained at R = 0.85 (147 MPa/173 MPa) for 20-minute hold time at 600°C. (A) Total curve. (B) Enlarged stress–strain profiles. CC, cyclic creep.**

Using the peak applied stress, the CC and monotonic creep, which were obtained at an identical temperature of 600°C, were comparatively examined to clarify any acceleration or deceleration in the creep rate or rupture life. Figs. 10 and 11 show the comparisons of the rupture time and creep rate between the monotonic creep and CC for each HT. The CC with each HT decreases significantly in rupture time compared with the monotonic creep without HT, and the creep rate accelerates more than the monotonic creep, although the measured data in the CC tests were scattered with HT and $\sigma_{max}$. Accordingly, the peak applied stress becomes an important factor affecting the CC properties. It is thus illustrated that, owing to accumulated creep damage during repeated tension–tension loading cycles, the CC tests induce more reduction in creep life than the monotonic creep.

From the results of the present CC tests, the CC behavior can be analyzed as follows. Parameters of superimposed cyclic loading may vary according to a wide change of stress and time such as HT, number of cycles, stress ratio, peak stress, stress range, etc. The influence of loading conditions on CC properties can either accelerate or decelerate, depending on the material structure, temperature, and stress conditions [11]. The observed influence of cyclic loading on creep deformation can be attributed to mechanisms involving inelastic

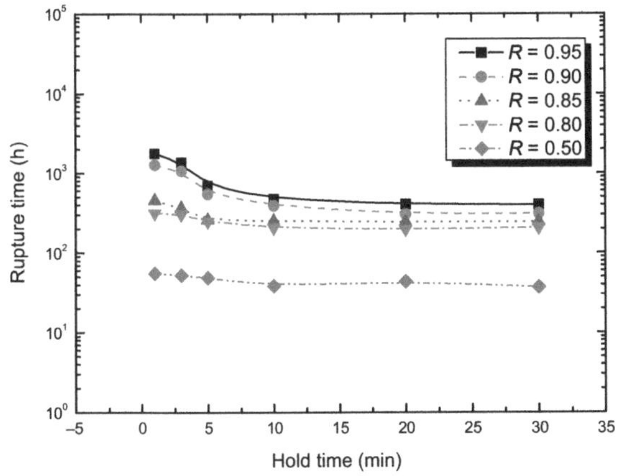


**Fig. 4 – Variations of cyclic rupture time with hold time at each stress ratio.**

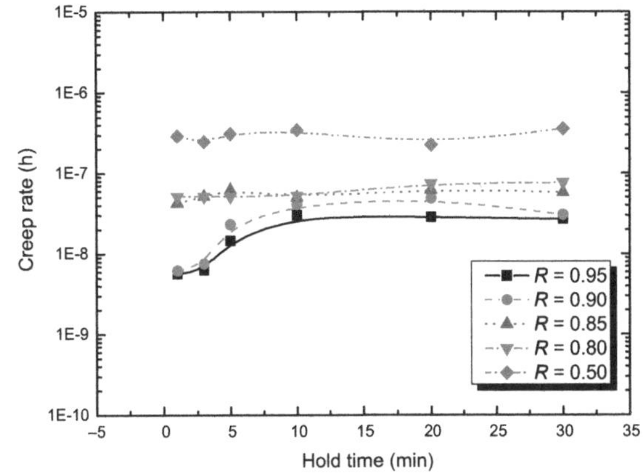


**Fig. 7 – Variations of creep rate with hold time at each stress ratio.**

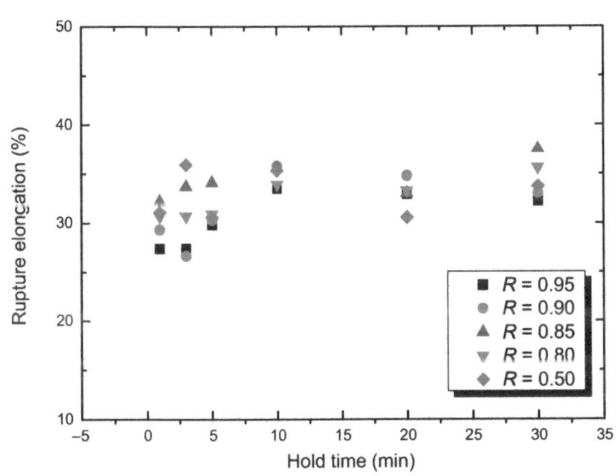


**Fig. 5 – Variations of rupture elongation with hold time at each stress ratio.**

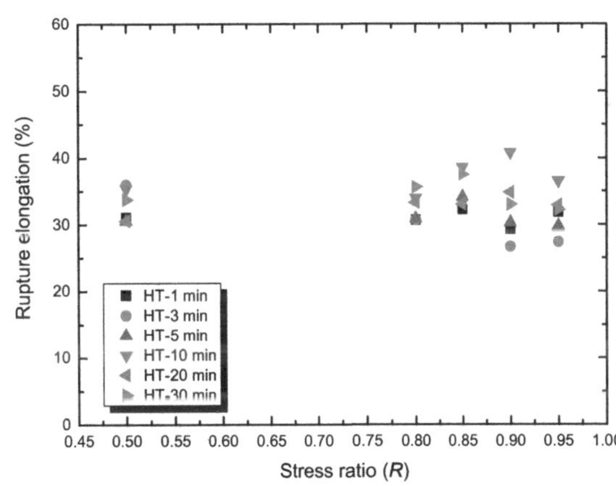


**Fig. 8 – Plot of rupture time versus stress ratio at each hold time. HT, hold time.**

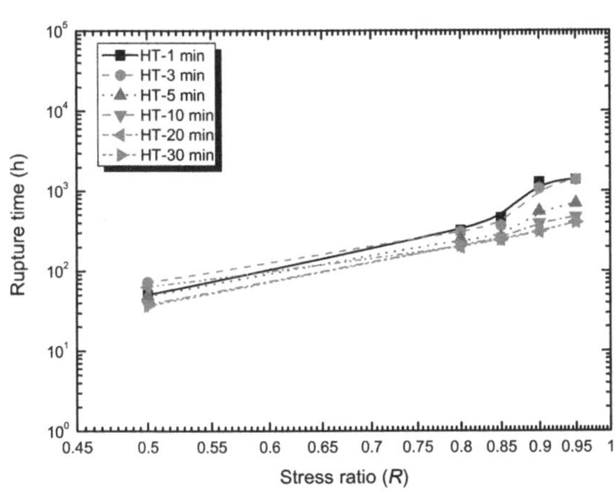


**Fig. 6 – Log–log plot of cyclic rupture time versus stress ratio at each hold time. HT, hold time.**

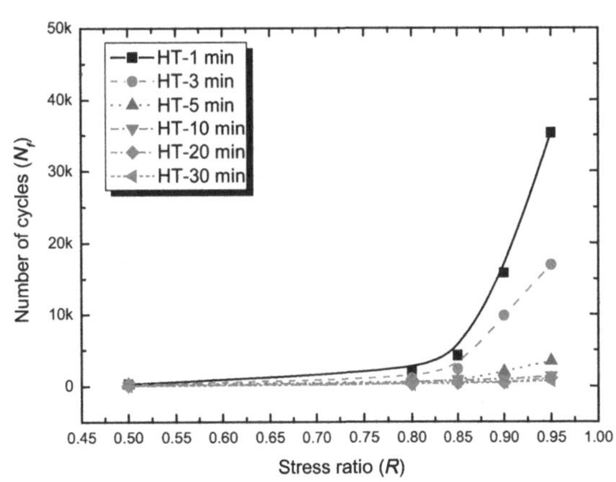


**Fig. 9 – Plot of the number of cycles versus stress ratio at each hold time. HT, hold time.**

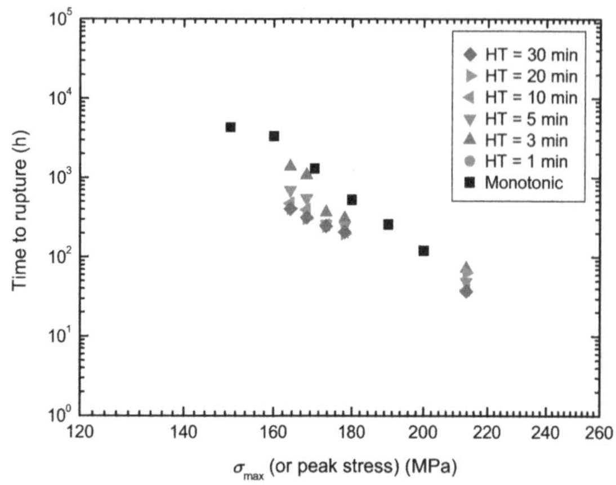


**Fig. 10 – Comparison of rupture time between monotonic creep without hold time and cyclic creep with each hold time when peak applied stress is used. HT, hold time.**

recovery, precipitation of carbides, and recovery of dislocation substructure. Under the tension–tension CC conditions employed in the present investigation, acceleration of the creep rate (or reduction of the rupture time) can be ascribed to stress factors such as a lower stress ratio, a higher stress range, and higher peak stress. The reason for this is that recovery of dislocation structure in higher fluctuation in the stress range is difficult during the CC tests. In addition, in the case of monotonic creep without HT under tension–tension cyclic loading, a higher frequency condition led to acceleration in the creep strain rate [11]. However, in the present investigation, regarding the increase in the rupture time at the higher frequency of short HT (HT = 1–3 minutes) at R = 0.90 and R = 0.95 (Fig. 4), this increase can be ascribed to the lower values of maximum stress ($\sigma_{max}$ = 168 at R = 0.90, $\sigma_{max}$ = 164 at R = 0.95) applied to the specimen, and also to the

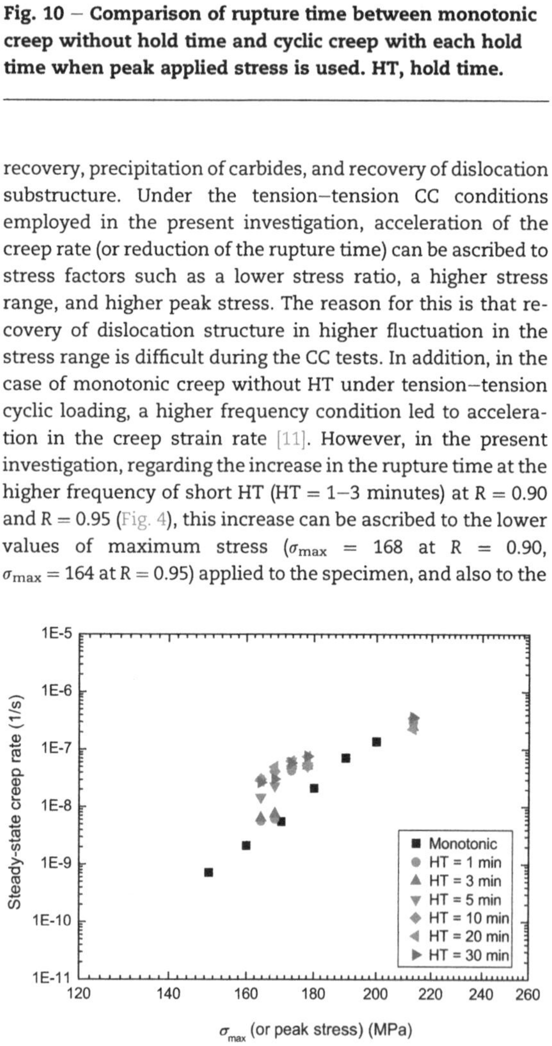


**Fig. 11 – Comparison of steady-state creep rate between monotonic creep without hold time and cyclic creep with each hold time when peak applied stress is used. HT, hold time.**

lower values in the stress range ($\sigma_R$ = 16 MPa at R = 0.90, $\sigma_R$ = 8 MPa at R = 0.95) when compared with those of other tested conditions, as shown in Table 2. Accordingly, it is found that the material responses of the employed tension (+)–tension (+) loading cyclic tests are different from those of tension (+)–compression (–) in the creep–fatigue loading cycles. It is thought that CC rupture properties depend dominantly on stress factors rather than on HT.

### 3.3. *Analysis of fracture micrographs*

Fig. 12 shows typical SEM photos taken in the cross-sectional and longitudinal directions; the figure also provides images of the fracture microstructures after the CC tests for HT = 10 minutes at R = 0.8. A ductile cup and cone fracture can easily be observed as shown in the failure region (b). The failure of Gr. 91 steel exhibits three regions: deformation region (a), slant region (c), and core region (d). The deformation region is due to plastic deformation during the CC tests, the slant region is due to shear stress in the middle area between the slant and the core regions, and the core region of the inside failed at the final step of the CC tests. The reason for this failure is that the creep deformation (or cracks) develops toward the inside after initiating at the outside of the surface. In the central equiaxed regions, the presence of many voids is observed. The CC specimens of Gr. 91 steel display typically transgranular fracture resulting from microvoid coalescence. It was observed that all of rupture specimens exhibit similar transgranular fractures regardless of the CC test conditions. In addition, because microvoids are closely related to creep properties such as rupture time and creep rate, fracture microstructures were observed and their quantity and size were measured using image analysis with the stress ratios of R = 0.5, 0.80, 0.85, 0.90, and 0.95.

Fig. 13 shows SEM photos taken in the CC tested specimens at the stress ratios of R = 0.5, 0.80, 0.85, 0.90, and 0.95. Distribution of voids was obtained by image analysis of the SEM photos shown in Fig. 13. Histograms showing the distribution of the void size (diameter) measured with the stress ratios of R = 0.5, 0.80, 0.85, 0.90, and 0.95 in the CC tested specimens are plotted in Fig. 14. The histograms clearly show an increase in the void area with an increase in the stress ratio. The void fraction to the total area increases from 2.86% to 6.69% with an increase of the stress ratio. The number of voids increases with the increase in the stress ratio, as listed Table 3. The presence of fine-sized voids at low stress dominates the nucleation process, while the presence of large-sized voids dominates the growth process at high stress. It is suggested that the microvoids originally nucleated at the sites of the precipitates. Then, the voids led to growth and increased in amount during the CC tests; finally, fracture occurred due to the coalescence of microvoids. A summary of void measurement with stress ratios in the CC test conditions is listed in Table 3.

In addition, microvoids with a variation of HT were also observed using SEM images. It is found that the quantity of microvoids was almost identical regardless of any increase in the HT. This means that the cyclic rupture time was dominantly affected by the stress ratio rather than by the HT, as shown in Fig. 4. Thus, it is suggested that the CC properties

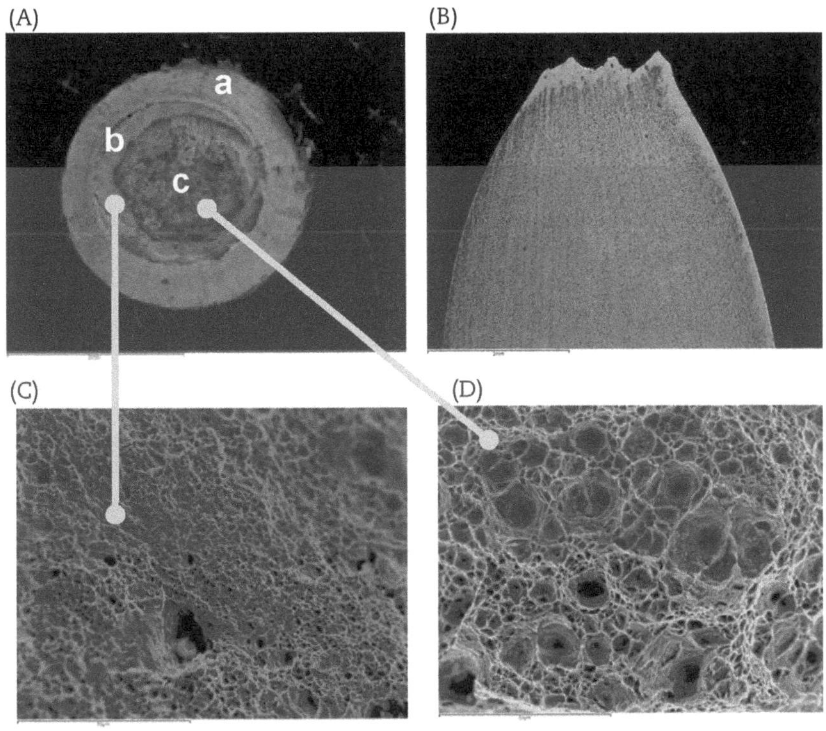


**Fig. 12 – Typical SEM photos taken after the CC test for HT = 10 minutes at R = 0.8. (A) Fracture surface. (B) Cross-sectional view. (C) Fracture surface in slant region. (D) Fracture surface in core region. CC, cyclic creep; HT, hold time; SEM, scanning electron microscope.**

under tension–tension loading cycles depend significantly on stress factors such as stress ratio, stress range, and maximum stress.

Fig. 15 shows initial microstructures of the "as received" material before the CC tests in the normalized and tempered (N + T) conditions of Gr. 91 steel. The microstructure is tempered martensite strengthened by carbides, and it has prior austenite grains containing martensite packets, blocks, and laths (Fig. 15B), as shown in the schematic diagram of the martensitic microstructure (Fig. 15C). Grain boundary carbides consisted almost entirely of $M_{23}C_6$. Round NbC and rod VC at the lath boundary enhanced lath stability: cuboidal $Cr_{23}C_6$ at the prior austenite grain boundary and Nb-rich MX precipitates were uniformly precipitated with sizes under 20 nm (Fig. 15D). As described above, in the N + T conditions before the CC tests, the microstructure was clearly composed of tempered lath martensite and precipitated at the prior austenite grain and martensite lath boundaries; the intralath matrix regions contained fine precipitates (Fig. 15B). The distinction between the grain and lath boundaries and the intralath matrix regions can clearly be observed. However, in the deformed CC specimen at 600°C (Fig. 16A), precipitates of the grain and lath boundaries, along with the matrix precipitates in the CC exposure, were coarsened and the lath width had increased compared with those characteristics of the initial Gr. 91 steel (N + T). The distinction between the grain and lath boundaries and the intralath matrix regions becomes unclear with precipitate coarsening due to exposure during the CC tests. By contrast, in the grip-section region (undeformed stress-free region), the precipitates are well connected along the prior austenite grain boundary, as shown clearly in Fig. 16B. However, in the deformed gage-section region, the prior austenite grain boundary was severely deformed with grain flow along the stress direction in the fracture zone. Owing to the cyclic stress, the CC tested specimens formed more precipitates when compared with the undeformed grip section regions (Fig. 16B).

It is reported that Gr. 91 steel in N + T conditions is strengthened by both solid solution and precipitation-strengthening mechanisms, with Cr, Mo, Nb, V, and N contributing to the formation of precipitates that act to improve the strength and limit the grain size. $M_{23}C_6$ precipitates occur at prior austenite and martensite lath boundaries together with finer V- and Nb-rich carbonitrides that precipitated within the grains ($M_2X$- and MX-type precipitates). At a higher temperature, $M_{23}C_6$ precipitates coarsen due to Oswald ripening reducing grain boundary pinning, which leads to recovery and grain growth, with corresponding reductions in dislocation densities [16,18]. Accordingly, enhanced coarsening of the precipitates in the CC specimens was due to Oswald ripening in terms of longer exposure and stress in the CC tests.

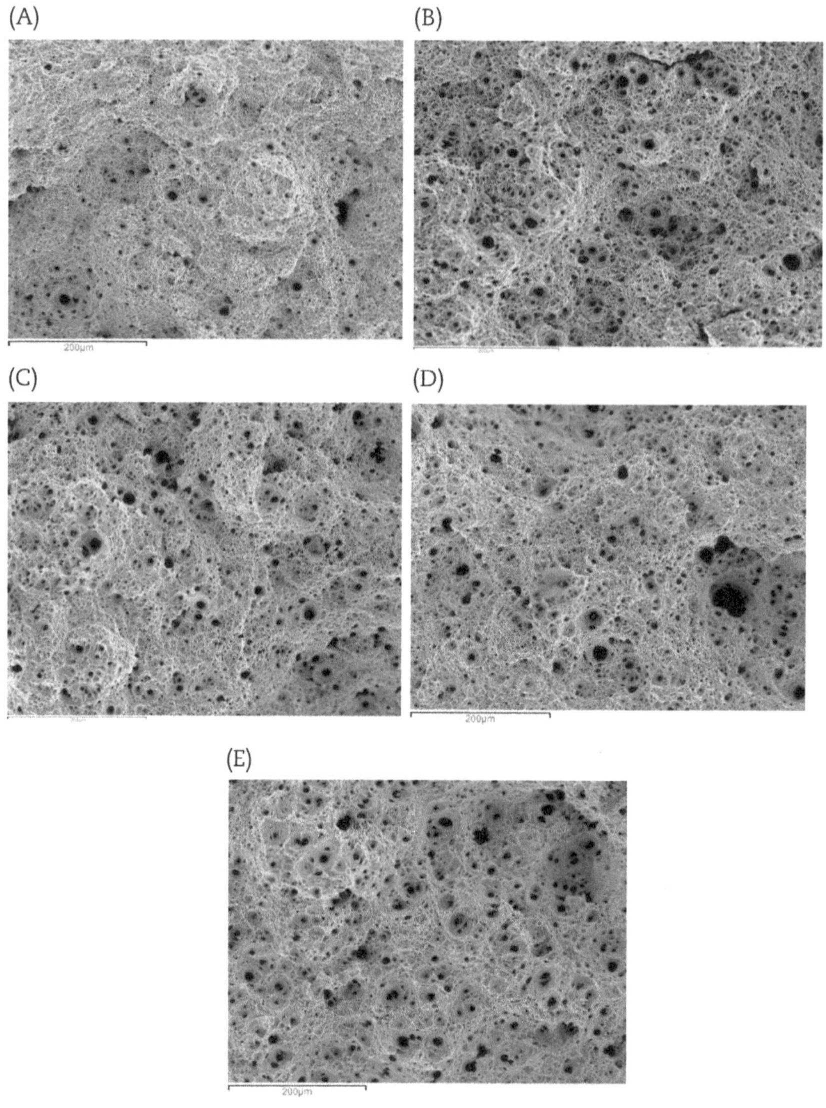

**Fig. 13 – SEM fractographs showing creep voids in the CC tested specimens at different stress ratios. (A) R = 0.5. (B) R = 0.80. (C) R = 0.85. (D) R = 0.90. (E) R = 0.95. CC, cyclic creep; SEM, scanning electron microscope.**

## 4. Conclusions

Influences of HT and stress ratio on the CC properties of Gr. 91 steel were systemically investigated in a wide range of CC tests, which were performed with various HTs and stress ratios under controlled tension–tension loading cycles at 600°C. Under the influence of HT, the rupture time increased for an HT of 10 minutes at R = 0.90 and R = 0.95, but no influence was found at the stress ratios of R = 0.50, 0.80, and 0.85. The creep rate was almost constant regardless of the increase in the HT except for the data obtained after the 10-minute hold at R = 0.90 and R = 0.95. Rupture elongation showed almost identical behavior regardless of the HT. Then, in terms of the influence of stress ratio, rupture time increased with an increase in the stress ratio, but the creep rate decreased, except for the data obtained at HT = 1 minutes and 3 minutes at R = 0.9 and R = 0.95. Rupture elongation exhibited no significant difference from the stress ratios. Compared with the monotonic creep, the CC showed a reduction in the rupture time and acceleration in the creep rate. It was suggested that CC properties depended dominantly on stress factors such as stress ratio, stress range, and maximum stress rather than any dependency on HT. Fracture surfaces in the CC tested specimens of Gr. 91 steel displayed transgranular fractures resulting from microvoid coalescence; the quantity of microvoids increased with an increase in the stress ratio. Enhanced coarsening of the precipitates in the CC test specimens was found under all conditions.

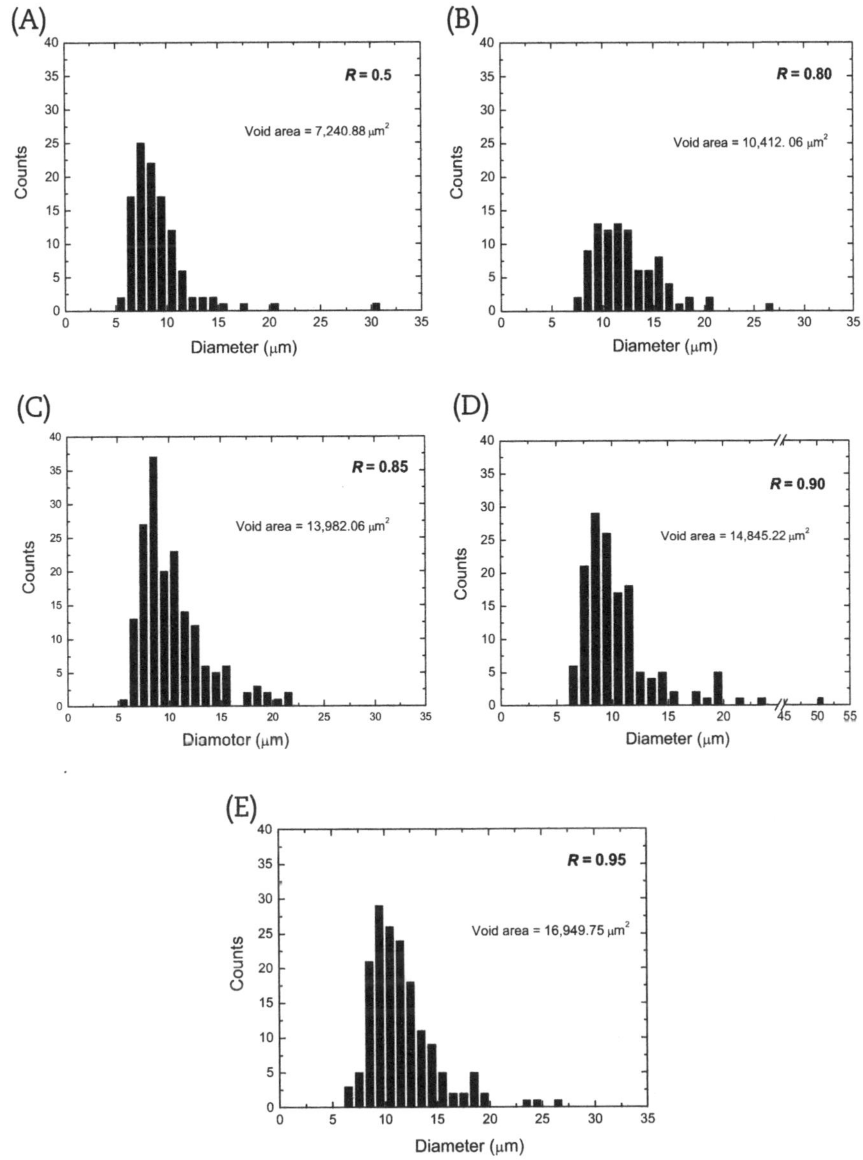


**Fig. 14 – Histogram showing distribution of diameter size in microvoids measured with various stress ratios in the CC tested specimens. (A) R = 0.5. (B) R = 0.80. (C) R = 0.85. (D) R = 0.90. (E) R = 0.95. CC, cyclic creep.**

**Table 3 – Summary of void measurement with stress ratios in the CC test conditions.**

| Stress ratio (R) | Total area ($\mu m^2$) | No. of void | Mean diameter of void (μm) | Void area ($\mu m^2$) | Fraction of void (%) |
|---|---|---|---|---|---|
| 0.50 | 253,350 | 111 | 9.17 | 7,240.88 | 2.86 |
| 0.80 | | 91 | 12.29 | 10,412.06 | 4.11 |
| 0.85 | | 178 | 10.28 | 13,982.06 | 5.52 |
| 0.90 | | 146 | 10.96 | 14,845.22 | 5.86 |
| 0.95 | | 165 | 11.60 | 16,949.75 | 6.69 |

CC, cyclic creep.

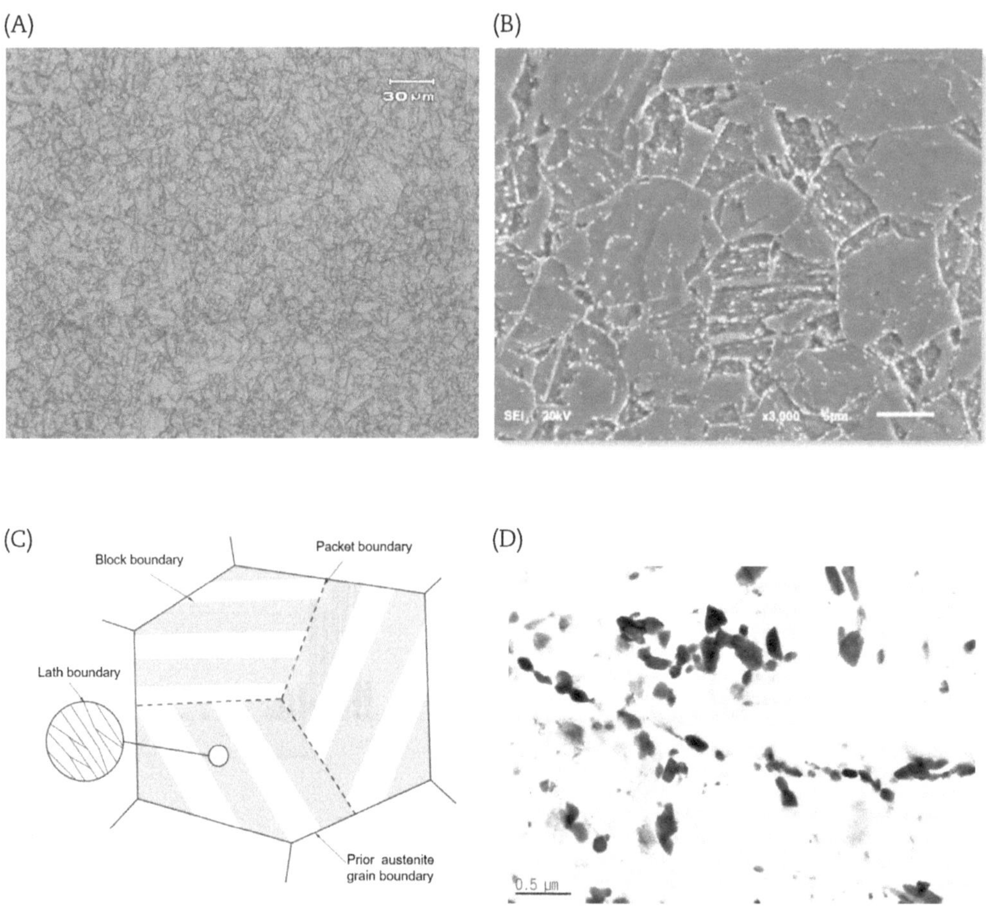


**Fig. 15 – Initial microstructures in the N + T conditions before the CC tests of Gr. 91 steel. (A) OM structure. (B) SEM microstructure for precipitates. (C) Schematic diagram for microstructure. (D) Formed precipitates (replica). CC, cyclic creep; Gr., grade; N + T, normalized and tempered; OM, optical microscope; SEM, scanning electron microscope.**

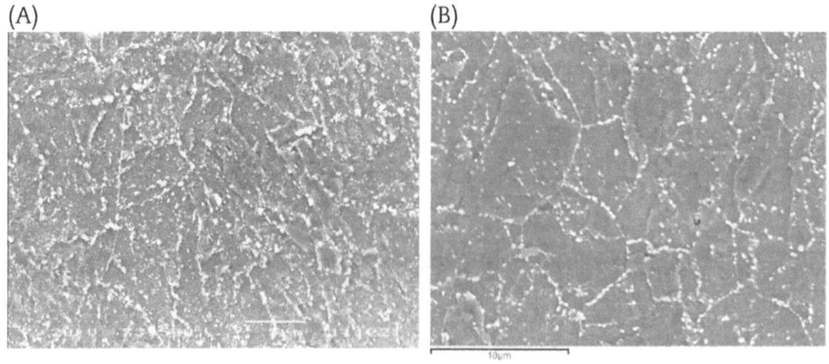


**Fig. 16 – Precipitates formed in the deformed gage-section and undeformed grip-section regions after the CC tests for HT = 1 minute at R = 0.5. (A) Deformed gage-section region. (B) Undeformed grip-section region. CC, cyclic creep; HT, hold time.**

## Conflicts of interest

No conflicts of interest.

## Acknowledgments

This study has been carried out as a part of the Innovative R&D Project of Korea Atomic Energy Research Institute.

## REFERENCES

[1] W.G. Kim, J.Y. Park, S.J. Kim, J. Jang, Reliability assessment of creeprupturelifeforGr.91steel,Mater.Des.51(2013)1045–1051.

[2] E. Issac Samual, B.K. Choudhary, K. Bhau Sankara Rao, Influence of post weld heat treatment on tensile properties of 9Cr-1Mo ferritic steel base metal, Mater. Sci. Technol. 23 (2007) 992–998.

[3] S.L. Mannan, S.C. Chetal, B. Raj, S.B. Bhoje, Selection of materials for prototype fast breeder reactor, Trans. Indian Inst. Met. 56 (2003) 155–178.

[4] B.K. Choudhary, V.S. Srinivasan, M.D. Mathew, Influence of strain rate and temperature on tensile properties of 9Cr–1Mo ferritic steel, Mater. High Temp. 28 (2011) 155–161.

[5] B.K. Choudhary, E. Issac Samual, Creep behavior of modified 9Cr–1Mo ferritic steel, J. Nucl. Mater. 412 (2011) 82–89.

[6] B. Raj, B.K. Choudhary, A perspective on creep and fatigue issues in sodium cooled fast reactors, Trans. Indian Inst. Met. 63 (2010) 75–84.

[7] K. Kimura, Y. Toda, H. Kushima, K. Sawada, Creep strength of high chromium steel with ferrite matrix, in: I.A. Shibli, S.R. Holdsworth (Eds.), Proceedings of the 2nd ECCC Creep Conference on Creep & Fracture in High Temperature Components, Zurich ETD, 2009, pp. 935–949.

[8] V.S. Srinivasan, B.K. Choudhary, M.D. Mathew, T. Jayakumar, Long-term creep-rupture strength prediction for modified 9Cr–1Mo ferritic steel and type 316L(N) austenitic stainless steel, Mater. High Temp. 29 (2012) 41–48.

[9] W.G. Kim, J.Y. Park, S.D. Hong, S.J. Kim, Probabilistic assessment of creep crack growth rate for Gr. 91 steel, Nucl. Eng. Des. 241 (2011) 3580–3586.

[10] D.K. Shetty, T. Mura, M. Meshii, Analysis of creep deformation under cyclic loading conditions, Mater. Sci. Eng. 20 (1975) 261–266.

[11] J. Zrnik, J.A. Wang, Y. Yu, L. Peijing, P. Hornak, Influence of cycling frequency on cyclic creep characteristics of nickel base single-crystal superalloy, Mater. Sci. Eng. A A234–A236 (1997) 884–888.

[12] J.H. Eom, D.H. Shin, S.W. Nam, Effects of stress amplitude and friction stress on cyclic creep deformation, J. Kor. Inst. Met. 20 (1982) 922–926.

[13] Y.K. Park, T.S. Kim, J.H. Choi, M.Y. Wee, A study on cyclic creep behavior of Zircaloy-4 at 0.3 Tm, J. Kor. Inst. Met. Mater. 38 (2000) 624–628.

[14] M. Boulbibane, A.R.S. Ponter, A method for the evaluation of design limits for structural materials in a cyclic state of creep, Eur. J. Mech. A Solid 21 (2002) 899–914.

[15] H.D. Chandler, S. Kwofie, A description of cyclic creep under conditions of axial cyclic and mean stresses, Int. J. Fatigue 27 (2005) 541–545.

[16] W.G. Kim, J.Y. Park, I.M.W. Ekaputra, S.J. Kim, J. Jang, Cyclic creep behavior under ten-tension loading cycles with hold time of modified 9Cr-1Mo steel, Mater. High Temp. 31 (2014) 249–257.

[17] ASTM International, Standard test method for conducting creep, creep-rupture, and stress-rupture tests of metallic materials, ASTM (2011). E139–11.

[18] C. Qin, S.V. Hainsworth, A. Strang, P.F. Morris, P.D. Clarke, A.P. Backhouse, TEM studies of microstructural evolution in creep exposed E911, in: I.A. Shibli, S.R. Holdsworth (Eds.), Proceedings of the 2nd ECCC Creep Conf. on Creep & Fracture in High Temperature Components, Zurich ETD, 2009, pp. 889–899.

Nuclear Engineering and Design 306 (2016) 177–185

Contents lists available at ScienceDirect

Nuclear Engineering and Design

journal homepage: www.elsevier.com/locate/nucengdes

# Investigation of creep rupture properties in air and He environments of alloy 617 at 800 °C

Woo-Gon Kim [a,*], I.M.W. Ekaputra [b], Jae-Young Park [b], Min-Hwan Kim [a], Yong-Wan Kim [a]

[a] Korea Atomic Energy Research Institute, 989-111 Daedeok-daero, Yuseong-gu, Daejeon 305-353, Republic of Korea
[b] Pukyong National University, Busan 608-739, Republic of Korea

ARTICLE INFO

Article history:
Received 31 March 2015
Received in revised form 22 March 2016
Accepted 29 April 2016
Available online 7 July 2016

ABSTRACT

Creep rupture properties for Alloy 617 were investigated by a series of creep tests under different applied stresses in air and He environments at 800 °C. The creep rupture time in air and He environments exhibited almost similar life in a short rupture time. However, when the creep rupture time reaches above 3000 h, the creep life in the He environment reduced compared with those of the air environment. The creep strain rate in the He environment was a little faster than that in the air environment above 3000 h. The reduction of creep life in the He environment was due to the difference of various microstructure features such as the carbide depleted zone, oxidation structures, surface cracking, voids below the surface, and voids in the matrix in air and He environments. Alloy 617 followed Norton's power law and the Monkman–Grant relationship well. As the stress decreased, the creep ductility decreased slightly. The thickness of the outer and internal oxide layers presented the trend of a parabolic increase with an increase in creep rupture time in both the air and He environments. The thickness in the He environment was found to be thicker than in the air environment, although pure helium gas of 99.999% was used in the present investigation. The differences in the oxide-layer thickness caused detrimental effects on the creep resistance, even in a low oxygen-containing He agent.



## 1. Introduction

A very high temperature reactor (VHTR) is one of the most promising Generation-IV reactor types for the economic production of electricity and hydrogen. Its major components are the reactor internals, reactor pressure vessel (RPV), piping, hot gas ducts (HGD), and intermediate heat exchangers (IHX), as shown in Fig. 1. The IHX among these is a key component. The heat of the primary helium (He) circuit transfers to the secondary helium loop through the IHX, because its component needs high-temperature creep resistance in the He environment, and requires good oxidation resistance, corrosion resistance, and phase stability at high temperatures (Kim et al., 2007, 2008, 2009, 2010, 2012a,b; Ennis et al.,1984; Huchtemann, 1989).

At present, the Korea Atomic Energy Research Institute (KAERI) is developing a nuclear hydrogen development and demonstration (NHDD) plan of a capacity with 200 $MW_{th}$ in the thermal and core outlet temperature of 950 °C. It has been carried out in a government funded project called "Development of Key Technologies for Nuclear Hydrogen" since 2006 (Chang et al., 2007; Kim et al., 2013). Since the VHTR system is designed for a life span of 60 years operating at 950 °C and 7 MPa in helium impurities, long-term creep and oxidation behaviors are considered as some of the most important material properties (Kim et al., 2014).

Of the existing alloys, nickel-based Alloy 617 is the leading candidate for use in next-generation nuclear plant (NGNP) heat exchangers because it has the highest creep strength of solid solution alloys under consideration for temperatures above 850 °C (Schneider et al., 1984; Schubert et al., 1983). However, until now, the ASME design code for Alloy 617 has not been developed for design use. Material works for developing the ASME NH code for Alloy 617 are ongoing according to a NGNP research and development plan. Various material data for Alloy 617 which will be used for design application have been uploaded and gathered to the GIF (Gen-IV Forum) Materials Handbook Database (DB) website arranged by GIF countries. Thus, to develop the ASME NH Code through this plan, a number of creep property data should be provided to complete the database, and the creep behavior should be investigated through a systematical analysis with creep temperature and stress conditions in air and He environments, because the creep rupture time varies widely with the environment interactions between He impurities and Alloy 617 (Jang et al., 2008; Dewson and Li, 2005; Cook, 1984; Christ et al., 1987; Lee et al.,

* Corresponding author. Tel.: +82 42 868 2493.
E-mail address: wgkim@kaeri.re.k (W.-G. Kim).

http://dx.doi.org/10.1016/j.nucengdes.2016.04.042
0029-5493/

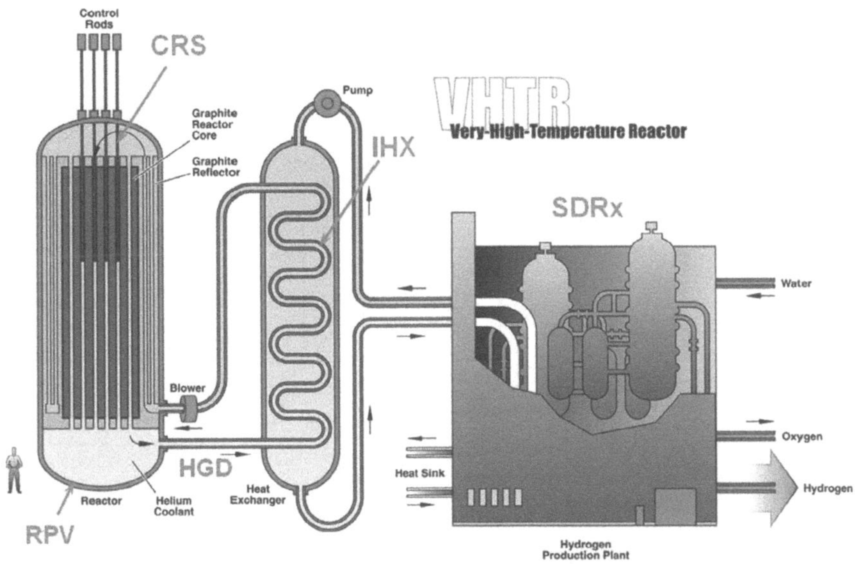


**Fig. 1.** Major components of VHTR system for nuclear hydrogen production.

2013; Schubert et al., 1984; Natesan et al., 2003). However, it has not been well established whether the test temperature and minor impure gases in He drastically affect the rupture time. The creep data for Alloy 617 in air and He environments are still insufficient for the design application, and the relationships between creep and oxidation behaviors are not yet well understood in terms of the He effect.

In this study, the creep rupture data tested in Alloy 617 in air and He environments were obtained through a series of creep tests conducted with different applied stress levels at 800 °C. The creep rupture properties in the air and He environments were comparatively investigated. In addition, the creep oxidation behavior was examined by measuring an oxide-layer thickness formed in the air and He environments during the creep rupture time.

## 2. Experimental procedures

Commercial grade nickel-based superalloy, Alloy 617 (Haynes 617) was used for this study, and was in a hot-rolled plate form with a thickness of 15.875 mm. The chemical composition is given in Table 1. All element compositions are included in the ASTM specification. Creep test specimens in the air and He environments were fabricated in the same cylindrical rigs with a 30 mm gauge length and 6 mm diameter. The gage section was parallel to the longitudinal rolling direction. Circular grooves were machined at both ends beyond the shoulder region of these specimens to attach extensometers for monitoring the elongation during creep tests.

The loading frames used in the constant-load creep tests had a lever arm ratio of 20:1. A split 3-zone furnace was used to heat the specimens in the air and He environments at 800 °C. K-type thermocouples were used to monitor the temperatures within the gage section of the specimens. Before testing, the thermocouple was calibrated using a reference thermometer. The materials of the pull rod and jig for the creep tests were used for a nickel-based superalloy to sufficiently endure the oxidation and thermal degradation during the creep test period. The creep tests were conducted under 100 MPa, 90 MPa, 80 MPa, 70 MPa, and 60 MPa in air and 100 MPa, 90 MPa, 80 MPa, 70 MPa, 60 MPa, and 50 MPa in He environment at an identical temperature of 800 °C. The overall creep rates were determined from elongations measured by extensometers equipped at the circular grooves. Creep strain data with elapsed times were taken automatically by a PC.

Before the creep tests in the He environment, a vacuum chamber made for a quartz tube was purged three or four times by a vacuum pump to remove some impurities in the chamber. During the creep tests, a pure He gas of 99.999% in the present investigation was supplied to the specimens equipped in the quartz tube. The impurity concentrations in the pure He gas were $O_2 < 1.0$ ppm, $N_2 < 5.0$ ppm and $H_2O < 1.0$ ppm. The flow rate of the He gas during the creep test period was controlled using a flow meter to under 20 $cm^3$/min. The control of the other impurities such as $CO_2$, CO, $CH_4$, and $H_2$ in He gas simulating a VHTR condition will be investigated in further tests. A schematic diagram of the He-gas flow for creep tests in the He environment is shown in Fig. 2. The creep testing apparatus in the He environment was given in detail in the author's previous study (Kim et al., 2014).

## 3. Results and discussion

### 3.1. Creep rupture properties

The creep rupture data for Alloy 617 such as the rupture time, minimum creep rate, rupture elongation, and reduction in area were obtained by the creep tests in the air and He environments

**Table 1**
Chemical composition of Alloy 617 (wt.%).

| | | C | Ni | Fe | Si | Mn | Co | Cr | Ti | P | S | Mo | Al | B | Cu |
|---|---|---|---|---|---|---|---|---|---|---|---|---|---|---|---|
| ASTM Spec. | Min. | 0.05 | Bal. | – | – | – | 10.0 | 20.0 | – | – | – | 8.00 | 0.80 | – | – |
| | Max. | 0.15 | Bal. | 3.00 | 0.50 | 0.50 | 15.0 | 24.0 | 0.60 | 0.015 | 0.015 | 10.0 | 1.50 | 0.006 | 0.50 |
| Haynes617 | | 0.08 | 53.11 | 0.949 | 0.084 | 0.029 | 12.3 | 22.2 | 0.41 | 0.003 | <0.002 | 9.5 | 1.06 | <0.002 | 0.0268 |

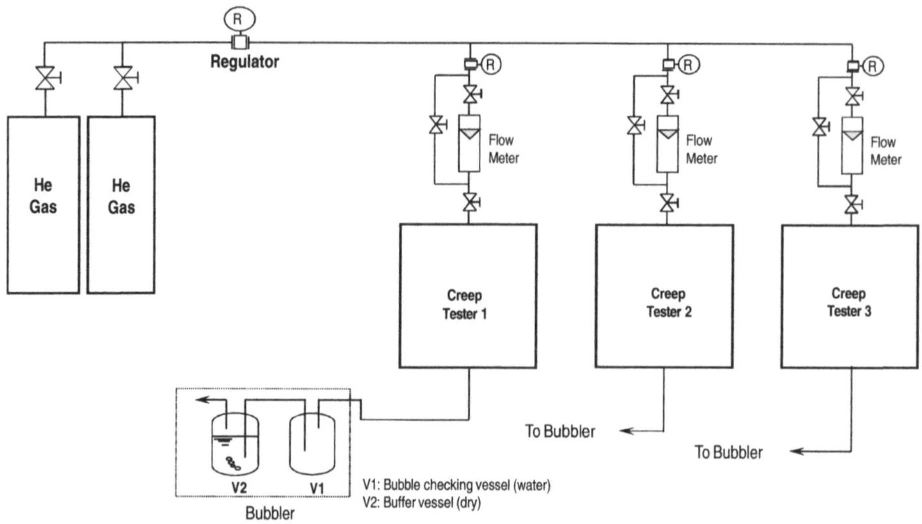


**Fig. 2.** A schematic diagram for creep tests in He environment.

at 800 °C. From the creep tests, long-term rupture data reaching 16,777 h (1.91y) under 60 MPa in air and 18,624 h (2.12y) under 50 MPa in the He environment were achieved. At present, a specimen for 50 MPa in air is running under 15,000 h in a creep machine. Using the creep experimental results, creep rupture properties between the air and He environments were comparatively investigated.

Fig. 3 shows a comparison of typical creep curves for stresses in the air and He environments at 800 °C. The creep curve of Alloy 617 does not show "textbook creep" behavior of well-defined primary, secondary, and tertiary creep stages. Alloy 617 exhibits little primary creep strain, an unclear secondary creep stage, and an unclear onset of the tertiary creep stage; in addition, the tertiary creep stage is initiated from a low strain level. No difference was found in the shapes of the creep curves between the air and He environments. Alloy 617 exhibits good ductility (>30%) in spite of the long duration creep time of 16,777 h at 800 °C.

Fig. 4 shows the plot of stress vs. rupture time in the air and He environments at 800 °C. The creep rupture time in the air and He environments was almost similar during a short rupture time. However, when the creep rupture time reaches above 3000 h, the creep stresses in the He environment are slightly reduced compared with those of the air environment. Also, the creep strain rate in the He environment is faster than that in the air environment, as shown in Fig. 5. The reduction of the creep life in the He environment is considered because of the difference of various microstructure features in air and He environments. Detail description is given in Section 3.3.

In this creep condition for Alloy 617, since the creep deformation corresponds to the power-law creep region, creep deformation is governed by the dislocation climb (Kim et al., 2010, 2014). Thus, the power-law relationship between the minimum creep rate and stress showed a good linearity, as shown in Fig. 6. The constants of Norton's power law, $\dot{\varepsilon}_m = A\sigma^n$, can be reasonably obtained. The $A$ values were obtained with about $3.25 \times 10^{-20}$ (MPa$^{-n}$ h$^{-1}$) in air, and $1.63 \times 10^{-15}$ (MPa$^{-n}$ h$^{-1}$) in He, and the $n$ values were 7.8 in air and 5.5 in He. There was a slight difference in the air and He environments for the $A$ and $n$ values. In addition, the Monkman–

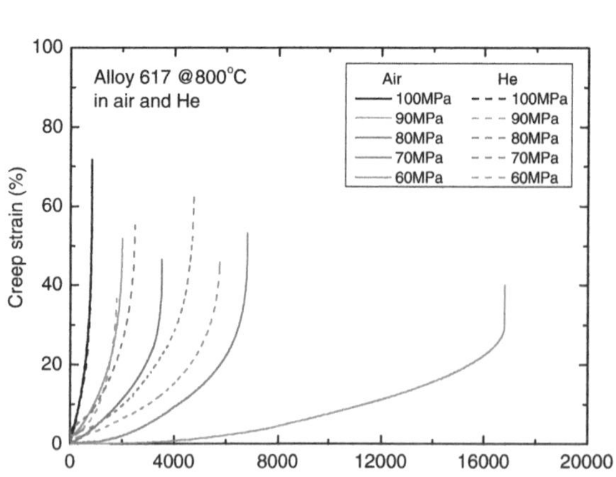


**Fig. 3.** Typical creep curves obtained in the air and He environments at 800 °C.

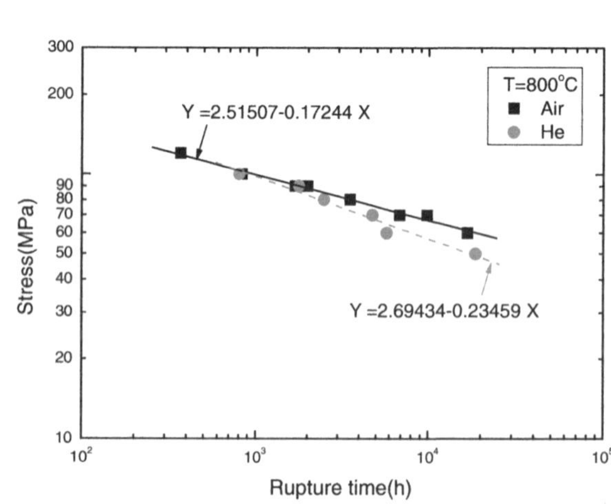


**Fig. 4.** Comparison of stress vs. rupture time in the air and He environments at 800 °C.

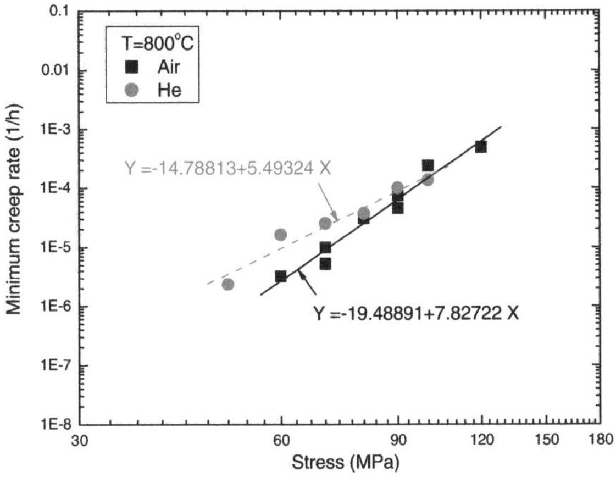


**Fig. 5.** Comparison of minimum creep rate vs. stress in the air and He environments at 800 °C.

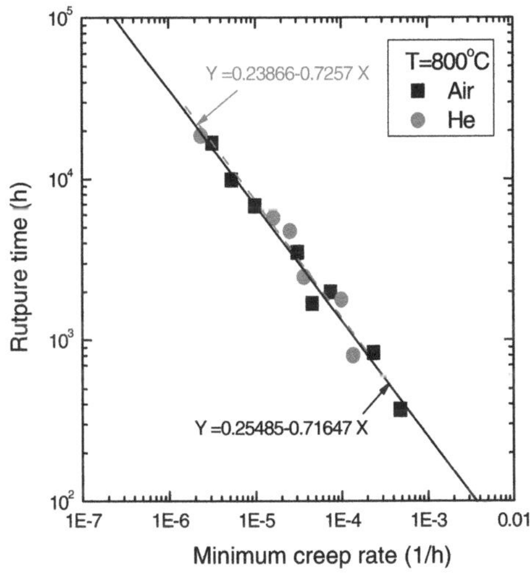


**Fig. 6.** Comparison of rupture time vs. minimum creep rate in the air and He environments at 800 °C.

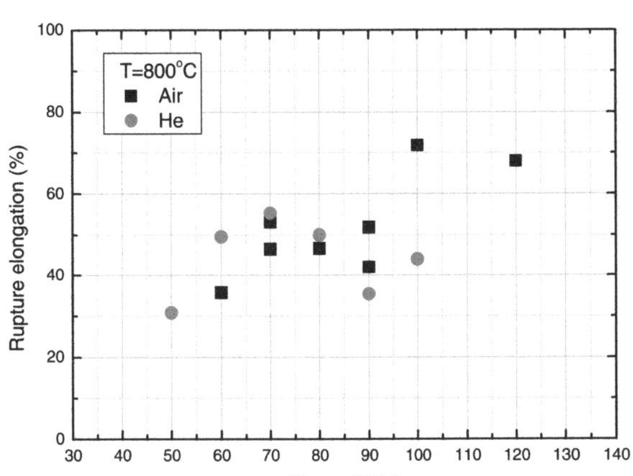


**Fig. 7.** Comparison of creep rupture elongation vs. stress in the air and He environments at 800 °C.

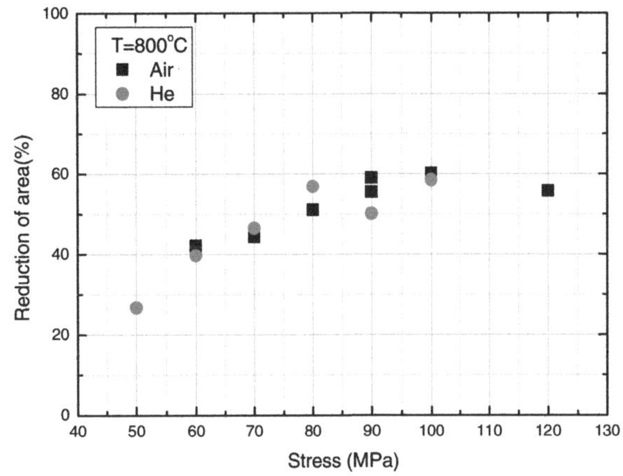


**Fig. 8.** Reduction of area vs. stresses of Alloy 617 in the air and He environments at 800 °C.

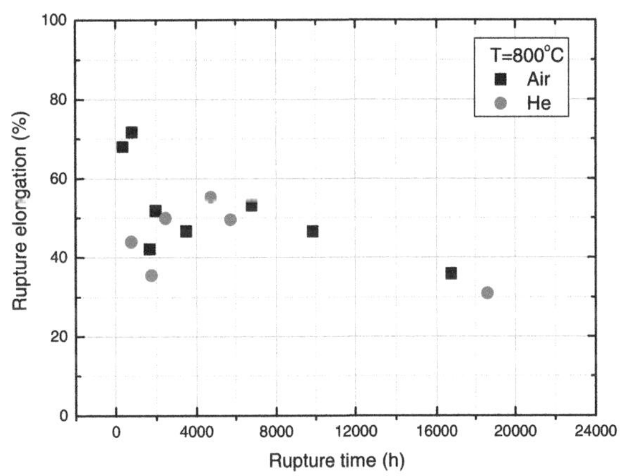


**Fig. 9.** Rupture elongation vs. rupture time of Alloy 617 in the air and He environments at 800 °C.

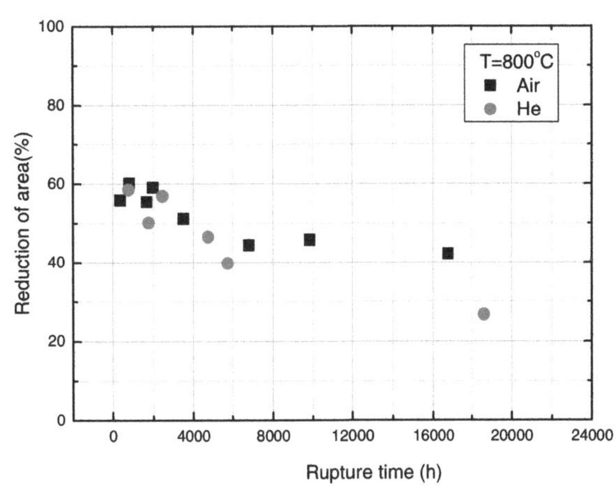


**Fig. 10.** Reduction of area vs. rupture time of Alloy 617 in the air and He environments at 800 °C.

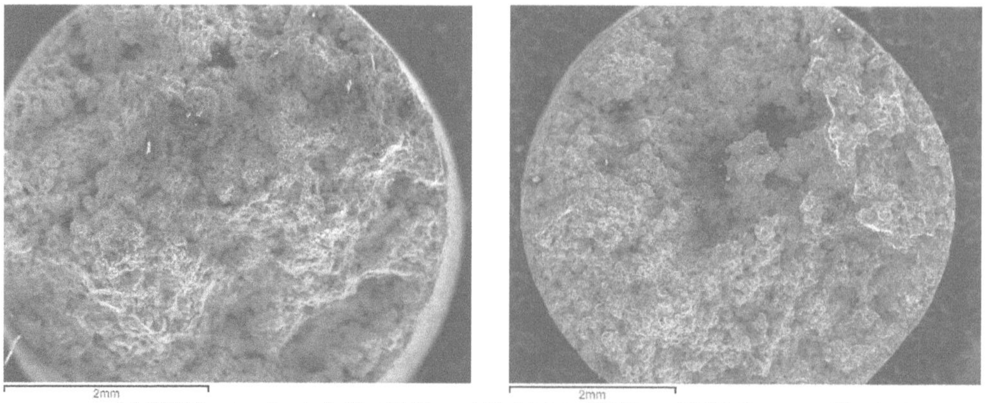


(a) SEM fracture in air (left) -60MPa and He (right) -50MPa at 800°C (low magnification)

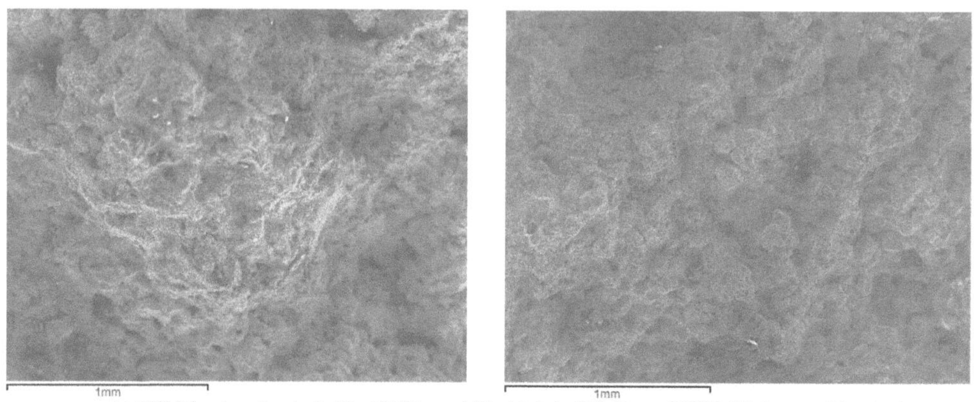


(b) SEM fracture in air (left) -60MPa and He (right) -50MPa at 800°C (high magnification)

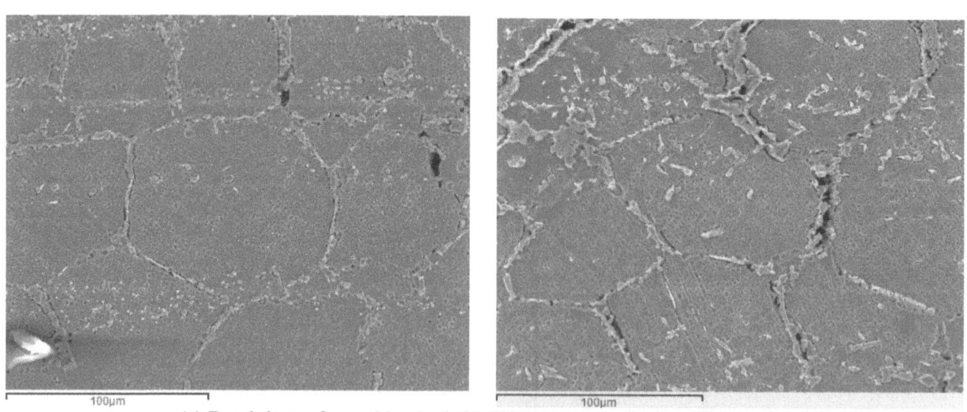


(c) Precipitates formed in air (left)-60MPa and He (right)-50MPa (matrix)

**Fig. 11.** Typical SEM fracture surface and precipitates formed in the crept specimens under 60 MPa in the air and 50 MPa in the He environment 800 °C. (Left side photos are for the air specimen (tr = 16,777 h, 60 MPa @800 °C), and right side photos are for the He specimen (tr = 18,624 h, 50 MPa @800 °C).

Grant (M–G) plot showing the plot of the minimum creep rate vs. rupture time presents good linearity, as shown in Fig. 6. In the M–G relationships, $\log t_r + m \log \dot{\varepsilon}_m = C_{MG}$, the plot exhibits no difference in the air and He environment. The $m$ and $C_{MG}$ values were almost the same at 0.71–0.72 and 0.23–0.25, respectively.

Figs. 7 and 8 show the plots of elongation and reduction of area during a rupture as a function of stress in the air and He environments at 800 °C. As the stress decreases, the creep ductility decreases slightly. It appears that the creep rupture ductility with respect to the stresses is almost similar between the air and He environments. The rupture elongation and reduction of area decreases with an increase in rupture time, as shown in Figs. 9 and 10. There was no difference in creep ductility between the air and He environments.

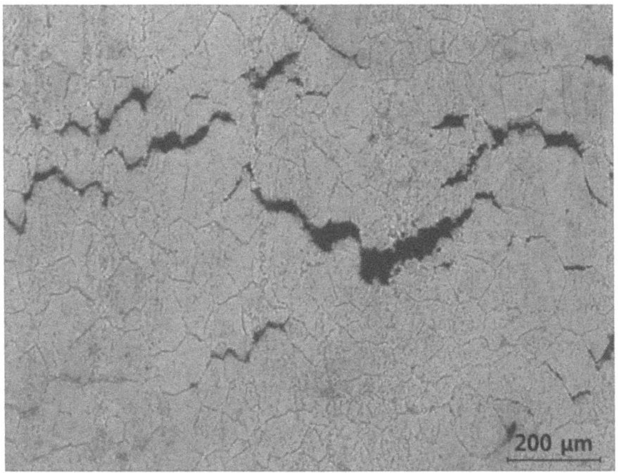


**Fig. 12.** A typical OM structure showing creep cracks along grain boundaries of a crept specimen (tr = 18,624 h) under 50 MPa in the He environment at 800 °C.

### 3.2. Fracture micrographs

After the creep tests, fracture micrographs were observed for all of the tested specimens in the air and He environments at 800 °C. Fig. 11 shows typical SEM fracture surfaces (Fig. 11a and b) and precipitates (c) formed in the crept specimens under 60 MPa in air and 50 MPa in He environment at 800 °C. It indicates that the left-side photos are for an air specimen (tr = 16,777 h, 60 MPa at 800 °C), and the right-side photos are for the He specimen (tr = 18,624 h, 50 MPa at 800 °C). The fracture surface shows a combination of dimple rupture and dominant intergranular fracture. In addition, as shown in Fig. 12, which presents a typical OM structure on creep cracks along the grain boundaries for a crept specimen (tr = 18,624 h) under 50 MPa in the He environment at 800 °C, the creep cavities and cracks developed along the grain boundaries perpendicular to the applied stress. This structure at 800 °C is similar to that of the author's previous study investigated in the air and He environments at 850, 900, and 950 °C [9]. The intergranular cracking is developed by the incorporation of minor cavities, formed on the grain boundary (GB) for prolonged creep duration. The cavities are widely distributed over the gage length of the specimen. Creep failure finally occurs by the incorporation of these cavities (or voids) introduced by the creep damage. These cavities are initially generated by precipitates formed along the GB. The precipitates become the sites a site initiating cavities. The fracture modes are not different in either environment.

In addition, as shown in Fig. 11c, the precipitates formed in air and He environments during creep rupture time occur mostly along the GB. He specimens form more coarse precipitates than the air ones. Cavities and cracks occur on the GBs where the carbides are formed. In the author's previous study (Kim et al., 2014), it was examined that cylindrically-shaped coarse carbides and Cr-rich $M_{23}C_6$ developed mainly on the GB. In the precipitate-denuded regions, the carbides developed near the GB. According to other studies on Alloy 617, $M_{23}C_6$ carbide was found to be abundant at temperatures of 649–1093 °C. Other carbides ($Cr_{23}C_6$), carbon nitrides [Cr Mo(C, N)], and nitride (TiN) were exhibited but were rare (Jang et al., 2008; Cook, 1984). Accordingly, it was noted that the fracture mode of Alloy 617 at 800 °C is clear to be the intergranular fracture along the GB, and the fracture modes are not different in both environments. The precipitates are mainly formed on the GB during creep exposure time, and they become site initiating cavities.

### 3.3. Oxidation structures in the air and He environments

To investigate the oxidation microstructures, all the rupture specimens after the creep tests were observed by SEM fracture micrographs.

Fig. 13 shows typical cross-sectional views of crept specimens under 60 MPa in air (a) (tr = 16,777 h, 1.91y) and (b) He (tr = 18,624 h, 2.12y) at 800 °C. For the air specimen, the creep rupture parts reveal a typical inter-granular fracture surface. There are a few large surface cracks along the specimen surfaces and crack voids. All specimens reveal a clear outer Cr-oxide layer with a smooth surface and density structure, which was observed at 850 °C, 900 °C, and 950 °C. For the He specimen, the fracture microstructures are similar to the air specimens: however, the outer oxide layer shows a rough surface and porous structure, and the internal oxide layer includes large cracks. Owing to high oxygen-partial pressure, there are thicker Cr-oxide layers along the ruptured and cracked surfaces compared with the air specimens, as shown in Fig. 13c and Fig. 13d of the enlarged magnifications. There were also thin Cr-oxide layers along the large cracked surfaces. This is thought to be from the low oxygen-partial pressure in the He environment. In these specimens, a delaminated layer of the Cr-oxide layer is not observed. In addition, it can be seen that outer-oxide layer thickness in the He environment is higher than that in air environment, as shown in Fig. 14a and b.

For the He sample, its structure is more porous and smooth than that in the air environment. The connection of the surface oxide layer in the He samples is broken by the cracks. On the other hand, for the air sample, the connection is continuous without cracks or breaks, and the structures are very dense over the gage length. The dense oxide layer is formed owing to a high oxygen concentration in the atmosphere, and it provides a proper protection layer from the surface oxidation. In the internal oxide layer, the He sample is deeper in Al-oxide intergrowth, and the interconnection between the surface oxide layer and matrix is weak. The structure is also porous and bared. Accordingly, the oxidation features in the air and He environments can be summarized as in Table 2.

Fig. 15 shows the changes of outer-oxide layer thickness measured for the crept specimens in the air and He environments 800 °C. For each sample, the average thickness was measured from more than 26 points from the surface images. As the creep rupture time increases, the thickness plots the trend of a parabolic increase. The increase of the outer-oxide layer thickness shows a smooth slope. Under about 3000 h in creep rupture time, the thickness in the air and He environment is almost the same, and above about 3000 h in creep rupture time, the thickness in the He environment is higher than that in the air environment, but the difference is within 5–6 μm.

Fig. 16 shows the change in internal-oxide layer depth measured for the crept specimens in the air and He environments at 800 °C. The thickness of the He environment has a slightly higher value than that of the air environment, but there is a small difference within about 5 μm. The thickness increases slowly with an increase in the creep rupture time. The internal-oxide layers are formed between the surface and matrix, and they are mainly composed of Al-oxide ($Al_2O_3$). The internal oxide acts like a chain between the outer oxide layer and matrix. Measuring the exact depth of the internal oxide layer is hard, but the depth increased with an increase in creep time. It seems that an excessive internal oxide layer with micro cracks affects the mechanical properties. From the change results of the internal- and outer-oxide layer thicknesses, it was found that the thickness in the He was thicker than in the air, although a pure helium gas of 99.999% was used in the present investigation. The differences in the oxide layer thickness caused detrimental effects on the creep resistance, even in the low oxygen-containing He agent used in this study.

(a) Cross-sectional view in an air specimen (tr=16,777h, 60MPa @800°C) (x200)

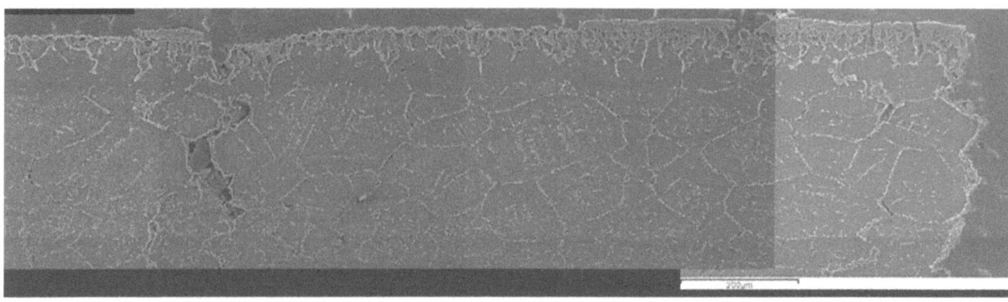

(b) Cross-sectional view in a He specimen (tr=18,624h, 50MPa @800°C) (x200)

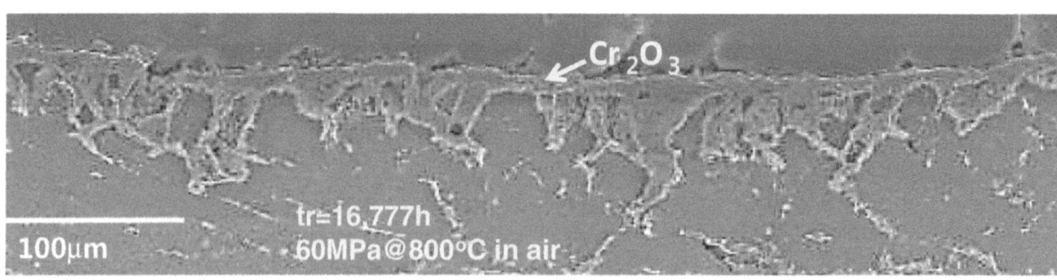


(c) Enlarged magnification in the air specimen (a)

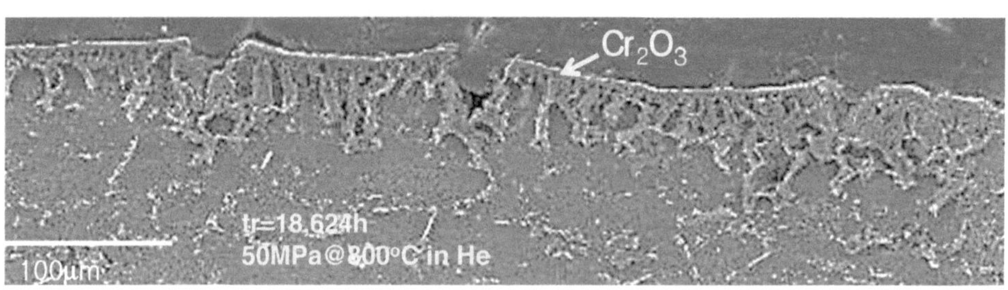


(d) Enlarged magnification of in the He specimen (b)

**Fig. 13.** SEM micrographs of the high and low magnifications for long-term ruptured specimens under stresses of 60 MPa (tr = 16,777 h) in air and 50 MPa (tr = 18,624 h) in He environments at 800 °C.

From the investigation of oxidation structures in the air and He environments for alloy 617 at 800 °C, it can be seen that in the He environment, the creep cavities and voids remained more severe on the grain boundary than that in the air environment. An oxidation layer ($Cr_2O_3$) was formed on the surface of alloy 617. The oxidation layer became thicker in accordance with a decrease in stress. The deterioration of the creep resistance in the He environment was due to the microstructural changes such as the thickness of the surface oxide layer, the depth of the internal oxidation, and the depth of the Cr-depleted zone. These changes were attributed to the enhanced diffusion of the oxidizing agent and the gaseous reaction products along the grain boundary. It is known that the creep properties of the material will have detrimental effects when the carbide depleted zone is enhanced, and they are dependent on various microstructure features such as surface cracking along the grain boundary, voids below the surface, and voids in the matrix (Jang et al., 2008). The reason why the surface oxide layer formed much deeper inside in the He than in air can be explained.

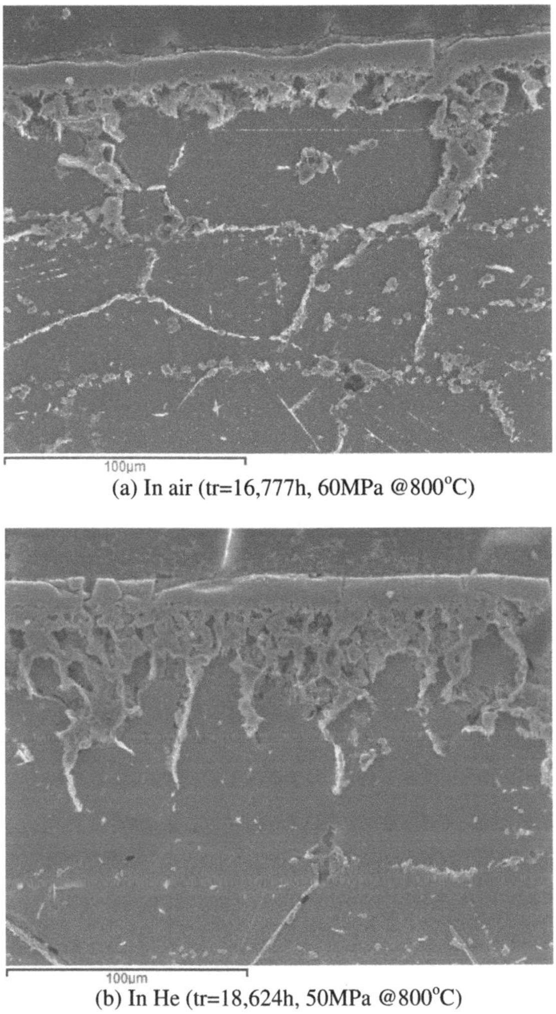


**Fig. 14.** SEM micrographs of the oxide layers formed during the creep time, in the air specimen (a) under 60 MPa and He specimen (b) under 50 MPa at 800 °C.

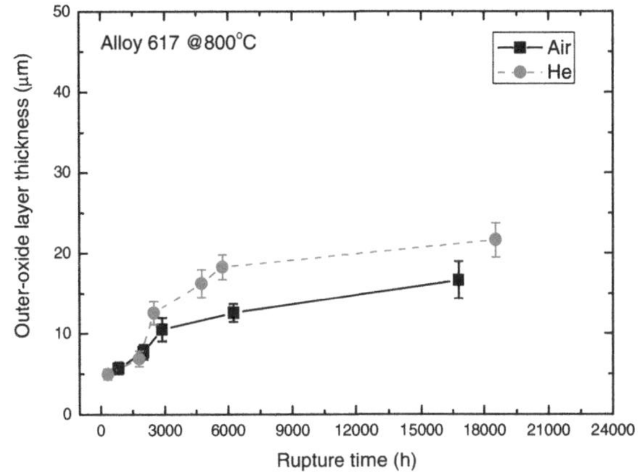


**Fig. 15.** Change of outer-oxide layer thickness with rupture times of the crept specimens in the air and He environments at 800 °C.

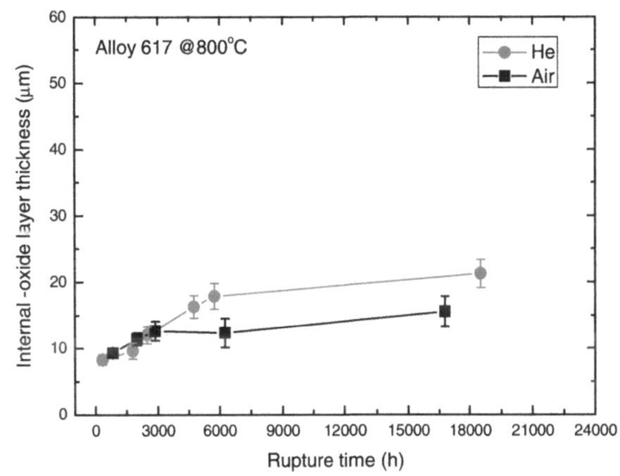


**Fig. 16.** Change of internal-oxide layer thickness with rupture times of the crept specimens in the air and He environments at 800 °C.

**Table 2**
Difference of oxidation structures in air and He environments.

| | Air | Helium |
|---|---|---|
| Outer surface oxide and cracks | Rough surface<br>Dense oxide<br>Without cracks | Smooth surface<br>Porous oxide<br>With broken cracks |
| Internal oxide | Uniform (good connection) | Discrete (weak connection) |
| Cavity size | Small and shallow | Large and deep |

Oxidation is a diffusion controlled process. Once the surface of a metal begins to oxidize, subsequent oxidation can take place only when oxygen ions diffuse through the oxide and react with the metal ions at the metal-oxide interface. As the oxide layer thickens, the diffusion distance increases and the rate of oxidation decreases. The density of oxide formed on Alloy 617 by annealing in air is much more than that formed in He, which tends to make the oxygen ion diffusion more difficult, while the much lower density of the oxide on the surface formed in the He ambience tends to make the oxygen ions diffuse easier, and increase the diffusion distance of the oxygen ions.

## 4. Conclusions

A series of creep tests was performed under different applied stresses for Alloy 617 in air and He environments at 800 °C, and experimental creep data were obtained. The results showed that the creep rupture time in the air and He environments was almost similar in the short rupture time. However, when the creep rupture time reaches above 3000 h, the creep stresses in the He environment were reduced compared with those of the air environment. The creep strain rate in the He environment was also a little faster than that in the air environment at above 3000 h. The reduction of creep life in the He environment was due to the difference of various microstructure features such as the carbide depleted zone, oxidation structures, surface cracking, voids below the surface, and voids in the matrix in air and He environments. Alloy 617 followed Norton's power law and the Monkman–Grant relationship well. As the stress decreased, the creep ductility also decreased slightly. The thickness of the outer and internal oxide layers showed a parabolic increase with an increase in creep rupture time in both the air and He environments. It was found that the thickness in the He environment was higher than in the air environment, although a pure helium gas of 99.999% was used in the present investigation. The differences of the oxide-layer thickness caused detrimental effects on the creep resistance, even in the low oxygen-containing He agent used in this study.

## Acknowledgements

This study was supported by Nuclear Research & Development Program of the National Research Foundation of Korea (NRF) grant funded by the Ministry of Science, ICT and Future Planning (2012M2A8A2025682).

## References

Chang, J.H. et al., 2007. A study of a nuclear hydrogen production demonstration plant. Nucl. Eng. Technol. 39 (2), 111.

Christ, H., Kunecke, U., Meyer, K., 1987. High temperature corrosion of the nickel-based alloy inconel 617 in helium containing small amounts of impurities. Mater. Sci. Eng. 87, 161.

Cook, R., 1984. Creep properties of Inconel 617 in air and helium at 800 to 1000°C. Nucl. Technol. 66, 283.

Dewson, S., Li, X., 2005. Selection criteria for the high temperature reactor intermediate heat exchanger. In: Proceedings of ICAPP 05, Paper No.5333, Seoul.

Ennis, P.J., Mohr, K.P., Schuste, H., 1984. Effect of carburizing service environments on the mechanical properties of high temperature alloys. Nucl. Technol. 66, 363.

Huchtemann, B., 1989. The effect of alloy chemistry on creep behavior in a helium environment with low oxygen partial pressure. Mat. Sci. Eng. A 121, 629.

Jang, C., Lee, D., Kim, D., 2008. Oxidation behavior of an alloy 617 in very high temperature air and helium environments. Int. J. Press. Vessels Pip. 85, 368.

Kim, W.G., Yin, S.N., Ryu, W.S., Chang, J.H., 2007. Analysis of the creep rupture data of alloy 617 for a high temperature gas cooled reactor. In: Proceedings of CREEP8, ASME PVP 2007-26834, Texas.

Kim, W.G., Yin, S.N., Kim, Y.W., Chang, J.H., 2008. Creep characterization of a Ni-based Hastelloy-X alloy by using theta projection method. Eng. Fract. Mech. 75, 4985.

Kim, W.G., Yin, S.N., Kim, Y.W., 2009. Long-term creep strain-time curve modeling of alloy 617 for a VHTR intermediate heat exchanger. J. Korean Inst. Met. Mater. 47 (10), 613 (in Korean).

Kim, W.G., Yin, S.N., Lee, G.G., Kim, Y.W., Kim, S.J., 2010. Creep oxidation behavior and creep strength prediction for alloy 617. Int. J. Press. Vessels Pip. 87, 289.

Kim, W.G., Yin, S.N., Park, J.Y., Hong, S.D., Kim, Y.W., 2012a. An improved methodology for determining tensile design strengths of alloy 617. J. Mech. Sci. Technol. 26 (2), 379.

Kim, W.G., Park, J.Y., Kim, S.J., Hong, S.D., Kim, Y.W., 2012b. Reliability evaluation on creep life prediction of alloy 617 for a very high temperature reactor. J. Korean Inst. Met. Mater. 50 (10), 721 (in Korean).

Kim, W.G., Park, J.Y., Ekaputra, I.M.W., Hong, S.D., Kim, S.J., Kim, Y.W., 2013. Comparative study on the high-temperature tensile and creep properties of alloy 617 base and weld metals. J. Mech. Sci. Technol. 27 (8), 2331.

Kim, W.G., Park, J.Y., Lee, G.G., Hong, S.D., Kim, Y.W., 2014. Temperature effect on the creep behavior of alloy 617 in air and helium environments. Nucl. Eng. Des. 271, 291.

Lee, G.G., Jung, S., Park, J.Y., Kim, W.G., Hong, S.D., Kim, Y.W., 2013. Microstructural investigation of alloy 617 creep-ruptured at high temperature in a helium environment. J. Mater. Sci. Technol. 29 (12), 1177.

Natesan, K., Puroit, A., Tam, S.W., 2003. Materials behavior in HTGR environments. NUREG/CR-6824, ANL-02/37.

Schneider, K., Hartnagel, W., Schepp, P., Ilschner, B., 1984. Creep behavior of materials for high temperature reactor application. Nucl. Technol. 66, 289.

Schubert, F., Bruch, U., Cook, H., Diehl, H., Ennis, P.J., Jakobeit, W., Penkalla, H.J., Heesen, E.T., Uilrich, G., 1984. Creep rupture behavior of candidate materials for nuclear process heat applications. Nucl. Technol. 66, 227.

Schubert, F., Seehafer, H.J., Bodmann, E., 1983. Status of design code work in Germany concerning materials and structural aspects for the heat exchanger components of advanced HTRs. J. Eng. Power 105, 713.

Engineering Failure Analysis 58 (2015) 441–451

Contents lists available at ScienceDirect

Engineering Failure Analysis

journal homepage: www.elsevier.com/locate/efa

# Creep deformation and rupture behavior of Alloy 617

Woo-Gon Kim [a,*], Jae-Young Park [b], I.M.W. Ekaputra [b], Seon-Jin Kim [b], Min-Hwan Kim [a], Yong-Wan Kim [a]

[a] Korea Atomic Energy Research Institute (KAERI), 989-111, Daedeok-daero, Yuseong-gu, Daejeon 305-353, Republic of Korea
[b] Pukyong National University, 365 Shinsunro, Nam-gu, Busan 608-739, Republic of Korea

ARTICLE INFO

*Article history:*
Received 19 December 2014
Received in revised form 15 July 2015
Accepted 28 July 2015
Available online 2 August 2015

*Keywords:*
Creep
Alloy 617
Monkman–Grant Relationship
Creep damage tolerance factor
Zenor–Holloman parameter

ABSTRACT

A series of creep data was obtained from creep tests at different applied stresses at the temperatures of 850 °C, 900 °C, and 950 °C for Alloy 617, which is a leading candidate material for high-temperature components in Gen-IV nuclear reactor systems. The creep deformation and rupture behavior were investigated in terms of Norton's power law, Monkman–Grant relation (MGR), modified Monkman–Grant relation (MMGR), creep damage tolerance factor (λ), Zener–Hollomon Parameter (Z), and fracture behavior. Alloy 617 did not exhibit textbook creep behavior and revealed somewhat differences from typical heat resistant steels. The MMGR provided improved correlation between creep rate and rupture life in Alloy 617. The Z parameter obeyed a good agreement for a function of $Z = 2.30 \times 10^{33} (\sigma/E)^{5.87}$, and the same creep mechanism was operative within the ranges tested in the present study. The value of λ for Alloy 617 was found to be 2.40, and this was in agreement with materials exhibiting typical cavitation damage. The creep failure analysis revealed a dominant intergranular fracture mode, which proceeds via initiation, linking, and incorporation of the cavities.



## 1. Introduction

A very high temperature reactor (VHTR) is one of the most promising Generation-IV reactor types for the economic production of electricity and hydrogen. The VHTR system is designed with a target life span of 60 years under operating conditions of 950 °C and 7 MPa in helium impurities. Its major components include the reactor internals, reactor pressure vessel (RPV), piping, hot gas ducts (HGD), and intermediate heat exchangers (IHX). The IHX is a key component that links the reactors to the hydrogen production plant. The IHX component is directly exposed to a primary coolant of 950 °C, making its very-high-temperature creep resistance particularly important as its integrity should be preserved during the design lifespan of the VHTR. Of the existing alloys, nickel-based Alloy 617 is the leading candidate for use in VHTR heat exchangers because it has been reported to have the highest creep allowable stress compared with other potential superalloys under consideration for $10^5$ h at 898 °C, as shown in Fig. 1 [1–4].

The Korea Atomic Energy Research Institute (KAERI) is developing a nuclear hydrogen development and demonstration (NHDD) plan with a capacity of 200 $MW_{th}$ and core outlet temperature of 950 °C. This work has been carried out in a government-funded project called "Development of Key Technologies for Nuclear Hydrogen" since 2006 [5–7]. However, till date, the ASME (American Society of Mechanical Engineering) design code for Alloy 617 was not developed for design use, and the draft Code Case for Alloy 617 [8] was only prepared for the design of high-temperature Alloy 617 nuclear components. The draft Code Case is a modification from ASME Section III Subsection NH [9] that was put forth by a special task force of the ASME subgroup that deals with elevated temperature

* Corresponding author.
*E-mail address:* wgkim@kaeri.re.kr (W.-G. Kim).

http://dx.doi.org/10.1016/j.engfailanal.2015.07.041
1350-6307/

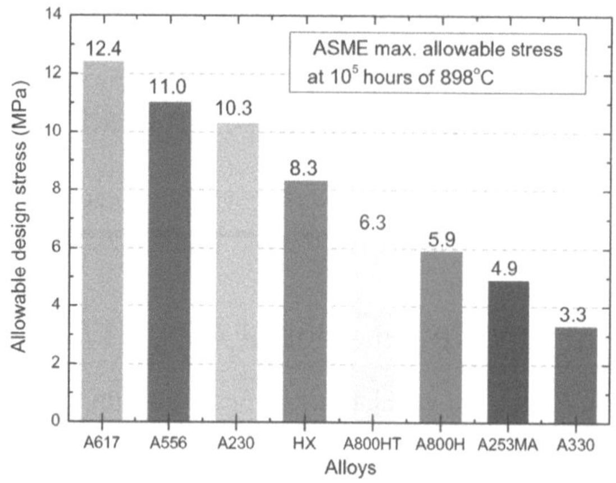


**Fig. 1.** Plot showing the highest creep strength of Alloy 617 compared with other candidate alloys.

design. The primary intended application of the draft Code Case is a VHTR. Currently, material database works for Alloy 617 are ongoing through Generation-IV forum (GIF) project plan of a next-generation nuclear plant research and development. As per this, a new Alloy 617 Code Case is planned to be approved by 2017. However, to do so, a number of creep data and creep constants should be obtained to help draft the new Code Case, and creep behavior should also be investigated through systematic analysis of a wide range of creep temperature and stress conditions.

In this study, creep rupture data for Alloy 617 were obtained through a series of creep tests under different applied stresses at three different temperatures of 850, 900, and 950 °C. A systematic analysis of the creep rupture data was performed using various creep equations and parameters, and the material constants used in the creep equations were determined using the experimental creep data. In addition, creep fracture microstructures were investigated for the crept specimens at the three temperatures, and the results are discussed.

## 2. Experimental procedure

Commercial grade nickel-based superalloy, Alloy 617 (Inconel 617), was used in this study. The material was a hot-rolled plate with a thickness of 15.875 mm. Creep specimens were fabricated in cylindrical form with a 30 mm gage length and 6 mm diameter, as shown in Fig. 2. The gage section was parallel to the longitudinal rolling direction. Circular grooves were machined at both ends beyond the shoulder region of these specimens to attach a high-precision extensometer for monitoring the elongation during creep tests.

The loading frames used in the creep tests had a lever arm ratio of 20:1. A split furnace was used to heat the specimens of Alloy 617 in dry air at 850, 900, and 950 °C. The pull rod and jig used for the creep tests were manufactured with a nickel-based superalloy material so that they could sufficiently endure oxidation and thermal degradation over the duration of creep tests at the very high temperatures employed. Creep strain data as a function of elapsed times were acquired automatically using a PC. K-type thermocouples were used to monitor temperatures within the gage section of the specimens. Creep strain-time curves were obtained, and the value of the minimum creep strain rate was obtained by calculating the secondary creep stage from the strain-time creep curves. After the creep tests, creep fracture micrographs were observed from the crept specimens using optical microscopy (OM) and scanning electron microscopy (SEM).

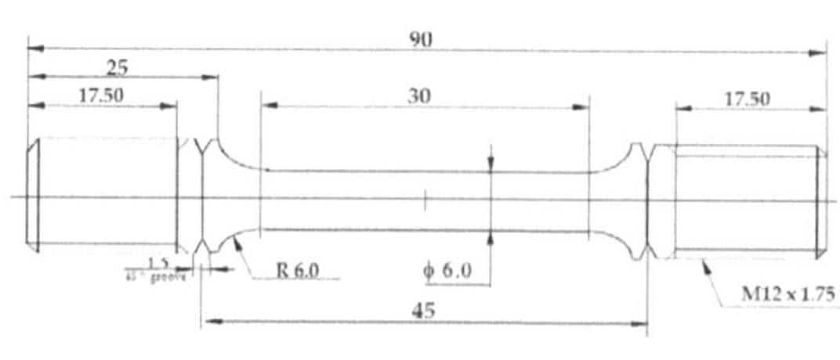


**Fig. 2.** Dimension of creep test specimen of Alloy 617.

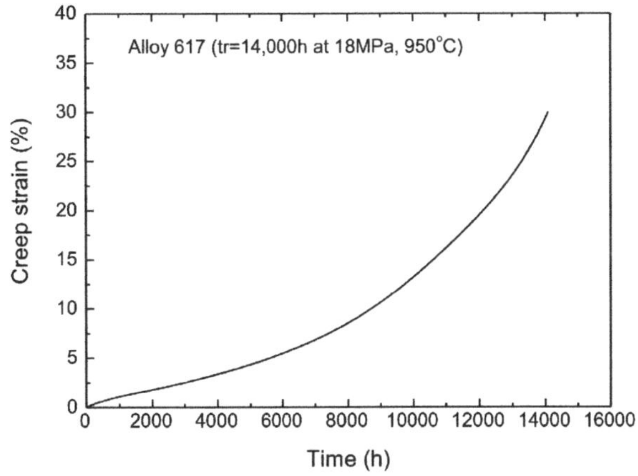


**Fig. 3.** Shape of typical long-term creep curve of Alloy 617.

## 3. Results and discussion

### 3.1. Analysis of creep rupture behavior

Data on the creep properties of Alloy 617, such as rupture time, minimum creep rate, rupture elongation, and reduction of area were obtained at 850, 900, and 950 °C. The creep rupture behavior was analyzed using various creep laws, and the constants in these creep laws were determined by a systematic analysis of the experimental data.

Fig. 3 shows a typical creep curve obtained for a rupture time of 14,000 h at 18 MPa and 950 °C for Alloy 617. The creep curve is somewhat different from those of typical heat-resistant low and high-Cr ferritic steels with distinct primary, secondary, and tertiary creep stages (referred to as "textbook creep curve"). In other words, Alloy 617 does not exhibit textbook creep behavior. The secondary creep stage is unclear, and it seems to exhibit a predominantly accelerated creep rate within a short time from the beginning of the tertiary creep stage. Owing to the unclear secondary creep stage, it is difficult to accurately determine the steady state creep rate as well as the onset of tertiary creep. Some specimens appeared to have a different value in the minimum creep rate and steady state creep rate, as shown in Fig. 4. It may introduce some scatters in the data of creep stress exponent and activation energy, which are directly affected by the value of creep rate. In order to maintain consistency in the determination of the creep rate, we determined the minimum creep rate for all specimens as the range of creep time in the secondary creep stage is unclear.

Fig. 5 shows a double-logarithmic plot of the minimum creep rate vs. the stress obtained at 850, 900, and 950 °C. The dependence of the minimum creep rate and rupture time on stress obeys the power-law, as follows.

$$\dot{\varepsilon}_m = A\sigma^n \quad (1)$$

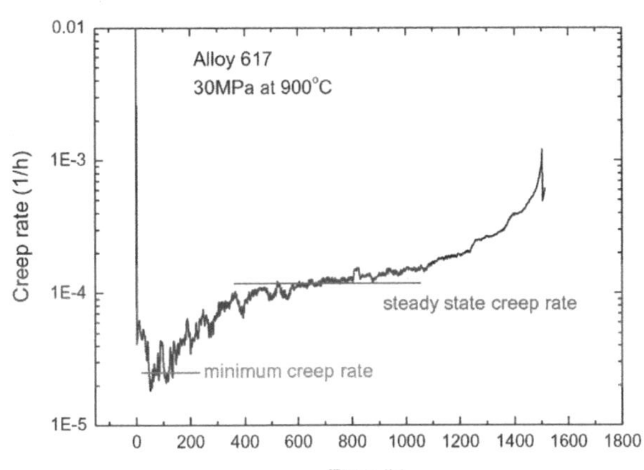


**Fig. 4.** Plot showing the difference between minimum creep rate and steady state creep rate of Alloy 617.

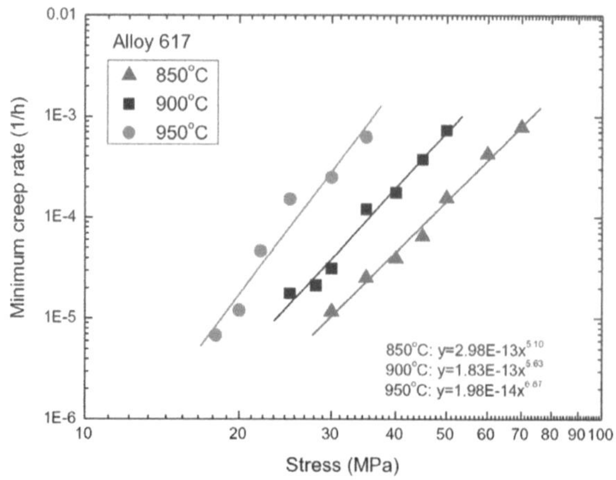


**Fig. 5.** Plot showing stress dependence of minimum creep rate at 850, 900, and 950 °C of Alloy 617.

$$t_r = A'\sigma^{n'}. \tag{2}$$

The slope is in the range of $n \cong 5.1$–6.9 at the three temperatures. Stress dependence of creep rupture time obeys the power-law at all three temperatures. The slope is in the ranges of $n' \cong 4.5$–6.0, as shown in Fig. 6. The four constants $A, A'$, $n$, and $n'$ are summarized in Table. 1. As seen in Table 1, the material constant values are nearly the same and independent of temperature. It can be inferred that the similar values of $n$ and $n'$ resulted from the same creep deformation and fracture mechanisms. Most metallic materials in the stress range $10^{-4} < \sigma / E < 10^{-2}$ follow the power-law creep with stress exponent $\approx 5$, and an approximate value of $n = 5$ implies dislocation glide and climb creep mechanism, which falls in the range of stress exponent typically observed for class II alloys [10,11].

The $n$ value obtained in the present study is similar to the values reported by Benz et al. for the temperature range 800–1000 °C ($n = 5.1$–6.9) [10]; however, in Benz's study, the reported value decreased from 6.9 to 5.1 with an increase in the temperature from 800 to 1000 °C; in contrast, the obtained value in the present study increases from 5.1 to 6.9 with an increase in the temperature from 850 to 950 °C. With regard to the temperature tendency of $n$, Sharma et al. has reported that the value of $n$ increased from 4.38 to 9.68 with an increase in the temperature from 800 to 900 °C, and then decreased to 6.29 with a further increase in the temperature to 1000 °C [11]. Shankar et al. has reported that the value of $n$ was in the range of 4.22–6.17 for temperatures between 843 and 950 °C, and it was unclear whether $n$ increased or decreased with the increase in temperature [11]. Thus, more tests will be needed to clarify the temperature tendency of $n$. However, for the temperature ranges investigated in this study, it can be strongly suggested that $n$ varies between 5 and 7 for Alloy 617 because the same creep mechanism is operative here, regardless of the temperature tendency of $n$.

Fig. 7 shows a plot of creep rupture elongation with rupture time at 850 °C, 900 °C, and 950 °C. The rupture elongation decreases slightly with an increase in rupture time although there are some scatters in the data.

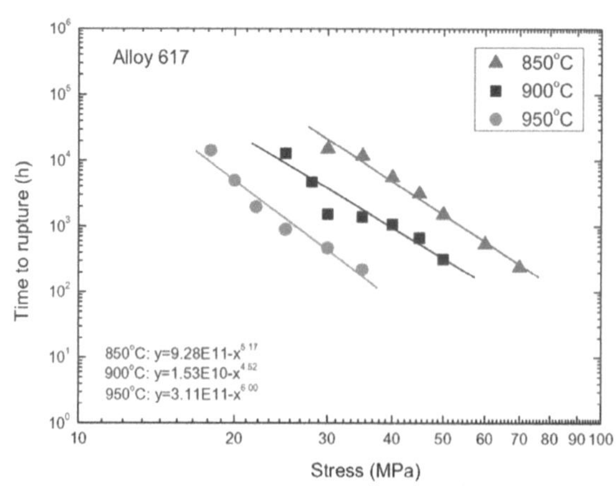


**Fig. 6.** Plot showing stress dependence of rupture time at 850, 900, and 950 °C of Alloy 617.

**Table 1**
Constants of $A$, $n$, $A'$, $n'$ obtained by a power law form at the test temperatures for Alloy 617.

| Temp. (°C) | $A$ | $n$ | $A'$ | $n'$ |
|---|---|---|---|---|
| 850 | $2.98 \times 10^{-13}$ | 5.10 | $9.28 \times 10^{11}$ | 5.17 |
| 900 | $1.83 \times 10^{-13}$ | 5.63 | $1.53 \times 10^{10}$ | 4.52 |
| 950 | $1.98 \times 10^{-14}$ | 6.87 | $3.11 \times 10^{11}$ | 6.00 |

The interrelation between creep deformation and damage is appropriately described by the empirical relationship proposed by Monkman and Grant, where the time to failure or rupture time $t_r$ is inversely proportional to the minimum or steady state creep rate. The Monkman–Grant relation (MGR) in the generalized form can be expressed as [13]:

$$\log t_r + m \log \dot{\varepsilon}_m = C \tag{3}$$

where $m$ is the slope of the double logarithmic plot of $t_r$ vs. $\dot{\varepsilon}_m$ and intercept $C$ is a constant. The modified Monkman–Grant relation (MMGR), proposed by Dobes and Milicka, in the generalized form can be expressed as [14]:

$$\log \left(t_r/\varepsilon_f\right) + m' \log \dot{\varepsilon}_m = C' \tag{4}$$

where $\varepsilon_f$ is the strain to failure and $m'$ is the slope of the double logarithmic plot of $t_r/\varepsilon_f$ vs. $\dot{\varepsilon}_m$ and intercept $C'$ is a constant.

The validity of the MGR and MMGR at 850, 900, and 950 °C for Alloy 617 was investigated, and the variations of rupture life with minimum creep rate are shown in Figs. 8 and 9. The validity of the MGR is observed as the slope of log $t_r$ vs. log$\dot{\varepsilon}_m$ plot equal to unity, and accordingly Eq. (3) can be expressed as:

$$\log t_r + \log \dot{\varepsilon}_m = C_{MG}. \tag{5}$$

When $m = 1$, the MGR constant $C$ becomes $C_{MG}$, which is a real constant independent of creep test conditions. When the slope is not unity, the product of creep rate and rupture life varies with creep test conditions. In the present investigation, the value of the MGR constant $C$ (Eq. (3)) is obtained as 0.517 for the slope of $m = 0.87$, as shown in Fig. 9. The validity in the MGR is weak because of $C \neq C_{MG}$. Similar to the MGR, the validity of the MMGR can also be observed as the slope of log $(t_r/\varepsilon_f)$ vs. log$\dot{\varepsilon}_m$ plot equals to unity [15–19]. When $m' = 1$, the MMGR constant $C'$ becomes $C_{MMG}$ and the MMGR can be expressed as:

$$\log \left(t_r/\varepsilon_f\right) + \log \dot{\varepsilon}_m = C_{MMG}. \tag{6}$$

The value of $C_{MMG}$ (Eq. (6)) is obtained as 0.420 for the slope of unity, as shown in Fig. 10. The MMGR has good validity because $C' = C_{MMG}$. The MMGR provides an improved correlation between the creep rate and rupture life in Alloy 617.

In addition, the creep damage tolerance factor (λ) provides a measure of the susceptibility of a material to localized cracking at locations of concentrated stress and strain. It is considered an important parameter for evaluating material performance. Based on

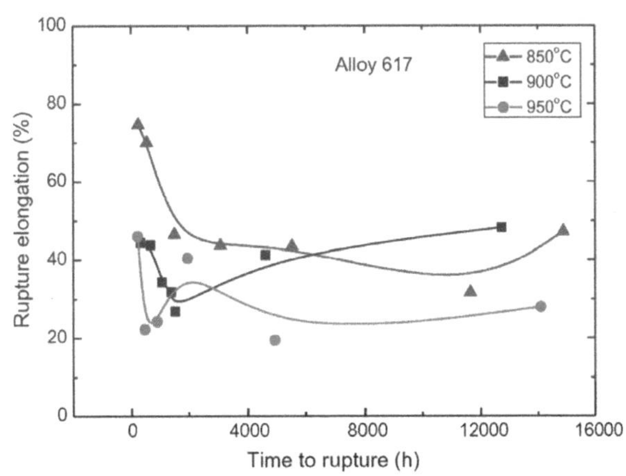


**Fig. 7.** Plot of creep rupture elongation with rupture time at 850, 900, and 950 °C of Alloy 617.

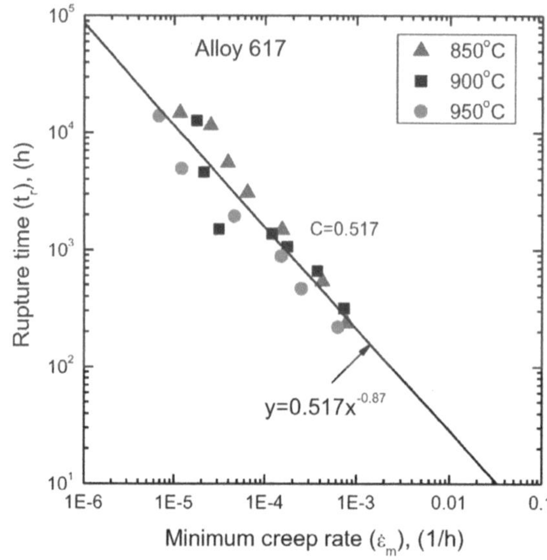


**Fig. 8.** The MGR plot of the creep data tested at 850, 900, and 950 °C of Alloy 617.

a continuum creep damage mechanics approach, λ is defined as the ratio of strain failure $\varepsilon_f$ to MGR, $\dot{\varepsilon}_m \cdot t_r$, i.e., the secondary creep strain. The λ equation can be expressed as:

$$\lambda = \frac{\varepsilon_f}{\dot{\varepsilon}_m \cdot t_r} = \frac{\dot{\varepsilon}_{avg}}{\dot{\varepsilon}_m}. \tag{7}$$

For Alloy 617, λ = 2.40, as shown in Fig. 10, which is the plot of the average creep rate (ACR) vs. the minimum creep rate (MCR) of the alloy tested at 850, 900, and 950 °C. Creep damage can occur by various mechanisms such as loss of external cross-section, loss of internal cross-section (formation, growth and linkage of cavities at grain boundaries), degradation of microstructure (thermal-coarsening of particles and/or substructure-induced acceleration of creep), and oxidation. It is known that each damage micromechanism, when acting

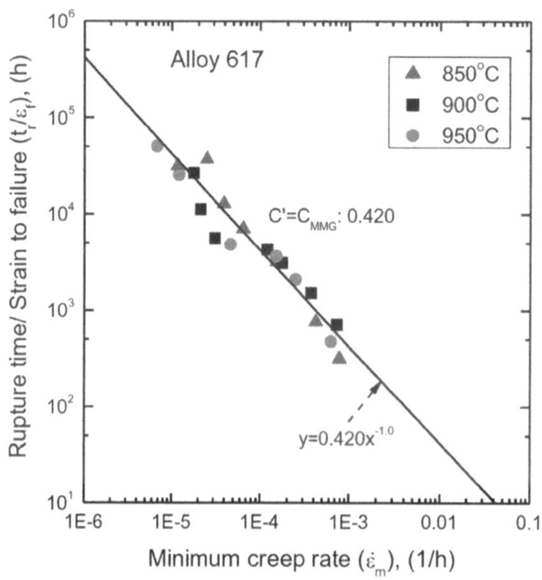


**Fig. 9.** The MMGR plot of the creep data tested at 850, 900, and 950 °C of Alloy 617.

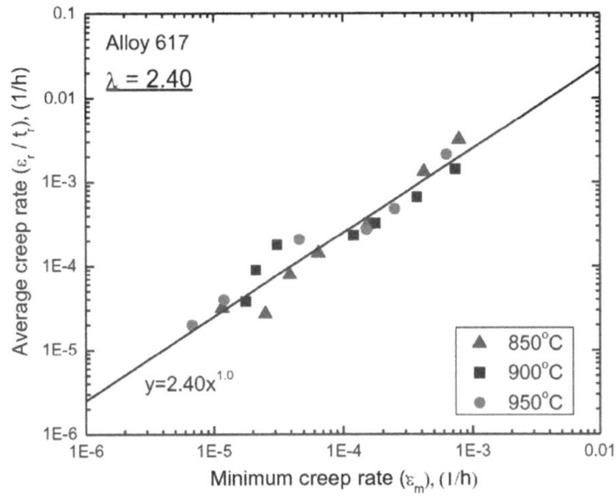


**Fig. 10.** Relationship of average creep rate and minimum creep rate tested of Alloy 617.

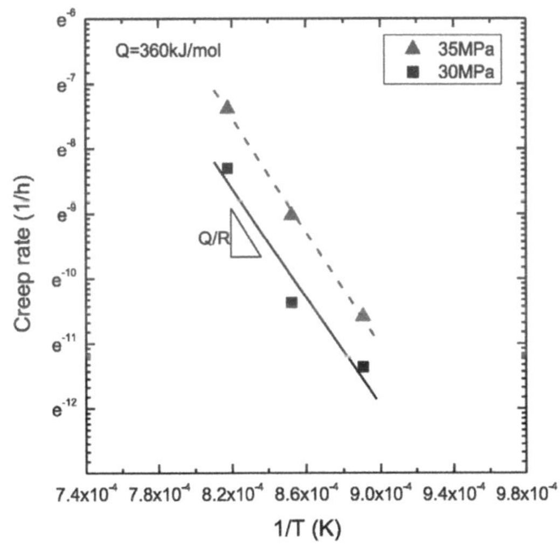


(a) Calculation of $Q$ value from the plot of log ( $\dot{\varepsilon}_m$ ) vs. 1/T

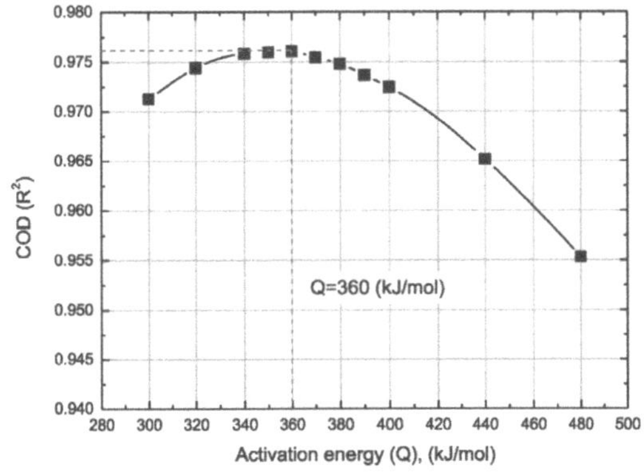


(b) Determination of $Q$ value from the best fit of Zener-Holloman plot

**Fig. 11.** Determination of activation energy at 850, 900, and 950 °C of Alloy 617.

alone, results in a distinctive shape of the creep curve and a corresponding characteristic value of λ [16]. For λ = 1.5–2.5, the creep damage results from growth of cavities by coupled diffusion and power-law creep; for λ = 5 or more, it results from microstructural degradation such as coarsening of the precipitates and/or dislocation substructural softening [20]. Thus, the creep damage mechanism for Alloy 617 can be reasonably inferred to be "cavitation" because for Alloy 617, λ = 2.4. It means that the average creep rate is 2.4 times the minimum creep rate. This is in agreement with materials exhibiting typical cavitation damage. It is thus evident that creep damage of Alloy 617 is accumulated by a cavitation process, and eventually creep rupture is developed by linking and coalescence of the cavities. The cavities form continuously throughout the creep process, but in the tertiary stage of creep the nucleation and growth of grain-boundary voids and cracks occurs at an accelerated rate. The details related to the creep fracture of Alloy 617 are described in Section 3.2.

The relationships of minimum creep rate with stress and temperature can be defined by a power-law as shown in Eq. (8), and this equation is frequently used to describe creep behavior.

$$\dot{\varepsilon}_m = A_1\sigma^{n_1}\exp(-Q/RT) \tag{8}$$

$$Z = \dot{\varepsilon}_m \exp(Q/RT) \tag{9}$$

$$Z = A_1\sigma^{n_1} \tag{10}$$

where $R$ is the universal gas constant (8.314 J mol$^{-1}$ K$^{-1}$). In Eq. (8), the parameter $A_1$, stress exponent $n_1$, and the apparent activation energy $Q$ are functions of stress and temperature themselves. The $Z$ parameter of Eq. (9) is referred to as the temperature-modified strain rate. Eq. (10) can be derived from Eqs. (8) and (9) in the power-law of $Z = f(\sigma)$. In Eq. (10), the constants $A_1$and $n_1$ are experimentally determined, and the quantity $Z$ is the "Zener–Hollomon parameter" [10,21].

To calculate the $Z$ parameter of Alloy 617, the activation energy was determined using two methods. One method determined $Q = 360$ (kJ/mol) from the semi-logarithmic plot of log $\dot{\varepsilon}_m$ vs. 1/T obtained under the constant stresses of 30 and 35 MPa at 850, 900, and 950 °C, as shown in Fig. 11(a). Another method determined $Q = 360$ kJ/mol from the best fit to the experimental data in the Zener–Hollomon plot in Fig. 12. As shown in Fig. 11(b), the coefficient of determination (COD, $R^2$), which is well known as a statistical parameter for error estimation, was the highest at $Q = 360$ kJ/mol. This value for $Q$ is higher than that for lattice-self diffusion of nickel, 284 (kJ/mol) [22,23], for Alloy 617 foils at 800 to 1000°C, 271 (kJ/mol) [11], and for bulk specimens tested in He + $CH_4$ environment, 309 (kJ/mol) [12]. On the contrary, the value of $Q = 360$ (kJ/mol) is lower than those for bulk specimens tested in pure helium environment, 408 (kJ/mol) and for Alloy 617 plate at 800 to1000°C, 410 (kJ/mol) [10]. However, $Q = 360$ (kJ/mol) obtained in the present study is consistent with the $Q$ value of 320 to 500 (kJ/mol) previously reported for Alloy 617 [10,24]. Wide variations of activation energy values for Alloy 617 are interpreted because the secondary creep stage was unclear as "non-textbook creep" behavior,

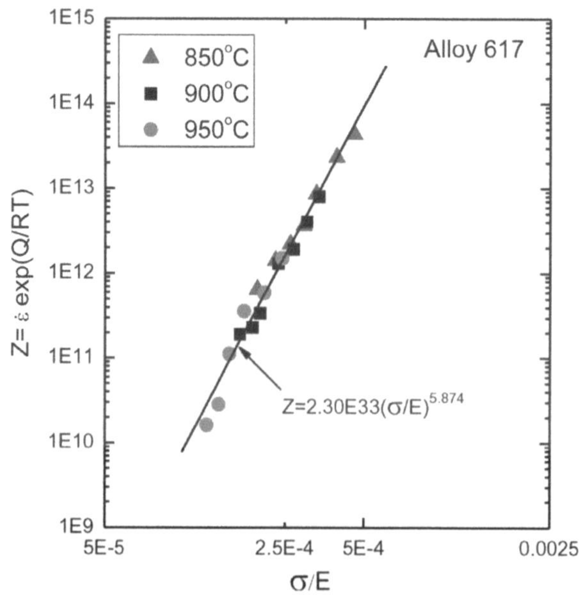


**Fig. 12.** Plot of the Zener–Holloman parameter as a function of stress (σ/E) at 850, 900, and 950 °C of Alloy 617.

**Fig. 13.** Shape of a crept specimen (tr = 14,896 h) under 30 MPa at 850 °C of Alloy 617.

as described in Figs. 3 and 4. This introduces some difficulty to correctly determine the value of the steady state creep rate. In addition, some scatters in the creep rate data may result in different values in the stress exponent $n$ and activation energy $Q$.

The $Z$ parameter was determined from Eq. (9). A plot of Z as a function of stress (σ/E) at 850, 900, and 950 °C for Alloy 617 is shown in Fig. 12. Here, the value of Young's modulus ($E$) was obtained from the reported data of Inconel 617 manufactured by Special Metals [25]. The data was fitted with a function of the form $Z = 2.30 \times 10^{33}\ (\sigma/E)^{5.87}$ regardless of the temperatures. Thus, it is obvious that the same mechanism was operative in the present temperature range.

### 3.2. Creep fracture microstructures

After the creep tests, creep fracture micrographs were observed for the crept specimens using OM and SEM images. A long-term crept specimen of Alloy 617 with $t_r$ = 14,896 h (1.7 years) ruptured under 30 MPa at 850 °C is shown in Fig. 13. It can be observed that minor and large cracks are widely distributed over the gage length, and creep failure occurred by the growth and interconnections of the cracks rather than a necking failure generated in the tertiary creep region. Creep deformation developed dominantly from the formation and growth of cavity during the creep damage process. The creep failure is due to initiation, linking, and incorporation of the cavities. The cavities and cracks develop along a direction perpendicular to the applied load, as shown in Fig. 14. It is obvious that the dominant failure mode of Alloy 617 was intergranular fracture. In addition, the precipitates formed coarsely and mainly on the grain boundaries and were finely scattered throughout the matrix, as shown in Fig. 15. Cylindrical coarse carbides, Cr-rich $M_{23}C_6$, developed mainly on the grain boundaries. According to other studies on Alloy 617, $M_{23}C_6$ carbide was found to be abundant at temperatures of 649 to 1093 °C. Other carbides ($Cr_{23}C_6$), carbon nitrides [Cr Mo(C, N)], and nitride (TiN) were exhibited but rarely [26,27].

Fig. 16 shows typical SEM micrographs taken at low and high magnifications for crept specimens at (a) 20 MPa ($t_r$ = 4943 h) and (b) 30 MPa ($t_r$ = 14,896 h) at 850 and 950 °C. The two specimens are almost similar in terms of fracture modes, regardless of rupture time. The fracture surface reveals dominant intergranular cracking, which is more prevalent in the specimen with the longer rupture time (Fig. 16b) than that with the shorter rupture time (Fig. 16a).

According to the report of Natesan et al. [28] on the SEM fractrography analysis for Alloy 617 base and weldment, Alloy 617 exhibited combined fracture modes consisting of dimple rupture and intergranular cracking, but intergranular fracture was more evident in the weldment. In the present study, Alloy 617 base metal revealed similar fracture behavior (i.e., combined fracture modes of dimple rupture and intergranular cracking), with the intergranular fracture being more evident than dimple rupture, especially, at a high temperature of 950 °C than at a low temperature of 850 °C for a longer rupture time. Dimple rupture observed in

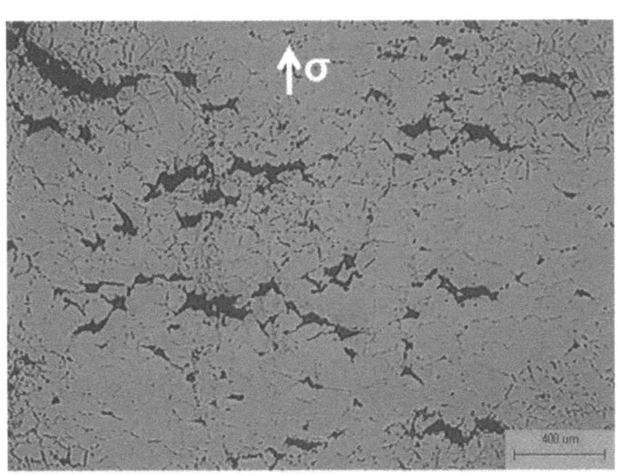


**Fig. 14.** Typical OM structures showing cavities in the crept specimen of Alloy 617 (tr = 14,896 h, 30 MPa, 850 °C).

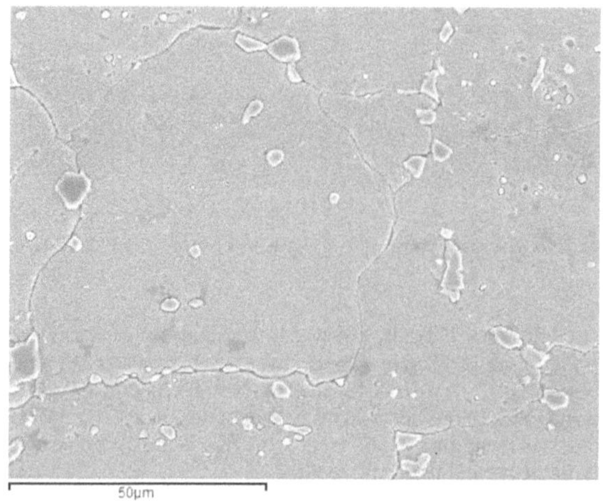


**Fig. 15.** Typical SEM micrograph of precipitates formed in a crept specimen of Alloy 617 (tr = 14,896 h, 30 MPa, 850 °C).

the facets of fracture surface was more evident in short-term rupture specimens at 850 °C than 950 °C. In addition, surface oxidation was evident owing to a prolonged exposure time $t_r$ = 14,896 h under 30 MPa at 850 °C. We believe that the intergranular cracking for Alloy 617 is initiated and developed by the incorporation of minor cavities formed on the grain boundary during the prolonged creep time.

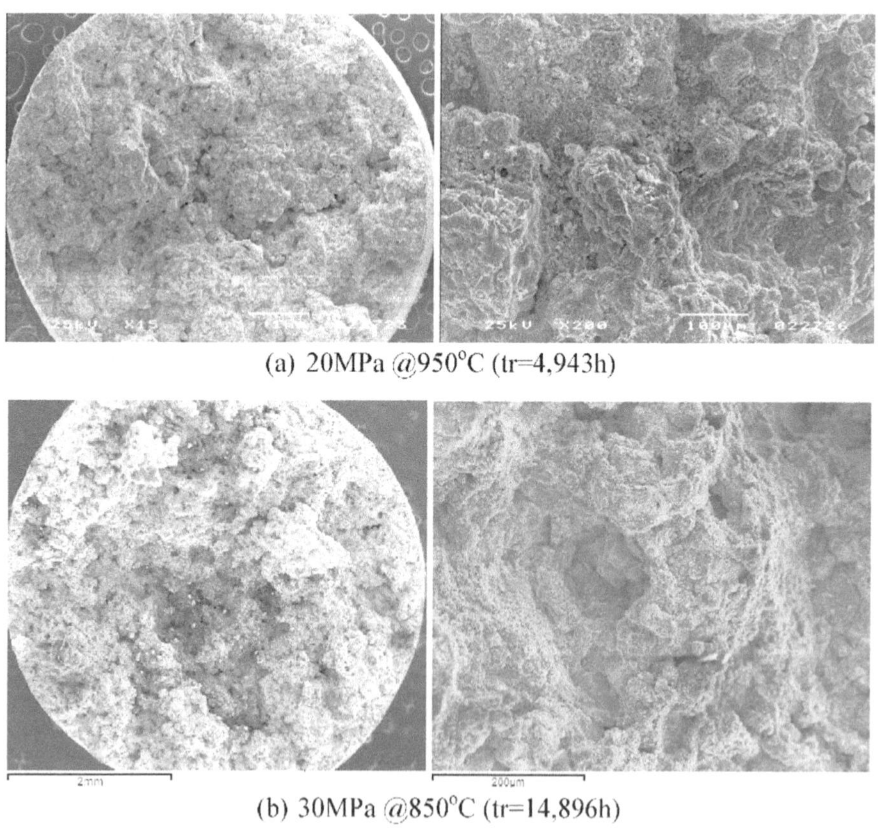


**Fig. 16.** SEM fracture surface of (a) short-term and (b) long-term ruptured specimen at 950 °C and 850 °C of Alloy 617.

## 4. Conclusions

A series of creep data was obtained from the creep tests at different applied stresses at temperatures of 850, 900, and 950 °C for Alloy 617. Creep deformation and rupture behaviors were analyzed using various creep laws on the basis of experimental creep rupture data. The creep curve of Alloy 617 did not show textbook creep behavior, and the secondary creep stage was unclear, and the tertiary creep stage was initiated during early stages of creep. The MMGR provided improved correlation between creep rate and rupture life in Alloy 617. The Z parameter obeyed a good agreement for a function of $Z = 2.30 \times 10^{33} (\sigma/E)^{5.87}$, and the same creep mechanism was operative within the ranges tested in this study. Creep damage tolerance factor of Alloy 617 was found to be 2.40, and this was in agreement with materials exhibiting typical cavitation damage. The creep failure of Alloy 617 occurred by growth and interconnections of cracks rather than a necking failure generated in the tertiary creep region. The creep deformation dominantly developed from the formation and growth of cavity during creep damage process. Creep failure analysis revealed intergranular fracture to be the dominant mode, which proceeds via initiation, linking, and incorporation of the cavities.

## Acknowledgment

This study was supported by Nuclear Research & Development Program of the National Research Foundation of Korea (NRF) grant funded by the Ministry of Science, ICT and Future Planning (2012M2A8A2025682).

## References

[1] R.H. Cook, Creep properties of Inconel-617 in air and helium at 800 to 1000 °C, Nucl. Technol. 66 (1984) 283–288.
[2] S. Dewson, X. Li, Selection criteria for the high temperature reactor intermediate heat exchanger, Proceedings of ICAPP 2005. Seoul; Paper No. 5333, 2005.
[3] W.G. Kim, et al., Comparative study on the high-temperature tensile and creep properties of alloy 617 base and weld metals, J. Mech. Sci. Technol. 27 (2013) 2331–2340.
[4] W.G. Kim, S.N. Yin, G.G. Lee, Y.W. Kim, S.J. Kim, Creep oxidation behavior and creep strength prediction for alloy 617, Int. J. Press. Vessels Piping 87 (2010) 289–295.
[5] J.H. Chang, et al., A study of a nuclear hydrogen production demonstration plant, Nucl. Eng. Technol. 39 (2010) 111–122.
[6] W.G. Kim, J.Y. Park, S.J. Kim, M.H. Kim, Improved methodology for determining tensile elastic and plastic strain components of alloy 617, J. Mech. Sci. Technol. 28 (2014) 3027–3032.
[7] J.Y. Park, et al., Development of materials for a high temperature gas cooled reactor in Korea, The 4th International Topical Meeting on High Temperature Reactor Technology-HTR 2008. Washington; Paper No. HTR 2008-58149, 2008.
[8] Draft Alloy 617 Code Case, Case N-47-28, Task Force Very High Temperature Design, ASME, 1988.
[9] ASME boiler and pressure vessel code, section III, rules for construction of nuclear power plant components, Div. 1, subsection NH, class 1 components in elevated temperature service, ASME, 2007.
[10] J.K. Benz, L.J. Carroll, J.K. Wright, R.N. Wright, T.M. Lillo, Threshold stress creep behavior of Alloy 617 at intermediate temperature, Metall. Mater. Trans. A 45A (2014) 3010–3022.
[11] S.K. Shrama, G.D. Ko, F.X. Li, K.J. Kang, Oxidation and creep failure of alloy 617 foils at high temperature, J. Nucl. Mater. 378 (2008) 144–152.
[12] P.S. Shankar, K. Natesan, Effect of trace impurities in helium on the creep behavior of alloy 617 for very high temperature reactor applications, J. Nucl. Mater. 366 (2007) 28–36.
[13] F.C. Monkman, N.J. Grant, An empirical relationship between rupture life and minimum creep rate in creep rupture tests, ASTM Proc. 56 (1956) 593–620.
[14] F. Dobes, K. Milicka, The relation between minimum creep rate and time to fracture, Met. Sci. 10 (1976) 382–384.
[15] W.G. Kim, S.H. Kim, W.S. Ryu, Evaluation of Monkman–Grant parameters for type 316LN and modified 9Cr–1Mo stainless steels, KSME Int. J. 16 (2002) 1420–1427.
[16] B.K. Choudhary, Tertiary creep behaviour of 9Cr–1Mo ferritic steel, Mater. Sci. Eng. A 585 (2013) 1–9.
[17] B.K. Choudhary, C. Phaniraj, Raj Baldev, Interesting relationships for creep deformation and damage and their applicability for 9Cr–1Mo ferritic steel, Trans. Indian Inst. Met. 63 (2010) 675–680.
[18] B.K. Choudhary, S. Saroja, K. Bhanu Sankara Rao, S.L. Mannan, Creep-rupture behaviour of forged thick section 9Cr–1Mo ferritic steel, Metall. Mater. Trans. A 30 (1999) 2825–2834.
[19] C. Phaniraj, B.K. Choudhary, K. Bhanu Sankara Rao, Raj Baldev, Relationship between time to reach Monkmon–Grant ductility and rupture life, Scr. Mater. 48 (2003) 1313–1318.
[20] B.K. Choudhary, E. Isaac Samuel, Creep behaviour of modified 9Cr–1Mo ferritic steel, J. Nucl. Mater. 412 (2011) 82–89.
[21] G.E. Deter, D. Bacon, Mechanical metallurgy, McGraw-Hill Book Company, London, 1988 306–307.
[22] M.F. Ashby, B.F. Dyson, in: Valluri, et al., (Eds.), Fracture Research, 1, Pergamon Press, Oxford 1984, pp. 3–30.
[23] H.J. Frost, M.F. Ashy, Deformation mechanism maps, Pergamon Press, Oxford, 1982 21.
[24] K. Schneider, W. Hartnagel, P. Schepp, B. Ilschner, Creep behavior of materials for high-temperature, Nucl. Technol. 66 (1984) 289–295.
[25] Special Metals Corporation, Inconel 617available at: www.specialmetals.com.
[26] C. Jang, D. Lee, D. Kim, Oxidation behavior of alloy 617 in very high temperature air and helium environments, Int. J. Press. Vessels Piping 85 (2008) 368–377.
[27] W.G. Kim, J.Y. Park, G.G. Lee, S.D. Hong, Y.W. Kim, Temperature effect on the creep behavior of alloy 617 in air and helium environments, Nucl. Eng. Des. 271 (2014) 291–300.
[28] K. Natesan, M. Li, W.K. Soppet, D.L. Rink, Creep rupture testing of Alloy 617 and A508/533 base metals and weldments, Argonne National Laboratory (ANL), Report; ANL/EXT-11/46, 2011.

## Glossary

$A, A', A_1$constants in power-law form$C, C', C_{MG}, C_{MMG}$intercept constants in Monkman–Grant and modified Monkman–Grant relations$E$modulus elasticity (GPa)$m, m'$ slope constants in Monkman–Grant and modified Monkman–Grant relations$n, n', n_1$creep exponent in power-law form$Q$activation energy (J/mol)$R$universal gas constant (8.3 J/mol K)$T$absolute temperature (K)$t_r$rupture time (h)$Z$Zener–Holloman parameter (or Z parameter)$\epsilon_f$strain in creep failure (%)$\dot{\epsilon}_m$minimum creep rate (1/s)$\sigma$stress (MPa)$\lambda$creep damage tolerance factor

Springer

KSME

Journal of Mechanical Science and Technology 28 (11) (2014) 4493~4501
www.springerlink.com/content/1738-494x
DOI 10.1007/s12206-014-1049-7

# Influence of data size on the reliability assessment of creep life of grade 91 steel†

Woo-Gon Kim[1,*], Jae-Young Park[2], B. K. Choudhary[3], Seon-Jin Kim[2], Min-Hwan Kim[1] and Jinsung Jang[1]

[1]*Nuclear Materials Division, Korea Atomic Energy Research Institute, 989-111 Daeduk-daero, Yuseong-gu, Daejeon, 305-353, Korea*
[2]*Dept. of Mechanical Design Engineering, Pukyong National University, 100 Yongdang-dong, Nam-gu, Busan, 608-739, Korea*
[3]*Metallurgy and Materials Group, Indira Gandhi Centre for Atomic Research, Kalpakkam - 603102, Tamil Nadu, India*

(Manuscript Received May 23, 2014; Revised July 31, 2014; Accepted August 24, 2014)

---

**Abstract**

Influence of data size on a reliability assessment of the long-term creep life of Grade 91 steel was investigated for two different sets of large body of creep-rupture data comprising 690 (Set-1) and 1072 (Set-2) data points. The Z-parameter on the Larson-Miller parameter of creep rupture data supported a normal distribution well. Based on the normal distribution, the reliability of the predicted creep life for the two data sets was assessed to obtain variations in reliability owing to fluctuations in the service conditions and the dispersion of the creep data using a service condition-creep rupture property interference (SCRI) model. A large number of random variables for Z-parameter were generated using Monte-Carlo simulation. A reliability assessment for the two data sizes under specific service conditions has been comparatively demonstrated. It was found that the size of the creep-rupture data influenced the reliability of the creep life prediction.

*Keywords*: Creep; Grade 91 steel; Creep life; Reliability; Z-parameter

---

## 1. Introduction

Grade 91 steel is a favored high temperature material for steam generator (SG) applications in fossil-fired thermal and nuclear power generating industries. The steel developed by the addition of strong carbide/nitride forming elements such as V and Nb along with the controlled addition of N in the plain 9Cr-1Mo steel, offers a good combination of high creep strength and ductility over long exposures at high temperatures [1-4]. In view of its superior creep strength, Grade 91 steel has been chosen as a candidate structural material for the steam generator, intermediate heat exchangers (IHX), and primary pipes of a generation-IV (Gen-IV) sodium-cooled fast reactor (SFR). In addition, Grade 91 steel will be a preferred choice for reactor pressure vessel (RPV) application of a very high temperature gas-cooled reactor (VHTR) [5]. The basic choice of Grade 91 steel for SFR applications is guided by its low thermal expansion coefficient and high resistance to stress corrosion cracking in water-steam systems compared to austenitic stainless steels, in addition to improved mechanical properties at elevated temperatures compared to its counterparts [6].

Under these circumstances, the Korea Atomic Energy Research Institute (KAERI) has recently inaugurated a long-term R&D program to develop the prototype Gen-IV SFR (PGSFR) system (inlet/outlet temperature: 390°C/545°C) of 150 MWe in electricity with enhanced safety, improved economics, efficient utilization of uranium resources, and efficient transmutation of light water reactor spent nuclear fuel. Key technologies for constructing the PGSFR system by 2028 are under development [7].

To achieve economical nuclear energy through the PGSFR route, the PGSFR system is being designed for 60 years. The French Nuclear Design Code RCC-MR [8] used for the design of fast reactors provides creep design curves of up to $3\times10^5$ h, i.e., ~35 years. To design the PGSFRs for 60 years, it is desirable to have the experimental creep-rupture data close to 20 years or, at least rupture lives significantly higher than $10^5$ h. This requirement arises from the fact that, as a rule of thumb for a creep design, an extrapolation factor of 3 times is considered to be appropriate. However, it is difficult to obtain experimental data close to 20 years because of expensive and time consuming works. In view of this, reliable life prediction and extrapolation techniques become necessary for a safe design life of 60 years.

Different time-temperature parametric (TTP) methods such as a Larson-Miller parameter (LMP), Manson-Haferd (MH) parameter, and Orr-Sherby-Dorn (OSD) parameter are being used frequently for the extrapolation of creep life [9-11]. However, without exception, all of these parametric methods suffer from the absence of detail pertaining to the appropriate microstructural degradation and damage from long-term creep exposures. Wilshire, et al. [12] introduced a theta (θ) projec-

*Corresponding author. Tel.: +82 42 868 2493, Fax.: +82 42 68 8549
E-mail address: wgkim@kaeri.re.kr
† Recommended by Editor Jai Hak Park

tion method as an extrapolation technique for the prediction of long-term creep curve by the combination of decaying transient creep and accelerating tertiary creep equations expressed using four θ parameters. Prager [13, 14] proposed the MPC-Omega method for the extrapolation of creep life incorporating the concept of continuum damage mechanics.

The various prediction and extrapolation methods described above are deterministic in nature and do not involve fluctuations in operating stress and temperature conditions. Although, the inherent scatter in experimental creep-rupture data is compressed into a master curve representing the mean value, the scatter in data still affects the predicted results. In addition to this, variations in the actual service conditions around the designed stress and temperature conditions in the operating lifetime are bound to provide a significant amount of dispersion in the creep life. In view of these facts, deterministic methods lack good agreement between the predicted and actual creep-rupture lives of components experienced during service. To overcome this problem, a probabilistic and statistical approach incorporating the fluctuations in the service conditions becomes necessary for a reliable prediction of creep life.

Zhao et al. [15] proposed a "service condition-creep rupture property interference" (SCRI) model based on a Z-parameter method to predict the reliability of the creep-rupture life for high-temperature components. In this approach, both the scatter in the experimental creep-rupture property and the dispersion of creep-rupture lives from the temperature and stress fluctuations during service are incorporated. This model was successfully used for the reliability assessment of creep-rupture lives of HK40 steel. It has been suggested that this approach can be extended to various heat-resistance steels for a reliability analysis and unified analysis of creep-rupture data under certain temperature and stress conditions. Recently, Kim, et al. [16] successfully used the SCRI model based on the Z-parameter to examine the reliability of creep life of Grade 91 steel for a proper size of 690 creep-rupture data.

In this study, a reliability assessment was performed in the framework of the SCRI model on a significantly increased size of 1072 creep-rupture data containing a large number of long-term creep test results for Grade 91 steel. The reliability of the predicted creep lives for two data sets has been investigated for various durations in the range $10^3$-$10^5$ h for different amount of fluctuations in the operating temperature and stress conditions during service. An attempt has been made to compare the influence of the size of the creep-rupture data on a reliability assessment of the creep life of Grade 91 steel.

## 2. Review of SCRI model

A basic concept and application procedures on the SCRI model were introduced in the previous studies by Zhao et al. and Kim et al. [15, 16]. Here, a brief review is given as follows.

A schematic diagram of the SCRI model is shown in Fig. 1.

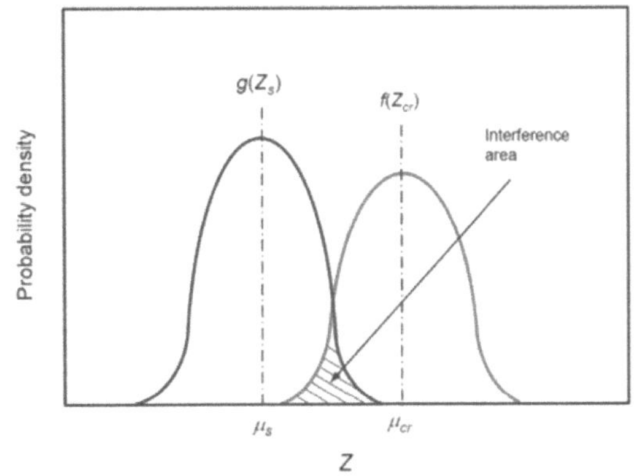


Fig. 1. A schematic diagram for a SCRI model showing the distribution of probability density function for $Z_s$ and $Z_{cr}$. The interference area shown by the hatched lines denotes a failure and unsafe region for condition $Z_s > Z_{cr}$ [16].

In this model, $Z_{cr}$ describes the experimental scatter in the creep-rupture data and $Z_s$ represents the dispersion in creep-rupture data owing to fluctuations in the service conditions. The distribution of Z-parameters supports a normal distribution well. The $Z_s$ component is used to describe the influence of the fluctuations in the service conditions such as temperature, stress and duration. In the SCRI model, $Z_{cr}$ and $Z_s$ are treated as two independent random variables. The probability density function of the distribution of $Z_s$ is expressed as

$$g\,(Z_S) = g\,(T,\ \Delta T, \sigma, \Delta\sigma, t_s)\,, \tag{1}$$

where $T$ is the service temperature, $\sigma$ is the applied stress, $t_s$ is the service duration and ΔT and Δσ are the fluctuations in service temperature and stress, respectively. The interference area, which is shown in the schematic concept of the SCRI model in Fig. 1, corresponds to the failure probability as a fracture state for condition $Z_s > Z_{cr}$ indicating an inability of the material to provide further service. This partial probability becomes the failure probability corresponding to certain operating conditions and service durations. The reliability '$R$' is obtained as

$$R = 1 - F \tag{2}$$

where $F$ is the cumulative failure probability and $R$ is the probability of no failure for a specific service condition.

## 3. Two sets of creep rupture data

To assess the reliability of creep life of Grade 91 steel, two sets of creep-rupture data were prepared. The first set, comprising 690 creep-rupture dataset-1 (Set-1) for temperature intervals of 50°C in the temperature range of 500-700°C, obtained for different heats of the plate, pipe, tube and forging product, is formed from the following three sources: (i) National Insti-

Table 1. Values of material constant '$C$', mean value (μ), standard deviation ($S_{cr}$) for creep-rupture dataset of Set-1 and Set-2 in Grade 91 steel.

| Creep-rupture data | Constant, $C$ | Mean value, μ | Standard deviation, $S_{cr}$ |
|---|---|---|---|
| Set-1 (690 points) | 38 | $3.2422 \times 10^{-5}$ | 0.33625 |
| Set-2 (1072 points) | 35 | $-2.6396 \times 10^{-5}$ | 0.31078 |

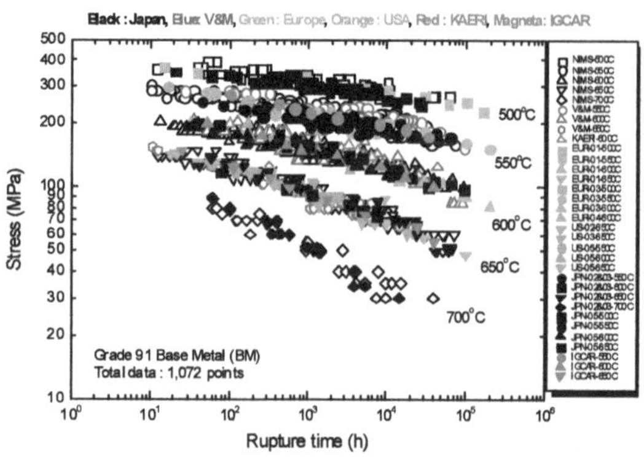

Fig. 2. Stress vs. rupture time plots at 500, 550, 600, 650 and 700°C for Set-2 comprising of 1072 data points for Grade 91 steel.

tute Materials Science (NIMS), Japan [17], (ii) Vallourec & Mannesmann (V&M), Europe [18] and (iii) Korea Atomic Energy Research Institute (KAERI), Republic of Korea, as reported in the author's pervious study [16]. In the second set (Set-2), the creep-rupture data consisted of 1072 data points. In addition to the data in Set-1, a significantly large amount of creep-rupture data was obtained from different sources from Japan [19-21], Europe [22-24], USA [25-28], and the Indira Gandhi Centre for Atomic Research (IGCAR), India [4]. The larger creep rupture database for Grade 91 steel was prepared by digitizing the published document data. Hence, the short-term data under 10 h were removed in this database.

In the collection of both sets of creep data, care has been taken regarding their reliability in terms of chemical compositions and heat treatments conforming to ASTM standards; C 0.08-0.12, Si 0.20-0.50, Mn 0.30-0.60, P ≤ 0.020, S≤0.010, Ni≤0.040, Cr 8.00-9.50, Mo 0.85-1.05, V 0.18-0.25, Nb 0.06-0.10, Al≤0.040, N 0.030-0.070 [29]. A typical log-log plot of stress vs. rupture time at 500, 550, 600, 650, and 700°C is shown in Fig. 2 for dataset-2 (Set-2) having 1072 creep-rupture data. It can be seen that the creep-rupture data are well distributed having reasonable scatter at each temperature in the range of 500-700°C.

## 4. Results and discussion

### 4.1 Master creep rupture curve

Both sets of creep-rupture data were analyzed using the Larson-Miller parameter (LMP), the most commonly used parametric approach for life extrapolation, expressed as

$$P = T\,(C + \log t_r) \tag{3}$$

where $T$ is the absolute temperature in Kelvin, $t_r$ is the rupture time (h), $C$ is a material constant, and $P$ is a stress functional parameter. The $C$ constant should be reasonably determined because it can affect sensitively the result of creep life prediction. The two sets of creep-rupture data were fitted separately using a third-degree polynomial non-linear regression analysis for the plots of Larson-Miller parameter, $P$ vs. log (stress) using log (stress) as a polynomial function. A detail investigation on the applicability of various time-temperature parametric approaches for long-term creep life prediction using 2nd, 3rd, and 4th order polynomials indicated that it is improper to use 2nd and 4th order polynomial equations, because the fitted master curves exhibit a sharp bend and inflections at a low stress regime owing to inherent mathematical characteristics in 2nd and 4th order polynomials [30]. Contrary to this, the master curve obtained using a 3rd order polynomial equation displayed a steady increase without inflection at low stresses. In view of this, the 3rd order polynomial was finally chosen to optimize the material constant '$C$' in the Larson-Miller parameter and to obtain master creep-rupture curve.

To optimize the material constant $C$ in the LMP, the $C$ value was varied from 20 to 50 and the best fit with the corresponding regression coefficient of determination (COD), $R^2$ for each of the $C$ values in the range from 20 to 50 was determined for each master curve. The $R^2$ has known as a statistical parameter, and its value has from 0 to 1. It means that a higher value (closer to 1) is better in fit. Based on the highest value of $R^2$, the optimum values of $C = 38$ and $C = 35$ were obtained for Set-1 and Set-2 consisting of 690 and 1072 creep-rupture data points, respectively. The optimum '$C$' values along with $R^2$ for both data sets are presented in Table 1. The high values of '$C$' for both the data sets obtained in the present investigation are in agreement with those reported for 9 % chromium steels [31, 32]. Separate creep-rupture master curves using a 3rd degree polynomial along with respective optimized material constants, i.e., $C = 38$ for Set-1 and $C = 35$ for Set-2 were obtained for the temperature range 500-700°C. The best polynomial fits for data Set-1 and Set-2 are given in Eqs. (4) and (5), respectively.

$$\text{Set-1:}\ P = 58.1427 - 23.4376\sigma + 12.2713\sigma^2 - 2.8413\sigma^3 \tag{4}$$

$$\text{Set-2:}\ P = 48.9560 - 14.4976\sigma + 8.0235\sigma^2 - 2.1451\sigma^3. \tag{5}$$

The respective master curves representing experimental creep-rupture data of Set-1 and Set-2 in the temperature ranges of 500-700°C are shown in Fig. 3. In the plot, the master curve of $C = 35$ is located lower than that of $C = 38$. Individual predicted master curves, which are obtained by Eq. (5)

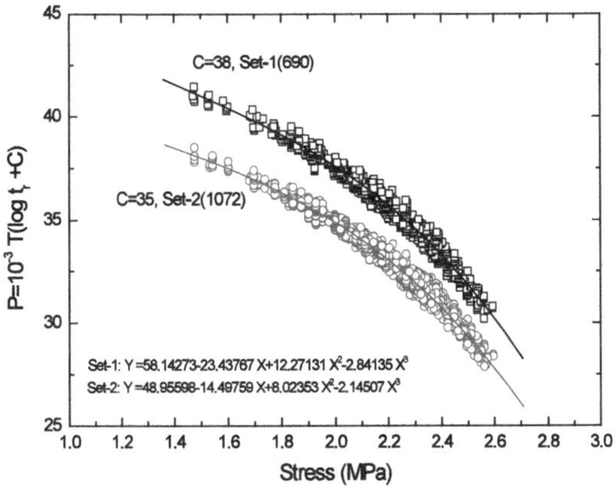


Fig. 3. Each master creep-rupture curve obtained by least square optimization for the Larson-Miller parameter using $C$ = 38 and 35 for Set-1 and Set-2 in Grade 91 steel.

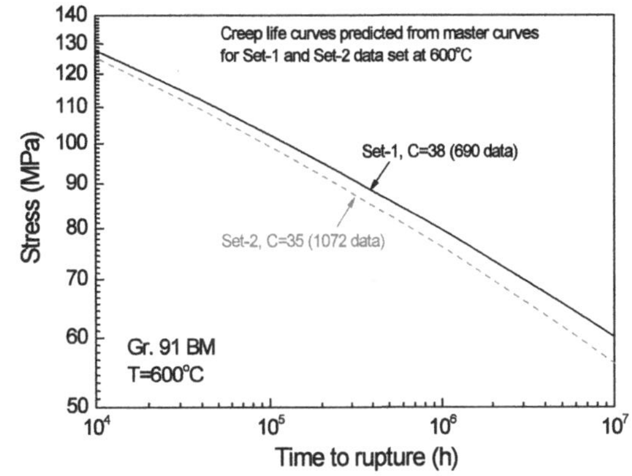


Fig. 5. Plot showing some differences of the predicted long-term creep curves for Set-1 (C = 38, 690data) and Set-2 (C = 35, 1072 data) at 600°C.

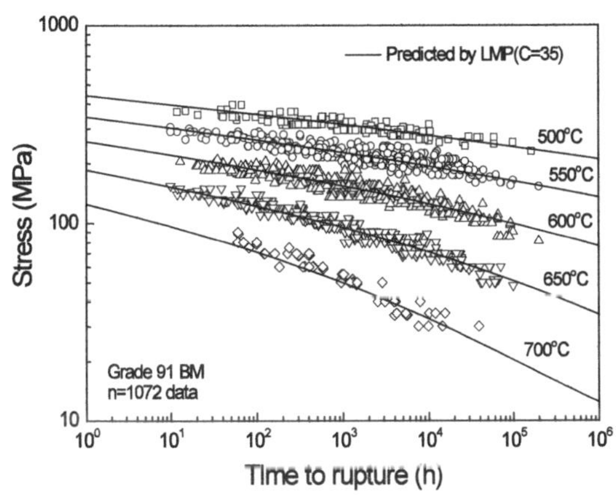


Fig. 4. Stress vs. rupture life plots at 500, 550, 600, 650, and 700°C for Set-2 in Grade 91 steel. Individual master curves obtained using the Larson-Miller parameter for different temperatures are shown by the full lines and the symbols represent the experimental data.

along with respective creep-rupture data at 500, 550, 600, 650 and 700°C for Set-2, are shown in Fig. 4. Reasonable agreement between the predicted master curves and experimental data obtained for the different temperatures can be identified for Set-2 in Fig. 4.

A decrease in the value of the material constant from $C$ = 38 for Set-1 to $C$ = 35 for Set-2 can be ascribed to the inclusion of a larger amount of creep-rupture data belonging to the long-term creep life regime in Set-2. The decrease in the $C$ values owing to increased size of creep-rupture data has resulted in a relative shift in the master curve towards lower stress values. For an example result, long-term predicted curves for Set-1 and Set data at 600°C are shown in Fig. 5. In the plot, the predicted curve of $C$ = 35 is lower in creep strength than that of $C$ = 38. It is suggested that the difference in the $C$ values between the two data sets may affect the reliability of the creep life prediction. Detail discussion for this is given in Secs. 4.4 and 4.5.

### 4.2 Master creep rupture curve

The creep-rupture data inherently exhibits some scatter owing to the different sources of data, heats and material variables (Figs. 2 and 4). Experimental variables and the accuracy of the measurements adopted world-wide in different laboratories add to the scatter in the creep-rupture data. The scatter in the experimental creep-rupture data can be defined as the Z-parameter describing the magnitude of data deviation from the obtained master curve using the following expression:

$$Z_i = P_{i,\exp} - P_{i,mean}, \tag{6}$$

where $Z_i$ corresponds to the deviation of experimental value $P_{i,exp}$ from the mean value $P_{i,mean}$ for the i$^{th}$ data ($T_i$, $\sigma_i$, $t_{ri}$). The values of $P_{i,mean}$ are obtained from the best fit polynomial equations for Set-1 and Set-2. Using respective optimized '$C$' for data Set-1 and Set-2, the values of $P_{i,exp}$ are evaluated using the Larson-Miller parameter, shown in Eq. (3). The distribution of the Z-parameter representing the dispersion in creep-rupture data for Set-2 is shown in Fig. 6. The statistical distribution of Z-parameter exhibits a normal distribution like a bell type, and this has been confirmed by the normal probability plot displaying good linearity [16]. The above observations also suggest that the reliability analysis for the creep life of Grade 91 steel can be performed on the basis of the normal distribution of the Z-parameter. For the normal distribution, the mean value ($\mu$) and standard deviation ($S_{cr}$) obtained for Set-1 and Set-2 are given in Table 1. A noticeable decrease in the value of standard deviation is observed as $S_{cr}$ = 0.310 for Set-2 compared with $S_{cr}$ = 0.336 for Set-1.

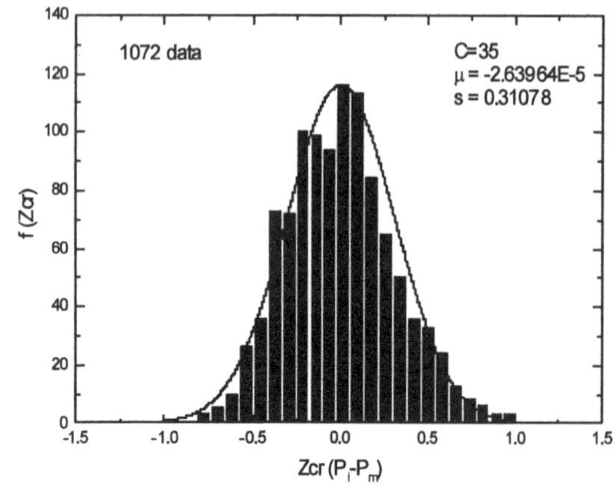


Fig. 6. Probability density plot showing $Z_{cr}$ distribution in Set-2 of Grade 91 steel.

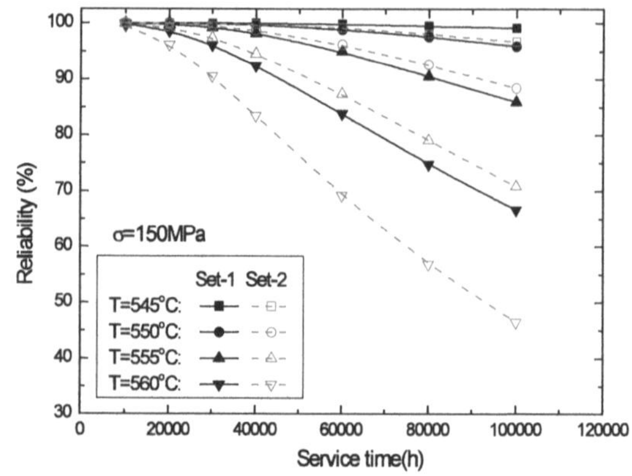


Fig. 7. Influence of increase in temperature on the reliability of creep life for different durations at σ = 150 MPa.

### *4.3 Generation of random variables through a Monte-Carlo simulation*

Reliability "*R*" can be obtained using Eq. (2), if the distribution under certain service condition is known. Following a normal distribution of the Z-parameter, as shown in Fig. 6, The evaluation of the distribution of $Z_s$ arising from the fluctuations in the service conditions becomes difficult when using an analytical method owing to the fluctuations in at least two factors, i.e., the variations in the operating temperature and stress. In view of this, the Monte-Carlo simulation technique is performed to evaluate the reliability of the SCRI model and the details of this simulation were described in the author's previous study [16].

Large numbers of random variables are generated using a Monte-Carlo simulation from the a normal distribution of the Z-parameter. For a set of service conditions ($T_i$, $r_i$, $t_{ri}$), the values of $Z_s$ are calculated using Eq. (6). Meanwhile, the values of $Z_{cr}$ are generated randomly in a normal distribution. The condition $Z_s < Z_{cr}$ defines the safe state of the material, whereas $Z_s > Z_{cr}$ indicates failure under the particular service condition. For the total number of simulations (*N*), reliability (*R*) is obtained as

$$R = \frac{n}{N} \tag{7}$$

where *n* is the number of safe states displaying $Z_s < Z_{cr}$ in the total number of simulation. In the present study, as an example, 30,000 random variables were generated for $Z_s$ and $Z_{cr}$ under certain conditions with fluctuations in the service temperatures and stresses, and the reliability percentage was assessed with service times for Set-1 and Set-2. The reliability of the predicted creep life is compared for the two data sets.

### *4.4 Reliability analysis for specific service conditions*

Temperature and stress are the most important factors influencing the creep life of high temperature materials. Any increase in the service temperature and/or stress greatly reduces the residual creep life and the reliability for the specific service duration. Therefore, it is appropriate to establish the reliability in terms of specific service duration. In view of this, the variations in reliability from fluctuation in the service temperature and stress has been evaluated for durations in the range $10^3$-$10^5$ h and presented in the current investigation. Following the Z-parameter concept, the deviations in the experimental creep-rupture data from the master curve are obtained using Eq. (6) and expressed in Eqs. (8) and (9) for Set-1 and Set-2, respectively.

Set-1:

$$\begin{aligned} Z &= T(38 + \log t_r) \times 10^{-3} - 58.1427 \\ &\quad + 23.4376\sigma - 12.2713\sigma^2 + 2.8413\sigma^3 . \end{aligned} \tag{8}$$

Set-2:

$$\begin{aligned} Z &= T(35 + \log t_r) \times 10^{-3} - 48.9560 \\ &\quad + 14.4976\sigma - 8.0235\sigma^2 + 2.1451\sigma^3 . \end{aligned} \tag{9}$$

The probability density function (pdf) is obtained for both Set-1 and Set-2 along with the respective mean value (μ) and standard deviation ($S_{cr}$) presented in Table 1. To account for the influence of increase in the operating temperature and stress conditions, a service temperature of *T* = 550°C and service stress of σ = 150 MPa are conveniently chosen to assess the reliability prediction up to $10^5$ h for Grade 91 steel. Following the procedure for the Monte-Carlo simulation described in Sec. 4.3, a reliability assessment was made using the chosen service temperature and stress conditions with different amounts of exclusions in temperature and stress.

The variations in the reliability percentage with a change in temperature in the range of 545-560°C for temperature intervals of 5°C at a constant applied stress of 150 MPa for different durations are shown in Fig. 7. In general, a larger decrease in reliability with an increase in the service durations is ob-

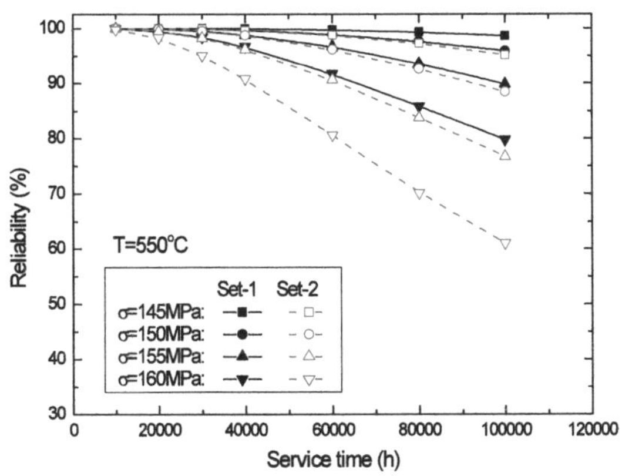


Fig. 8. Influence of increase in stress on the reliability of creep life for different durations at T = 550°C.

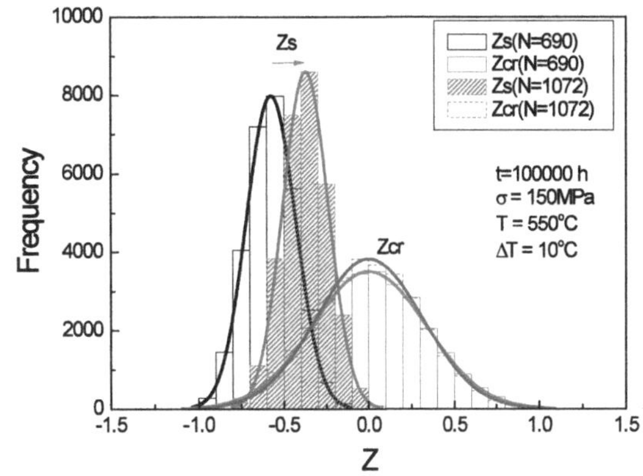


Fig. 9. Distribution of temperature fluctuation, $\Delta T$ = 10°C, on the relative distribution of $Z_s$ and $Z_{cr}$ showing interference area in the SCRI model for $10^5$ h for Set-1 and Set-2.

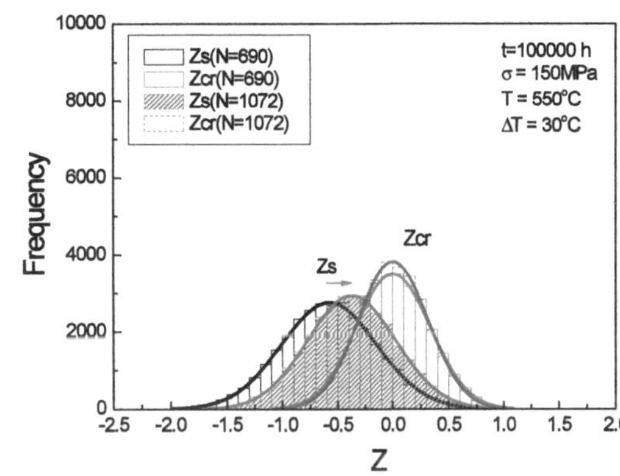


Fig. 10. Distribution of temperature fluctuation, $\Delta T$ = 30°C, on the relative distribution of $Z_s$ and $Z_{cr}$ showing an increased interference area in the SCRI model for $10^5$ h for Set-1 and Set-2.

served for all conditions. Further, a larger increase in temperature results a larger decrease in reliability. Similarly, for service temperature of 550°C, the reliability decreases with an increase in applied stress from 145 to 160 MPa (for stress intervals of 5 MPa) with a more dominant reduction for a larger increase in stress at longer durations, as shown in Fig. 8. The observed decrease in the reliability with respect to an increase in the operating temperature and stress demonstrates clearly that the reliability of creep life is sensitive to the variations in operating conditions. Further, a systematic and significant decrease in reliability percentage for Set-2 having a larger amount of creep-rupture data can be seen for both an increase in temperature as well as stress conditions. In addition, at longer service durations, a larger decrease in reliability can be seen in Figs. 7 and 8. During the actual operation of a plant, it is difficult to maintain the service temperature and stress at constant design values giving rise to variations in the operating temperature and stress conditions. This becomes all the more severe during start-ups and shut-downs of the plants. In the following, the variations in reliability owing to the fluctuations in temperature and/or stress conditions by specific predetermined amounts are presented as examples.

Before presenting the results on the variations in reliability owing to fluctuations in temperature ($\Delta T$) and/or stress ($\Delta\sigma$), it is appropriate to examine the relative distribution of Z-parameters, i.e., the distribution of $Z_{cr}$ owing to scatter in the creep-rupture property, and $Z_s$ due to fluctuations in operating temperature and stress conditions in the framework of SCRI model for some specific service conditions. Fig. 9 shows the distributions of $Z_{cr}$ and $Z_s$ obtained for Set-1 and Set-2 for temperature fluctuation, $\Delta T$ of 10°C for a chosen service temperature, T = 550°C, and constant stress, $\sigma$ = 150 MPa.

The distributions of $Z_{cr}$ and $Z_s$ evaluated for temperature fluctuation, $\Delta T$ of 30°C for service temperature T = 550°C and constant stress, $\sigma$ = 150 MPa for both the data sets are presented in Fig. 10. As in $Z_{cr}$, the normal distribution obtained for $Z_s$ can be seen in Figs. 9 and 10. A comparative evaluation of the interference area clearly suggests a significant increase in the interference area specifying a failure and unsafe zone with an increase in temperature fluctuation from $\Delta T$ = 10°C to $\Delta T$ = 30°C. A similar increase in the interference area indicated with an increase in the stress fluctuations is observed. Here, a comment on the relative distributions of both $Z_{cr}$ and $Z_s$ with respect to the creep-rupture data size is appropriate. It can be clearly seen that the distribution of $Z_{cr}$ is almost superimposed regardless of the data size. However, a marginal increase in the frequency of $Z_{cr}$ owing to a decreased standard deviation for larger creep-rupture dataset (Set-2) is discernible in Figs. 9 and 10. On the contrary, a noticeable shift in $Z_s$ (hatched lines) towards the right side with an increase in Set-2 is observed for all temperatures and stress fluctuations, as shown in Figs. 9 and 10. This shift in $Z_s$ towards $Z_{cr}$ results in a larger interference area specifying a failure and unsafe area with increase in the size of creep-rupture data in Set-2.

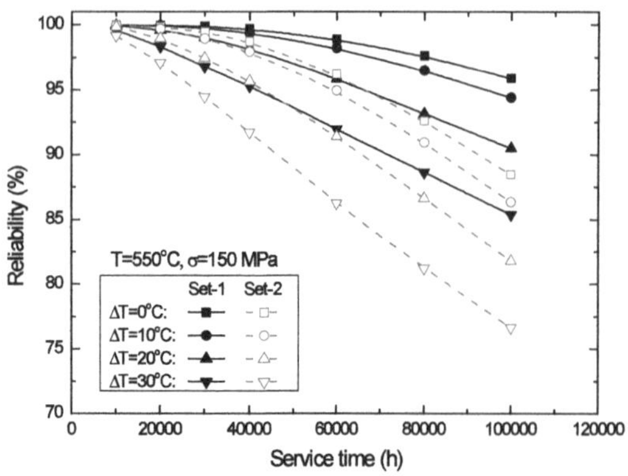


Fig. 11. Influence of increasing temperature fluctuations (ΔT) in the range of 0-30°C on the reliability of creep life for various durations for operating temperature, T = 550°C and stress, σ = 150 MPa, for data Set-1 and Set-2.

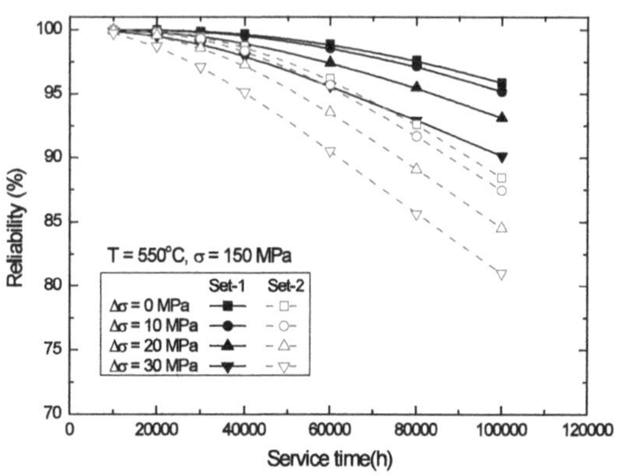


Fig. 12. Influence of increasing stress fluctuations (Δσ) in the range 0-30 MPa on the reliability of creep life for various durations for operating temperature, T = 550°C, and stress, σ = 150 MPa, for Set-1 and Set-2.

The origin of such a shift in the distribution of $Z_s$ can arise from the change in the material constant $C$ values obtained for the two data sets. Owing to an increase in data size, a decrease in $C$ value from 38 (Set-1) to 35 (Set-2) indicates a reduced creep-rupture strength, as is well shown in Fig. 5 predicted for an example at 600°C. Accordingly, a shift in the distributions of $Z_s$ towards $Z_{cr}$ has led to a decrease in the reliability percentage by providing more interference area for $Z_s > Z_{cr}$.

The variations in reliability owing to temperature and stress fluctuations for operating temperature, T = 550°C and stress, σ = 150 MPa, for different durations in the range of $10^3$-$10^5$ h are presented in Figs. 11 and 12, respectively. Fig. 11 clearly demonstrates a rapid decrease in reliability with an increase in the temperature fluctuations in the range ΔT = 10°C to ΔT = 30°C for an operating constant stress of σ = 150 MPa. A more pronounced reduction in reliability from an increase in the fluctuations in temperature is obtained at longer durations. Similarly, a systematic decrease in the reliability with an increase in the stress fluctuations from Δσ = 10 MPa to Δσ = 30 MPa for a constant operating temperature of T = 550°C with a more pronounced decrease at longer durations is evident in Fig. 12. The influence of the increased amount of creep-rupture data is reflected in the systematic decrease in reliability percentage with more pronounced effects at longer durations as shown in Figs. 11 and 12.

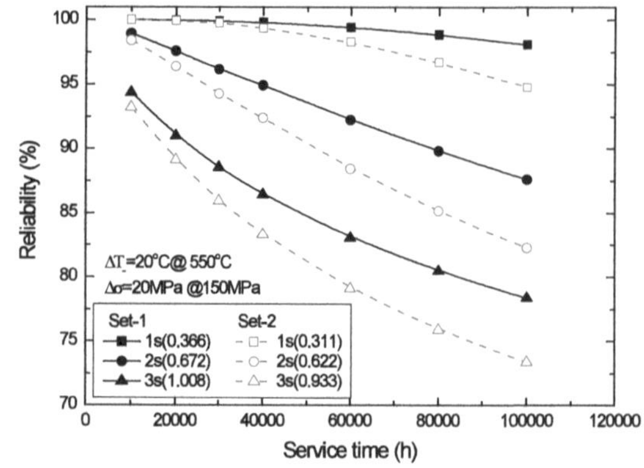


Fig. 13. Influence of standard deviation on the reliability of creep life predicted in the fluctuations of ΔT = 20°C and Δσ = 20 MPa for operating conditions of T = 550°C, and σ = 150 MPa, for Set-1 and Set-2.

To examine the influence of standard deviation (s) under the combined operating conditions of both fluctuations in a temperature and stress, the reliability of predicted creep life was assessed. Fig. 13 shows the influence of standard deviation on the reliability of creep life, which is predicted under the fluctuations of ΔT = 20°C and Δσ = 20 MPa for operating conditions of T = 550°C and σ = 150 MPa for Set-1 and Set-2, respectively. The reliability percentage decreases with an increase in the standard deviation from 1 s to 3 s. The reliability percentage of Set-2 is lower than that of Set-1, although Set-2 having s = 0.311 was lower than Set-1 having s = 0.366. The reason for this is described in Sec. 4.5.

In addition, there is a larger decrease in the reliability percentage for the conditions when there are fluctuations in both temperature and stress together than the condition when there is a fluctuation in either temperature or stress alone, as shown in Figs. 11 and 12.

### *4.5 Influence of creep-rupture data size on reliability*

As presented above, in spite of the lower value of s = 0.311 in the larger dataset (Set-2), the Set-2 revealed worse reliability than the Set-1. It can be described that the influence of creep-rupture data size is significantly reflected in the decreased values of the material constant '$C$' obtained for the Larson-Miller parameter and a noticeable decrease in the value standard deviation '$S_{cr}$' (Table 1). A decrease in the value of '$C$' from 38 (for Set-1 with 690 data) to 35 (for Set-2 with 1072 data) with an increase in the size of the creep data results mainly from the inclusion of a higher number of long-term creep data in Set-2. A larger amount of creep-rupture data with a reduced $C$ value is expected to provide more realistic and better creep life predictions. A larger amount of creep data is also expected to introduce higher scatter and a higher standard deviation. As a result, for the observed decrease in the standard deviation for a larger amount of creep data in Set-2, we may misjudge that the master curve obtained with $C$ = 35 for Set-2 can provide an improved representation of the creep-rupture data. However, in the present investigation, owing to the lower creep strength predicted by the lower $C$ value for larger creep-rupture data Set-2, the shift in $Z_s$ towards $Z_{cr}$ resulted in a larger interference area for $Z_s > Z_{cr}$. This has facilitated a larger failure area and a decrease in the reliability percentage. In addition, the influence of the standard deviation for Set-1 and Set-2 is not significant with an increase in service time, as shown in Fig. 13. However, there are some differences in the reliability percentage for the two data sets. This difference is due to the shift in $Z_s$ towards $Z_{cr}$, as shown in Figs. 9 and 10. The reason for the shift in $Z_s$ is due to the decreased creep-rupture strength originated from the lower $C$ value in Set-2 rather than the difference in the standard deviation values between the two data sets.

## 5. Conclusions

A reliability assessment for the predicted creep life steel has been performed for the two dataset of Set-1 and Set-2 with a large body of creep-rupture data for Grade 91. For the SCRI model, using a Monte-Carlo simulation and probabilistic and statistical analyses, the reliability of the predicted creep life has been evaluated from the relative distribution of probability density function of the Z-parameter, $Z_s$ describing fluctuations in the service conditions, and $Z_{cr}$ defining the dispersion in the creep-rupture data. Using two different amounts of creep-rupture data, variations in reliability have been comparatively examined. The reliability decreased with increased fluctuations in service temperature and stress conditions in Grade 91 steel, and a larger decrease in reliability was observed for longer service durations. The reliability of the predicted creep life decreased noticeably in the large amount of 1072 data points (Set-2) compared with a smaller size of 690 data points (Set-1) of creep-rupture data. The reason for this was arisen from the difference of the $C$ values in LMP between Set-1 and Set-2. A decrease in reliability of Set-2 was originated from the decreased creep strength predicted from the lower $C$ values in LMP, and the shift in $Z_s$ towards $Z_{cr}$ resulted in a larger interference area for $Z_c > Z_{cr}$.

## Acknowledgment

This study has been carried out as a part of the Innovative R&D Project of Korea Atomic Energy Research Institute.

## References

[1] W. G. Kim, S. H. Kim and W. S. Ryu, Evaluation of Monkman-Grant parameters for type 316LN and modified 9Cr-1Mo stainless steels, *J. Mech. Sci. Technol.*, 16 (11) (2002) 1420-1427.

[2] B. Raj and B. K. Choudhary, A perspective on creep and fatigue issues in sodium cooled fast reactors, *Trans. Indian Inst. Met.*, 63 (2010) 75-84.

[3] W. G. Kim, S. N. Yin and G. H. Koo, Generation of isochronous stress-strain curves with a nonlinear least square fitting method for modified 9Cr-1Mo steel, *Met. Mater. Int.*, 15 (2009) 727-732.

[4] B. K. Choudhary and E. Isaac Samuel, Creep behavior of modified 9Cr-1Mo steel, *J. Nucl. Mater.*, 412 (2011) 82-89.

[5] W. G. Kim, J. Y. Park, S. N. Yin, D. W. Kim and S. J. Kim, Reliability prediction of long-term creep strength of Gr. 91 steel for next generation reactor structure materials, *Kor. J. Met. Mater.*, 29 (2011) 275-280.

[6] S. L. Mannan, S. C. Chetal, B. Raj and S. B. Bhoje, Selection of materials for prototype fast breeder reactor, *Trans. Indian. Inst. Met.*, 56 (2003) 155-178.

[7] H. Y. Lee, J. H. Eoh and Y. B. Lee, High temperature design of finned-tube sodium-to-air heat exchanger in a sodium test loop, *Nucl. Eng. Des*, 265 (2013) 833-840.

[8] Design and construction rules for mechanical components of nuclear installations, *French Nuclear Design Code, RCC-MR, Section 1, Subsection Z, Appendix A3: Characteristics of Materials*, 1S & 18S (2007).

[9] F. R. Larson and J. Miller, A time-temperature relationship for rupture and creep stresses, *Trans. ASME*, 74 (1952) 765-771.

[10] R. L. Orr, O. D. Sherby and J. E. Dorn, Correlations of rupture data for metals at elevated temperature, *Trans. ASME*, 46 (1954) 113-118.

[11] S. S. Manson and A. M. Haferd, A linear time-temperature relation for extrapolation of creep and stress rupture data, *NASA Technical Note 2890*, Mar (1952).

[12] R. W. Evans, J. D. Parker and B. Wilshire, The theta projection concept-A model-based approach to design and life extension of engineering plant, *Int. J. Pres. Ves. Piping*, 50 (1992) 147-160.

[13] M. Prager, Development of MPC omega method for life assessment in the creep range, *The Materials Properties Council Inc.*, USA (1994).

[14] M. Prager, The omega method- an engineering approach to life assessment, *J. Pres. Ves. Technology*, 122 (2000) 273-280.

[15] J. Zhao, D. Li, J. Zhang, W. Feng and Y. Fang, Introduction of SCRI model for creep rupture life assessment, *Int. J. Pres. Ves. Piping*, 86 (2009) 599-603.

[16] W. G. Kim, J. Y. Park, S. J. Kim and J. Jang, Reliability assessment of creep rupture life for Gr. 91 steel, *Mater. Des.*, 51 (2013) 1045-1051.

[17] K. Kimura, Y. Toda, H. Kushima and K. Sawada, Creep strength of high chromium steel with ferrite matrix, *The 2nd ECCC Creep Conference*, Zurich, Swiss (2009) 935-949.

[18] K. Haarmann, J. C. Vaillant, B. Vandenberghe, W. Bendick and A. Arbab, *The T91/P91 book-ferritic pipe for high temperature use in boilers and petrochemical applications*, Vallourec & Mannesmann Tubes Inc. (2002) 156.

[19] T. Asayama and Y. Tachibana, Collection of available creep-fatigue data and study existing creep-fatigue evaluation procedures for grade 91 and hastelloy XR, *JAEA Task 5 Report* (2007).

[20] R. P. Chen, H. G. Armaki, K. Yoshimi, K. Maruyama, Y. Minami and M. Igarashi, Premature creep rupture and overestimation of rupture life in modified 9Cr-1Mo steel, *Testu-to-Hagane*, 96 (2010) 564.

[21] R. P. Chen, H. G. Armaki, K. Maruyama and M. Igarashi, Long-term microstructure degradation and creep strength in Gr. 91 Steel, *Mater. Sci. Eng. A*, 528 (2011) 4390-4394.

[22] V. Sklenicka, K. Kucharova, M. Svoboda, L. Kloc, J. Burik and A. Kroupa, Long-term creep behavior of 9-12%Cr power plant steels, *Mater. Charact.*, 51 (2003) 35-48.

[23] *ECCC Data Sheet*, www.vgb.org/gbg wug mpa pres-dfid-23836.html (2009).

[24] T. Masse and Y. Lejeail, Creep behavior and failure modeling of modified 9Cr 1Mo steel, *Nucl. Eng. Des.*, 246 (2012) 220-232.

[25] V. K. Sikka, C. T. Ward and K. C. Thomas, Modified 9Cr1Mo steel - An improved alloy for SG application, *ASM Conf.*, USA (1981).

[26] V. K. Sikka, M. G. Cowgill and B. W. Roberts, Creep properties of modified 9Cr 1Mo steel, *Proc. of Topical Conf. on Ferritic Alloys for Use in Nuclear Energy Technologies*, USA (1983).

[27] R. W. Swindeman, V. K. Sikka and P. J. Maziasz, Evaluation of T91 after 130,000 hours in service, *ASME Conf. on PVP*, California, USA (1998).

[28] T. C. Totemeier, J. A. Simpson, R. Viswanathan, D. Gandy and K. Colman, Effect of off-normal austenization on creep strength of ferritic-martensitic steels, *Advances in Materials Technology for Fossil Power Plants,* South Carolina, USA (2004) 1242-1255.

[29] Standard specification for seamless ferritic and austenitic alloy-steel boiler, superheater, and heat-exchanger tubes, *ASTM Standards A213/A213M-11a*, ASTM International, Pennsylvania, USA (2011).

[30] V. S. Srinivasan, B. K. Choudhary, M. D. Mathew and T. Jayakumar, Long-term creep-rupture strength prediction for modified 9Cr-1Mo ferritic steel and type 316L(N) austenitic stainless steel, *Mater. High Temp.*, 29 (2012) 41-48.

[31] W. G. Kim, J. Y. Park, J. Jang, B. K. Choudhary, S. J. Kim, Y. W. Kim and Y. H. Jeong, Methodologies for long-term creep life extrapolation of Gr. 91 steel, *Proc. the 8th Pacific Rim Int. Congress on Advanced Materials and Processing-TMS*, Hawaii, USA (2013) 397.

[32] F. Masuyama, Creep rupture life and design factors for high-strength ferritic steels, *Int. J. Pres. Ves. Piping*, 84 (2007) 53-61.

**Woo-Gon Kim** received Ph.D. at Chungbuk National University at Dept. of Mechanical Engineering in 1998. He is a principal researcher at Korea Atomic Energy Research Institute. His specialty is mechanical assessment, analysis, modeling and database establishment of high temperature nuclear materials for Generation-IV reactor systems, and especially, time-dependent creep and crack growth behaviors at elevated temperature.

Springer

KSME

Journal of Mechanical Science and Technology 28 (8) (2014) 3027~3032
www.springerlink.com/content/1738-494x
DOI 10.1007/s12206-014-0709-y

# Improved methodology for determining tensile elastic and plastic strain components of alloy 617 †

Woo-Gon Kim[1,*], Jae-Young Park[2], Seon-Jin Kim[2] and Min-Hwan Kim[1]

[1]*Korea Atomic Energy Research Institute, 989-111 Daeduk-daero, Yuseong-gu, Daejeon, 305-353, Korea*
[2]*Dept. of Mechanical Design Engineering, Pukyong National University, Busan, 608-739, Korea*

(Manuscript Received November 20, 2013; Revised February 23, 2014; Accepted March 25, 2014)

**Abstract**

This paper presents an improved methodology for calculating tensile elastic and plastic strain components in total creep strain, which are needed in the generation of the isochronous stress-strain curves (ISSC). The procedures for the RCC-MR code and Blackburn's methods to determine the total strain are analyzed in detail, and a new method is proposed to determine the tensile elastic and plastic strain components. To practically obtain the tensile elastic and plastic strain components for Alloy 617, high-temperature tensile tests were conducted at 800, 850, 900, and 950°C. Using the tested data, three methods of the RCC-MR code, Blackburn, and the newly proposed method were applied and compared. In the elastic regime, the three methods were identical in the tensile curves, but in the plastic regime, Blackburn's method was higher in the stress-strain curves than the RCC-MR code. The difference gap for the two methods was larger with an increase in the plastic strain. In addition, the proposed new method revealed almost midway values for the two methods. It was suggested that the proposed method is useful to determine tensile and plastic strain components.

*Keywords*: Alloy 617; Plastic strain; Isochronous stress-strain curves; Tensile; RCC-MR

## 1. Introduction

Strain depends on both the stress level and time at a given temperature within a creep range. For a design application, the limitation of strain may be an important factor for preserving the structural integrity or functional capability of a component. Such strain limitations are more readily incorporated into a design process by first translating them into design stress limits, which will preclude an excessive distortion of the intended service. Thus, a design for creep materials should reflect the essential interrelationship among the stress, strain, and time. One method of depicting these interrelationships is through the isochronous stress-strain curve (ISSC), which is presented in a plot of the stress against strain, the locus of the total strains accumulated when different constant stresses are applied for a fixed time. It is thus necessary to develop an ISSC for up to $10^5$ h to design the materials [1, 2].

In addition, a very high temperature reactor (VHTR) is one of the most promising Generation-IV reactor types for the economic production of electricity and hydrogen. At present, the Korea Atomic Energy Research Institute (KAERI) is developing a nuclear hydrogen development and demonstration (NHDD) plan with a capacity of 200 $MW_{th}$ in the thermal and core outlet temperature of 950°C. It has been carried out on a government funded project called "Development of Key Technologies for Nuclear Hydrogen" since 2006 [3-5]. Its major components are the reactor internals, reactor pressure vessel (RPV), hot gas ducts (HGD), and intermediate heat exchangers (IHX), as shown in Fig. 1. Among the components, the IHX is an essential component to link the reactors to the hydrogen production plant. It is directly exposed by a primary coolant of 950°C. Thus, high-temperature creep properties are very important, because its integrity should be preserved during the design life for 60 years in VHTR coolant. Alloy 617 for IHX material is a prime candidate material owing to its superior creep resistance above 800°C compared with other potential superalloys [6-9]. For the design application of Alloy 617, the ISSC is needed to avoid an excessive deformation of the material during an intended service life, because they are designed with a target life based on a specified amount of allowable strain and stress. However, until now, the ISSCs for Alloy 617 have not been developed although they were prepared in in the ASME draft code case, which is not insufficient and perfect for a design use [10].

In this study, the procedures of the French code RCC-MR and Blackburn's methods were described in detail to develop the ISSCs. To practically determine the tensile elastic and

*Corresponding author. Tel.: +82 42 868 2493, Fax.: +82 42 868 8549
E-mail address: wgkim@kaeri.re.kr
† Recommended by Editor Jai Hak Park

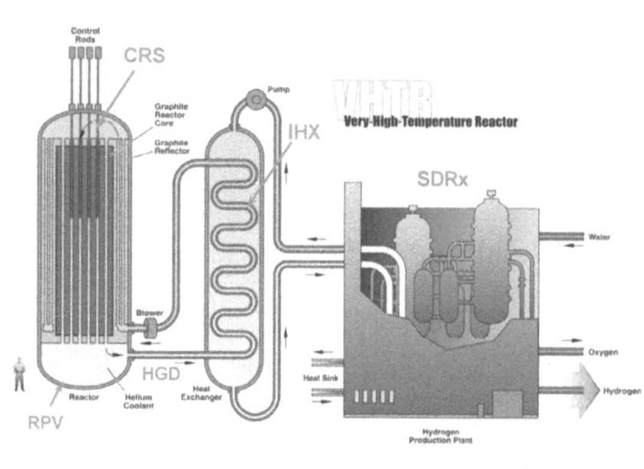


Fig. 1. Major components of VHTR system for nuclear hydrogen production.

plastic strain components for Alloy 617, tensile tests were conducted at 800, 850, 900, and 950°C. A new method was proposed in addition to two methods, and the tensile elastic and plastic strain components were calculated and compared for the three methods of the RCC-MR code, Blackburn's methods, and the method newly proposed in this study.

## 2. Analysis of RCC-MR and Blackburn's method

### *2.1 RCC-MR code*

For the design application of high-temperature materials, it is necessary to construct the ISSC, which can be developed using Young's modulus, the average tensile hardening rule and creep strain laws. The strain used for a determination of the ISSC is the total strain, $\varepsilon_{tot}$, given as

$$\varepsilon_{tot} = \varepsilon_{el} + \varepsilon_p + \varepsilon_c, \quad (1)$$

where $\varepsilon_{tot}$ is the total strain as a percentage (%) corresponding to stress $\sigma$, $\varepsilon_{el}$ is the elastic strain (%), $\varepsilon_p$ is the plastic strain (%), and $\varepsilon_c$ is the creep strain (%). The elastic, plastic and creep strain components of the total strain can be derived using the following procedures [11, 12].

(1) Tensile elastic strain ($\varepsilon_{el}$).

The elastic strain in percentage is calculated by,

$$\varepsilon_{el} = \sigma / E \times 100, \quad (2)$$

where $E$ is Young's modulus in MPa.

(2) Tensile plastic strain ($\varepsilon_p$)

In the RCC-MR code, the plastic strain can be derived from the tensile work hardening equation given by the following formula [7].

$$\varepsilon_p = D\left(\frac{\sigma}{\sigma_y}\right)^m, \quad (3)$$

where $\sigma_y$ is the average yield strength at a 0.2% offset, given as a function of temperature (°C) in MPa, $\varepsilon_p$ is the plastic strain induced by stress $\sigma$, and $D$ and $m$ are the material constants. The plastic strain is limited to 1.5% in the above equation for the purpose of the calculation of the design curves. The MCC-MR code is applied to calculate the plastic strain for X10CrMoVNb9-1 normalized-tempered (or quenched-tempered) steel and X2CrNiMo17-12-2(N) solution annealed austenitic stainless steel. It shows that the tensile stress-strain curves for design use can be derived using Eq. (3) at a given temperature [11].

(3) Creep strain ($\varepsilon_c$)

The creep strain is the sum of the primary and secondary creep strains, as creep strain ($\varepsilon_c$) = primary creep strain ($\varepsilon_{cp}$) + secondary creep strain ($\varepsilon_{cs}$). The primary creep strain can be determined by fitting the transient creep data to the following equation [11, 12],

$$\varepsilon_{cp} = C_1(t)^{C_2}(\sigma)^{n_1}. \quad (4)$$

In this formula, $\varepsilon_{cp}$ (%) designates the primary creep strain attained under stress $\sigma$ (MPa) after time $t$ (h) at temperature (°C). The coefficients $C_1$, $C_2$, and $n_1$ are functions of temperature. This formula is valid for $\varepsilon_c < 2\%$ and $t \leq t_{cp}$, where $t_{cp}$ is the time at the end of the primary creep.

Obtaining the time at the end of the primary creep is necessary to identify the transient creep data for fitting into Eq. (4) and to determine the secondary creep strain in Eq. (7). The time at the end of the primary creep is estimated graphically by locating the point of deviation of creep curve from the straight line representing the secondary stage. A graph is plotted between the stress values and corresponding time at the end of the primary creep at a constant temperature. These data can be fitted into a power law relationship, which is given as follows:

$$t_{cp} = C_3(\sigma)^{n_3}. \quad (5)$$

The coefficients $C_3$ and $n_3$ are functions of temperature. The steady state creep rate data are plotted against stress at constant temperature, and the data are fitted into the power law relation as follows:

$$\dot{\varepsilon}_s = C(\sigma)^n. \quad (6)$$

In this formula, (%/h) designates the secondary creep strain rate attained under stress $\sigma$ at temperature. The $C$ and $n$ are temperature-dependent material constants. This formula is valid for $\varepsilon_c > \varepsilon_{cfp}$, where $\varepsilon_{cfp}$ is the strain at the end of the primary creep. The secondary creep strain at time $t$ is given by

$$\varepsilon_{cs}(\%) = C\sigma(t - t_{cp}). \quad (7)$$

The total strain can be estimated by adding the elastic, plastic and creep strain values at different stress levels and required time intervals using Eq. (1). These data give an average value of stress, and a minimum value equal to 0.8 times the average will be used. The average isochronous stress-strain curves are obtained by plotting the graph between average stress (MPa) and total strain (%) at a constant time. The minimum stresses inducing 1% total strain can be obtained from the minimum isochronous stress- strain curves.

### *2.2 Blackburn's method*

(1) Tensile plastic strain ($\varepsilon_p$)

The elastic strain can be identically obtained by Eq. (2). Blackburn proposed the plastic strain equation, which is derived from the tensile work hardening equation by the following formula [1].

$$\varepsilon_p = D_1(\sigma - \sigma_p)^{m_1} \text{ (for } \sigma \geq \sigma_p\text{)}, \quad (8)$$

where $\varepsilon_p$ is the plastic strain induced by stress $\sigma$, and $D_1$ and $m_1$ are material constants. In addition, $\sigma_p$ is the true stress at the proportional elastic limit. There is a difference in the forms between Eqs. (3) and (8). Blackburn developed the ISSCs using Eq. (8) in the calculation of the tensile plastic strain for type 304 and 316 stainless steels.

Note that Eq. (8) is derived in terms of a true stress and a true strain, while all the equations for a creep strain are in engineering strain units. Therefore, when significant plastic strains are involved, it is necessary to convert the stress and strain values from Eq. (8) to engineering units before combining the elastic and plastic strains with the creep strain.

(2) Creep strain ($\varepsilon_c$)

As described above, the creep strain is the sum of primary and secondary creep strain. To define the primary and secondary creep strain, Garofalo's equation can be widely used by the following form [1, 13],

$$\varepsilon_c \ (= \varepsilon_{tot} - \varepsilon_{el} - \varepsilon_p) = \varepsilon_t[1 - \exp(-rt)] + \dot{\varepsilon}_m t \ , \quad (9)$$

where $\varepsilon_t$ is the limiting magnitude of a transient creep strain, $r$ is a constant related to the rate of an exhaustion of a transient creep, $\dot{\varepsilon}_m$ is the secondary creep rate (or minimum creep rate), and $t$ is the time. It is modified by the following form:

$$\varepsilon_c = \varepsilon_x[1 - \exp(-st)] + \varepsilon_t[1 - \exp(-rt)] + \dot{\varepsilon}_m t. \quad (10)$$

It is a modification of Eq. (9) in which the first term, representing a rapid initial transient strain contribution, was added to obtain a better fit to the experimental strain-time creep data. The second term represents a slower and longer transient strain contribution, as reported by Garofalo and coworkers [13].

The parameters $\varepsilon_p$, $\varepsilon_x$, $s$, $\varepsilon_t$, $r$, and $\dot{\varepsilon}_m$ are all functions of stress and temperature. The ISSCs can be generated using all the parameters defined from both the tensile test and creep test data. A good result can be achieved when the ISSCs computed from this relation can reflect properly the magnitude of time-independent elastic and plastic strain, the magnitude and time dependence of transient creep strain, and the magnitude and time dependence of secondary creep strain.

## 3. Experimental procedures

To practically determine the tensile elastic and plastic strain components for Alloy 617, high-temperature tensile tests were conducted at 800, 850, 900, and 950°C. Commercial grade nickel-based superalloy, Alloy 617 (Inconel 617), was used for this study, which is a hot-rolled plate type with a 16mm thickness. The composition of each element was identified and was to be included well within the ASTM specification ranges. Tensile test specimens were machined using an electric discharge machine (EDM) with a rectangular cross section of 25.0 mm in gage length, 6.2mm in width, and 2.0 mm in thickness. The specimens were taken along a rolling direction. The strain rate in the tensile tests was conducted with the identical strain rate of 6.666E-4 /sec at each temperature of 800, 850, 900, and 950oC. The jigs and pull rod were made using a Ni-based superalloy to endure a high-temperature deformation of the material during the high-temperature tensile tests.

The tensile tests were started with the duration time of 30min after reaching a target temperature. The temperature was constantly maintained under ± 2°C. Three thermocouples of a K type were attached on the gage length of the specimen.

The tensile load-displacement data were automatically collected through a PC. The tensile properties were obtained, and the material constants, $D$ and $m$ were analyzed using the tensile test data.

## 4. Results and discussion

### *4.1 Proposition of a new method*

For the total strain (elastic + plastic + creep) needed to generate the ISSC, we did not consider the creep strain component. To determine the tensile plastic strain component, a new equation is proposed that is a modification of Eq. (3) used in the RCC-MR code. Similarly, the plastic strain is derived from the tensile work hardening equation given by the following formula.

$$\varepsilon_p = D_2\left(\frac{\sigma}{\sigma_p}\right)^{m_2} \quad \text{(for } \sigma \geq \sigma_p\text{)}, \quad (11)$$

where $\varepsilon_p$ is the plastic strain induced by stress $\sigma$, and $D_2$ and $m_2$ are the material constants, which are determined from the tensile test dada. In addition, $\sigma_p$ is the true stress at the proportional elastic limit. The stress term in the equation uses the proportion elastic limit instead of the 0.2% offset yield stress ($\sigma_y$) used in the MCC-MR code. This modified equation can be considered because the experimental $\sigma_p$ values have frequently been used as a typical material parameter in the evaluation and analysis of high-temperature materials. The three equations of Eqs. (2), (3), (8) and (11) were applied to Alloy 617 to determine the values of the elastic and plastic strain components, and the results,

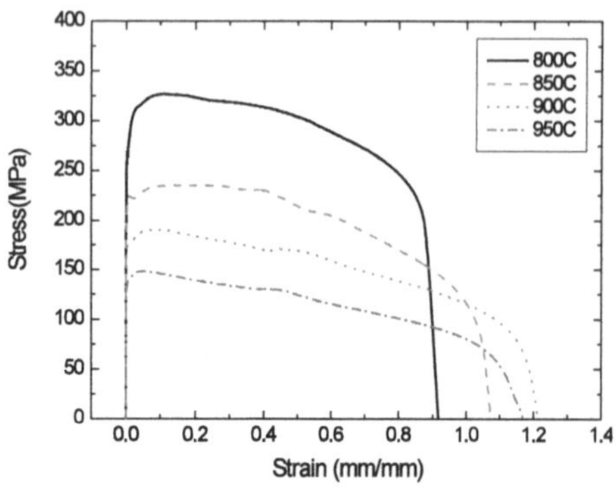


Fig. 2. Tensile curves at 800, 850, 900, and 950°C of Alloy 617.

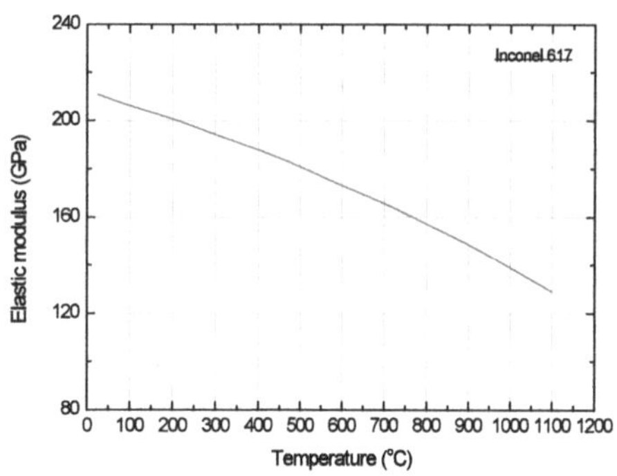


Fig. 3. Variations of elastic modulus with temperatures of Alloy 617.

which were calculated using the experimental tensile curves for the three methods of the RCC-MR, Blackburn, and new method, were obtained and compared.

### 4.2 Determination of material constants

Tensile properties were obtained at 800, 850, 900, and 950°C of Alloy 617. The four tensile curves at each temperature are shown in Fig. 2. In the tensile curves, as the temperatures were increased, the stress values were decreased. Hence, the values of elastic modulus $E$ were not obtained in this experiment. It is noted that the $E$ values cited the reported data of Inconel 617 provided in Special Metals [14]. The value with the temperatures is shown in Fig. 3.

Using the experimental tensile curves, the 0.2% offset yield stress values were determined at each temperature. The tensile properties of the 0.2% offset yield stress $\sigma_y$, proportional elastic limit $\sigma_p$, and ultimate tensile stress (UTS) were determined, and the results are listed in Table 1.

As the temperatures increased, material properties of E, $\sigma_y$, $\sigma_p$, and UTS values decreased. The material constants $D$, $m$, $D_1$, $m_1$, $D_2$, and $m_2$ were determined graphically from the slope and intercept on the plots of log stress ($\sigma/\sigma_y$, $\sigma$-$\sigma_p$, and $\sigma/\sigma_p$) against log strain ($\varepsilon_p$), i.e., using RCC-MR Eq. (3), Blackburn's Eq. (8), and a new method Eq. (11). The three methods determined for the material constants at 800°C are shown in

Table 1. Elastic modulus and tensile properties obtained at each temperature of Alloy 617.

| Temp. (°C) | E (GPa) | Yield stress, $\sigma_y$ (MPa) | Proportional elastic limit, $\sigma_p$ (MPa) | UTS (MPa) |
|---|---|---|---|---|
| 800 | 157 | 259.2 | 242.0 | 327.0 |
| 850 | 153 | 212.8 | 181.0 | 235.5 |
| 900 | 149 | 168.9 | 139.3 | 190.7 |
| 950 | 144 | 132.3 | 107.3 | 148.7 |

Table 2. Material constants determined for the three methods at each temperature of Alloy 617.

| Temp. (°C) | RCC-MR | | Blackburn | | New method | |
|---|---|---|---|---|---|---|
| | $D$ | $m$ | $D_1$ | $m_1$ | $D_2$ | $m_2$ |
| 800 | 2.47E-3 | 11.264 | 1.45E-07 | 2.809 | 2.28E-03 | 9.44 |
| 850 | 1.34E-2 | 10.055 | 1.33E-06 | 2.573 | 3.23E-03 | 9.06 |
| 900 | 7.48E-3 | 11.700 | 9.01E-08 | 3.250 | 9.07E-04 | 11.37 |
| 950 | 3.50E-3 | 15.662 | 2.19E-09 | 4.343 | 2.78E-04 | 13.37 |

Fig. 4. A summary of the material constants at 800, 850, 900, and 950°C is listed in Table 2.

### 4.3 Comparison for three methods

As discussed above, with the obtained material constants, the tensile elastic and plastic strain values can be obtained by calculating the equations of the RCC-MR, Blackburn, and the new method proposed in this study. In the calculation of each equation, a stress range of 10 to 450 MPa was properly used. Comparative plots of the three methods for stress vs. strain (elastic + plastic) curves at 800, 850, 900, and 950°C are shown in Fig. 5. In the comparisons for the three methods of the RCC-MR, Blackburn, and the newly proposed methods, in the elastic regime, they have the same curves without any differences, but in the plastic regime, the three methods show some differences. The different gap for the three methods was larger with an increase in the strains. Blackburn's method was higher in the stress-strain curves than the RCC-MR code. In particular, the newly proposed method in this study has midway curves between the two curves of Blackburn > New method > RCC-MR. The reason for this is that the proportion elastic limit $\sigma_p$ was lower than the yield stress $\sigma_y$. It is thus thought to be better to use the proposed method.

However, in spite of this difference in the plastic regime for the three methods, it was suggested that the three methods can be utilized to calculate the elastic and plastic strain of Alloy 617, because the creep stresses of Alloy 617 at 800, 850, 900, and 950°C are generally applied for a lower elastic stress regime rather than a higher plastic stress regime. Accordingly, in order to determine the tensile strain (elastic + plastic) values which are needed to construct the ISSCs of Alloy 617, it is suggested that the new method proposed in this study is better than the other two methods.

From this result, thus, the tensile elastic and plastic strain components in total strain which is needed to generate the

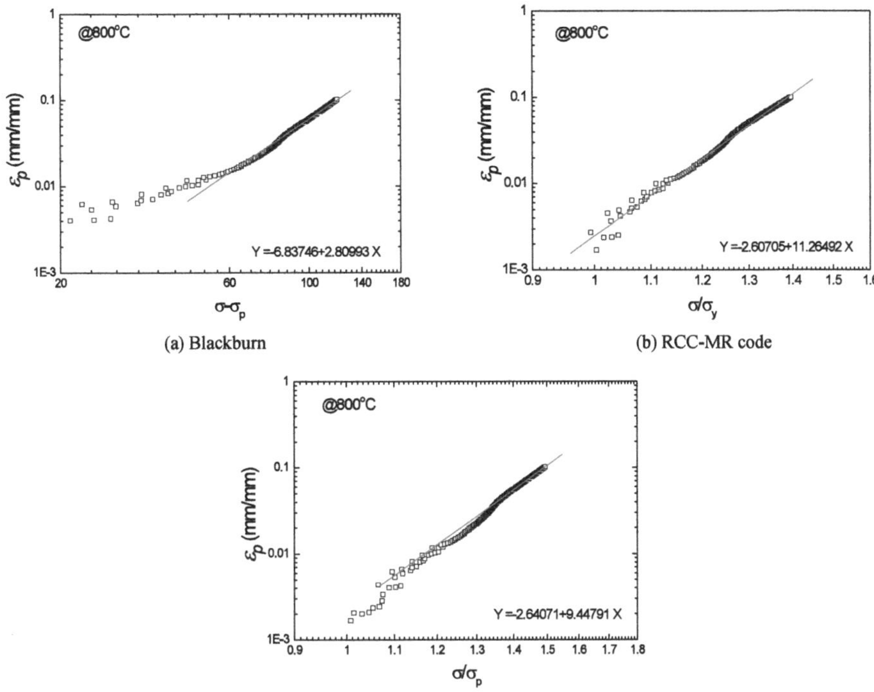


(a) Blackburn

(b) RCC-MR code

(c) Proposed new method

Fig. 4. Tensile plastic curves at 800°C for the three methods.

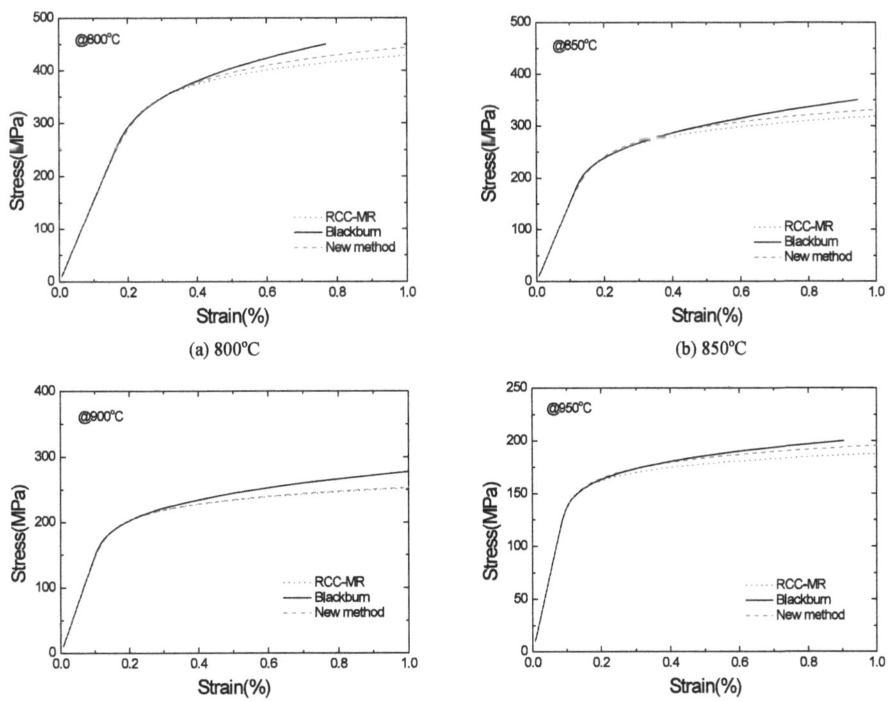


(a) 800°C

(b) 850°C

(c) 900°C

(d) 950°C

Fig. 5. Comparisons of the stress-strain (elastic+plastic) curves obtained for the three methods at 800, 850, 900, and 950°C of Alloy 617.

ICCS can be mathematically obtained using the proposed equations without creep tests. This calculation method is more accurate than the experimental method, which is obtained by creep tests, because the value of the initial loading strain corresponding to the elastic and plastic strain is difficult to accurately measure as an instantaneous strain in the early creep stages. In addition, the jig and pull rod attached to a creep specimen is easily elongated owing to a high temperature during creep tests, and it can generate some errors in an accurate measurement of the initial loading strain. Thus, if the high-temperature tensile properties of material ($E$, $\sigma_y$, $\sigma_p$) and material constants ($D$ and $m$) are known, we can obtain accurately the elastic and plastic strain components using the proposed equations above.

## 5. Conclusions

To determine the total strain for developing the ISSCs, the RCC-MR code and Blackburn's method were preliminarily analyzed. In addition, a new method was proposed to practically determine the tensile elastic and plastic strain components of Alloy 617. Tensile tests were conducted at 800, 850, 900, and 950°C. Using the tensile tested data, the materials constants $D$, $m$, $D_1$, $m_1$, $D_2$, and $m_2$ for the three methods were determined. Following conclusions were drawn.

(1) In the elastic regime, the three methods were identical in the stress-strain curves, but in the plastic regime, Blackburn's equation was higher in the stress-strain curves than the RCC-MR code.

(2) The new method showed midway values in turn Blackburn > New method > RCC-MR. The reason for this was that the proportion elastic limit $\sigma_p$ was lower than the yield stress $\sigma_y$.

(3) It was found that the tensile elastic and plastic strain components in total strain values could be mathematically obtained using the proposed equations without creep tests. This method was found to be better than the experimental method of the creep tests, because it is difficult to accurately measure the value of an initial loading strain corresponding to the elastic and plastic strain.

## Acknowledgment

This study was supported by Nuclear Research & Development Program of the National Research Foundation of Korea (NRF) grant funded by the Ministry of Science, ICT and Future Planning (2012M2A8A2025682).

## References

[1] L. D. Blackburn, *Isochronous stress strain curves for austenitic steels*, ASME, New York (1972) 15-48.

[2] W. G. Kim, S. N. Yin and G. H. Koo, Generation of isochronous stress-strain curves with a nonlinear least square fitting method for modified 9Cr-1Mo steel, *Met. Mater. Int.*, 15 (5) (2009) 727-732.

[3] H. Y. Lee, Y. W. Kim and K. N. Song, Preliminary application of the draft code case for alloy 617 for a high temperature component, *J. of Mechanical Science and Technology*, 22 (2008) 856-863.

[4] J. Y. Park, D. W. Kim, D. J. Kim, E. S. Kim, H. P. Kim, S. H. Kim, W. J. Kim, W. G. Kim, W. S. Ryu and Y. D. Kim, Development of materials for a high temperature gas cooled reactor in Korea, *The 4th International Topical Meeting on High Temperature Reactor Technology-HTR 2008*, Washington, USA (2008) Paper HTR 2008-58149.

[5] J. H. Chang, Y. W. Kim, K. Y. Lee, Y. W. Lee, W. J. Lee, J. M. Noh, M. W. Kim, H. S. Lim, Y. J. Shin, K. K. Bae and K. D. Jung, A study of a nuclear hydrogen production demonstration plant, *Nuclear Engineering and Technology*, 39 (2) (2007) 111-122.

[6] S. J. Dewson and X. Li, Selection criteria for the high temperature reactor intermediate heat exchanger, *Proc. of ICAPP 05*, Seoul, Korea, May (2005) Paper 5333.

[7] W. G. Kim, S. N. Yin, G. G. Lee, Y. W. Kim and S. J. Kim, Creep oxidation behavior and creep strength prediction for Alloy 617, *Int. J. of Pressure Vessels and Piping*, 87 (2010) 289-295.

[8] W. G. Kim, S. N. Yin, W. S. Ryu and J. H. Chang, Analysis of the creep rupture data of Alloy 617 for a high temperature gas cooled reactor, *The 8th Int. Conf. on Creep and Fatigue at Elevated Temperatures*, Texas, USA, July (2007) Paper PVP 2007-26834.

[9] W. G., Kim, S. N., Yin, J. Y., Park, S. D., Hong and Y. W. Kim, An improved methodology for determining tensile design strengths of Alloy 617, *J. of Mechanical Science and Technology*, 26 (2012) 379-387.

[10] J. M. Corum and J. J. Blass, Rules for design of alloy 617 nuclear components to very high temperatures, Attachment 2, *Draft Code Case for Alloy 617*, USA (1988).

[11] RCC-MR, *Design and construction rules for mechanical components of nuclear installations*, Section 1, subsection. Z. AFCEN, Paris (2007).

[12] J. Ganesh Kumar, M. Chowdary, V. Ganesan, R. K. Paretkar, K. Bhanu Sankara Rao and M. D. Mathew, High temperature design curves for high nitrogen grades of 316LN stainless steel, *Nuclear Engineering and Design*, 240 (2010) 1363-1370.

[13] F. Garofalo, *Fundamentals of creep and creep rupture in metals*, New York: Macmillan (1966).

[14] *Special Metals Corporation*, Inconel 617, www.specialmetals.com.

**Woo-Gon Kim** earned his Ph.D. at Chungbuk National University in Mechanical Engineering in 1998. He is a principal researcher at Korea Atomic Energy Research Institute. Dr. Kim's specialty is mechanical assessment, analysis, modeling and database establishment of high temperature nuclear materials for Generation-IV reactor systems, and especially, time-dependent creep and crack growth behaviors at elevated temperature.

Nuclear Engineering and Design 271 (2014) 291–300

Contents lists available at ScienceDirect

Nuclear Engineering and Design

journal homepage: www.elsevier.com/locate/nucengdes

# Temperature effect on the creep behavior of alloy 617 in air and helium environments

Woo-Gon Kim [a,*], Jae-Young Park [b], Gyeong-Geun Lee [a], Sung-Deok Hong [a], Yong-Wan Kim [a]

[a] Korea Atomic Energy Research Institute, Daejeon 305-353, Republic of Korea
[b] Pukyong National University, Busan 608-739, Republic of Korea

A B S T R A C T

The temperature effect on creep and oxidation behaviors in air and helium (He) environments was investigated at 950, 900 and 850 °C for Alloy 617, which is considered as a prime candidate material for VHTR components. Creep data were obtained with different stress levels at the three temperatures. Oxidation microstructural features such as the surface oxide layer, internal oxidation and decarburization were analyzed by observing each crept specimen. At 950 and 900 °C, the creep rupture time in He environment was shorter than that in air, and the thickness of the surface oxide layer was thicker than in air. The deterioration of creep resistance in the He environment was due to a thicker oxide-layer thickness, which reduced the effective area carrying the creep load. On the other hand, at 850 °C, the creep rupture time was almost the same regardless of both environments. The surface oxide-layer thickness was reduced by about 50% compared to the temperatures of 950 and 900 °C. The temperature of 850 °C could be assumed as a boundary temperature at which the He effects disappeared. It was found that the relationship between the creep data and surface oxide-layer thickness was in accordance with both environments.



## 1. Introduction

A very high temperature gas reactor (VHTR) is one of the Gen-IV reactors aiming at safe, long-lived, proliferation-resistant and economical nuclear power plants. Its high operating temperature of over 800 °C enables a high energy efficiency and the production of hydrogen gas using Sulfur–Iodine process. The heat of the primary helium (He) circuit transfers to the secondary helium loop through the intermediate heat exchanger (IHX). The IHX component needs high-temperature creep resistance in the He environment, and requires good oxidation resistance, corrosion resistance, and phase stability at high temperatures. Among them, creep and oxidation properties are importantly considered, as the integrity of the components should be preserved during a prolonged period in a VHTR coolant (Kim et al., 2007, 2008, 2009, 2010, 2012a,b).

Of the existing alloys, nickel-base Alloy 617 is the leading candidate for use in the next generation nuclear plant (NGNP) heat exchangers because it has the highest creep strength of solid solution alloys under consideration for temperature above 850 °C. It was reported that the creep rupture time varied widely with the environments interactions between He impurities and the alloy (Jang et al., 2008; Dewson and Li, 2005; Cook, 1984). The impurities in He coolant during steady-state operation of the VHTRs were reported as $H_2O$, $H_2$, CO, $CO_2$, $CH_4$, $O_2$ and $N_2$. However, it has not been well established on the temperature effect in He environment to be drastically affected to the rupture time. The relationships between creep and oxidation features are not well understood. Therefore, it is necessary to investigate the temperature effect on the creep and oxidation behaviors in air and He environments of Alloy 617.

In this study, the creep behaviors of Alloy 617 were comparatively investigated by creep tests with different stresses in air and He environments at 950, 900, and 850 °C. The temperature effect on the creep and oxidation behaviors was evaluated, and relationships between these behaviors were found and discussed.

## 2. Experimental procedures

Commercial grade nickel-based superalloy, Alloy 617 (Inconel 617), was used in this study. The material was a hot-rolled plate with a thickness of 15.875 mm. Creep test specimens in the air and He environments were fabricated in cylindrical form with a 30 mm gauge length and 6 mm diameter. The gage section was parallel to the longitudinal rolling direction. Circular grooves were machined at both ends beyond the shoulder region of these specimens to attach extensometers for monitoring the elongation during creep testing.

The loading frames used in the creep tests had a lever arm ratio of 20:1. The overall creep rates were determined from elongations measured by extensometers equipped at the circular grooves. Creep strain data with elapsed times were taken automatically by

* Corresponding author at: Korea Atomic Energy Research Institute, 989-111 Daedeok-daero, Yuseong-gu, Daejeon 305-353, Republic of Korea.
Tel.: +82 42 868 2493.
E-mail address: wgkim@kaeri.re.kr (W.-G. Kim).

0029-5493/$ – see front matter © 2013 Elsevier B.V. All rights reserved.
http://dx.doi.org/10.1016/j.nucengdes.2013.11.050

**Fig. 1.** Creep testing apparatus in He environment.

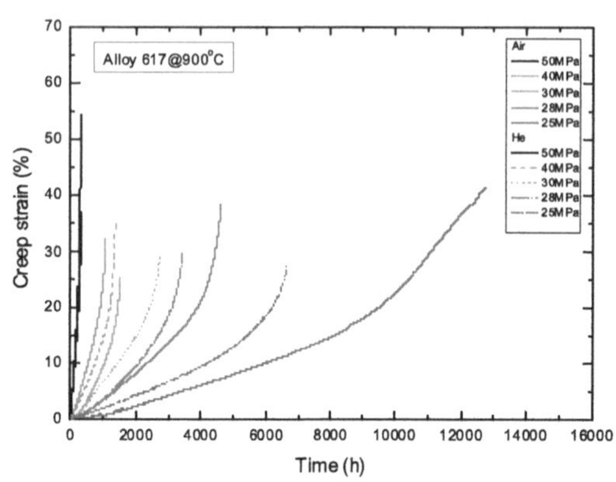


**Fig. 2.** Typical creep curves obtained in air and He environments at 900 °C.

a PC. K-type thermocouples were used to monitor temperatures within the gage section of the specimens. A split furnace was used to heat the specimens of Alloy 617 in the air and He environments to 950, 900, and 850 °C. The pull rod and jig used for the creep tests were manufactured with a nickel-based superalloy material to sufficiently endure oxidation and thermal degradation during creep.

Before creep tests in the He environment, a vacuum chamber made for the quartz tube was purged three or four times by a vacuum pump to remove some impurities in the chamber. During the creep tests, pure He gas of 99.999% was supplied to the specimens equipped in the quartz tube. Impurity concentration in pure He gas was $O_2 < 1.0$ ppm, $N_2 < 5.0$ ppm and $H_2O < 1.0$ ppm. Flow rate of the He gas during the creep test was controlled under 20 $cm^3$/min. The control of the other impurities such as $CO_2$, CO, $CH_4$, and $H_2$ in He gas simulating a VHTR condition will be investigated in the further tests. Creep testing apparatus in the He environment is shown in Fig. 1. Photos showing each component such as the quart tube, He flow meter, strain indicator, specimen and jigs, furnace controller, and data acquisition PC are presented.

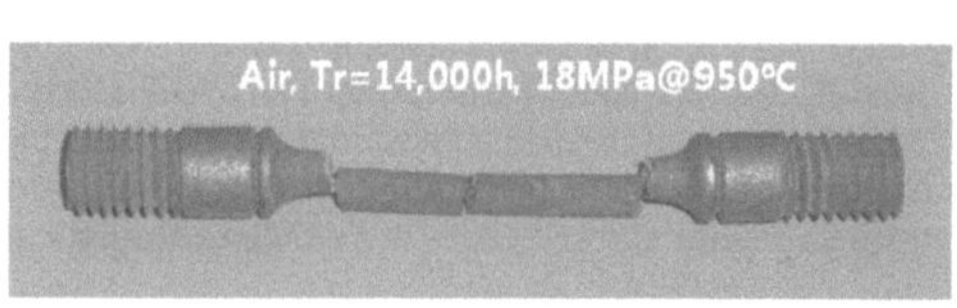


**Fig. 3.** A photo of crept specimen under 18 MPa at 950 °C in air of Alloy 617.

## 3. Results and discussion

### 3.1. Comparison of creep behaviors in air and He environments at 950, 900 and 850 °C

The creep property data of Alloy 617 such as the rupture time, minimum creep rate, rupture elongation, and reduction of area were obtained in the air and He environments at 950, 900, and 850 °C. The creep results in both environments were compared at the three temperatures.

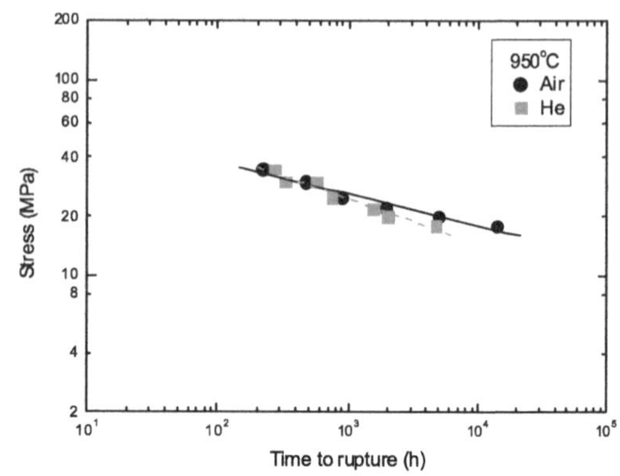


(a) 950°C

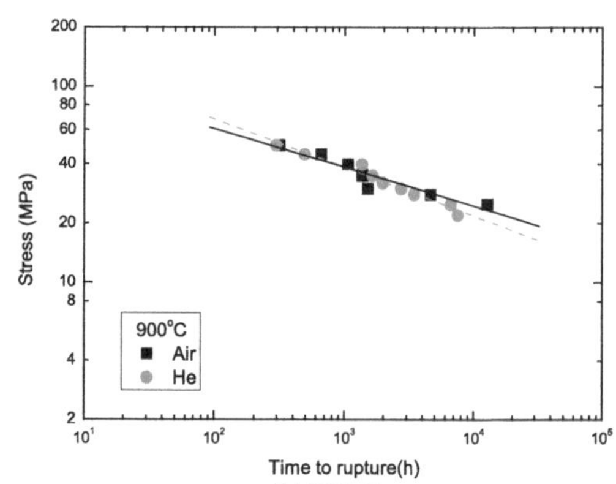


(b) 900°C

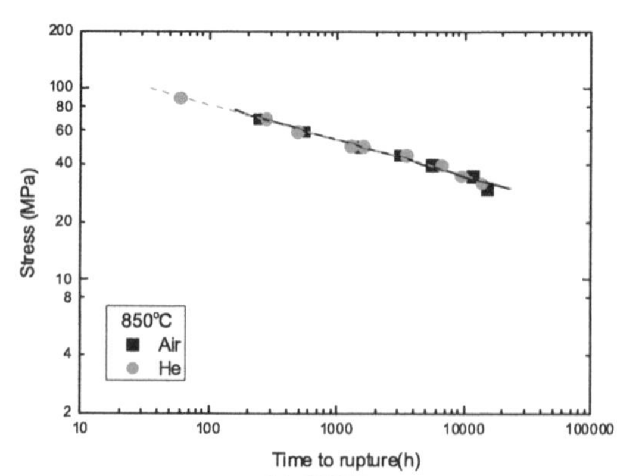


(c) 850°C

**Fig. 4.** Comparison of stress vs. rupture time in the air and He environments at (a) 950 °C, (b) 900 °C, and (c) 850 °C.

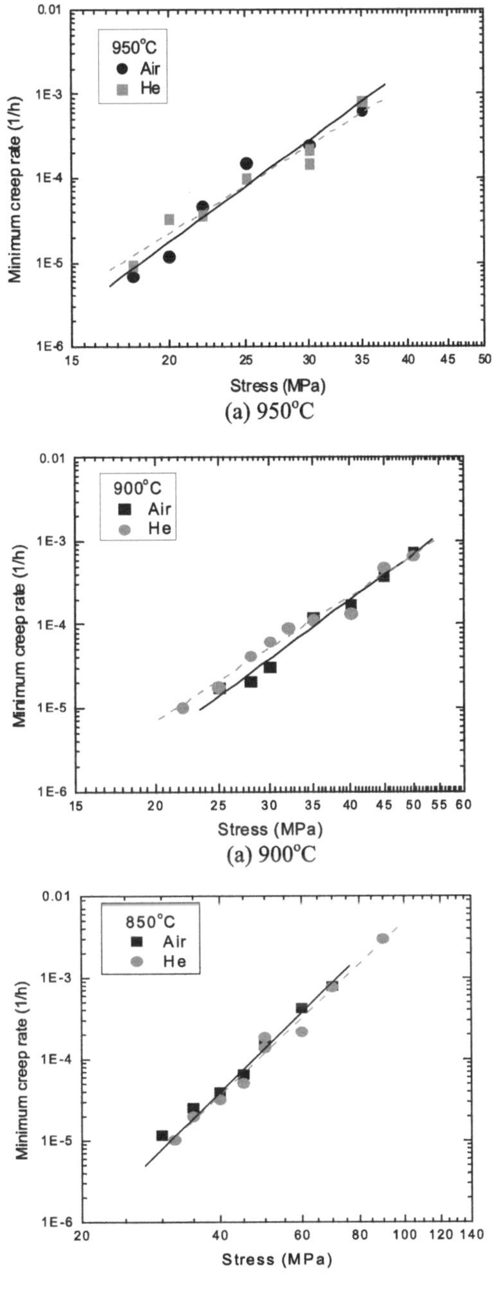


**Fig. 5.** Comparison of minimum creep rate vs. stress in air and He environments at 950, 900, and 850 °C.

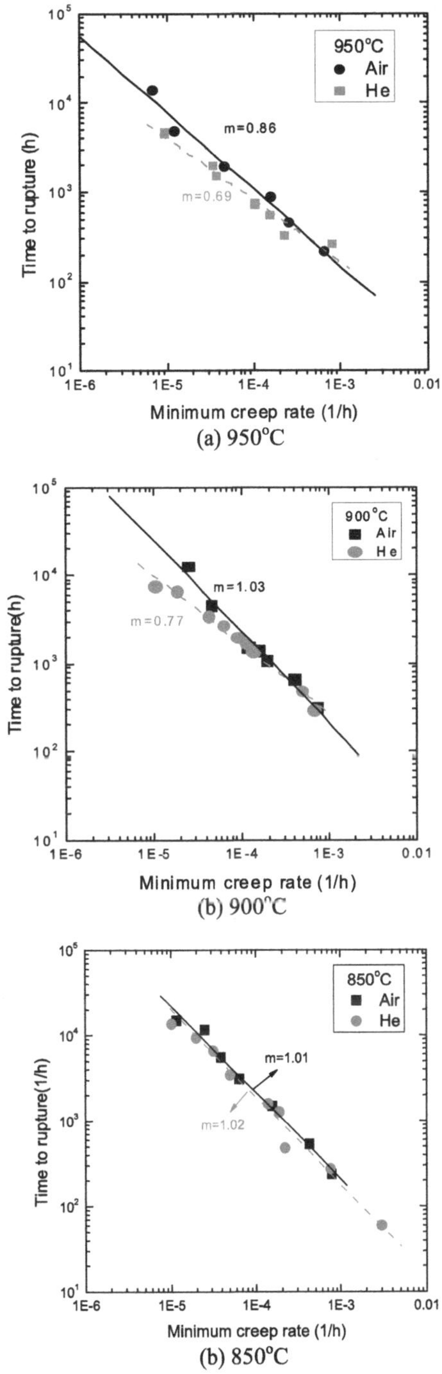


**Fig. 6.** Comparison of rupture time vs. minimum creep rate in air and He environments at 950, 900, and 850 °C.

Fig. 2 shows typical creep curves for stresses in the air and He environments at 900 °C. Alloy 617 showed little primary creep strain, and a well-defined secondary creep stage was not exhibited in the full creep curves. The onset of a tertiary creep was unclear, and a tertiary creep stage was initiated from a low strain level. There were no differences in the shapes of the creep curves between the air and He environments. Alloy 617 revealed good ductility (>30%) in spite of the long duration above $10^4$ h at 950, 900, and 850 °C.

From a series of creep tests, long-term creep data reaching 14,000 h at 950 °C were achieved in the air, as shown in Fig. 3. Creep rupture occurred at three positions within the gage length. The rupture elongation was about 30%, and the reduction of area was as small as 22%. It was identified that the creep rupture was developed by the incorporation of the many small cavities formed during the creep, rather than the necking of the specimen.

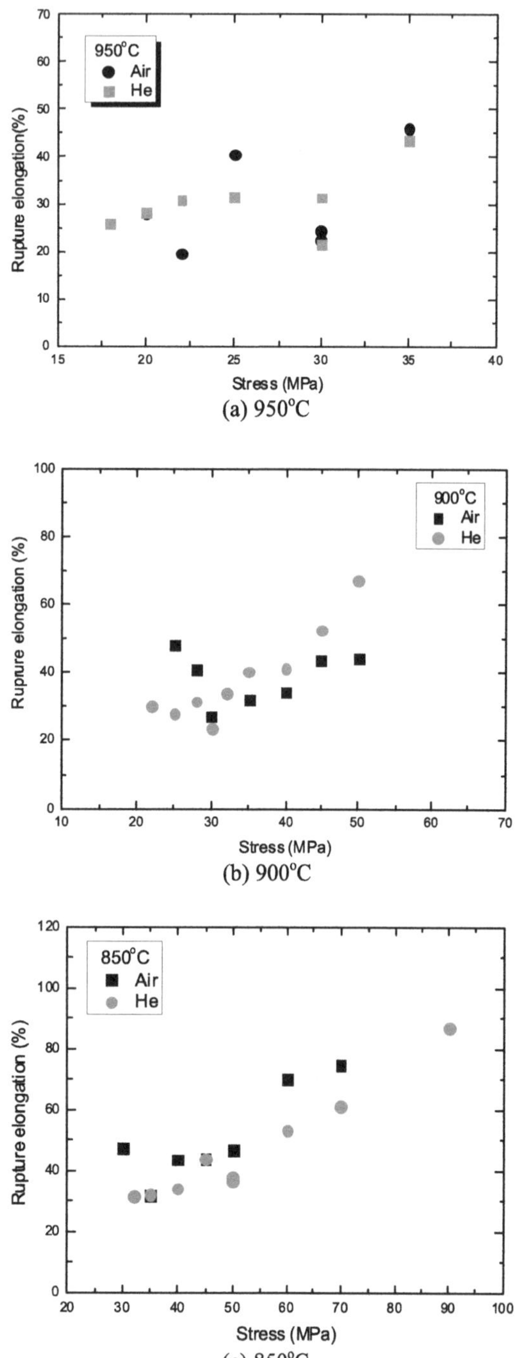


**Fig. 7.** Comparison of creep rupture elongation vs. stress in air and He environments at 950, 900, and 850 °C.

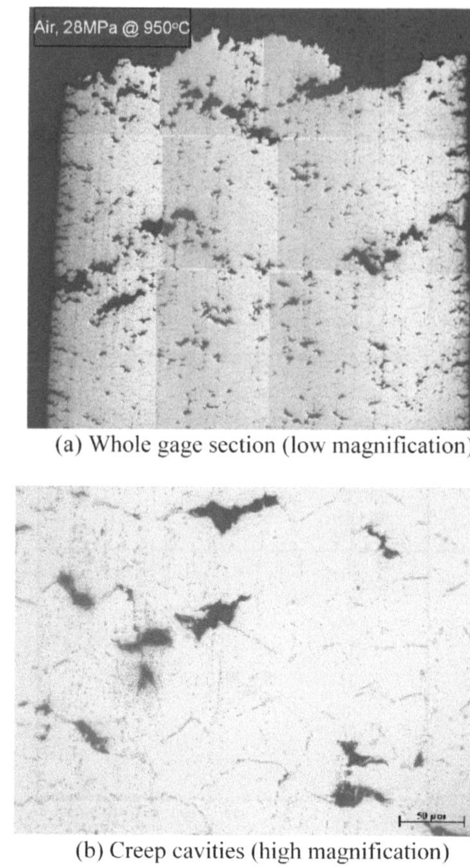


**Fig. 8.** Typical OM micrographs of a crept specimen under 28 MPa in air at 950 °C.

Fig. 4 shows a comparison of the log stress vs. log rupture time in the air and He environments at 950, 900 and 850 °C. For 950 and 900 °C, in the high stress ranges above about 30 MPa, the creep stress between the air and He environments was almost similar. However, in the low stress ranges below about 30 MPa, there were differences in the creep stress (or creep rupture time) between the two environments. Namely, the creep stress in air was higher than that in He environment, but for 850 °C, the creep data of the log–log plot was almost superimposed regardless of both environments, as shown well in the figure. Therefore, it is obvious that the creep properties in air and He environments have a temperature effect. The reason for this is closely related to the oxide layer thickness formed during the creep. The oxide-layer thicknesses in both environments are measured at the three temperatures, and the results are discussed in a later section.

Fig. 5 shows a comparison of the minimum creep rate vs. stress at 950, 900, and 850 °C in air and He environments. The relationships between the minimum creep rate and stress revealed a good linearity regardless of both environments at the three temperatures. It is clear that Alloy 617 followed Norton's power law well under this creep condition because the deformation map corresponded to the power-law creep region. The creep mechanism is governed by a climb of dislocation (Kim et al., 2010). The creep rates between both environments at 950 and 900 °C plot a significant difference at low stress levels. The creep rate in He was faster than that in air. However, the creep rates at 850 °C were almost the same regardless of both environments.

Fig. 6 shows a plot of Monkman–Grant (M–G) relationships. In the equation of M–G relationship, $\log t_r + m\dot{\varepsilon}_m = C$, the $m$ values of the slope were obtained in the air and He environments at the three temperatures of 950, 900, and 850 °C. At 950 and 900 °C, it means that the creep rupture time of the He specimens was shorter than that of the air one. But at 850 °C, their values were almost equal to

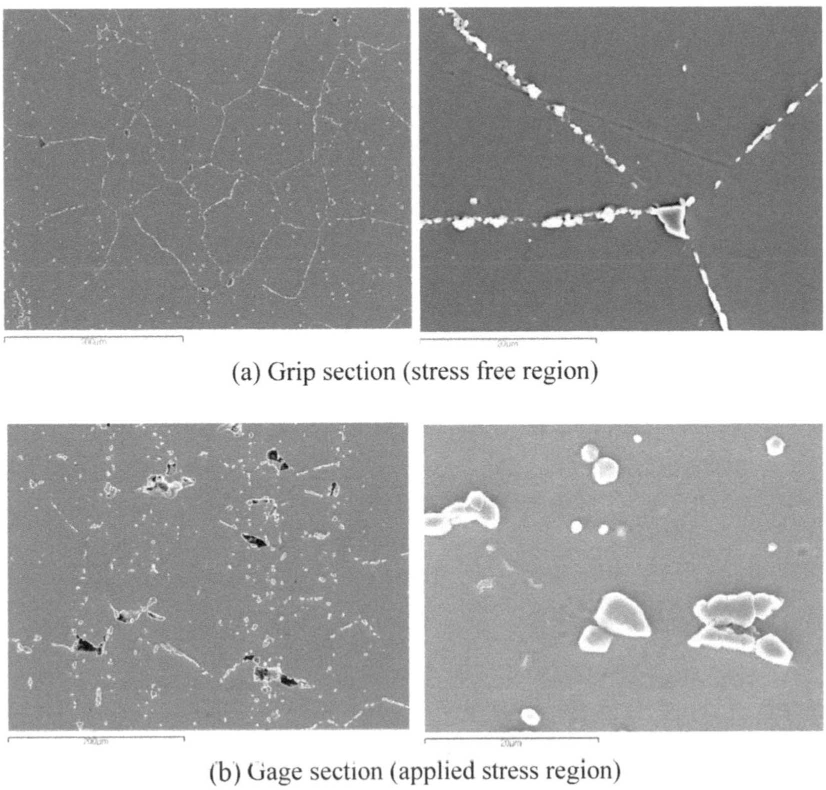


**Fig. 9.** Precipitates formed in the grip-section region and gage-section region of the specimen crept in He environment under 20 MPa at 950 °C.

$m = 1.0$ regardless of the environment. This means that the creep properties were not changed at 850 °C under the He environment. It is obvious that the He effects on the creep properties disappeared at 850 °C. This is assumed to be a temperature effect on the creep properties under this He-agent condition. Thus, the temperature of 850 °C can be regarded as a boundary temperature at which the He effects disappeared. The data on the creep rupture elongation were similar in both environments, although the data were scattered, as shown in Fig. 7.

### 3.2. *Creep fracture micrographs*

Fig. 8 shows typical OM micrographs at two magnifications of the crept specimen under 28 MPa in air at 950 °C. Photo (a) is a low magnification of the whole gage section. While photo (b) is a high magnification showing the creep cavities formed inside the region. The creep cavities and cracks developed along the grain boundaries perpendicular to the applied stress. The cavities were widely distributed over the gage section. Creep failure finally occurred by the incorporation of these cavities (or voids) introduced from the creep damage. These cavities were initially generated by precipitates formed along the grain boundary. The precipitates become site initiating cavities.

Fig. 9 shows the precipitates observed in the (a) grip section (stress free region) and (b) gage section (applied stress region) of the specimens crept in the He environment under 20 MPa (rupture time of 2020 h) at 950 °C. The grip-section sample of the stress free region, taken from the grip end, precipitated through the carbides in the matrix and on the grain boundaries, as clearly shown by the carbides.

On the contrary, in the gage-section sample (near the failure region), which was applied by creep stress, the carbides were very scattered throughout matrix, and the grain boundaries were not clear. Also, the carbides were coarse and blocky, and the cavities occurred on the grain boundaries where the carbides were formed. Cylindrically-shaped coarse carbides, Cr-rich $M_{23}C_6$, developed mainly on the grain boundaries. In the precipitate-denuded regions, the carbides developed near the grain boundaries. According to other studies on Alloy 617, $M_{23}C_6$ carbide was found to be abundant at temperatures of 649 to 1093 °C. Other carbides ($Cr_{23}C_6$), carbon nitrides [Cr Mo(C, N)], and nitride (TiN) were exhibited but rare (Kim et al., 2012a; Jang et al., 2008). After the creep failure, the fracture surface of each tested specimen, conducted under the high and low stress levels in the air and He environments at the three temperatures of 950, 900, and 850 °C, was examined by scanning electron microscopy (SEM).

Fig. 10 shows typical SEM micrographs taken at two magnifications showing fracture modes under 70 MPa and 35 MPa in the air and He environments at 850 °C. The fracture modes are similar in both environments. The plane of the surfaces is not flat at the high stress level of 70 MPa. The fracture surface shows a dominant dimple rupture and secondary cracking with the indication of intergranular cracking. However, the fracture surface at the low stress level of 35 MPa shows a combination of dimple rupture and dominant intergranular fracture. The plane of the surfaces is very even when compared with the high stress of 70 MPa. In particular, surface oxidation is evident owing to a prolonged exposure time under 35 MPa. The intergranular cracking is initiated and developed by incorporation of minor cavities, formed on the grain boundary for prolonged creep duration.

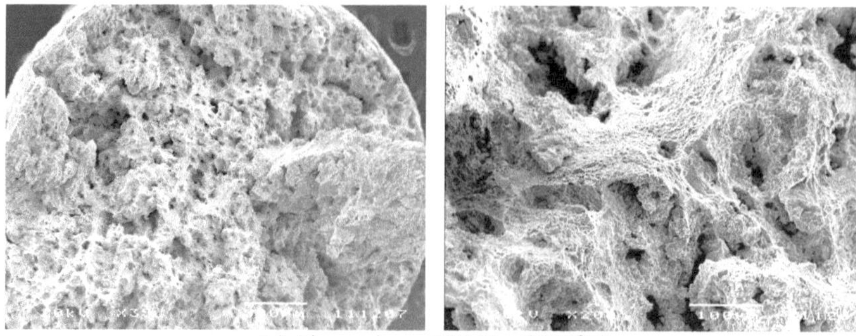

(a) Air specimen (70MPa @850°C)

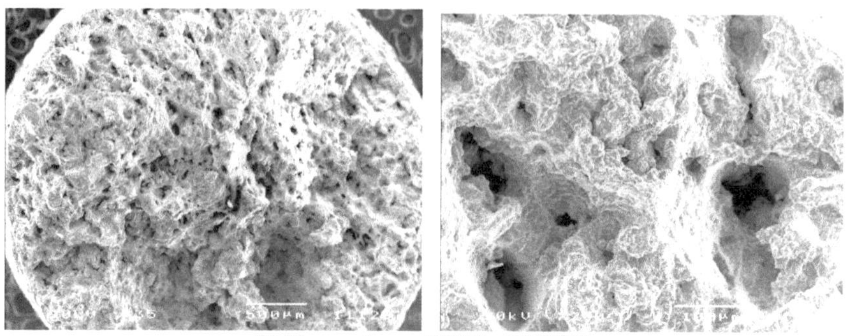

(b) He specimen (70MPa @850°C)

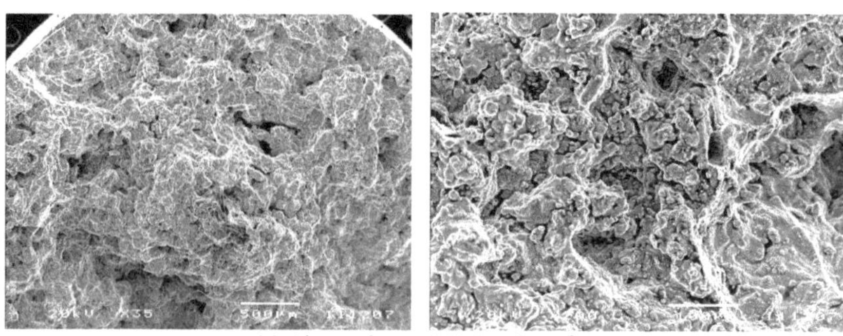

(c) Air specimen (35MPa @850°C)

(d) He specimen (35MPa @850°C)

**Fig. 10.** Typical SEM micrographs of the fracture surface under the high and low stress levels of 70 and 35 MPa in air and He environments at 850 °C.

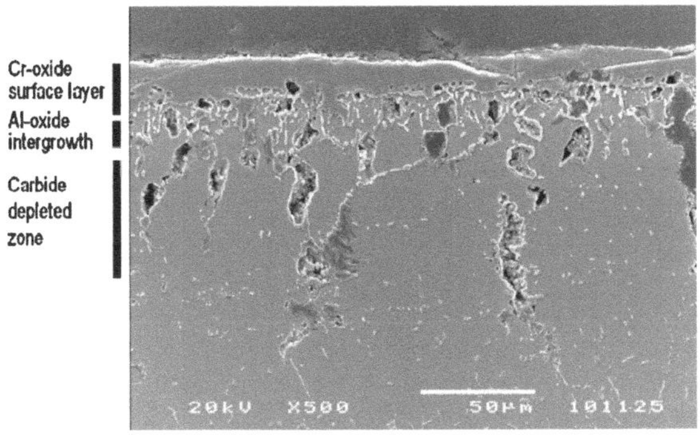


**Fig. 11.** Typical oxide layers formed for Alloy 617 (28 MPa in He environment at 900 °C).

### 3.3. Oxidation features in air and He environments with temperatures

Three microstructure features are investigated in this study related to oxidation behavior during creep exposure time of Alloy 617. They are the surface oxide-layer thickness, the depth of the internal oxidation and the depth of the carbide depleted zone (or Cr depleted zone). These three regions are labeled in the micrographs, as shown in Fig. 11. By quantifying the time dependence of the change in the microstructure features, the mechanism of degradation can be ascertained in the air and He environments at the three temperatures of 950, 900 and 850 °C. An energy dispersive X-ray (EDX) analysis was used to identify the metallic element constituents at different locations.

On the outer surface, a Cr-rich (mainly $Cr_2O_3$) oxide layer was formed owing to Cr diffusion from the matrix during the creep exposure time. Just below the $Cr_2O_3$ layer, an Al-rich (mainly $Al_2O_3$) discrete internal oxide layer was formed. The internal oxidation was formed by the diffusion of oxygen through the surface oxide and matrix. Then, below the internal oxidation layer, a thick carbide-depleted zone (or Cr depleted zone) was formed by depletion of the carbides at the grain boundary near the surface oxides as the oxidation progressed gradually. The depletion of grain boundary carbides is caused by a reaction between the grain boundary carbides and oxygen provided by diffusion from the environment or the oxidation reaction of the surface oxide layers. Carbide dissolution occurs from chromium migration to support oxide formation on the surface, as Alloy 617 is a chromium-forming alloy. It is known that the optimal distribution of grain boundary carbide improves the creep resistance as the grain boundary carbide can prevent grain boundary migration and sliding (Christ et al., 1987).

Fig. 12 shows comparisons of typical SEM micrographs of the oxide layers formed for creep failure specimens in the (a) air and (b) He environments under an identical stress of 20 MPa at 950 °C. As shown in (a) and (b), different oxidation features were found between the two environments, as follows.

First, for the surface oxide layer, the thickness in He was thicker than in air, although the helium gas used in this study was at high purity at 99.999% containing small amount of impurities, such as $H_2O < 1.0$ ppm, $O_2 < 1.0$ ppm, $N_2 < 5.0$ ppm. The structure was more porous and smooth than that in the air samples. The connection of the surface oxide layer in the He samples was broken by cracks. In the air samples, the connection is continuous without cracks or breaks, and the structures were very dense over the gage length. The dense oxide layer is formed owing to a high oxygen concentration in the atmosphere, and it provides a proper protection layer from surface oxidation. Then, for the internal oxide layer, the He samples were deeper in Al-oxide intergrowth, and the interconnection between the surface oxide layer and matrix was weak. The structure was also porous and bared. As the temperature and creep rupture time increased, the depth showed roughly an increasing trend. Finally, for the Cr-depleted layer, the depth of the He environment was not different compared to that of the air. As the temperature and creep rupture time increased, the depth was enhanced on the micrographs, although the depth was difficult to quantitatively measure. The amount of carbide precipitates at the grain

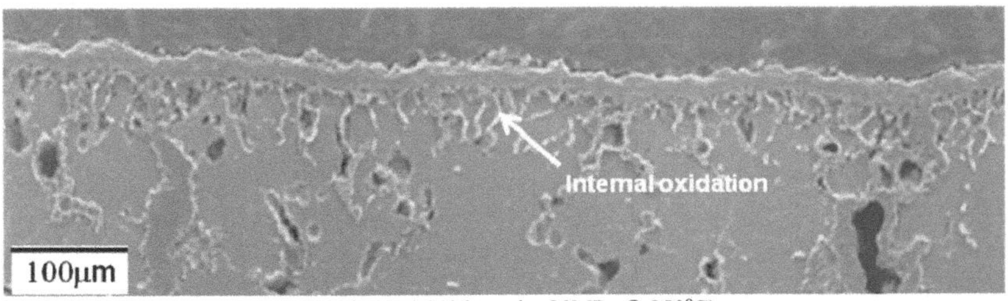


(a) Air (tr=4,943 h, under 20MPa @ 950°C)

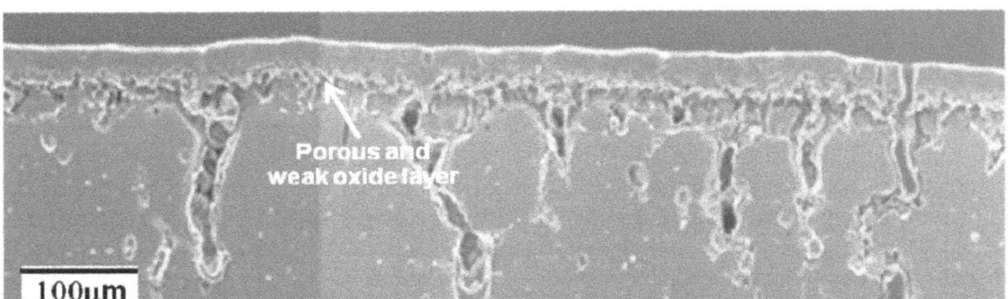


(b)He (tr=2,021 h, under 20MPa @ 950°C)

**Fig. 12.** Comparison of SEM micrographs of the oxide layers formed during creep rupture time in (a) air and (b) He environments under an identical applied stress of 20 MPa at 950 °C.

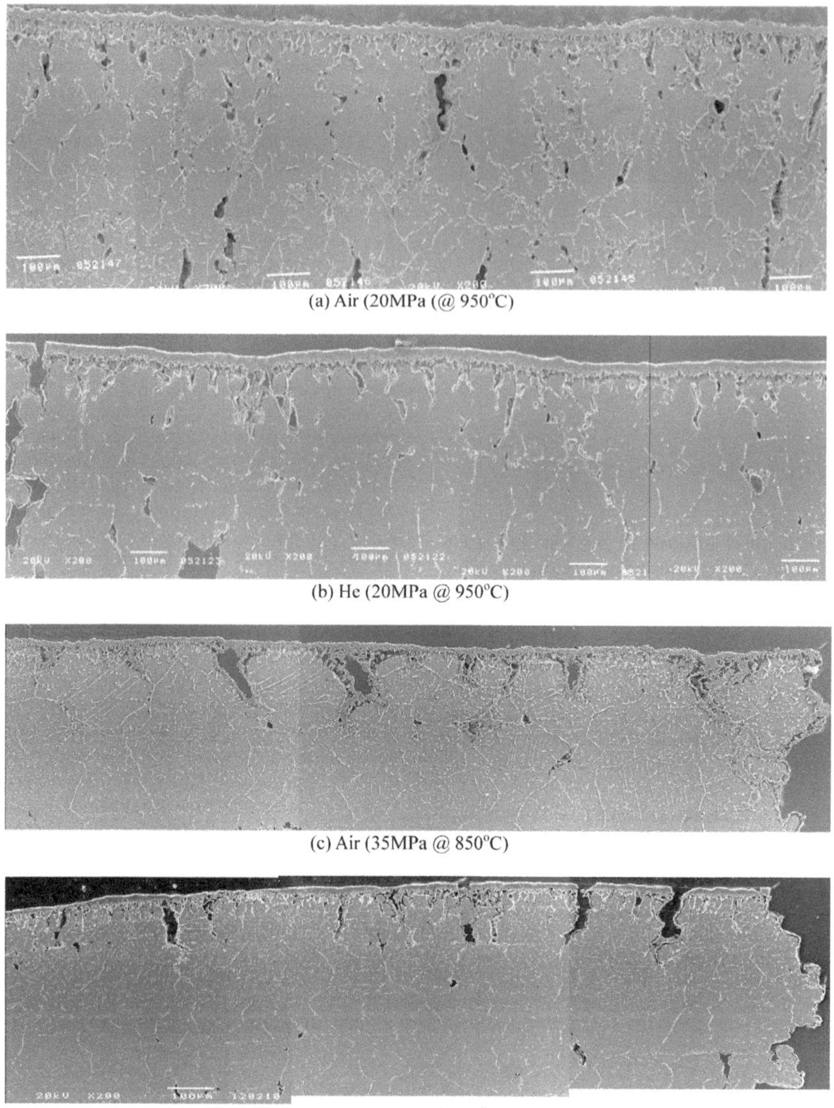


**Fig. 13.** Typical SEM image observed for longitudinal section of the specimens crept in air and He environments under 20 MPa at 950 °C and 35 MPa at 850 °C.

boundary was much smaller than that in the air, as shown in Fig. 13. Accordingly, it is thought that the differences of the above three microstructure features caused detrimental effects on the creep resistance, even in the low oxygen-containing He agent used in this study.

In addition, in this examination, it was difficult to quantitatively measure the depths of the Al-oxide intergrowth and Cr-depleted zone with creep rupture time. Thus, it is focused on the surface oxide-layer thickness, and its thickness was measured by SEM taken for the crept specimens in both environments at 950, 900, and 850 °C. Typical micrographs of longitudinal section are shown in Fig. 13. The samples were polished and etched for each crept specimen. The surface oxide-layer thickness was measured for more than 20 points with an interval of ~80 μm along the longitudinal direction of samples. The average value and standard deviation were obtained.

Fig. 14 shows the measured results of the surface oxide-layer thickness with the creep rupture time at (a) 950 °C, (b) 900 °C, and (c) 850 °C in the air and He environments. The surface oxide-layer thickness increased as the rupture time increased for both environments at the three temperatures. The He samples revealed higher thickness values compared to the air samples as the enhanced oxide-layer thickness in the He environment at 950 and 900 °C reduced the effective area of the specimens carrying creep load. The deterioration of the creep resistance was due to the microstructural changes such as the thickness of the surface oxide layer, the depth of the internal oxidation, and the depth of the Cr-depleted zone. These changes were attributed to the enhanced diffusion of the oxidizing agent and the gaseous reaction products along the grain boundary. It is known that the creep properties of the material will have detrimental effects when the carbide depleted zone is enhanced, and they are dependent on various

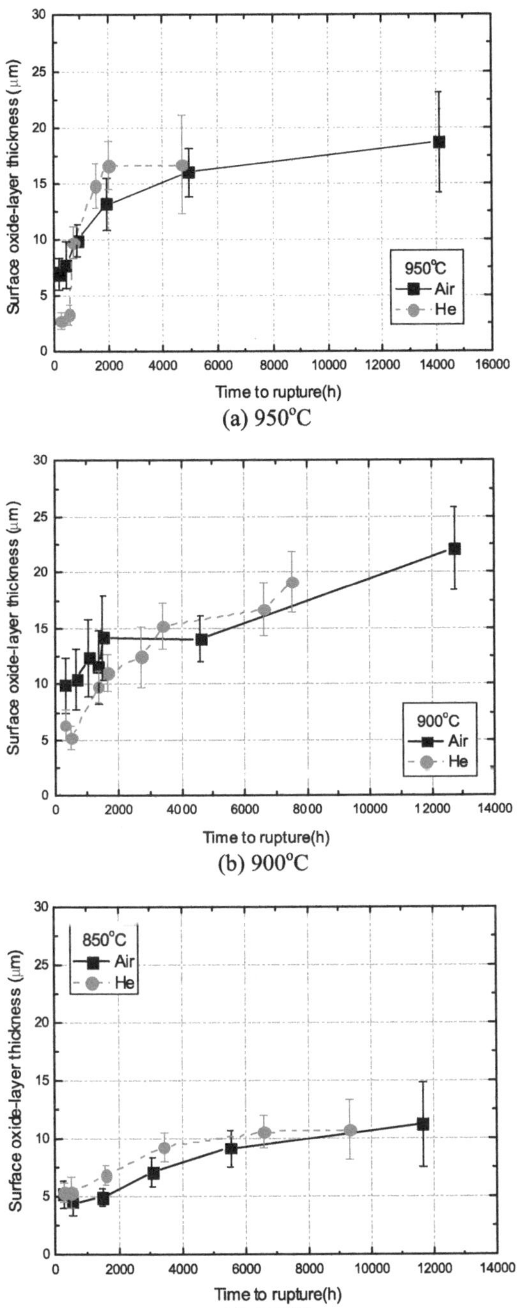


**Fig. 14.** Comparisons of surface oxide-layer thickness with creep rupture time at (a) 950 °C, (b) 900 °C, and (c) 850 °C in air and He environments.

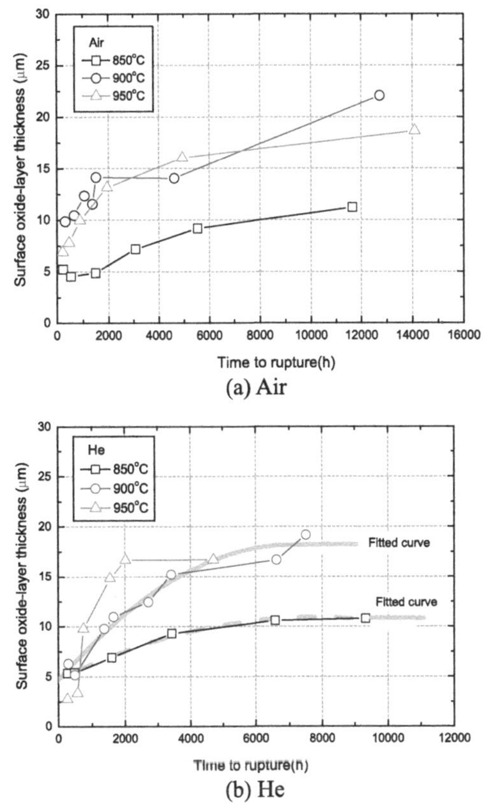


**Fig. 15.** Comparisons of surface oxide-layer thickness with creep rupture time in (a) air and (b) He environments at 950, 900, and 850 °C.

microstructue features such as surface cracking along the grain boundary, voids below the surface, and voids in the matrix (Dewson and Li, 2005).

At 950 and 900 °C, the surface oxide-layer thickness of the He samples was thinner at short rupture time than that of air, but after the time passed for 1000 h at 950 °C, and 3000 h at 900 °C, the thickness increased when compared to in air. This result at the short rupture time is due to the low oxygen-containing He environment used in this study. However, at 850 °C, the thickness was as low as 10 μm in spite of the long-term rupture reaching 12,000 h. The thickness for both environments was almost same. The thickness changes with time variations revealed a roughly parabolic trend, as shown in Fig. 15. It is believed that the parabolic trend was determined by the diffusion rate when the supply of oxygen atom penetrates a dense oxide layer formed for long-term creep. At 950 and 900 °C, the surface oxide-layer thickness was almost similar in both environments, but at 850 °C, the thickness was reduced by about 50% when compared to the temperatures of 950 and 900 °C.

The reason why the surface oxide layer formed much deeper inside in the He than in air can be explained as following. Oxidation is a diffusion controlled process. Once the surface of a metal begins to oxidize, subsequent oxidation can take place only when oxygen ions diffuse through the oxide and react with the metal ions at the metal-oxide interface. As the oxide layer thickens, the diffusion distance increases and the rate of oxidation decreases. The density of oxide formed on Alloy 617 by annealing in air is much more than that formed in He, which tends to make the oxygen ions diffuse more difficult, while the much less density of the oxide on the surface formed in the He ambience tends to make the oxygen ions diffuse easier and then to increase the diffusion distance of the oxygen ions.

From the results, it is found that the results of the surface oxide thickness were in accordance with those of the creep properties, as shown in Figs. 4–6. In particular, at 850 °C, the creep and oxidation behaviors were similar regardless of both environments. Therefore,

850 °C is assumed to be a boundary temperature at which the He effects disappeared.

## 4. Conclusions

A series of creep tests was performed at 950, 900, and 850 °C in air and He environments to investigate the temperature effect on creep and oxidation behaviors of Alloy 617. The creep properties were analyzed, and oxidation microstructural features were analyzed by observing the crept specimens tested in both environments. The surface oxide-layer thickness was measured to define oxidation behavior with the rupture time. Based on the creep data and oxidation microstructural analysis, the following conclusions were drawn.

At 950 and 900 °C, the creep rupture time in the He environment was shorter than that in air, and the thickness of the surface oxide layer was thicker than in air. A significant deterioration of the creep resistance in He environment was due to the thicker surface oxide-layer formed by the reaction of surface oxygen and Cr diffusion in the matrix, even in a low oxygen-containing He agent. The thicker surface oxide-layer reduced the effective area carrying the creep load, and the creep resistance was reduced. On the other hand, at 850 °C, the rupture time was almost the same regardless of both environments. The surface oxide-layer thickness was reduced by about 50% compared to that of 950 and 900 °C. 850 °C could be assumed as a boundary temperature at which the He effects disappeared. The relationships between the creep properties and surface oxide-layer thickness were in accordance in both environments.

## Acknowledgments

This study was supported by Nuclear Research and Development Program of the National Research Foundation of Korea (NRF) grant funded by the Korean government (MSIP) (Grant code: NRF-2012MZA8A2025682).

## References

**Christ, H., Kunecke, U., Meyer, K., 1987.** High temperature corrosion of the nickel-based alloy Inconel 617 in helium containing small amounts of impurities. Mater. Sci. Eng. 87, 161.

**Cook, R., 1984.** Creep properties of Inconel-617 in air and helium at 800 to 1000 °C. Nucl. Technol. 66, 283.

**Dewson, S., Li, X., 2005.** Selection criteria for the high temperature reactor intermediate heat exchanger. In: Proceedings of ICAPP 05, Paper No. 5333, Seoul.

**Jang, C., Lee, D., Kim, D., 2008.** Oxidation behavior of an alloy 617 in very high temperature air and helium environments. Int. J. Press. Vessels Piping 85, 368.

**Kim, W.G., Yin, S.N., Ryu, W.S., Chang, J.H., 2007.** Analysis of the creep rupture data of alloy 617 for a high temperature gas cooled reactor. In: Proceedings of CREEP8, ASME PVP 2007-26834, Texas.

**Kim, W.G., Yin, S.N., Kim, Y.W., Chang, J.H., 2008.** Creep characterization of a Ni-based Hastelloy-X alloy by using theta projection method. Eng. Fract. Mech. 75, 4985.

**Kim, W.G., Yin, S.N., Kim, Y.W., 2009.** Long-term creep strain-time curve modeling of alloy 617 for a VHTR intermediate heat exchanger. J. Korean Inst. Met. Mater. 47 (10), 613 (in Korean).

**Kim, W.G., Yin, S.N., Lee, G.G., Kim, Y.W., Kim, S.J., 2010.** Creep oxidation behavior and creep strength prediction for alloy 617. Int. J. Press. Vessels Piping 87, 289.

**Kim, W.G., Yin, S.N., Park, J.Y., Hong, S.D., Kim, Y.W., 2012a.** An improved methodology for determining tensile design strengths of alloy 617. J. Mech. Sci. Technol. 26 (2), 379.

**Kim, W.G., Park, J.Y., Kim, S.J., Hong, S.D., Kim, Y.W., 2012b.** Reliability evaluation on creep life prediction of alloy 617 for a very high temperature reactor. J. Korean Inst. Met. Mater. 50 (10), 721 (in Korean).

# Cyclic creep behaviour under tension–tension loading cycles with hold time of modified 9Cr–1Mo steel

**W.-G. Kim*[1], J.-Y. Park[2], I. M. W. Ekaputra[2], S.-J. Kim[2] and J. Jang[1]**

Cyclic creep behaviour of modified 9Cr–1Mo steel was investigated by a series of cyclic creep (CC) tests at 600°C, which were performed under controlled tension–tension loading cycles with the magnitude of stress ranges in a constant stress ratio ($R$=0·1). Hold time was applied for a 10 min hold at the maximum stress ($\sigma_{max}$) and minimum stress ($\sigma_{min}$). The CC properties were compared with the static creep (SC) using Norton's power law, Larson–Miller plot, and Monkman–Grant relation, and the microstructure was examined. For the test conditions employed in the present investigation, retardation in the CC behaviour in terms of a lower creep rate and longer rupture time compared to those in the SC was obtained. The retardation was ascribed to the effects associated with anelastic recovery during the 10 min hold time at the minimum load of the cyclic loading. The creep rupture ductility decreased with a general decrease in stress, and there was no difference in the creep ductility between the CC and SC. The steel displayed a transgranular fracture characterised by the presence of dimples resulting from micro-void coalescence. Carbide precipitation was more coarsened with increasing in exposure time in the CC tests.

**Keywords:** Static creep, Cyclic creep, Modified 9Cr–1Mo steel, Stress range, Creep rate, Sodium cooled fast reactors

## Introduction

Modified 9Cr–1Mo steel (ASME Grade 91, hereafter referred to as Gr. 91 steel) is regarded as a promising candidate for structural materials such as steam generators (SG), intermediate heat exchangers (IHX), and hot pipes in sodium cooled fast reactors (SFR). The selection of Gr. 91 steel is mainly based on its high creep and low cycle fatigue resistance than those exhibited by its counterparts such as 9Cr–1Mo and 2·25 -1Mo steels.[1–8]

Presently, Korea Atomic Energy Research Institute (KAERI) has established an R&D program to develop a Prototype Gen-IV SFR (PGSFR) by 2028, and key technologies for constructing its reactor are being developed step by step through a long term R&D program.[9,10] Two main materials of the PGSFR consist of Gr. 91 and 316 stainless steels, as shown in Fig. 1. Its structural components are designed for service of up to 60 years at elevated temperatures and expected to experience complex loading conditions for long durations requiring high resistance to creep, fatigue and superimposed creep fatigue. Both 'cyclic creep' (CC) and general static creep (hereafter referred to as SC) are important considerations for the design of the components operating at high temperatures because the high temperature structural components are exposed to cyclic loadings.

In view of this, the CC has been evolved for describing the material behaviour under different loading conditions, where test under load control mode is involved and the applied cycling frequency simulates the creep/fatigue interaction. The parameters of cyclic

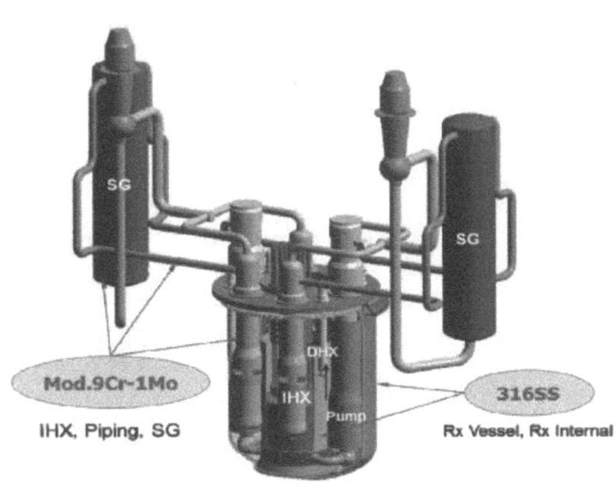


1 **Two main materials in Prototype Gen-IV sodium cooled fast reactor**

[1]Korea Atomic Energy Research Institute, 989-111 Daedeokdaero, Yuseong-gu, Daejeon 305-353, Korea
[2]Pukyong National University, 365 Shinsunro, Nam-gu, Busan 608-739, Korea

*Corresponding author, email wgkim@kaeri.re.kr


Received 21 May 2014; accepted 7 June 2014
DOI 10.1179/1878641314Y.0000000021

**Table 1 Chemical compositions of Gr. 91 steel/wt-%**

| C | Si | Mn | P | S | Ni | Cr | Mo | Cu | V | Al | N | Nb |
|---|---|---|---|---|---|---|---|---|---|---|---|---|
| 0·115 | 0·23 | 0·415 | 0·012 | 0·0014 | 0·22 | 8·9 | 0·869 | 0·038 | 0·194 | 0·020 | 0·0513 | 0·073 |

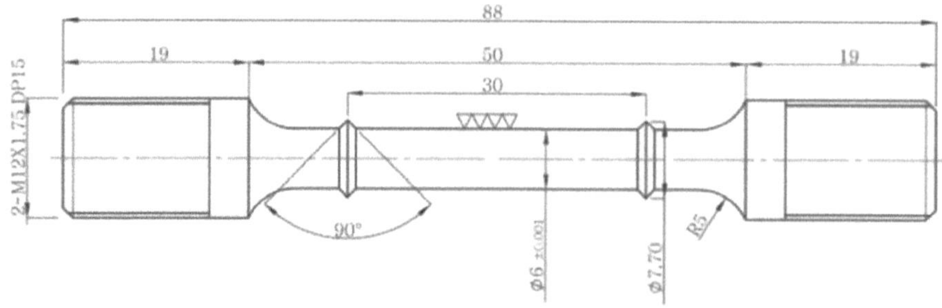


**2 Specimen dimensions of CC test specimen**

**Table 2 Employed CC test conditions of Gr. 91 steel at 600°C**

| Stress ratio, $R$=0·1 ($\sigma_{min}/\sigma_{max}$) | Stress range ($\sigma_R=\sigma_{max}-\sigma_{min}$) (MPa) | Mean stress $\sigma_m=(\sigma_{max}+\sigma_{min})/2$ (MPa) | Hold time (min/min) |
|---|---|---|---|
| 22/220 | 198 | 121·0 | 10/10 |
| 21/210 | 189 | 115·5 | 10/10 |
| 20/200 | 180 | 110·0 | 10/10 |
| 19/190 | 171 | 104·5 | 10/10 |
| 18/180 | 162 | 99·0 | 10/10 |
| 17/170 | 153 | 93·5 | 10/10 |

loading can vary in broad regions of stress and time. There can be considerable interaction between creep and cyclic loading. The effect of cyclic loading can either accelerate or retard creep deformation depending on the material structure, temperature and stress conditions.[11–16] In the case of static creep, the response of the material is simple as a static state. However, in the case of CC, it is complex because of the dynamic nature of the loading, and furthermore, its behaviour has been rarely reported. It is thus important to understand whether the CC leads to an acceleration or retardation in the evolution of creep strain leading to change in creep rate and the rupture time compared to the SC. The influence of frequency, stress range ($\sigma_R$), stress ratio ($R$) and hold duration on the creep strain evolution affecting the creep rate and rupture time should be examined. In view of this, it is necessary to examine the CC behaviour in terms of these experimental variables influencing the creep deformation and fracture behaviour compared to those observed during static creep.

In the present investigation, a series of controlled CC tests was performed with the different stress ranges ($\sigma_R$) in a constant stress ratio ($R$=0·1) under controlled tension–tension loading cycles at 600°C. Hold time was applied for a 10 min hold time at the maximum stress ($\sigma_{max}$) and minimum stress ($\sigma_{min}$). A comparison between the results obtained for the SC and CC properties is presented in terms of stress dependence of creep rate and rupture life, and Monkman–Grant relation. Fracture micrographs after the CC tests were observed and discussed.

## Experimental procedure

### Materials and test conditions

A commercial grade hot rolled Gr. 91 steel plate with a thickness of 16 mm was used for the SC and CC tests. The chemical composition is listed in Table 1. The heat treatment condition of the steel was normalized at 1050°C/1 min mm$^{-1}$ followed by tempering at 770°C/3 min mm$^{-1}$. The cylindrical specimens with a 30 mm gauge length and 6 mm diameter were identically used for the SC and CC test specimens, as shown in Fig. 2. The SC tests were performed with different stress levels of 200, 190, 180, 170, 160, 150 MPa at 600°C using constant load creep machine. In the SC tests, the

**3 Complete view of CC testing apparatus**

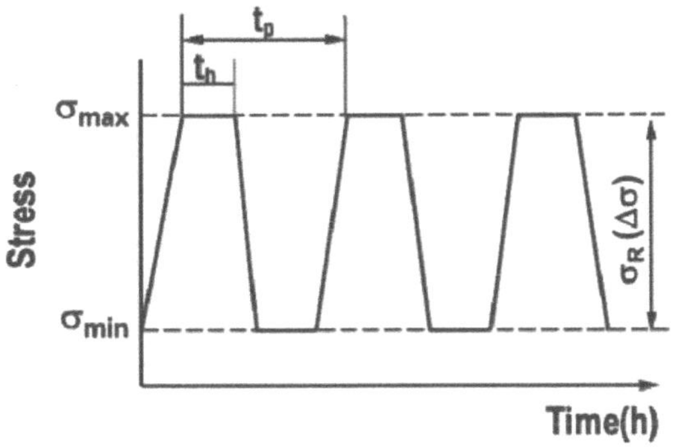


**4 Loading schedules of controlled CC tests**

experimental procedures, typical creep curves, and creep tested results were in detail given in the author's previous study.[17]

In addition, the CC tests were performed with different applied stress ranges ($\sigma_R$) at a constant stress ratio of $R$=0·1 (i.e. $R=\sigma_{min}/\sigma_{max}$) of 20 MPa ($\sigma_{min}$)/ 200 MPa ($\sigma_{max}$.), 19 MPa/190 MPa, 18 MPa/180 MPa, and 17 MPa/170 MPa at 600°C. In the CC tests, hold time (or dwell time) was applied for 10 min ($t_h$) at the maximum stress ($\sigma_{max}$) and minimum stress ($\sigma_{min}$), respectively. The stress range ($\sigma_R=\sigma_{max}-\sigma_{min}$) was chosen as a test variable. The stress ranges at $R$=0·1 were 180 MPa ($\sigma_{R1}$), 171 MPa ($\sigma_{R2}$), 162 MPa ($\sigma_{R3}$) and 153 MPa ($\sigma_{R4}$) as listed in Table 2. The $\sigma_R$ values were chosen as a lower value than the yield stress of 247·5 MPa obtained from the tensile tests for Gr. 91 steel at 600°C. In all cyclic creep tests, the crosshead speed to the maximum stress value from the minimum stress was fixed at 3920 N min$^{-1}$.

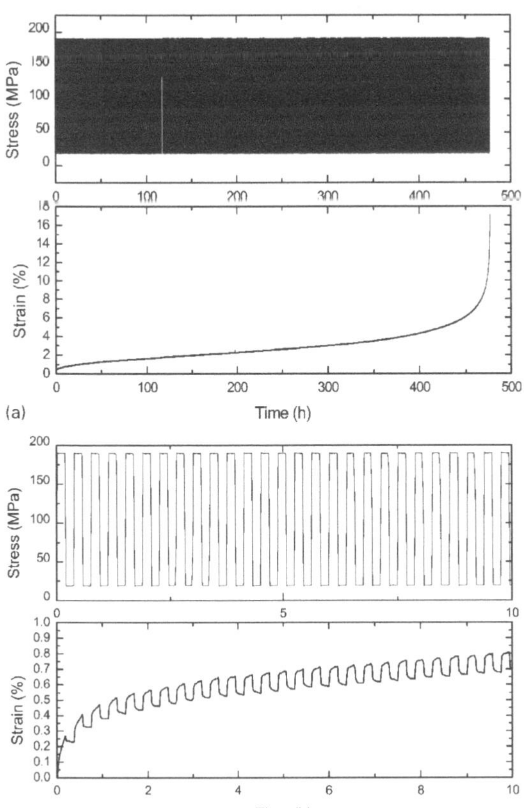


**5 *a* total CC curves and *b* enlarged stress–strain profiles obtained at $\sigma_R$=171 MPa (19 MPa/190 MPa) for 10 min hold time at 600°C**

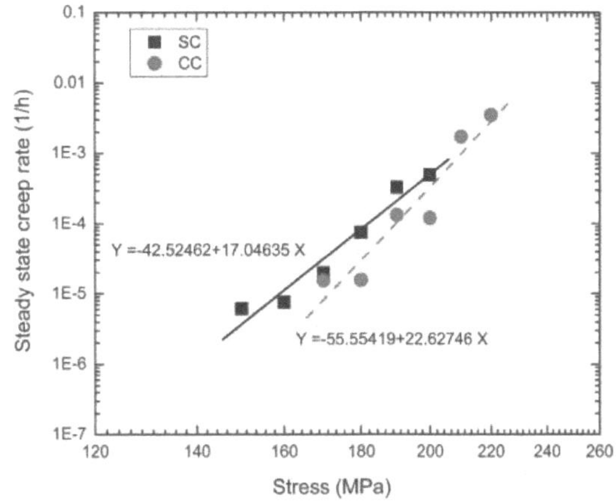


**6 Plot of creep rate versus stress obtained for CC and SC conditions**

## Cyclic test method

Load controlled tension–tension cyclic tests were performed using a universal testing machine with a 100 KN capacity (model no.: RB Unitech-M), manufactured by an R&B company in South Korea. A photo of the total view is shown in Fig. 3. The cyclic loading of the maximum and minimum stresses was applied to a specimen using an AC servomotor type, which can be periodically repeated in the clockwise and counter-clockwise rotations. The main components are composed of a three-zone heating furnace, temperature controller, extensometer, strain gage, data acquisition system (PC and monitor) and program controller.

The cyclic creep tests were conducted following the procedure recommended in ASTM E139-11.[18] The

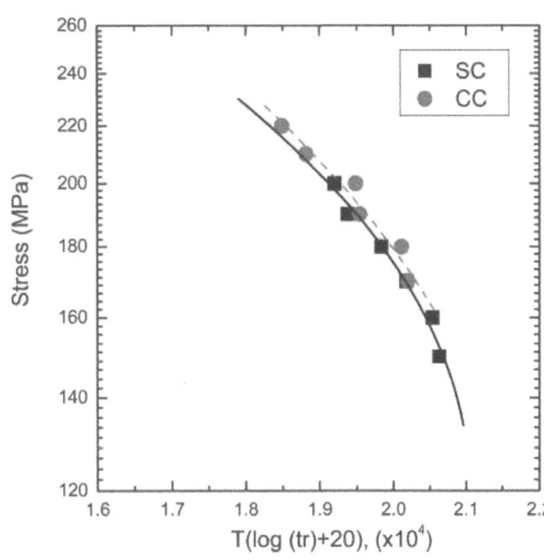


**7 Plot of Larson–Miller plot obtained for SC and CC tests**

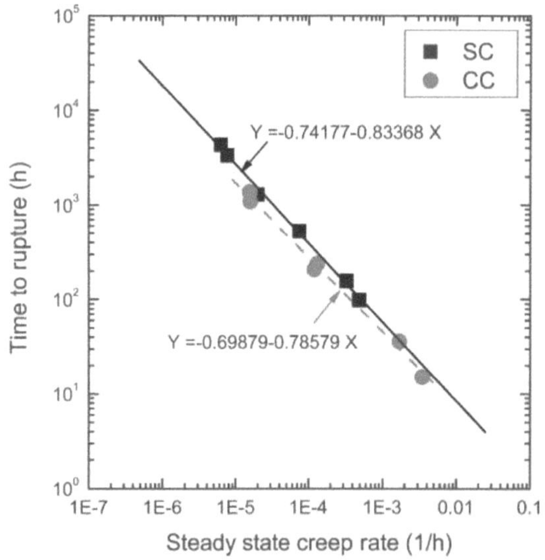


**8 Plot of rupture time versus creep rate obtained for SC and CC tests**

loading schedules for the CC tests are shown in Fig. 4. The real time data of the strain and stress at the elapsed times were monitored and collected by a PC through a high precision extensometer attached to the specimen. The temperature during the CC tests was maintained to within ±2°C. The steady state creep rate (SSCR) in the cyclic creep was taken as a mean value of the secondary creep strain time data. In order to compare the creep properties between the SC and CC appropriately, the SSCR and rupture time were determined by taking the half value of the total elapsed time. This is based on the fact that the unloading time during a minimum stress of cyclic loading does not contribute practically to creep deformation.

## Results and discussion

### Cyclic creep curves

Figure 5 shows typical cyclic creep curves obtained for the stress range of $\sigma_R$=171 MPa ($\sigma_R$=190 MPa−19 MPa) under a stress ratio $R$=0·1 for a 10 min hold at 600°C. The creep curve shows well defined primary creep, secondary creep and tertiary creep stages similar to that generally obtained during the static creep. The secondary creep stage occupied almost the entire creep life. It seems that the cyclic creep curves were no different in their shapes compared with the static creep curves. The black part in figure (Fig. 5*a*) indicates a number of stress profiles with time variations. Enlarged magnification of the primary and secondary creep stages is clearly shown in figure (Fig. 5*b*). From the curves, it is thus justified that the stress and strain profiles with time variations were accurately monitored. In addition, the profiles of typical cyclic creep curves were obtained for a stress range of $\sigma_R$=153 MPa ($\sigma_R$=170 MPa−17 MPa) under a stress ratio of $R$=0·1 for a 10 min hold at 600°C. It appeared that in the cyclic strain time profiles, the width of the strain range increased gradually to the secondary creep stage from the primary creep stage. However, the width was constant for the secondary creep stage. This constancy in width during the secondary creep stage resulted from a balance between hardening and softening.

### Comparison of SC and CC behaviour

Creep rupture data in the CC tests such as the rupture time, SSCR, rupture elongation and a reduction of area were reasonably obtained for Gr. 91 steel at 600°C. Herein, the results of the CC tests were comparatively given with those of the SC tests. The SSCR in the CC curves was determined from the mean value of the strain profile range in the secondary creep stage, and the stress value was used for maximum stress. It is noted that the values of the creep rupture time and SSCR are taken for the half values of elapsed time, because the unloading time corresponding to a minimum stress during cycling does not practically contribute to creep deformation.[14]

Figure 6 shows a comparison of Norton's power law plot obtained for the SC and CC tests. The material constants of Norton's power law, $\dot{\varepsilon}_{ss}=A\sigma^n$, were obtained for Gr. 91 steel. In the CC, the $A$ value was $2\cdot79\times10^{-56}$ $MPa^{-n}$ $h^{-1}$), and the $n$ value of the slope was 22·62, and in the SC, the $A$ value was $3\cdot01\times10^{-43}$ $MPa^{-n}$ $h^{-1}$, and the $n$ value of the slope was 17·04. From the comparison between the SC and CC in log SSCR versus log stress, it was observed that

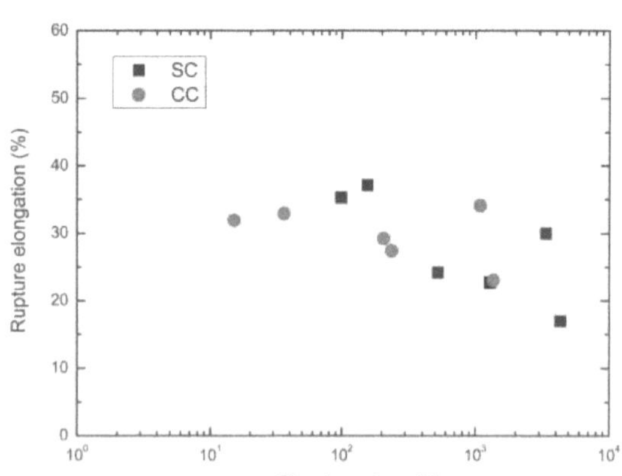


**9 Plot of rupture elongation versus rupture time obtained for SC and CC tests**

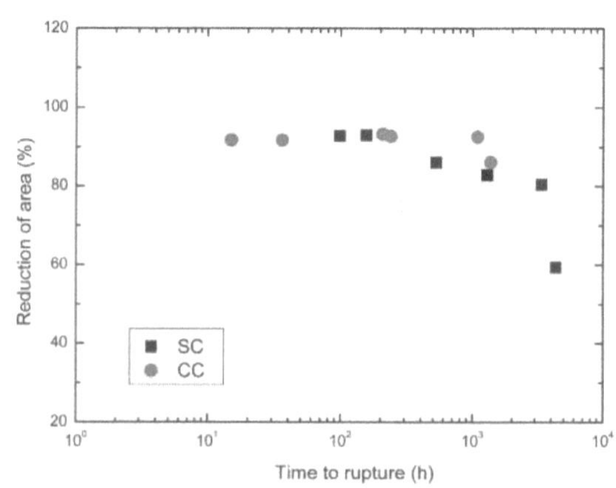


**10 Plot of reduction of area versus rupture time obtained for SC and CC tests**

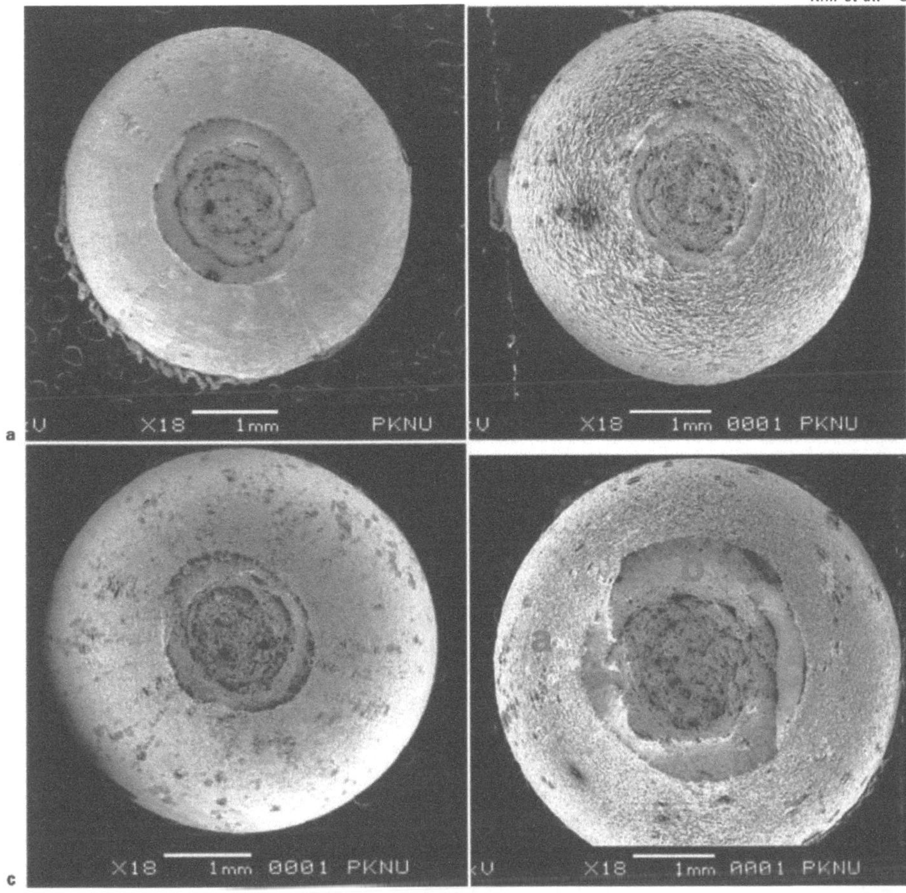


*a* $\sigma_R$=180 MPa (20 MPa/200 MPa); *b* 19MPa/190MPa; *c* 18MPa/180MPa; *d* $\sigma_R$=153 MPa (17 MPa/170 MPa)

**11 Fracture surface after CC tests for 10 min hold time at 600°C**

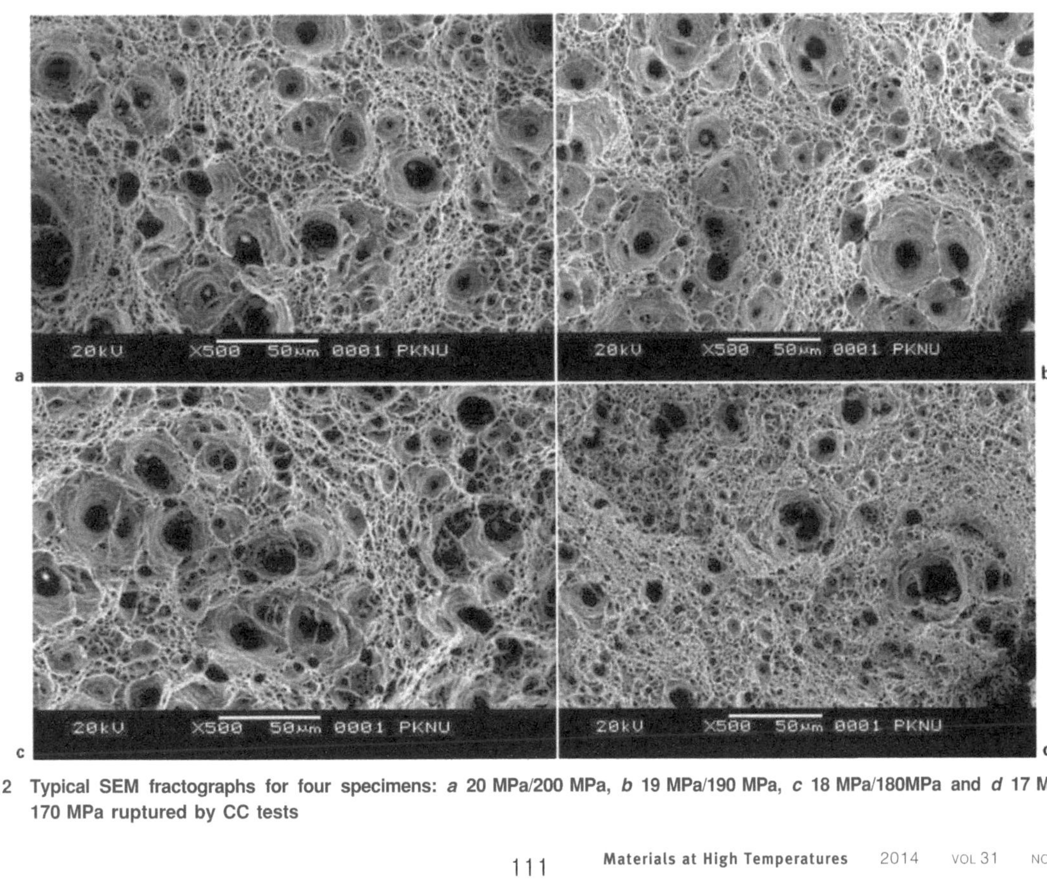


**12 Typical SEM fractographs for four specimens: *a* 20 MPa/200 MPa, *b* 19 MPa/190 MPa, *c* 18 MPa/180MPa and *d* 17 MPa/170 MPa ruptured by CC tests**

Gr. 91 steel exhibits a lower creep rate in the CC compared to that in the SC. The observed lower creep rate (Fig. 6) is reflected from the longer rupture time in this CC test condition, as shown in the plot of stress versus Larson–Miller Parameter (LMP) plot in Fig. 7. In the LMP equation, $T(\log t_r + C)$, $T$ is the absolute temperature (K), $t_r$ is the rupture time (h), and $C$ is material constant. Here, the average value of $C=20$ in the LMP plot has been used for comparison purpose.

Figure 8 shows a comparison of the results for the SC and CC using a Monkman–Grant (M–G) plot indicating the relationship between the creep rupture time and SSCR. The M–G plot revealed good linearity for both the SC and CC conditions. In the M–G relation

$$\log t_r + m \log \dot{\varepsilon}_{ss} = C_{MG}$$

the $m$ values obtained from the linear slope were similar at about 0·78–0·83. The deceleration in the SSCR and longer rupture life obtained for cyclic creep can be ascribed to the anelastic recovery occurring during a 10 min hold time employed at the minimum stress of the cyclic loading. A comparison of the rupture elongation and reduction of area as a function of rupture life for the SC and CC tests is shown in Figs. 9 and 10, respectively. Gr. 91 steel indicates a decrease in creep ductility with an increase in rupture life for the static and cyclic loadings. No significant difference in creep ductility between the SC and CC tests was noticed.

The retardation of the creep rate and creep rupture life by the CC tests can be described, as follows. The parameters of superimposed cyclic loading may vary in an abroad region of the stress and time such as frequency, stress ratio, stress range, and hold time. The effect of cyclic loading onto the creep stress can either accelerate or decelerate, and it depends on material structure, temperature, and stress conditions. The creep effect on the cyclic loading can be attributed to different mechanisms involving anelastic strain storage and recovery, carbide precipitation, repeated effect of primary creep deformation, and recovery of dislocation structure. The beneficial effects of cyclic stress, which are expressed in an increase of the rupture life and/or a decrease of the minimum creep strain rate, have been reported for nickel superalloys and ODS alloys.[12] Therefore, the retardation of the creep rate for the CC conditions employed in the present investigation was ascribed due to the 10 min hold and lower minimum stress ($\sigma_{min}$) of the stress ranges ($\sigma_R$). The reason for this is attributed to the recovery of dislocation structure during the CC tests owing to the effect of the hold time and lower value in $\sigma_{min}$. It has been reported that a higher frequency condition without a hold time in the case of tension–tension cyclic loading was accelerated in the creep strain rate.[12]

## Fracture micrographs in cyclic creep

Figure 11 shows typical SEM fractographs observed for four specimens ruptured by the CC tests under 20 MPa/200 MPa (200 MPa), 19 MPa/190 MPa (190 MPa), 18 MPa/180 MPa (180 MPa), and 17 MPa/170 MPa

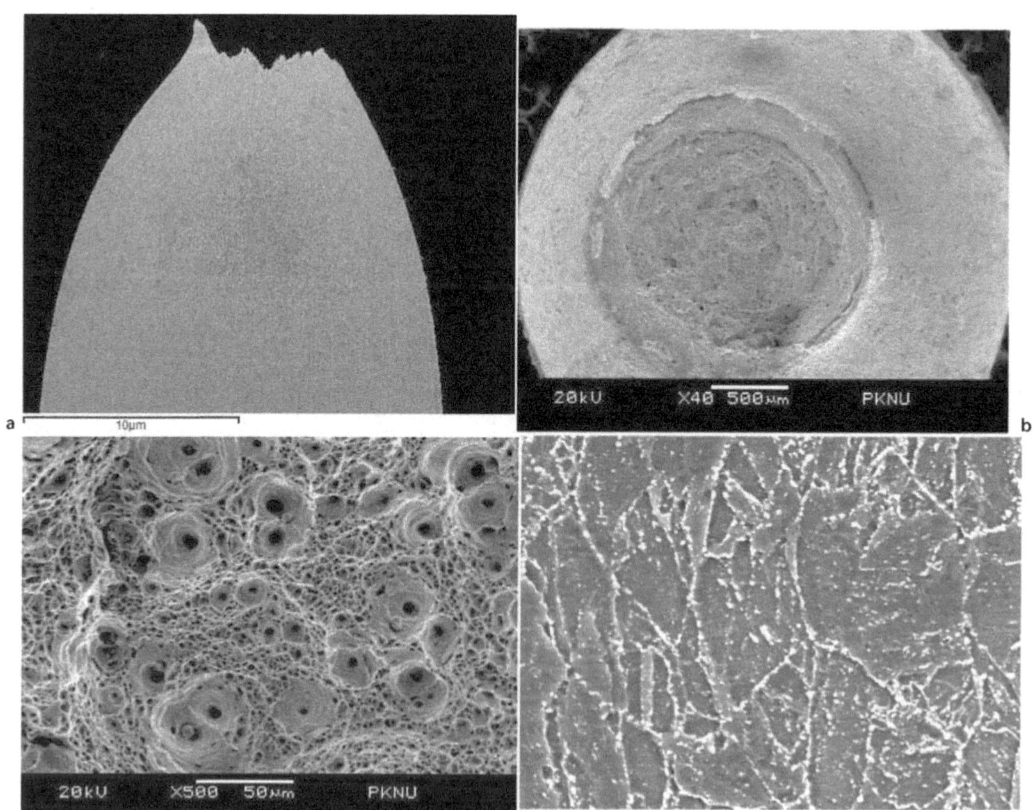


*a* cross-section view; *b* plain view; *c* enlarged view of core region *b*; *d* precipitates

**13 Typical SEM fractographs and precipitates ruptured by SC test under 190 MPa at 600°C**

(170 MPa). A ductile cup and cone fracture for all the CC tested specimens can be seen well. The reduction of the area increases with a decrease in the stress ranges. Namely, it means that the reduction of the ruptured diameter size was not large with decrease in the CC stress. The failure of Gr. 91 steel exhibits three regions; deformation region (Fig. 11*a*), slant region (Fig. 11*b*), and core region (Fig. 11*c*), as shown in Fig. 11*d*. The deformation region is due to plastic deformation during the CC tests, the slant region is due to shear stress in the middle area between the slant and core regions, and the core region reveals equiaxed dimples as failed at the final step of the CC tests. In the central equiaxed regions (Fig. 11*c*), the presence of many dimples in the ruptured specimens can be seen in Fig. 12. All of the ruptured specimens exhibit a similar transgranular fracture characterised by the presence of dimples resulting from microvoid coalescence. No difference in dimple failure between the CC and SC conditions is observed as shown in Fig. 13, which was ruptured in the SC test of 190 MPa at 600°C. Figure 13*a* is the cross-section view of the ruptured specimen, Fig. 13*b* is the plain view, Fig. 13*c* is the enlarged photograph for the core region of the photo (Fig. 13*b*), and Fig. 13*d* indicates the precipitates formed by the SC creep under 190 MPa at 600°C.

Figure 14 shows the variations in the size and density of the dimples measured with the stress ranges of four ruptured specimens in the CC tests. It appeared to be a decrease in the size of dimples with a decrease in the stress range. The presence of fine sized dimples at low stress suggests the dominance of the nucleation process, while the presence of large sized dimples is dominating owing to the dominance of the growth process at high stresses.

Figure 15 shows the precipitates formed in the matrix after the CC tests for Gr. 91 steel at 600°C. The microstructures in the normalised and tempered (N+T) condition (in the grip region of stress free) (Fig. 15*a*) and the CC tested condition (Fig. 15*b–d*) are shown in the figure, respectively. The microstructure was composed of tempered lath martensite and precipitates at the prior austenite grain boundary (PAGB) and martensite lath boundary in the grip region of stress free (Fig. 15*a*). The intralath matrix regions display fine precipitates. Higher coarsening of grain and lath boundaries along with matrix precipitates with increasing creep exposure (or lower applied stress) at 600°C can be seen in Fig. 15*b–d*. As annotated on the microstructure in Fig. 15*d*, the PAGB containing martensite packets, blocks and laths can be seen. It is also seen that more precipitates remain in the CC test (Fig. 15*c*) compared with the SC (Fig. 13*d*) under the same stress condition of 190 MPa at 600°C.

It is known that in the normalised and tempered conditions of Gr. 91 steel, the microstructure consists of tempered martensite, and the microstructure consists of prior austenite grains containing martensite packets,

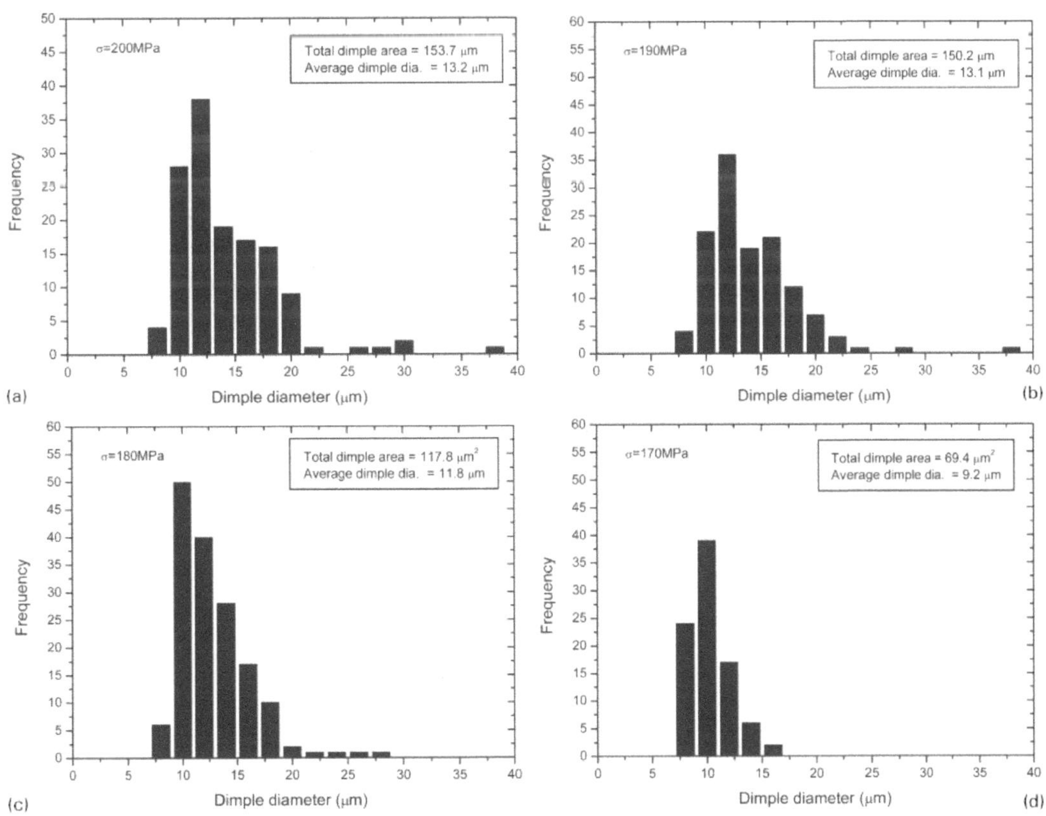


**14 Variations in dimple size and density measured with stress range ranges *a* $\sigma_R$=180 MPa (20 MPa/200 MPa), *b* $\sigma_R$=171 MPa (19 MPa/190 MPa), *c* $\sigma_R$=162 MPa (18 MPa/180MPa) and *d* $\sigma_R$=153 MPa (17 MPa/170 MPa) in CC tested specimens**

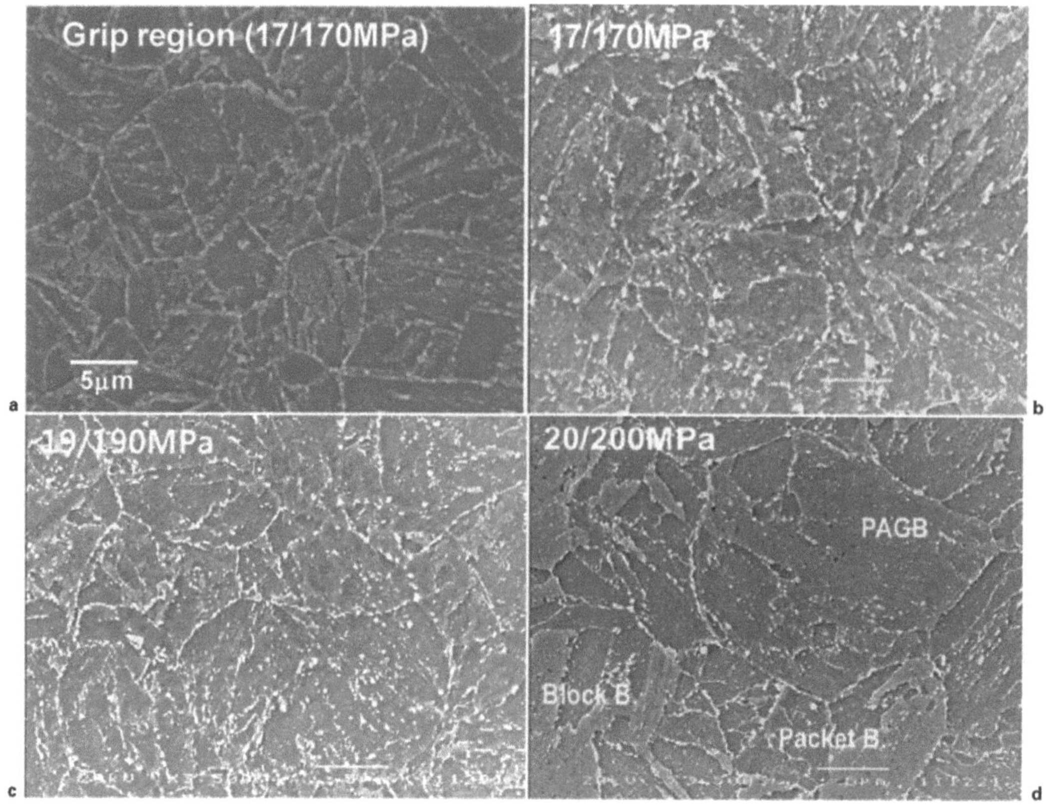


*a* N+T (Grip section region); *b* Gage section region (17 MPa/170 MPa); *c* Gage section region (19 MPa/190 MPa); *d* Gage section region (20 MPa/200 MPa)

**15 SEM photos showing precipitates formed in CC tests at 600°C and prior austenitie grain boundaries, packet boundaries and block boundaries**

blocks and laths, as schematically shown in Fig. 16. Gr. 91 steel is strengthened by both solid solution and precipitation strengthening mechanisms with Cr, Mo, and Nb, V and N contributing to the formation of precipitates which act to improve the strength and limit grain size. $M_{23}C_6$ precipitates occur on the prior austenite and martensite lath boundaries together with finer V and Nb rich carbonitrides precipitated within the grains ($M_2X$ and MX type precipitates). At higher temperature, the $M_{23}C_6$ precipitates coarsen due to Oswald ripening reducing grain boundary pinning which leads to recovery and grain growth with corresponding reductions in dislocation densities.[19] Accordingly, the enhanced coarsening of the precipitates in the CC tests remained, and this was due to Oswald ripening in terms of longer exposure time in the CC tests at 600°C.

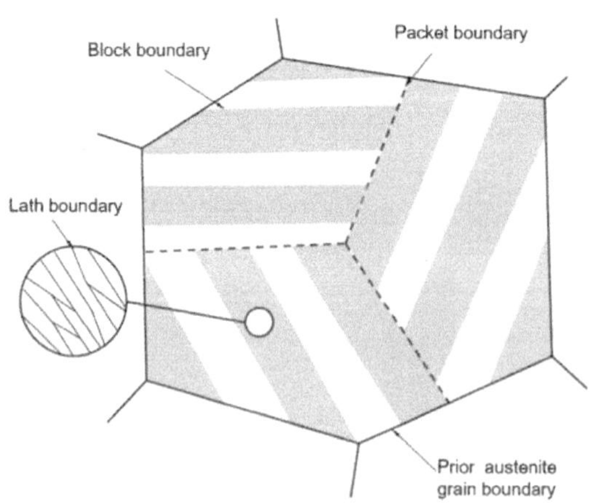


**16 Schematic diagram microstructures of ferritic/martensitic steel**

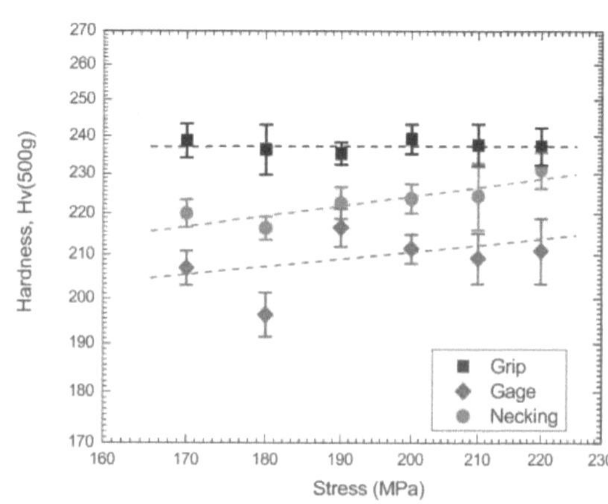


**17 Variations of microhardness measured with stresses for grip, gage and necking regions in CC tests**

In addition, Vickers microhardness values were measured in the grip region, gage length region, and necking region of the CC ruptured specimens with different stresses after the CC tests, respectively, as shown in Fig. 17. In the microhardness measurement, the mean value and standard deviation value were obtained by 10 measurements at each region. In the grip part of the stress-free region, the hardness was constant regardless of the stresses and the highest. However, in the gage and necking area regions, the hardness decreased with a decrease in stress. The hardness of the necking region was higher than that of the gage region. The reason for this is due to the hardening effect caused in the necking area under the loading cycles in the CC tests. Owing to the hardening effect, the cyclic creep exhibited a lower creep rate and longer life when compared with the SC.

## Conclusions

Tension–tension cyclic creep behaviour was successfully examined by a series of controlled cyclic creep tests performed at 600°C of Gr. 91 steel, and the results of the CC properties were compared with those of the SC properties. For the present cyclic loading conditions with different stress ranges of 10 min hold at $R=0{\cdot}1$, retardation of the lower creep rate and longer rupture time compared with those in the SC conditions was obtained. The retardation was ascribed to the effects associated with anelastic recovery during a 10 min hold time at the lower minimum stress of the loading stress. The results of the CC and SC properties followed the well known Norton's power law and Monkman–Grant (M–G) relation. The creep rupture ductility decreased with a decrease in stress, and no difference in creep ductility between the CC and SC conditions was noticed. Fracture micrographs exhibited a transgranular fracture as a typical ductile fracture of a cup-cone type of dimple morphology characterised by the presence of dimples originated by microvoid coalescence. Enhanced carbide precipitates remained with increasing in exposure time (or lower applied stress) in the CC tests.

## Acknowledgement

This study has been carried out as a part of the Innovative R&D Project of Korea Atomic Energy Research Institute.

## References

1. B. K. Choudhary, V. S. Srinivasan and M. D. Mathew: 'Influence of strain rate and temperature on tensile properties of 9Cr–1Mo ferritic steel', *Mater. High Temper.*, 2011, **28**, 155–161.
2. B. K. Choudhary and E. Issac Samual: 'Creep behavior of modified 9Cr-1Mo ferritic steel', *J. Nucl. Mater.*, 2011, **412**, 82–89.
3. S. L. Mannan, S. C. Chetal, B. Raj and S. B. Bhoje: 'Selection of materials for prototype fast breeder reactor', *Trans. Ind. Inst. Met.*, 2003, **56**, 155–78.
4. B. Raj and B. K. Choudhary: 'A perspective on creep and fatigue issues in sodium cooled fast reactors', *Trans. Ind. Inst. Met.*, 2010, **63**, 75–84.
5. E. Issac Samual, B. K. Choudhary and K. Bhau Sankara Rao: 'Influence of post weld heat treatment on tensile properties of 9Cr-1Mo ferritic steel base metal', *Mater. Sci. Technol.*, 2007, **23**, 992–998.
6. W. G. Kim, J. Y. Park, S. J. Kim and J. Jang: 'Reliability assessment of creep rupture life for Gr. 91 steel', *Mater. Des.*, 2013, **51**, 1045–1051.
7. K. Kimura, Y. Toda, H. Kushima and K. Sawada: 'Creep strength of high chromium steel with ferrite matrix', Proc. 2nd ECCC Creep Conf. on 'Creep & fracture in high temperature components', (ed. I. A. Shibli and S. R. Holdsworth), 935–949; 2009, Zurich, ETD.
8. V. S. Srinivasan, B. K. Choudhary, M. D. Mathew and T. Jayakumar: 'Long-term creep-rupture strength prediction for modified 9Cr–1Mo ferritic steel and type 316L(N) austenitic stainless steel', *Mater. High Temper.*, 2012, **29**, 41–48.
9. W. G. Kim, J. Y. Park, S. N. Yin, D. W. Kim, J. Y. Park and S. J. Kim: 'Reliability prediction of long-term creep strength of Gr. 91 steel for next generation reactor structure materials', *Kor. J. Met. Mater.*, 2011, **29**, 275–280.
10. W. G. Kim, J. Y. Park, S. D. Hong and S. J. Kim: 'Probabilistic assessment of creep crack growth rate for Gr. 91 steel', *Nucl. Eng. Design*, 2011, **241**, 3580–3586.
11. D. K. Shetty, T. Mura and M. Meshii: 'Analysis of creep deformation under cyclic loading conditions', *Mater. Sci. Eng.*, 1975, **20**, 261–266.
12. J. Zrnik, J. A. Wang, Y. Yu, L. Peijing and P. Hornak: 'Influence of cycling frequency on cyclic creep characteristics of nickel base single-crystal superalloy', *Mater. Sci. Eng. A*, 1997, **A234–A236**, 884–888.
13. J. H. Eom, D. H. Shin and S. W. Nam: 'Effects of stress amplitude and friction stress on cyclic creep deformation', *J. Korean Inst. Met.*, 1982, **20**, 922–926.
14. Y. K. Park, T. S. Kim, J. H. Choi and M. Y. Wee: 'A study on cyclic creep behavior of Zircaloy-4 at 0·3 Tm', *J. Kor. Inst. Met. Mater.*, 2000, **38**, 624–628.
15. M. Boulbibane and A. R. S. Ponter: 'A method for the evaluation of design limits for structural materials in a cyclic state of creep', *Eur. J. Mech. A/Solid*, 2002, **21**, 899–914.
16. H. D. Chandler and S. Kwofie: 'A description of cyclic creep under conditions of axial cyclic and mean stresses', *Int. J. Fatigue*, 2005, **27**, 541–545.
17. W. G. Kim, S. H. Kim and C. B. Lee: 'Long-term creep characterization of Gr. 91 steel by modified creep constitutive equations', *Met. Mater. Int.*, 2011, **17**, 497–504.
18. 'Standard test method for conducting creep, creep-rupture, and stress-rupture tests of metallic materials', ASTM E139-11, ASTM International, 2011.
19. G. Qin, S. V. Hainsworth, A. Strang, P. F. Morris, P. D. Clark and A. P. Backhouse: 'TEM studies of microstructural evolution in creep exposed E911', (ed. I. A. Shibli and S. R. Holdsworth), Proc. 2nd ECCC Creep Conf. on 'Creep & fracture in high temperature components', 889–899; 2009, Zurich, ETD.

International Journal of Pressure Vessels and Piping 110 (2013) 66–71

Contents lists available at SciVerse ScienceDirect

International Journal of Pressure Vessels and Piping

journal homepage: www.elsevier.com/locate/ijpvp

# Time-dependent crack growth behavior for a SMAW weldment of Gr. 91 steel

Woo-Gon Kim [a,*], Jae-Young Park [b], Hyeong-Yeon Lee [a], Sung-Deok Hong [a], Yong-Wan Kim [a], Seon-Jin Kim [c]

[a] Korea Atomic Energy Research Institute, 989-111, Daedeok-daero, Yuseong-gu, Daejeon 305-353, Republic of Korea
[b] Department of Mechanical Design Engineering, Pukyong National University, Busan 608-739, Republic of Korea
[c] Department of Mechanical & Automotive Engineering, Pukyong National University, Busan 608-739, Republic of Korea

*Keywords:*
Creep
$C^*$ parameter
Creep crack growth
Gr. 91 steel
Shield metal arc weld
Weld metal
Heat affected zone

A B S T R A C T

This paper presents the experimental results on creep crack growth (CCG) behavior for BM, WM and HAZ in the weldment of Gr.91 steel, which was prepared by a shield metal arc weld (SMAW) process. A series of CCG data was obtained by the creep and creep crack growth tests under different applied loads for BM, WM and HAZ in Gr. 91 welded joint at 600 °C. The CCG behaviors were characterized by the empirical equation of the da/dt vs. $C^*$ fracture parameter, and the CCG laws for the BM, WM and HAZ were constructed and compared, respectively. Results showed that for a given value of $C^*$, the WM and HAZ were almost equal, but they were about two-times faster in the CCGR curves than the BM. This reason is that the CCGRs were closely attributed to their creep strength and creep strain rates. It is thus supposed that creep cracks will dominantly initiate in the HAZ, because lower creep strength existed in the HAZ compared with BM.



## 1. Introduction

High-chromium steels are widely used in recent fossil power plants because of their excellent high temperature strength. In particular, modified 9Cr–1Mo steel (ASME grade 9Cr–1Mo, hereafter Gr.91) is regarded as a promising candidate for structural materials of Generation-IV reactor types such as steam generators, intermediate heat exchangers and hot pipes in sodium-cooled fast reactors (SFR), and pressure vessels in very high temperature reactors (VHTR) [1,2].

Their structural components are designed to last for up to 60 years at elevated temperatures and are often subjected to non-uniform stress and temperature distribution during service. These conditions may generate localized creep damage and propagate the cracks and ultimately may cause a fracture. A significant portion of their lives will be spent during crack propagation [3–7]. It is therefore necessary to evaluate creep crack growth (CCG) behavior during creep loading for design and safety assessment of the components, especially for Type-IV cracking of a heat affected zone (HAZ) in weldments of high-Cr ferrite and martensite (FM) steels. It is known that 9Cr or 12Cr steel tends to fail with Type-IV cracking at welded joints, resulting in shorter plant life than expected. To prevent such unexpected failures, an accurate residual life assessment method for the welded joints is required. Welded joints are considered as the composite structure of different materials consists of base metal (BM), weld metal (WM), and HAZ. Since the inhomogeneity among those materials affects the state of stress field or strain rate field near the weld fusion line and near the BM/HAZ interface, it is not easy to estimate the crack propagation behavior at the welded joint. Also, it is concerning that the difference in creep deformation properties among the BM, WM, and HAZ may bring about a difficulty in the crack assessment process [8–11]. Therefore, it is necessary to clarify the creep crack growth (or creep crack propagation) laws through experimental CCG tests for the weld joint of Gr. 91 steel.

This paper is a comparative investigation on time-dependent creep crack growth behaviors for BM, WM, and HAZ in the weldment of Gr. 91, which is prepared using a shielded metal arc weld (SMAW) process. Creep crack growth laws for the BM, WM, and HAZ are constructed using a $C^*$ fracture parameter and the results are discussed.

## 2. Experimental procedures

### 2.1. Material and weld method

The Gr. 91 steel used in this study was a commercial type hot rolled plate 32 mm in thickness. Heat treatment conditions were

* Corresponding author.
*E-mail address:* wgkim@kaeri.re.kr (W.-G. Kim).

0308-0161/$ – see front matter © 2013 Elsevier Ltd. All rights reserved.
http://dx.doi.org/10.1016/j.ijpvp.2013.04.024

**Table 1**
Chemical composition of the G91 steel (wt.%).

| C | Si | Mn | P | S | Ni | Cr | Mo | Cu | V | Al | N | Nb |
|---|---|---|---|---|---|---|---|---|---|---|---|---|
| 0.115 | 0.23 | 0.415 | 0.012 | 0.0014 | 0.22 | 8.9 | 0.869 | 0.038 | 0.194 | 0.020 | 0.0513 | 0.073 |

normalized and tempered at 1050 °C/1 mim/mm and 770 °C/3 mim/mm. Chemical composition is listed in Table 1. The groove shape of the welding of two plates was designed as a single V-groove with 60°. Welded blocks were prepared by using the shielded metal arc welding process. The filler metal, CM-9Cb (brand name), was manufactured by Kobe steel as AWS Class, E9016-G (3.2–4.0 mm). The post weld heat treatment was maintained for 255 min at 750 °C.

A photo of a piece of weld block is shown in Fig. 1. The variations of Micro-Vickers hardness for the BM, WM and HAZ regions in the welded joint were measured, as shown in Fig. 2. The thickness of the HAZ region was around 3.0–5.0 mm. The hardness of the WM region showed higher values than that of the BM and HAZ regions, and in particular, the hardness in the HAZ decreased significantly by softening. This HAZ will be very weak from cracks compared with the BM and WM, and it will initiate Type-IV crack in fine-grained zone which is located in the HAZ adjacent to the parent metal. It is known that the Type-IV cracks are parallel or offset from the fusion line.

### 2.2. Tensile, creep and CCG tests

To obtain the material properties for the BM, WM, and HAZ regions, the tensile and creep tests were performed at 600 °C. The HAZ specimens were cut out toward the transverse direction against the welding direction (longitudinal direction) in the welded block. The tensile specimens had a rectangular cross section of 2 mm in thickness and 6.25 mm in width, with a 25 mm gauge length. The strain rate was conducted with $1 \times 10^{-4}$/s at 600 °C. The creep specimens had a cylindrical shape with a 30 mm gauge length and 6 mm diameter. The HAZ specimens position the HAZ location at the center of gage length. Using constant load creep machines of a dead-weight type with lever ratio of 20:1, creep tests were carried out with different stress levels at 600 °C. Creep strain data with elapsed times were taken automatically by a PC through a high precision LVDT. The steady state creep rate was measured from the secondary creep region of experimental creep curves. The experimental procedures for the creep tests followed the recommendations of the ASTM standard E139 [12]. From these tensile and creep tests, the material constants of *D*, *m*, *A*, and *n* were obtained for the BM, WM, and HAZ samples.

The CCG tests for the BM, WM, and HAZ samples were carried out at a constant load with different applied load levels at 600 °C. Compact tension (CT) specimens had a width (*W*) of 25.4 mm, a thickness (*B*) of 12.7 mm, and side grooves of a 10% depth. The initial crack ratio (*a*/*W*) was about 0.5, and the pre-cracking size was 2.0 mm and was machined by an electric discharge machining (EDM) technique to introduce a sharp crack tip starter for the BM, WM, and HAZ regions. In the welded joint, sharp pre-cracks for the HAZ were taken to conform to one of the HAZ locations, as shown in Fig. 3, which is an overview on the BM, WM and HAZ regions of a welded joint in a CT specimen. Load-line displacement was measured using a linear gauge assembly attached to the specimen, and the crack length was determined using the direct current potential drop (DCPD) method. Crack extension data was continuously collected using a data acquisition system. All of the experimental procedures followed the recommendations of the ASTM standard E1457 [13].

After the CCG testing, the CT specimens were broken open at liquid nitrogen temperature to measure the actual crack length, as shown in Fig. 4. The actually measured final crack length ($a_{mf}$) was calculated from measurements made on the fracture surface at nine equally spaced points (so-called "nine points method") using the enlarged photo of the fractured surfaces, because the individual measurements on the fracture surface vary due to crack front irregularities.

In addition, the predicted crack length ($a_p$) by DCPD technique was calculated by Johnson's formula:

$$\frac{a_p}{W} = \frac{2}{\pi}\cos^{-1}\left[\frac{\cosh(\pi Y_o/2W)}{\cosh\left[\frac{V}{V_o}\cosh^{-1}\left\{\frac{\cosh\pi Y_o/2W}{\cosh\pi a_o/2W}\right\}\right]}\right] \tag{1}$$

where $a_o$ = initial crack size (reference crack size for the reference voltage $V_o$), $Y_o$ = the half distance between the output voltage loads, $V$ = the output voltage, and $W$ = the width of specimen. The $a_p$ was compensated by the potential measurement error and became the corrected crack length $a$:

$$a = \frac{a_{mf} - a_o}{a_{pf} - a_o}(a_p - a_o) + a_o \tag{2}$$

where $a_{pf}$ is the predicted value of the final crack length.

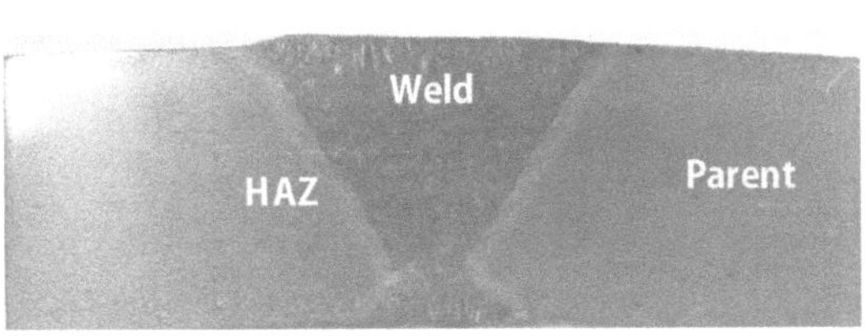


**Fig. 1.** Photo of a welded block.

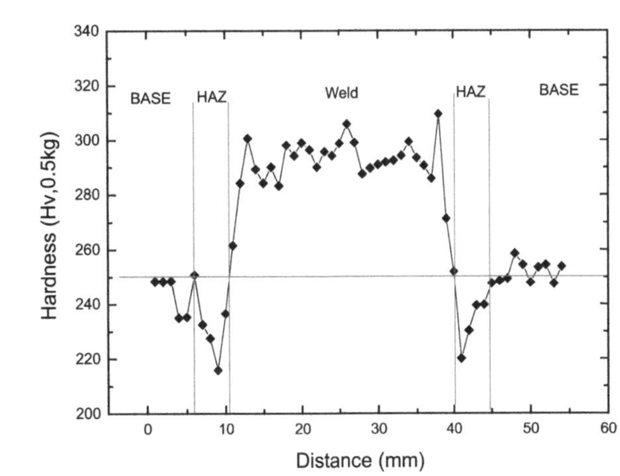


**Fig. 2.** Vickers hardness for weld joint of Gr. 91 steel.

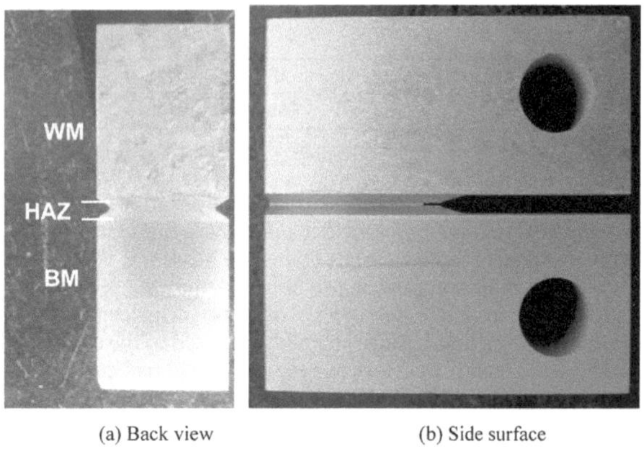


**Fig. 3.** Photo showing BM, WM, and HAZ in 1/2" CT specimens. (a) Back view. (b) Side surface.

Hence, the difference values between the $a_{mf}$ and $a_{pf}$ were investigated whether their values were fairly included within the requirements of ASTM E1457 [13], or not. Namely, the CCG data are valid for further processing if

$$0.85 \leq \left(\Delta a_{pf}/\left(a_{mf} - a_o\right) \leq 1.15 \tag{3}$$

where the predicted crack extension, $\Delta a_{pf}$, was calculated by subtracting the initial crack length, $a_o$, from the finally predicted crack length, $a_{pf}$ [14]. From these procedures, the validity for the CCG tests was identified for all the tested CT specimens.

In the BM, the CCG test data were obtained for eight samples with the applied load ranges of 3800N–5200N. In the WM, the CCG test data were obtained for seven samples with the applied load ranges of 3500N–4500N. And in the HAZ, the CCG test data were obtained for nine samples with the applied load ranges of 3500N–4500N, respectively.

## 3. Results and discussion

### 3.1. Material constants

The $D$ and $m$ constants for the BM, WM, and HAZ were obtained from the tensile tests at 600 °C. Plastic constants were obtained using $\varepsilon_p = D(\sigma/\sigma_y)^m$, where $\varepsilon_p$ = true plastic strain, $\sigma$ = true stress, and $\sigma_y$ = yield stress. The relationship between plastic strain and true stress followed good linearity. From this linear relation, the $D$ and $m$ values for the BM, WM, and HAZ could be obtained.

In addition, the creep constants of $A$ and $n$ were obtained by a series of the creep tests for the BM, WM, and HAZ at 600 °C. Fig. 5 shows the relationships between steady state creep rate and applied stress, which followed good linearity based on Norton's power law. While the WM and HAZ were almost similar, they were higher in $A$ and $n$ values than the BM. The creep strain rates of the WM and HAZ were found to be the difference of one order approximately to the BM. It is believed that the WM and HAZ will be significantly attributed to a faster crack propagation rate. Also, in the creep rupture stress, the HAZ was similar to the WM, and the WM and HAZ were lower than the BM, as shown in Fig. 6, which was obtained by a series of creep tests at 600 °C. It is clearly identified that lower creep strength exists in the HAZ of welded joints. Thus, it is strongly supposed that the creep damages and cracks will initiate in the HAZ, which is called as a Type-IV crack decreasing the creep life of weldment at higher temperatures.

Accordingly, through the tensile and creep tests at 600 °C, the material constants of $D$, $m$, $A$, and $n$ were obtained for the BM, WM, and HAZ, respectively. A summary of these material constants, which were used to calculate the $C^*$ fracture parameter later, is given in Table 2.

### 3.2. Calculation of $C^*$ parameter

The fracture mechanics parameter, $C^*$-integral, has been widely used to characterize the creep crack growth rate in metals undergoing a steady state creep [15]. The steady state creep rate can be written using Norton's power law:

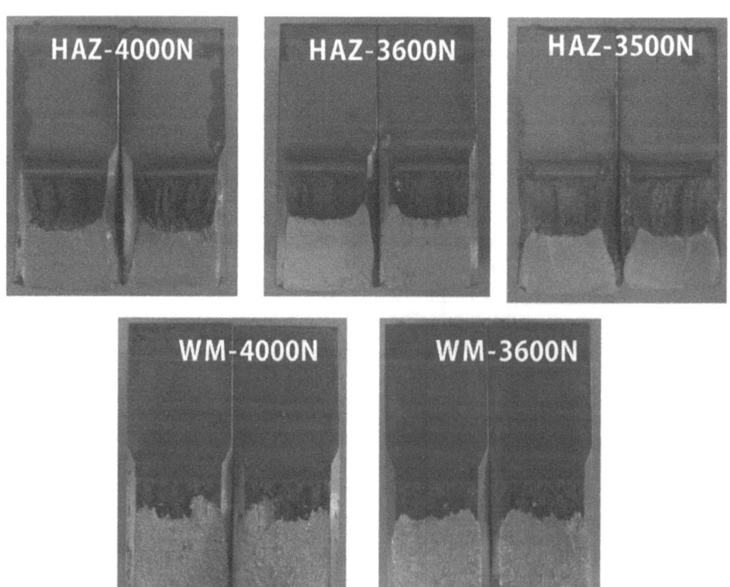


**Fig. 4.** Photos of typical fracture surfaces on the HAZ and WM samples after the CCG tests.

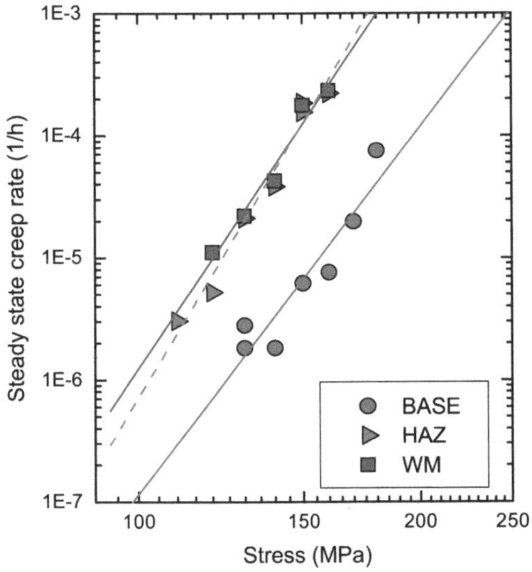


**Fig. 5.** Comparison of the steady state creep rate for the BM, WM, and HAZ at 600 °C.

$$\dot{\varepsilon}_{ss} = A\sigma^n \tag{4}$$

The general form between the creep crack growth rate (d$a$/d$t$) and the $C^*$ can be expressed by

$$\mathrm{d}a/\mathrm{d}t = B\left[C^*\right]^q \tag{5}$$

where $n$ is the creep exponent, and the $B$ and $q$ coefficients are material constants, which are generally obtained from a regression line of the CCGR data. They are related to the intercept and slope, respectively, of the d$a$/d$t$ vs. $C^*$ relationship on a log–log plot. To calculate the d$a$/d$t$ in Eq. (5), material data obtained in Table 2 were used to calculate the $C^*$ values. The relationship between d$a$/d$t$ and $C^*$ was obtained for all samples.

In the CT specimen, the $C^*$ value was calculated using Eq. (6), and load-line displacement rate ($\dot{V}_c$) due to creep strain was calculated by Eq. (8).

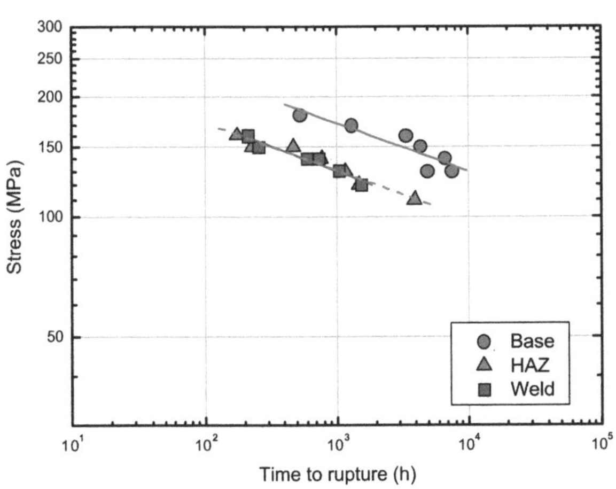


**Fig. 6.** Comparison of creep properties for the BM, WM and HAZ at 600 °C.

**Table 2**
Material constants obtained for the BM, WM, and HAZ of G91 steel at 600 °C.

| Material | $\sigma_y$ (MPa) | $\varepsilon_p = D(\sigma/\sigma_y)^m$ | | $\dot{\varepsilon}_s = A\sigma^n$ | |
|---|---|---|---|---|---|
| | | $D$ | $m$ | $A$ (MPa$^{-n}$h$^{-1}$) | $n$ |
| BM | 247.5 | 0.0017 | 6.4 | 1.28E−27 | 9.98 |
| WM | 298.2 | 0.00191 | 12.99 | 2.28E−29 | 11.27 |
| HAZ | 282.5 | 0.00184 | 11.13 | 3.26E−32 | 12.63 |

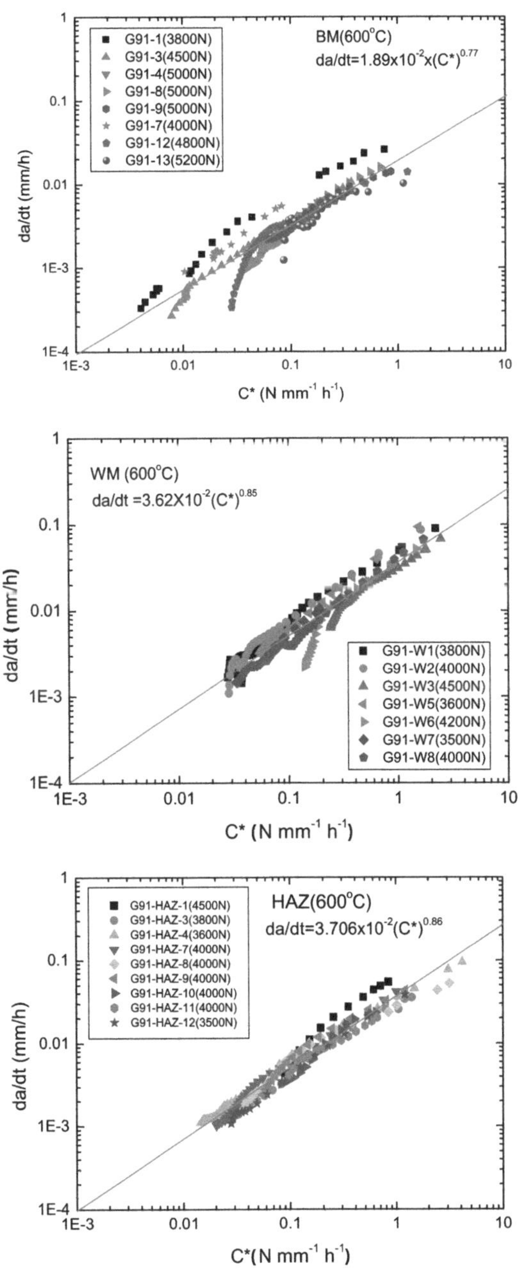


**Fig. 7.** Plot of $C^*$ vs. CCGR for the BM, WM, and HAZ.

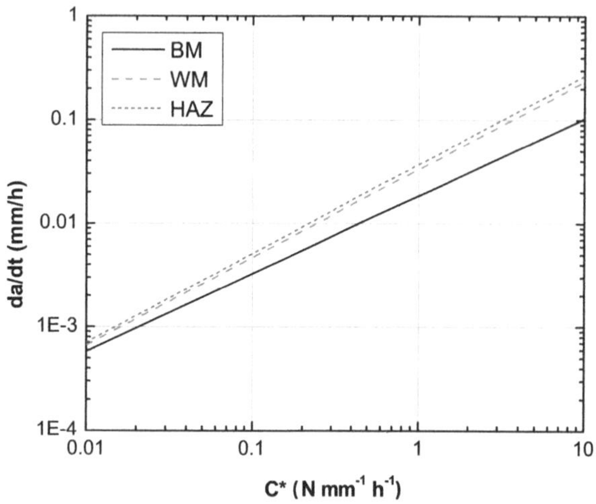


**Fig. 8.** Comparison of CCGR curves for the BM, WM, and HAZ.

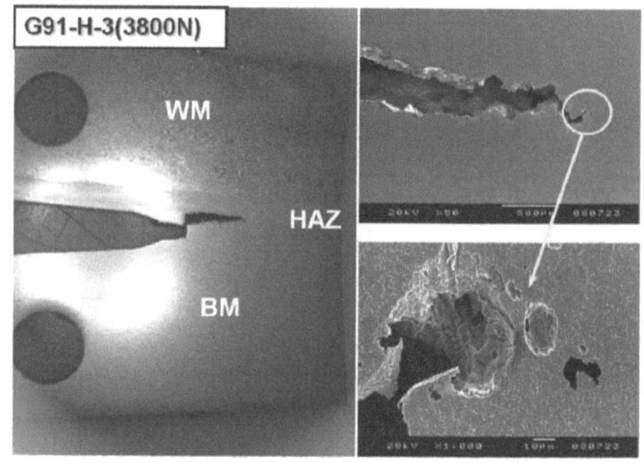


a) Creep crack propagation in HAZ (b) High magnification in front of crack tip

**Fig. 9.** CCG micrographs along Type-IV crack in typical HAZ sample (3800N). (a) Creep crack propagation in HAZ. (b) High magnification in front of crack tip.

$$C^* = \frac{P\dot{V}_c}{B_N(W-a)}\eta\left(\frac{a}{W},n\right) \tag{6}$$

$$\eta\left(\frac{a}{W},n\right) = \frac{1}{(1-a/W)}\frac{n}{n+1}\left(\gamma-\frac{\delta}{n}\right) \tag{7}$$

$$\dot{V}_c = \dot{V} - \frac{\dot{a}B_N}{P}\left(\frac{2K^2}{E'}+\left(m+1\right)J_p\right) \tag{8}$$

where $P$ = applied load, $a$ = crack size, $W$ = width of the specimen, $\dot{V}$ = total load-line displacement rate, $B_N$ = net thickness of specimen, $E'$ = elastic modulus for plane strain, $K$ = stress intensity factor, $\dot{a}$ = crack growth rate, and $m$ = stress exponent in the Ramberg–Osgood stress versus strain relationship. Also, the $J_p$ contribution for the CT specimen is given by

$$J_p = \frac{D_1h_1(a/W,m)}{(\sigma_{ys}(W-a))^m}\left(\frac{P}{1.455B_N\alpha}\right)^{m+1} \tag{9}$$

where $h_1$ is a function of $a/W$ and $m$, and it was obtained using the table result of ASTM E1457.

### 3.3. CCG laws for the BM, WM and HAZ

Fig. 7 shows the plots of $C^*$ vs. $da/dt$ obtained for the BM, WM, and HAZ of Gr. 91 steel at 600 °C. The solid lines show the regression curves obtained using the least squares fit method for each CCG data. The creep crack growth laws for the BM, WM and HAZ were defined, as follows:

$$da/dt = 1.89\times10^{-2}(C^*)0.77(\text{for BM}) \quad (\text{validity range}: 0.003 < C^* < 1.5\ \text{N/mm h}) \tag{10}$$

$$da/dt = 3.62\times10^{-2}(C^*)0.85\ (\text{for WM}) \quad (\text{validity range}: 0.025 < C^* < 2.5\ \text{N/mm h}) \tag{11}$$

$$da/dt = 3.71\times10^{-2}(C^*)0.86\ (\text{for HAZ}) \quad (\text{validity range}: 0.012 < (C^*) < 4.0\ \text{N/mm h}) \tag{12}$$

As given in the equations above, in the creep crack growth laws, WM and HAZ were almost equal, but and were 2.0 times faster than BM. Also, when the three CCGR curves for the BM, WM, and HAZ were overlapped together, as shown in Fig. 8, it appeared clearly that the WM and HAZ were almost the same, but that they were faster than BM. The reason for this is thought to be largely attributed to the higher creep rate in the WM and HAZ.

Fig. 9 shows CCG micrographs fractured for an applied load (3800N) of a typical HAZ sample. As shown in the left side photo, the crack was developed along a fine grain zone near the BM/HAZ interface. Micro voids in front of the crack tip were grown due to creep damage accumulated under high temperature and stress conditions. The cracks were initiated by linking and incorporating with the micro voids, and were then propagated along the voids.

## 4. Conclusions

To evaluate the CCG behavior of the SMAW weldment of Gr. 91 steel, a series of creep and creep crack growth tests was carried out under different applied loads at 600 °C. Creep crack growth laws for the BM, WM, and HAZ were constructed using the empirical equation of $da/dt$ vs. the $C^*$ parameter and compared. Results showed that in the CCGRs, the BM and HAZ were similar, but the BM was lower than WM and HAZ. The WM and HAZ were found to be about 2.0 times faster in the CCGR than the BM. This reason was closely related to the higher creep strain rate in the WM and HAZ. It was identified that lower creep strength existed in the HAZ of welded joints. Thus, in welded joints of Gr. 91 steel, creep cracks will dominantly initiate in the HAZ, because the creep life (or strength) in the HAZ was significantly decreased compared with the BM.

## References

[1] Choudhary BK, Bhanu Sankara Rao K, Mannan SL. Steady state creep deformation behavior of 9Cr–1Mo ferritic steel forging in quenched and tempered condition. Transactions of the Indian Institute of Metals 1999;52:327–36.
[2] Kim WG, Yin SN, Koo GH. Generation of isochronous stress–strain curves with a nonlinear least square fitting method for modified 9Cr–1Mo steel. Metals and Materials Internation 2009;15:727–32.
[3] Kim WG, Yin SN, Ryu WS, Kim SJ, Yi W. Probabilistic analysis of the creep crack growth rate of type 316LN stainless steel by the Monte Carlo simulation. Journal of ASTM International 2006;3:71–80.
[4] Nikbin KM, Smith DJ, Webster GA. An engineering approach to the prediction of creep crack growth. Transactions of the ASME 1986;108:186–91.

[5] Kim WG, Park JY, Hong SD, Kim SJ. Probabilistic assessment of creep crack growth rate for Gr. 91 steel. Nuclear Engineering and Design 2011;241:3580–6.
[6] Kim WG, Kim HH, Yoon KB, Ryu WS. Application of creep ductility model for evaluating creep crack growth rate of type 316SS series. Materials Science Forum 2005;475–479:1433–6.
[7] Takahasi Y, Igari T, Kawashima F, Date S, Isobe N, Itoh T, et al. High-temperature crack growth behavior of high-chromium steels. SMiRT 18, SMiRT18-G11-7 20051904–15.
[8] Kim WG, Park JY, Yin SN, Kim YW, Kim SJ. Creep crack growth rates on the base and welded metals of modified 9Cr–1Mo steel. Materials Science Forum 2010;654–656:524–7.
[9] Tabuchi M, Watanabe T, Kubo K, Matsui M, Kinugawa J, Abe F. Creep crack growth behavior in the HAZ of Weldments of W containing high Cr steel. International Journal of Pressure Vessels and Piping 2001;78:779–84.
[10] Sugiura R, Yokobori J, Tabuchi M, Yokobori T. Comparison of creep crack growth rate in heat affected zone of welded joint for 9%Cr ferritic heat resistance steel based on $C^*$, $d\delta/dt$, $K$ and $Q^*$ parameter. Engineering Fracture Mechanics 2007;74:868–81.
[11] Yamamoto M, Miura N, Ogata T. Applicability of $C^*$ parameter in assessing type-IV creep cracking in mod. 9Cr–1Mo steel welded joint. Engineering Fracture Mechanics 2010;77:3022–34.
[12] ASTM E139-07. Standard test method for conducting creep, creep-rupture and stress rupture tests of metallic materials 2007.
[13] ASTM E1457-07. Standard test method for measurement of creep crack growth rates in metals 2007.
[14] Dogan B, Petrovski B. Creep crack growth of high temperature weldments. International Journal of Pressure Vessels and Piping 2001;78:795–805.
[15] Saxena A. Nonlinear fracture mechanics for engineers. New York: CRC Press; 1997363–77.

Materials and Design 51 (2013) 1045–1051

Contents lists available at SciVerse ScienceDirect

Materials and Design

journal homepage: www.elsevier.com/locate/matdes

# Reliability assessment of creep rupture life for Gr. 91 steel ☆

Woo-Gon Kim [a,*], Jae-Young Park [b], Seon-Jin Kim [b], Jinsung Jang [a]

[a] Korea Atomic Energy Research Institute, 989-111 Daedeok-daero, Yuseong-gu, Daejeon 305-353, Republic of Korea
[b] Pukyong National University, 100 Yongdang-dong, Nam-gu, Busan 608-739, Republic of Korea

ARTICLE INFO

*Article history:*
Received 28 February 2013
Accepted 7 May 2013
Available online 18 May 2013

*Keywords:*
Creep
Creep rupture life
Creep life prediction
Reliability
Monte-Carlo simulation

ABSTRACT

This paper presents reliability assessment of the long-term creep life of Gr. 91 steel, which is a major structural material for high temperature structural components of Generation-IV reactor systems. A number of creep rupture data for Gr. 91 steel were collected through literature surveys, and the long-term creep life was predicted by Larson–Miller parameter. A "Z parameter" method was used to describe the magnitude of the deviation of the creep rupture data to a master curve. A "Service Condition-creep Rupture property Interference (SCRI) model" based on the Z parameter was used to simultaneously consider the scattering of the creep rupture data of materials and the fluctuations of service conditions in reliability assessment. A statistical analysis of the creep rupture data was conducted by the Z parameter. To carry out the SCRI model, a number of random variables for $Z_s$ describing service conditions and $Z_{cr}$ describing the dispersion of the creep rupture data were generated using a Monte-Carlo simulation technique. As examples for application, the creep rupture life under a certain service conditions of Gr. 91 steel was reasonably drawn from the viewpoints of reliability.



## 1. Introduction

High-chromium steel is widely used in current fossil power plants because of their excellent high-temperature strength [1–3]. In particular, modified 9Cr–1Mo steel (ASME Grade 9Cr–1Mo, hereafter Gr. 91) is regarded as a promising candidate for structural materials such as steam generators, intermediate heat exchangers (IHX), and primary pipes of a sodium-cooled fast reactor (SFR), which is a the Generation-IV nuclear reactor. The SFR system is designed for up to 60 years of use at an elevated temperature, as shown in Fig. 1, which is a concept drawing of a demonstration reactor developing at the Korea Atomic Energy Research Institute (KAERI) [1]. To design their components, the long-term creep life (or strength) should be reasonably assessed under service temperature and stress conditions. However, there are difficulties to actually obtain the long-term creep rupture data for more than $10^5$ h, because it is a time-consuming and expensive process. It is thus necessary to accurately predict the long-term creep life using short-term creep experimental data. In author's previous work, some attempts were made to accurately predict the long-term creep life for the nuclear materials such as type 316LN stainless steel, modified 9Cr–1Mo steel, and Alloy 617 [4–7].

In addition, so far, various methods for predicting the long-term creep life of components operating at high temperatures have been developed: the Larson–Miller (LM) parameter, Manson–Harped (MH) parameter, Dorn parameter, Monkman–Grant (MG) method, Theta projection method, and Omega method, etc [8–22]. Wilshire, Evans, and Brown et al. [8–10,13,21,22] introduced a theta ($\theta$) projection method, as an extrapolation method for long-term creep curves and creep lifetime behaviors. This model describes normal creep curves in terms of a decaying primary creep and accelerating tertiary creep equations expressed using four $\theta$ parameters. A lack of a physical meaning in the model has driven some studies to determine the relationship between the $\theta$ parameters and micro-mechanisms of creep, but these approaches have not been fully exploited. Also, Prager [15,16] proposed the omega ($\Omega$) method, which has been known to extend the concepts of the Larson–Miller plot, the Monkman–Grant (M–G) relationship, Robinson, $\theta$ projection method, Kachanov's damage mechanics, and Norton's creep law. Among them, the Larson–Miller (LM) method [11], known as a typical time-temperature parameter (TTP), has been most commonly used to predict long-term creep life from short-term creep data due to its simplicity and ease of application.

However, most of these methods have used deterministic methods, and their methods do not incorporate uncertainties in the parametric values affecting creep life, because the material rupture data exhibit some dispersion originally, and the use of limited data lowers their accuracy. Although a master curve for life prediction



* Corresponding author. Tel.: +82 042 868 249; fax: +82 042 868 8549.
*E-mail address:* wgkim@kaeri.re.kr (W.-G. Kim).

0261-3069/$ - see front matter © 2013 Elsevier Ltd. All rights reserved.
http://dx.doi.org/10.1016/j.matdes.2013.05.013

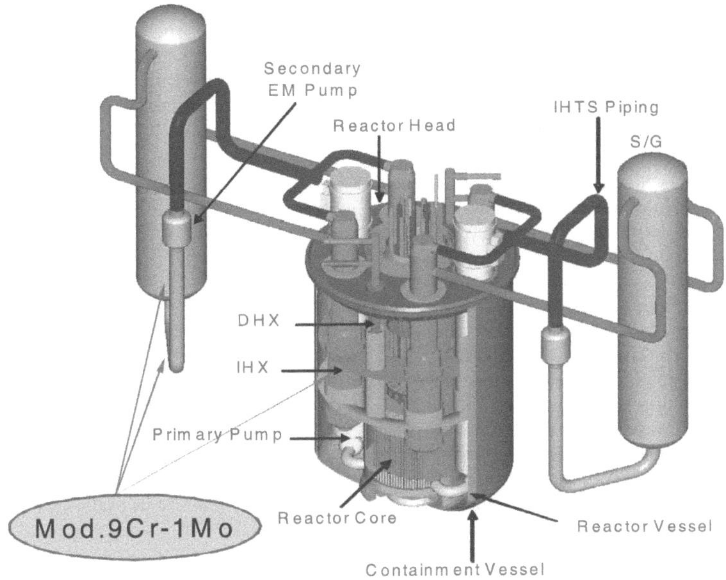


**Fig. 1.** SFR major components developing at KAERI.

can be well compressed into a curve or line, data scattering still exists and influences the predicted results. Furthermore, the actual service conditions would fluctuate around the designed criteria in the operating history, which led to some dispersion of creep life. Owing to these facts, deterministic methods do not have sufficient agreement for establishing the creep rupture life. Thus, a statistical approach will be used to improve the accuracy in the creep life prediction.

Under these circumstances, Zhao et al. [23] recently proposed a "Service Condition-creep Rupture property Interference (SCRI) model" based on Z parameter method to predict the reliability of creep rupture life for high-temperature components. In this model, they considered the dispersion of creep rupture property and the fluctuation of service temperature and stress conditions. As an example, HK40 steel has been used, and the reliability assessment has been successfully conducted under different temperature and stress fluctuations at very high temperature ranges of 871–900 °C. Using this model, they reported that it was possible to achieve unified analysis of creep rupture data in certain temperature and stress conditions for various heat-resistance steels. Therefore, it is required to verify an applicability of this model for Gr. 91 steel, which is considered as a prime structural material of Generation-IV reactor systems for use up to 60 years.

In this paper, a number of creep rupture data were collected and the creep life was predicted using the LM parameter. The scattering of the creep rupture data was analyzed on the basis of the Z parameter. A Monte-Carlo simulation was carried out to generate a number of random variables for $Z_s$ to describe of service conditions and $Z_{cr}$ to describe the dispersion of the creep rupture data. Quantitative results in the assessment of the creep rupture life were obtained and discussed from the viewpoints of reliability.

## 2. Concept of SCRI model

The two types of factors can affect the accuracy of creep life prediction: experiment factors and operating condition factors. The former includes dispersion in the material composition and structure, experimental temperature or stress error, and measuring the device precision. The latter includes a fluctuation of the operating conditions such as service temperature and service stress. In a fatigue life prediction, a stress-strength interference model has been used for a life evaluation by considering the strength data scattering and stress level fluctuation. Zhao et al. [23] introduced this idea for the prediction of creep rupture life. They applied to HK40 heat resistance steel, and reasonable results were drawn in the reliability assessment of creep rupture life. Here, the concept of the SCRI model is preliminarily analyzed to apply Gr. 91 steel.

Fig. 2 shows a schematic diagram of the SCRI model. In the model, $Z_{cr}$ is used to describe the rupture property portion, which represents the dispersion of creep rupture data, and $Z_s$ is used to describe the fluctuation of the service conditions. As the distribution of parameter Z follows a normal distribution, the probability density function of $Z_{cr}$ can be given by

$$f(Z_{cr}) = \frac{1}{s_{cr}\sqrt{2\pi}} \exp\left(-\frac{Z_{cr}^2}{2s_{cr}^2}\right), \quad (1)$$

where $s_{cr}$ is the standard deviation of the creep rupture data.

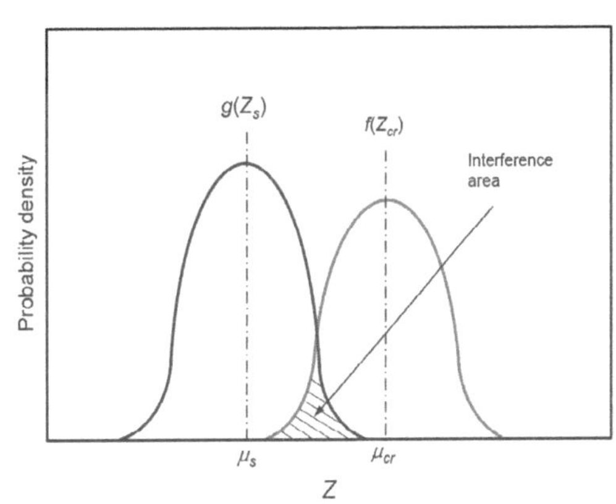


**Fig. 2.** A schematic diagram of the SCRI model.

In addition, $Z_s$ is used to describe the fluctuation of the service conditions, such as the service temperature, applied stress and service time. As $Z_{cr}$ and $Z_s$ are two independent random variables, the probability density function of the distribution of $Z_s$ in the corresponding service conditions can be written by

$$g(Z_s) = g(T, \Delta T, \sigma, \Delta\sigma, t_s), \quad (2)$$

where $T$ is the service temperature, $\sigma$ is the applied stress, $t_s$ is the service time, $\Delta T$ is the service temperature fluctuation and $\Delta\sigma$ is the applied stress fluctuation and $Z_s$ is the value of the $Z$ parameter corresponding to the fluctuation of the service conditions.

In the SCRI model, as shown in Fig. 2, the interference area corresponds to the fracture probability as a fracture state. If $Z_s > Z_{cr}$, it means that the creep rupture property of material cannot offer enough capability for service. Also, the material cannot be used any more. This partial probability becomes the failure probability of material corresponding to certain operating conditions and servicing time. A rupture failure can occur only in the interference area. Namely, the distribution of $Z_{cr}$ is always fixed as the experimental distribution of the creep rupture data. If the distribution of $Z_s$ is overlapped with the distribution $Z_{cr}$, it indicates that failure occurs in the service conditions, because the service conditions such as stress, temperature, and time are laid within the creep rupture values, as $Z_s > Z_{cr}$. Thus, the reliability, $R$ can be obtained by

$$R = 1 - F, \quad (3)$$

where $F$ is the cumulative failure probability, and $R$ is the probability without failure at a specific service condition.

## 3. Results and discussion

### 3.1. Statistical distribution of Z parameter

To assess the reliability of Gr. 91 steel, a number of the creep rupture data were collected through world-wide literature surveys for plate, pipe, forging and tube types, and the data was digitized. Owing to multiple sources, the creep properties may be different on the heat to heat variations. To avoid the differences of the creep properties, it is noted that the database was finally created from the reliable data within the ASME specification in the chemical compositions of Gr. 91 steel. A total of 690 data points was obtained for the temperature ranges of 500–700 °C, produced by National Institute Materials Science (NIMS) [10,24], Vallourec & Mannesmann (V&M) [25], and KAERI.

The plot of log stress vs. log time to rupture for the total data is shown in Fig. 3. Using these data, long-term creep life was predicted by the LM parameter, which was most commonly used in time–temperature parametric (TTP) methods.

The equation of the LM parameter ($P$) [9] is given by

$$P = T(\log t_r + C), \quad (4)$$

where $T$ is the absolute temperature, $t_r$ is the time to rupture, and $C$ is a material constant. $P$ is a synthetic parameter of temperature and time to rupture. In the LM parameter, the master curve can be fitted by a polynomial expression which is expressed as a stress function, $f(\sigma)$. A master curve in the LM parameter was obtained, as shown in Fig. 4. In the plot of $P$ vs. log stress, the master curve was proper for a cubic equation. In the LM parameter, the $C = 38$ was determined as an optimum value. This value was applied for the creep life prediction of Gr. 91 steel.

In addition, as noted above, the creep rupture data are generally scattered because of the influence of some uncertain factors such as the inhomogeneity of the material and experiment errors. The scattering can be defined as the $Z$ parameter describing the magnitude of data deviation from the master curve. It is obtained by the following equation:

$$Z_i = P_i - a_1(\log\sigma_i) - a_2(\log\sigma_i)^2 - a_3(\log\sigma_i)^3 - Z_o \quad (5)$$

where $Z_i$ corresponds to the deviation from the master curve at $i$ data ($T_i$, $\sigma_i$, $t_{ri}$). The distribution of the $Z$ parameter can represent the dispersion of creep rupture property for certain materials. The $Z$ parameter for Gr. 91 steel followed a normal bell type distribution, as shown in Fig. 5. Also, the normal probability of the $Z$ parameter can be plotted, as shown in Fig. 6. It is thus obvious that the distribution of the $Z$ parameter supports a normal distribution well, because the distribution showed a good linearity. In the normal distribution, a mean value ($\mu$) was obtained for 3.2422E–5, and a standard deviation ($s$) was obtained for 0.33625. Therefore, the reliability of the creep rupture life for Gr. 91 steel can be achieved using a normal distribution of the $Z$ parameter.

### 3.2. Generation of random variables by Monte-Carlo simulation

If the distribution under certain service conditions is determined, the reliability $R$ can be obtained by Eq. (3). $Z_{cr}$ representing the scattering of rupture property of the material can be calculated using Eq. (1) because the distribution of $Z_{cr}$ is generally supported by a normal distribution. However, it is difficult to confirm the

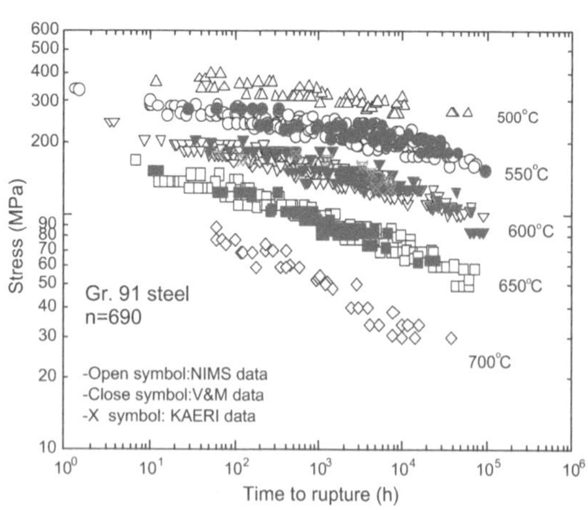


**Fig. 3.** Creep rupture data collected for Gr. 91 steel.

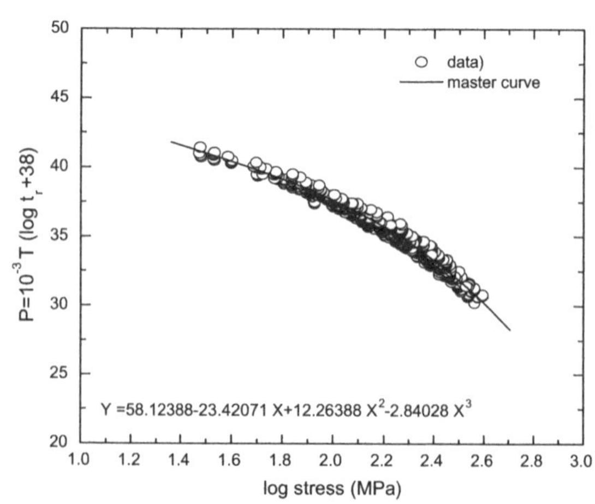


**Fig. 4.** A master curve obtained by LMP for Gr. 91 steel.

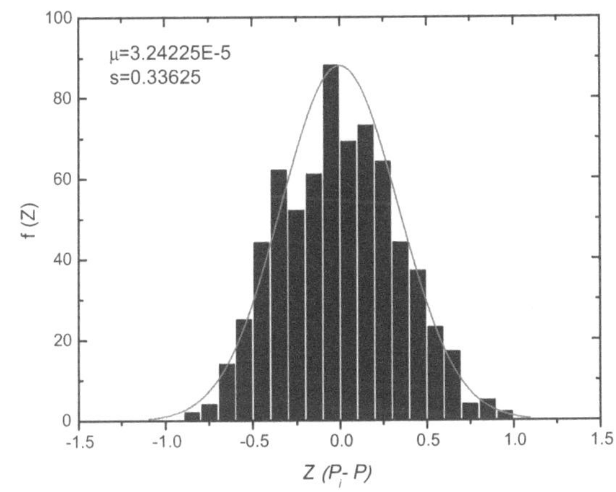


**Fig. 5.** Distribution of the *Z*-parameter for Gr. 91 steel.

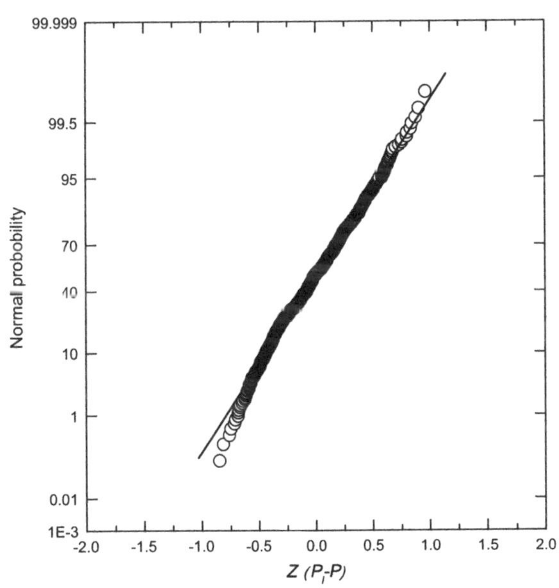


**Fig. 6.** Normal probability of Gr. 91 steel.

distribution of $Z_s$ using an analytic method, because there are at least two factors of operating temperature and stress that fluctuate during the service time. Thus, the Monte-Carlo simulation technique can be used to evaluate the safety reliability in a SCRI model.

Fig. 7 shows a flowchart of the Monte-Carlo simulation to evaluate the safety reliability in the SCRI model. A number of random variables are generated using the Monte-Carlo simulation from the normal distribution of the *Z* parameter. For a set of service conditions ($T_i$, $\sigma_i$, $t_r$), the value of $Z_s$ can then be calculated using Eq. (5). Meanwhile, the values of $Z_{cr}$ are generated randomly in a normal distribution or computed using Eq. (1). If $Z_s < Z_{cr}$, it means the material is in a safety state. If $Z_s > Z_{cr}$, it means a failure has occurred under this service condition. For total number of simulation, *N*, the safety reliability can be obtained by

$$R = \frac{n}{N} \tag{6}$$

where *n* is the number of safety states where $Z_s < Z_{cr}$ in total number *N* of simulation. Here, the random variables of 30,000 ea for $Z_s$ and

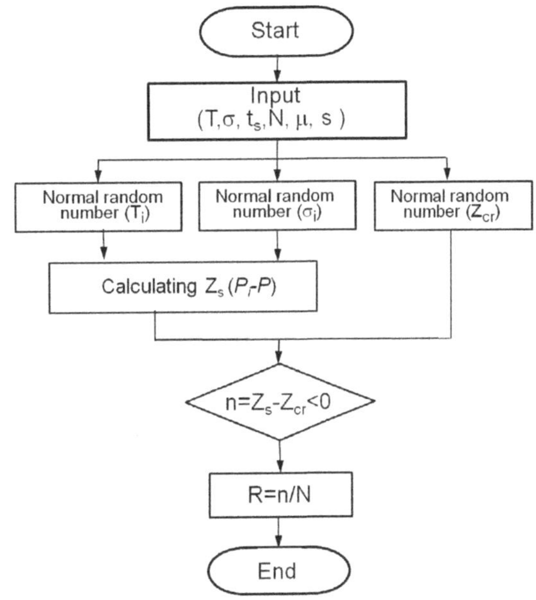


**Fig. 7.** Flowchart of Monte-Carlo simulation for a reliability analysis.

$Z_{cr}$ were generated under certain conditions with a fluctuation of service temperatures and stresses, and its safety reliability was evaluated.

### *3.3. Reliability results under specific service conditions*

As described in detail above, if using the interference model of the SCRI, it is anticipated that a reliability assessment of the creep rupture life in a set of service condition can be applied to Gr. 91 steel. Herein, as some examples for application, the assessment of the creep rupture life under specific service conditions is conducted and obtained from the viewpoints of reliability.

According to common sense, temperature and stress are regarded as the most important factors influencing the residual life of high temperature materials. Namely, since an increase in service temperature and stress reduces the reliability of the specific service time, the reliability for a service time should be established. Thus, on the basis of the Z parameter, the master curve of creep rupture data for Gr. 91 steel can be obtained by

$$Z = 10^{-3}T(\log t_r + 38) + 23.420(\log \sigma) - 12.263(\log \sigma)^2 - 2.840(\log \sigma)^3 - 58.123 \tag{7}$$

Also, the probability density function (pdf) can be obtained by Eq. (1), where scr = 0.3326.

In addition, for a safe reliability assessment, the service temperature and stress values can be freely chosen. In this study, for a living example, a service temperature choose *T* = 550 °C, because the SFR system developing at KAERI is designed at around 550 °C. Also, a service stress choose $\sigma$ = 150 MPa to assess the reliability prediction for up to around $10^5$ h. The results of the reliability assessments under specific service conditions for Gr. 91 are given in detail as follows.

Fig. 8 shows the results of a reliability change on the temperature influences with the service time variations, when the temperature varies from 545 °C to 560 °C under a constant applied stress of 150 MPa. In each case, the reliability was observed to decrease an increase in service time. A higher temperature reveals a faster

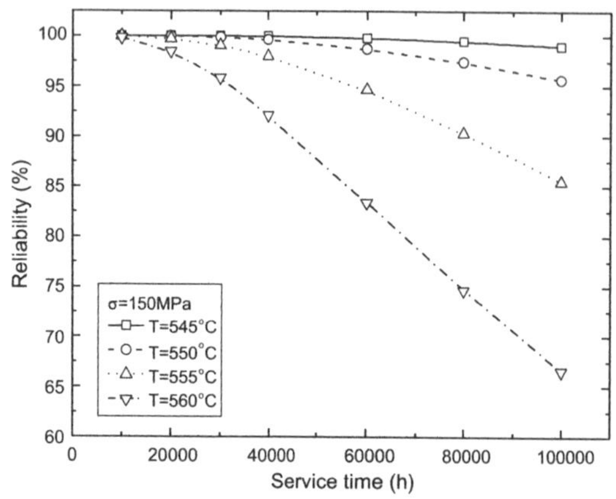


Fig. 8. The influence of temperature on the reliability of residual life ($\sigma$ = 150 MPa).

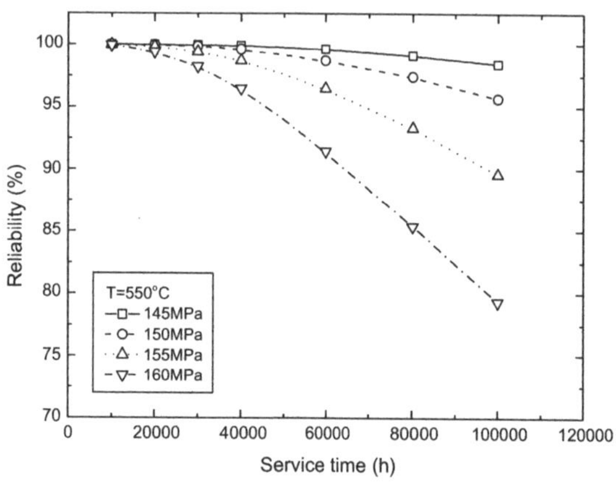


Fig. 9. The influence of stress on the reliability of residual life ($T$ = 550 °C).

decrease. Also, Fig. 9 shows the results of a reliability change on the stress influenced with the service time variations, when the stress varies from 145 MPa to 169 MPa under a constant temperature of 550 °C. This trend is similar to the influence of temperature. With increase in stresses, the decrease was accelerated. During the operation of power plants, the service temperature and stress are varied steadily against the controlling parameter because it is difficult to maintain them at constant values. Furthermore, power-on and off processes may generate a great change in temperature and stress. Thus, some examples of a reliability analysis were carried out under fluctuations of service temperature and stress, and the results were given as follows.

Figs. 10 and 11 reveal the reliability change on the temperature fluctuation of $\Delta T$ = 10 °C and $\Delta T$ = 30 °C (under $T$ = 550 °C, $\sigma$ = 150 MPa, and service time $t = 10^5$ h) in the interference region using SCRI model, respectively. The two figures show well that the interference area was large at a high fluctuation temperature of $\Delta T$ = 30 °C when compared with the fluctuation temperature of $\Delta T$ = 10 °C.

Figs. 12 and 13 reveal the reliability change in the temperature and stress fluctuations under $T$ = 550 °C and $\sigma$ = 150 MPa. In Fig. 12, it is clear that the reliability decreases rapidly at a higher temperature fluctuation amplitude. Also, in Fig. 12, similarly, the reliability decreases with an increase in the stress fluctuation amplitude. As shown in the two examples, it is verified that both

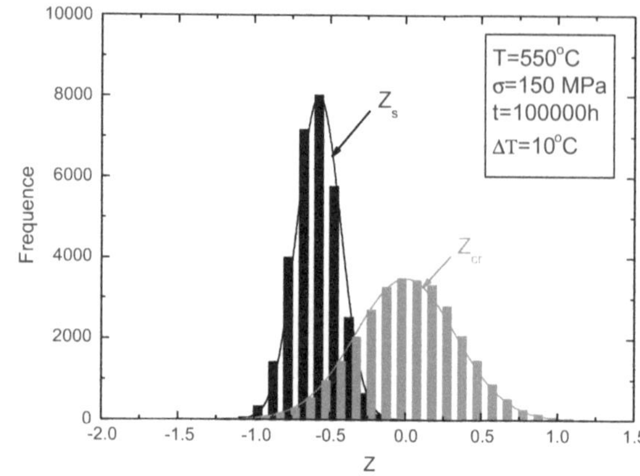


Fig. 10. The effect of temperature fluctuation on the area of interference region by the SCRI model ($\Delta T$ = 10 °C).

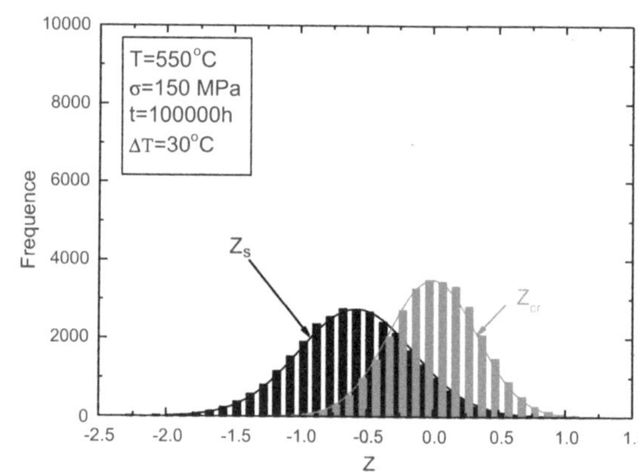


Fig. 11. The effect of temperature fluctuation on the area of interference region by the SCRI model ($\Delta T$ = 30 °C).

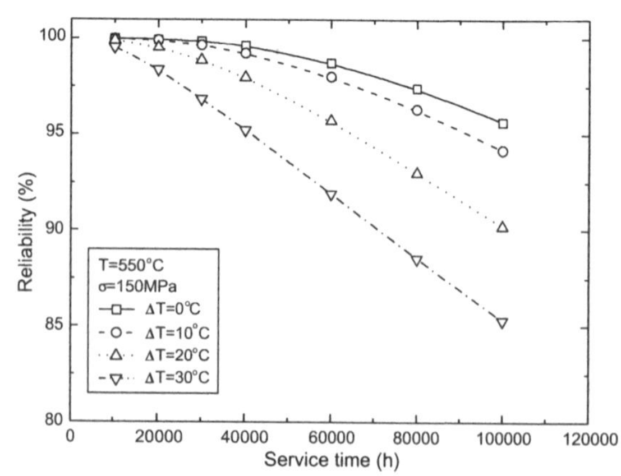


Fig. 12. The influence of temperature fluctuations on the reliability of residual life ($T$ = 550 °C, $\sigma$ = 150 MPa).

temperature and stress fluctuations were closely related to the trend of the reliability change.

Fig. 14 shows the probability density distribution corresponding to the change in a standard deviation ($s$) under $\Delta T$ = 20 °C at

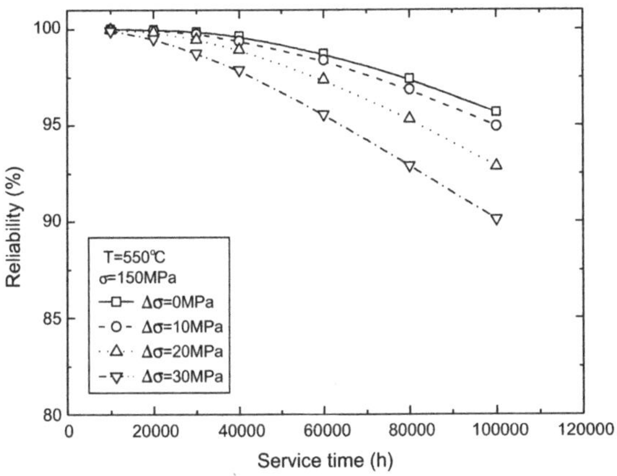


**Fig. 13.** The influence of stress fluctuations on the reliability of residual life ($T$ = 550 °C, $\sigma$ = 150 MPa).

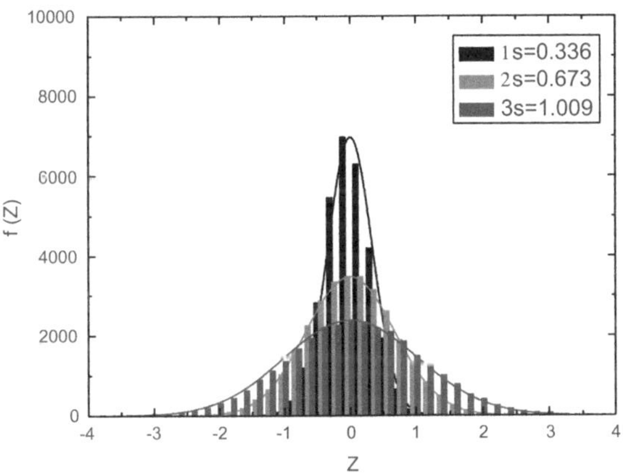


**Fig. 14.** The influence of standard deviation of the $Z$-parameter for the probability density ($\Delta T$ = 20 °C at $T$ = 550 °C, $\Delta\sigma$ = 20 MPa at $\sigma$ = 150 MPa).

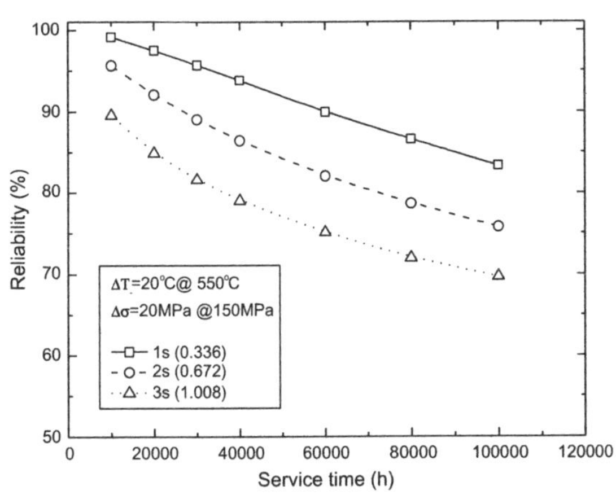


**Fig. 15.** The influence of standard deviation of the $Z$-parameter for a reliability prediction ($\Delta T$ = 20 °C at $T$ = 550 °C, $\Delta\sigma$ = 20 MPa at $\sigma$ = 150 MPa).

$T$ = 550 °C and $\Delta\sigma$ = 20 MPa at $\sigma$ = 150 MPa. In the SCRI model, the standard deviation ($s$) is considered as a factor effecting to the reliability because the creep rupture data are dispersed originally. Thus, herein, the dispersion of creep rupture property was involved in the reliability analysis in the SCRI model. Also, Fig. 15 shows the influence of the standard deviations, 1$s$, 2$s$, and 3$s$ on the reliability under $\Delta T$ = 20 °C at $T$ = 550 °C and $\Delta\sigma$ = 20 MPa at $\sigma$ = 150 MPa. It is shown in the two figures that as the values of the standard deviations became largely 3$s$ from 1$s$, the dispersion of the $Z$ parameter was widely changed in the probability density plot, and the results of the reliability also decreased significantly.

## 4. Conclusions

A reliability assessment of the creep rupture life of Gr. 91 steel was carried out using the SCRI model based on the $Z$ parameter. A number of creep rupture data were collected through literature surveys, and using these data, the long-term creep life was predicted by the LM parameter. The scattering of the rupture data based on the $Z$ parameter followed a normal distribution well. From the normal distribution, a number of random variables for $Z_s$ and $Z_{cr}$ were generated using a Monte-Carlo simulation technique. As examples for application of Gr. 91 steel, the results for the creep rupture life under specific service conditions were reasonably drawn from the viewpoints of reliability. It was found that the reliability of the creep rupture life decreased with increase in service time, and a higher temperature and stress conditions caused a faster deterioration. Also it decreased rapidly at the higher temperature/stress fluctuations and the large scattering of the creep rupture data. It is considered that the SCRI model can be utilized to various heat resistance steels for a reliability assessment of the creep rupture life.

## Acknowledgments

This work was performed under the Nuclear R&D Program sponsored by the Korean Ministry of Education, Science and Technology.

## References

[1] Kim WG, Park JY, Yin SN, Kim DW, Park JY, Kim SJ. Reliability prediction of long-term creep strength of Gr. 91 steel for next generation reactor structure materials. Kor J Met Mater 2011;29:275–80.
[2] Spigarelli S, Quadrini E. Analysis of the creep behavior of modified P91 (9Cr–1Mo–NbV) welds. Mater Des 2002;23:547–52.
[3] Samantaray D, Mandal S, Bhaduri AK. Constitutive analysis to predict high-temperature flow stress in modified 9Cr–1Mo (P91) steel. Mater Des 2010;31:981–4.
[4] Kim WG, Yin SY, Kim SH, Ryu WS, Lee CB, Kim SJ. Numerical approach to determine creep constants for time-temperature parametric methods. Met Mater Int 2009;15:559–64.
[5] Kim WG, Yin SN, Koo GH. Generation of isochronous stress–strain curves with a nonlinear least square fitting method for modified 9Cr–1Mo steel. Met Mater Int 2009;15:727–32.
[6] Kim WG, Park JY, Kim SJ, Hong SD, Kim YW. Reliability evaluation on creep life prediction of alloy 617 for a very high temperature reactor. Kor J Met Mater 2012;50:721–8.
[7] Kim WG, Yin SN, Park JY, Hong SD, Kim YW. An improved methodology for determining tensile design strength of alloy 617. J Mech Sci Technol 2012;26:379–87.
[8] Wilshire B, Scharning PJ. A new methodology for analysis of creep and creep fracture data for 9–12% chromium steels. Int Mater Rev 2008;53:91–104.
[9] Wilshire B, Scharning PJ. Prediction of long term creep data for forging 1Cr–1Mo-0.25V steel. Mater Sci Technol 2008;24:1–9.
[10] Wilshire B, Bache MR. Cost effective prediction of creep design data for power plant steels. In: Shibli IA, Holdsworth SR, editors. Creep & fracture in high temperature components-2nd ECCC creep conference. Zurich, Switzerland: DES Tech Publication Inc; 2009. p. 44–55.
[11] Larson FR, Miller J. A time-dependence relationship for rupture and creep stresses. Trans ASME 1952;74:765–75.
[12] Yeom JT, Kim JY, Na YS, Park NK. Creep strain and creep-life prediction for alloy 718 using the Omega method. Met Mater Int 2003;9:555–60.
[13] Evans RW, Parker JD, Wilshire B. The theta projection concept – a model-based approach to design and life extension of engineering plant. Int J Pres Ves Piping 1992;50:147–60.

[14] Park KS, Bae DS, Lee SK, Lee GH, Kim JH, Endo T. Creep modeling for life evaluation and strengthening mechanism of tungsten alloyed 9–12% Cr steels. Met Mater Int 2006;12:385–91.
[15] Prager M. Development of MPC Omega method for life assessment in the creep range. The Materials Properties Council Inc; 1994.
[16] Prager M. The Omega method-an engineering approach to life assessment. J Pres Ves Technol-Trans ASME 2000;122:273–80.
[17] Kim WG, Yin SY, Kim YW, Chang JH. Creep characterization of Ni-based Hastelloy-X alloy by using theta projection method. Eng Fract Mech 2008;75:4985–95.
[18] Maruyama K, Oikawa H. Comments on exponential descriptions of normal creep curves by SGR Brown, RW Evans and B Wilshire. Scripta Metall 1987;21:233–7.
[19] Wolf H, Mathews MD, Mannan SL. Prediction of creep parameters of type 316 stainless steel under service conditions using the θ-projection concept. Mater Sci Eng A 1992;159:199–204.
[20] Koul AK, Castillo R. A critical assessment of θ-projection concept for creep life prediction of Nickel-based superalloy components. Mater Sci Eng A 1991;138:213–9.
[21] Brown SGR, Evans RW, Wilshire BA. Comparison of extrapolation techniques for long-term creep strain and creep life prediction based on equations designed to represent creep curve shape. Int J Pres Ves Piping 1986;24:251–68.
[22] Brown SGR, Evans RW, Wilshire BA. Exponential descriptions of normal creep curves. Scripta Metall 1986;20:855–60.
[23] Zhao J, Li D, Zhang J, Feng W, Fang Y. Introduction of SCRI model for creep rupture life assessment. Int J Pres Ves Piping 2009;86:599–603.
[24] Kimura K, Toda Y, Kushima H, Sawada K. Creep strength of high chromium steel with ferrite matrix. In: Shibli IA, Holdsworth SR, editors. Creep & fracture in high temperature components-2nd ECCC creep conference. Zurich, Switzerland: DES Tech Publication Inc; 2009. p. 935–49.
[25] Haarmann K, Vaillant JC, Vandenberghe B, Bendick W, Arbab A. The T91/P91 book. 2nd ed. Vallourec & Mannesmann Tubes; 2002.

Springer

KSME

Journal of Mechanical Science and Technology 27 (8) (2013) 2331~2340
www.springerlink.com/content/1738-494x
DOI 10.1007/s12206-013-0616-7

# Comparative study on the high-temperature tensile and creep properties of Alloy 617 base and weld metals †

Woo-Gon Kim[1,*], Jae-Young Park[2], I. M. W. Ekaputra[2], Sung-Deok Hong[1], Seon-Jin Kim[2] and Yong-Wan Kim[1]

[1]*Korea Atomic Energy Research Institute, 989-111 Daeduk-daero, Yuseong-gu, Daejeon 305-353, Korea*
[2]*Dept. of Mechanical Design Engineering, Pukyong National University, Busan, 608-739, Korea*

(Manuscript Received February 12, 2013; Revised March 22, 2013; Accepted April 1, 2013)

**Abstract**

This paper presents a comparative investigation on the high-temperature tensile and creep properties of Alloy 617 base metal (BM) and weld metal (WM) fabricated by a gas tungsten arc weld process. The WM had higher yield strength and lower ultimate tensile strength than the BM does; however, its elongation was significantly lower than that of the BM. The creep curve of the BM and WM was somewhat different from that of typical heat-resistance steel, and did not show a textbook creep. The WM exhibited a longer creep rupture life, lower creep rate, and lower rupture ductility than the BM. However, as the creep rupture time reached approximately 36,800 h, the creep life of the WM was expected to be almost similar to that of the BM; and after 36,800 h, its creep life was expected to be worse than the BM. Loner creep tests is needed to investigate the long-term creep life of the WM. The creep failure mode of the BM and WM was obviously an intergranular cracking of the cavity formation and growth mechanisms, although it was more evident in the WM. The BM had a more ductile fracture surface than the WM.

*Keywords*: Alloy 617; Tensile; Creep; Weld metal; GTAW

## 1. Introduction

The very high temperature reactor (VHTR) is one of the most promising Generation-IV reactor types for the economic production of electricity and hydrogen. Its major components are the reactor internals, reactor pressure vessel, piping, hot gas ducts, and intermediate heat exchangers (IHX), as shown in Fig. 1 [1-4]. The IHX is a key component, and Alloy 617 is a prime candidate material compared with other potential superalloys due to its superior creep resistance above 800°C [5-7].

At present, the Korea Atomic Energy Research Institute is developing a nuclear hydrogen development and demonstration plan with a capacity of 200 $MW_{th}$ with thermal and core outlet temperature of 950°C. The plan has been carried out on a government-funded project called "development of key technologies for nuclear hydrogen" since 2006 [8-10]. Because the components are designed for a life span of 40~60 years operation at 950°C and 3~8 MPa in helium impurities, one of the most important properties of the components is the creep behavior, especially for the welded structures.

Generally, because the weld metal (WM) may have original some defects, the creep strength can be worse than that of the base metal (BM). Thus, in the design rules of high-temperature components, the allowable stresses in the welds are restricted to the lower value of the following: (i) the allowable stress in the BM is based on 67% of the stress to rupture, 80% of the stress to cause onset of a tertiary strain, and 100% of the stress to cause a 1% total strain, and (ii) the product of $0.8\ \sigma_{min} \times R$, where $\sigma_{min}$ is the expected minimum stress of the BM and $R$ is the appropriate ratio of the creep rupture strength of the WM compared with that of the BM [11, 12]. The allowable deformation in the welds is also restricted to half of the deformation permitted for the BM, because the ductility of the welds at elevated temperatures is generally low. For the design, the data of the tensile and creep properties of Alloy 617 WM should be sufficiently provided, and a design code for Alloy 617 WM should be developed. However, the data for the WM have been very rare and limited until now, although the data of the BM are available in a draft Alloy 617 code case, which was suspended at the end of the 1980s because of a lack of support and interest [1].

This study comparatively investigated the high-temperature tensile and creep properties of Alloy 617 BM and WM, which were fabricated by a gas tungsten arc welding (GTAW) procedure. The BM and WM properties were obtained and com-

*Corresponding author. Tel.: +82 42 868 2493, Fax.: +82 42 68 8549
E-mail address: wgkim@kaeri.re.kr
† Recommended by Editor Jai Hak Park

Table 1. Chemical composition of Alloy 617 plate (wt. %).

| | | C | Ni | Fe | Si | Mn | Co | Cr | Ti | P | S | Mo | Al | B | Cu |
|---|---|---|---|---|---|---|---|---|---|---|---|---|---|---|---|
| ASTM spec. | Min | 0.05 | Bal. | - | - | - | 10.0 | 20.0 | - | - | - | 8.00 | 0.80 | - | - |
| | Max | 0.15 | Bal. | 3.00 | 0.50 | 0.50 | 15.0 | 24.0 | 0.60 | 0.015 | 0.015 | 10.0 | 1.50 | 0.006 | 0.50 |
| Haynes 617 | | 0.08 | 53.11 | 0.949 | 0.084 | 0.029 | 12.3 | 22.2 | 0.41 | 0.003 | < 0.002 | 9.5 | 1.06 | < 0.002 | 0.0268 |

Table 2. Chemical composition of filler metal (wt. %).

| | C | Ni | Fe | Si | Mn | Co | Cr | Ti | P | S | Mo | Al | Cu |
|---|---|---|---|---|---|---|---|---|---|---|---|---|---|
| AWS spec. | 0.05 | Rem | MAX | MAX | MAX | 10.0 | 20.0 | MAX | MAX | MAX | 8.00 | 0.80 | MAX |
| | 0.15 | Rem | 3.0 | 1.0 | 1.0 | 15.0 | 24.0 | 0.60 | 0.03 | 0.015 | 10.0 | 1.50 | 0.50 |
| KW-T617 filler metal | 0.07 | Bal | 0.33 | 0.3 | 0.3 | 11.29 | 22.5 | 0.42 | <0.003 | 0.001 | 8.8 | 1.1 | < 0.01 |

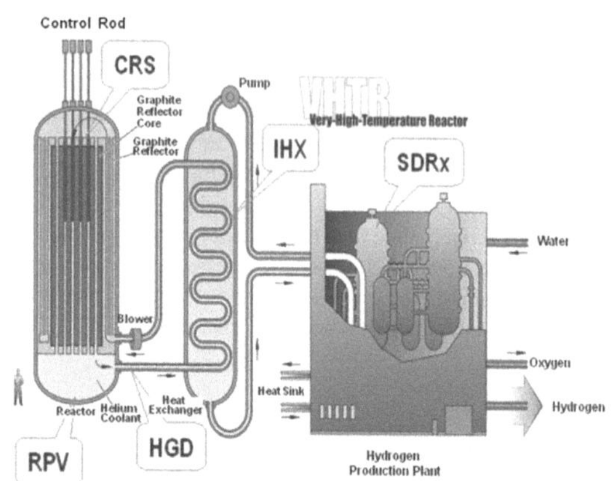


Fig. 1. Main components of the VHTR system for nuclear hydrogen production.

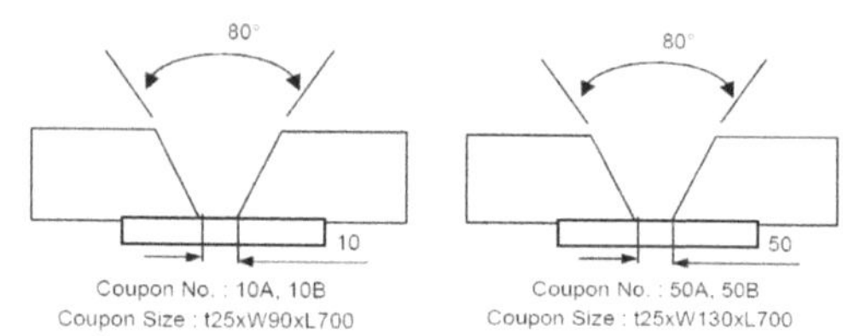


(a) 10 mm root gap (b) 50 mm root gap

Fig. 2. Single V-groove dimensions for fabricating the Alloy 617 weldments.

pared, and the fracture morphologies from the fracture specimens were also examined.

## 2. Experimental procedures

### *2.1 Welding method*

Commercial-grade nickel-based superalloy, Alloy 617 (Haynes 617) was used in this study, which was a hot-rolled plate with a thickness of 25 mm. The chemical composition is given in Table 1, and the amount of each element was well within the ASTM specifications. The shape of the weld joint (WJ) has a single V-groove with an angle of 80°.

A filler metal was used for KW-T617 (brand name), manufactured by KISWEL Company. It was prepared according to the American Welding Society (AWS) specifications, AWS A 5.14-05 ERNiCrCoMo-1 (UNS N06617), and its diameter was 2.4 mm. The chemical composition is given in Table 2. The amount of each element was well within the AWS specifications. To shield the welding pool from the atmosphere during welding, a 99.99% pure argon gas from 10 L/min to 15 L/min was supplied.

Two types of root gaps of 10 and 50 mm were fabricated. The 10 mm root gap specimens were prepared to evaluate the WJ, and the 50 mm root gap specimens were prepared to evaluate the WM. Here, they are designed as 10A and 10B for the 10 mm root gap and 50A and 50B for the 50 mm root gap, as shown in Fig. 2. A backing gas to prevent oxidation to the back side of the WM was not supplied. The GTAW conditions are given as follows:

- Polarity: DC straight polarity.
- Current range: from 180 A to 200 A.
- Voltage range: from 14 V to16 V.
- Welding speed: from 18 cm/min to 22 cm/min.
- Pre-heating temperature: minimum of 18°C.
- Temperature between passes: under 177°C.

In the GTAW process, 14 welding layers and 41 passes were applied for the 10A and 10B specimens, and 15 layers and 133 passes were applied for the 50A and 50B specimens. To prevent a bending deformation of the weldments, some passes at the back side were added for each specimen, as shown in Figs. 3(a) and 3(b). Post heat-treatment was not conducted because the Ni-based superalloy was not normally applied. After welding, the soundness of the weldments was investigated by ultrasonic, tensile, and bending tests.

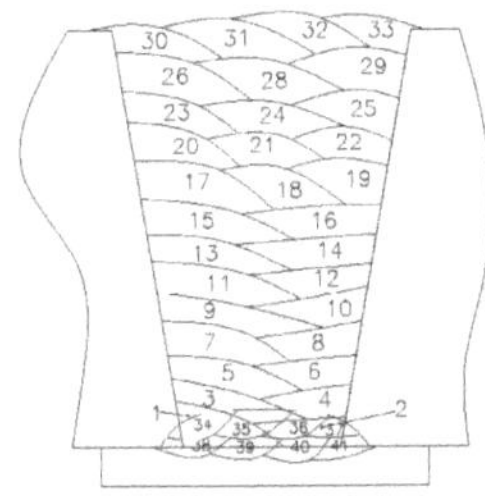

(a) 10A and 10B weldments (10 mm root gap)

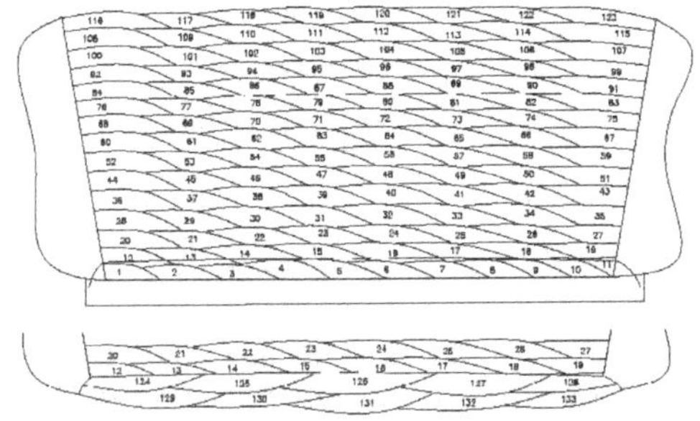

(b) 50A/50B wledments (50 mm root gap)

Fig. 3. Profiles of the welding layers and pass number for 10A/10B and 50A/50B weldments.

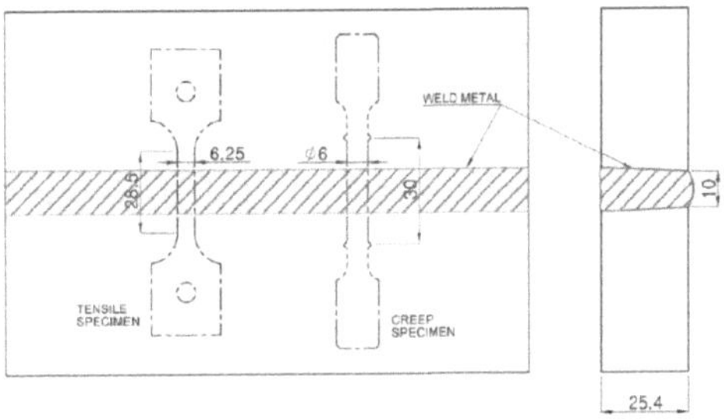


(a) WJ

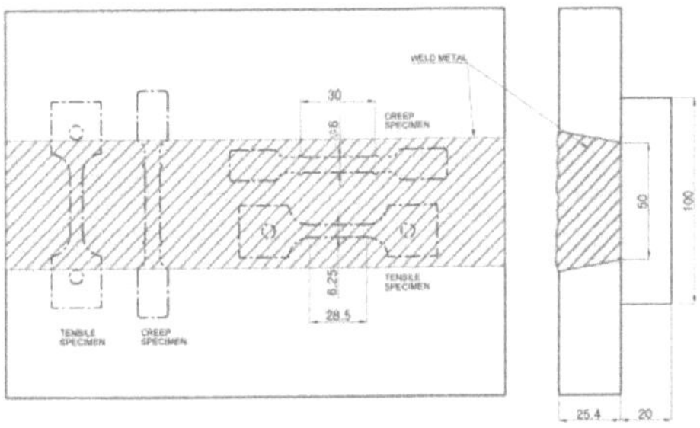

(b) WM

Fig. 4. Cutting method of the tensile and creep test specimens for the Alloy 617 weldments.

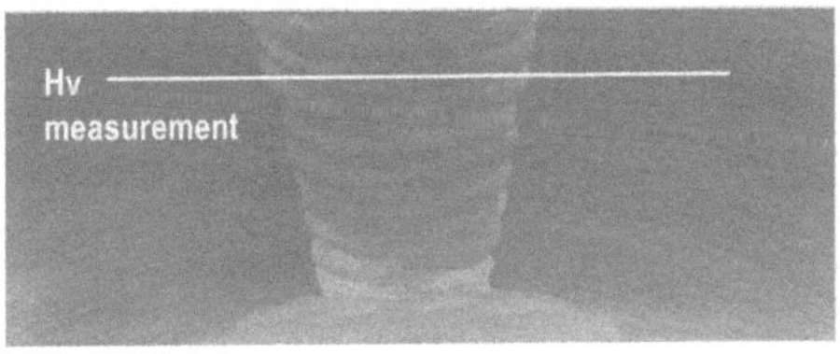


Fig. 5. OM structure of the Alloy 617 weldments.

### 2.2 Test method

Figs. 4(a) and 4(b) show the cutting methods used in the WJ and in the WM to prepare the tensile and creep test specimens. In the WJ specimens, the WM and WJ regions were cut within the gage length for inclusion. These specimens were prepared to evaluate the joint strength in the WJ region. In the WM specimens, the tensile and creep specimens were taken from fully welded metal. The specimens were machined into the transverse direction (TD) and longitudinal direction (LD) against the welding direction, and the tensile properties in the two directions were compared.

The tensile test specimens were machined into a rectangular cross section with a gage length of 28.5 mm, width of 6.25 mm, and thickness of 1.5 mm. The strain rate in the tensile tests was conducted with a slow strain rate of $5.85 \times 10^{-4}$ 1/s at temperature ranges from room temperature to 950°C. Further, the creep test specimens were machined into a cylindrical form with a gage length of 30 mm and diameter of 6 mm.

A series of creep tests was performed under different stress levels of 35, 30, 25, 22, 20, and 18 MPa for the BM and 35, 30, 24, and 20 MPa for the WM at an identical temperature of 950°C. The pull rod and jig used for the creep tests were manufactured using Ni-based superalloy materials to endure sufficiently the oxidation and thermal degradation during the creep tests. Creep strain data with elapsed times were taken automatically by a PC through a high-precision extensometer. The steady-state creep rate was measured from the experimental creep curves.

A photograph of the Alloy 617 weld block taken with a 10 mm root gap after welding is shown in Fig. 5. The profiles of the multi layers can be observed. A Vickers micro-hardness survey was performed across the Alloy 617 weldment, as indicated by the white straight line in the photo.

After testing, to observe the optical microscope (OM) and scanning electron microscopy (SEM) micrographs, an electro etching was conducted for 12~15 s using two voltages in a mixed solution of $H_2O$ (90%) + HCl (10%).

## 3. Results and discussion

### 3.1 Tensile properties

Figs. 6(a)-6(c) show the optical microstructure of the BM, WM, and heat affected zone (HAZ), respectively. In the photograph of the BM structure, the grain boundary (GB) is uniform and stable; however, in the WM structure, the GB is meanderingly non-uniform as a cell-type-dendrite solidification structure. This weld structure appeared to have been formed from the solidification when the welding material is

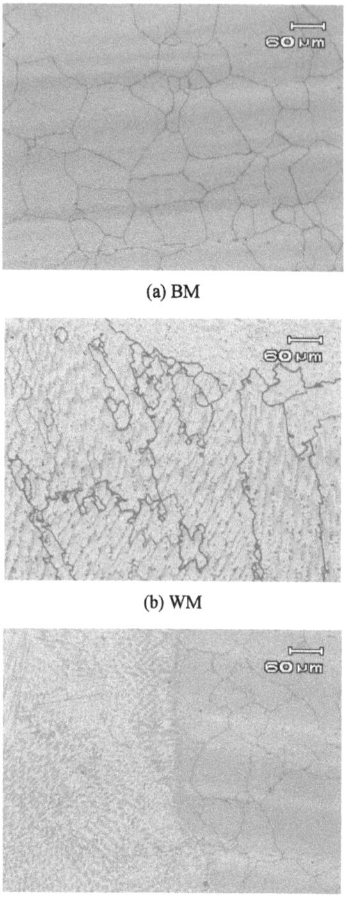

(a) BM

(b) WM

(c ) HAZ

Fig. 6. Optical microstructure showing the solidified grain boundaries of the Alloy 617 weldment.

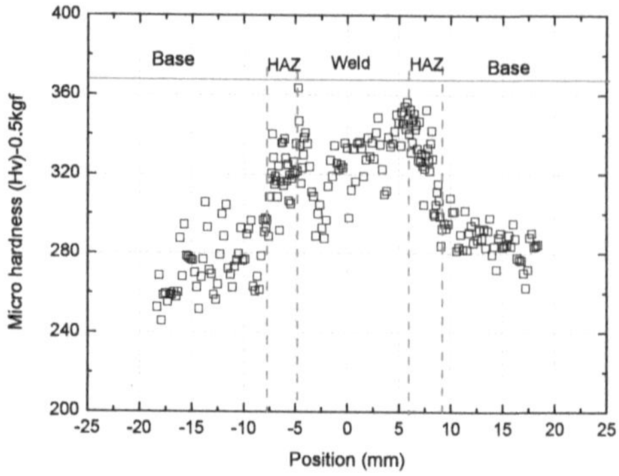


Fig. 7. Micro-hardness variation (Hv) across the WJ.

solidified between the solid and liquid interfaces. Then, in the HAZ structure, a fusion line between the BM and WM is observed. The WM solidification appeared to have been started from this fusion line, and its crystallization was an epitaxial growth that grew directly along the same directions as that in the coarse GBs in the HAZs. The HAZ is believed to be different in terms of the mechanical properties compared with the BM, because of the number of thermal cycles in the multi-layer welding process used in this study.

Fig. 7 shows the results of the micro-hardness measurements of the Alloy 617 weldment, with the dashed lines delineating the approximate location of the HAZs on each side of the WM. The BM hardness was approximately from 280 to 290, rising to approximately 350 in the HAZ; it was approximately from 320 to 340 in the WM. The WM had a higher value (approximately from 40 to 50) than the BM. The hardness in the HAZ sharply changed in the transition region between the BM and WM. The HAZ width was very narrow at approximately from 3 mm to 5 mm, and the grains coarsened.

Fig. 8 shows the results of the high-temperature tensile property measurements. Figs. 8(a)-8(c) shows 0.2% yield stress (YS), ultimate tensile strength (UTS), and tensile failure elongation for the BM, WJ, and WM (LD and TD) of the Alloy 617 weldments, respectively. The YS value of the BM was lower than that of the WM; however, at above 800°C, the YS values of the BM, WJ, and WM were almost the same. A large difference in the stress values between the TD and LD of the WM appeared. Meanwhile, the UTS value of the BM was higher than those of the WM and WJ. At above 800°C, the UTS values of the BM, WJ, and WM were almost the same. In terms of the tensile elongation, the WM ductility dropped significantly compared with that of the BM. That of the WJ was the lowest.

Fig. 9 shows the typical SEM fracture surfaces of the BM and WM at room temperature (RT), 400, 600, 800, and 950°C. At RT, the WM surface exhibited clearly an equal-axial dendrite structure, which mainly solidified along the GBs. It was formed at the final cooling phase of the solidification of the WM. Some differences existed in the fractures between the BM and WM. The WM showed a significantly cleave failure surface, but the BM showed a more ductile fracture than the WM. In addition, at 950°C, dimples were significantly observed in the BM compared with the WM.

### 3.2 Creep rupture properties

To understand the creep behavior of BM and WM of Alloy 617, the creep properties such as the creep rupture time, creep strain rate (steady-state creep rate), rupture elongation, and reduction in area were investigated through a series of creep tests at an identical temperature of 950°C. Long-time rupture data that reached approximately 14,000 h (1.6 years) for the BM and 10,172 h (1.16 years) for the WM were experimentally obtained. After the creep testing, the fracture micrographs of each tested specimen were examined using OM and

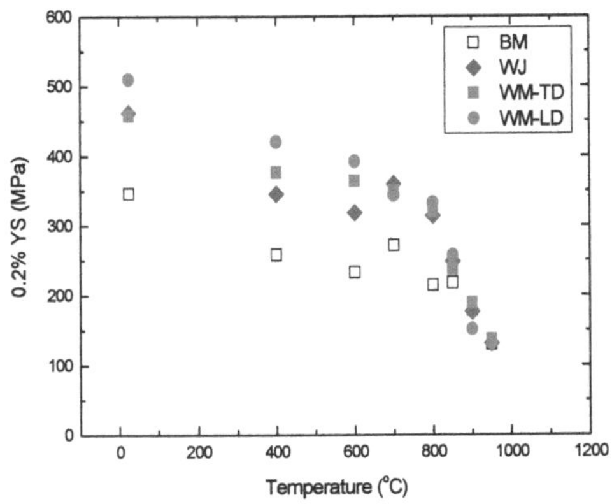


(a) 0.2% YS

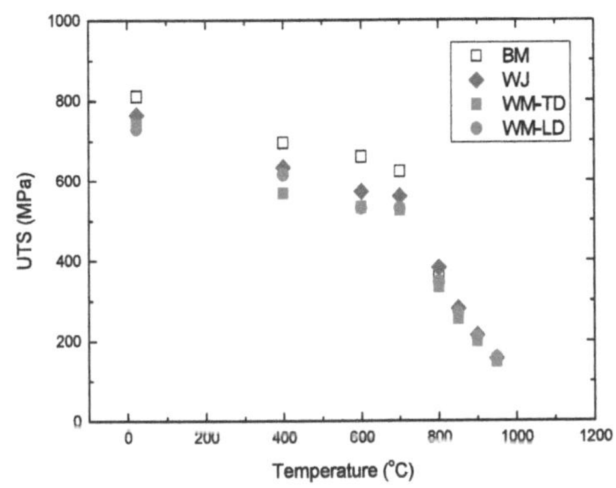


(b) UTS

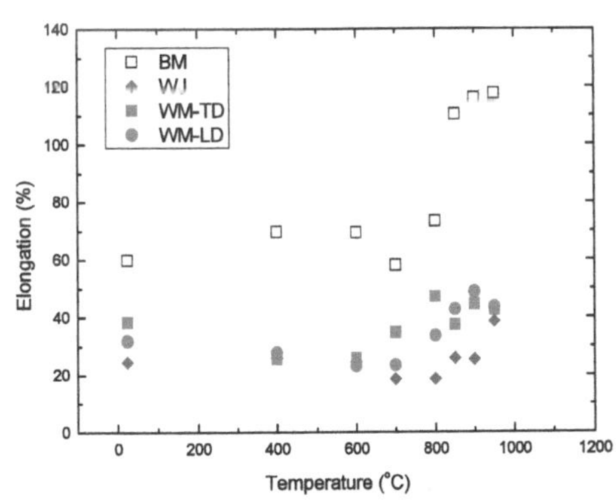


(c) Elongation

Fig. 8. High-temperature tensile properties of the Alloy 617 weldments.

SEM. In this study, the creep test results for the BM and WM were comparatively presented in detail.

Figs. 10 and 11 show the comparisons of the creep curves of the BM and WM of Alloy 617 tested at 950°C. Fig. 10 shows that of the 30 MPa and Fig. 11 shows that of the 20 MPa. The shape of the creep curves for the BM and WM was not different in the primary, secondary, and tertiary creep re-

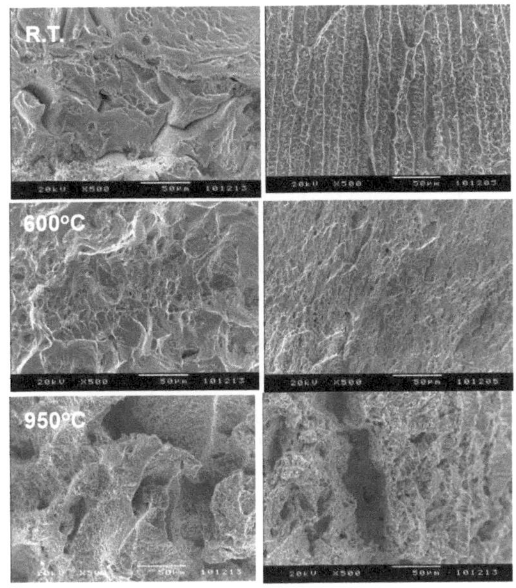


(a) BM (b) WM

Fig. 9. SEM micrographs of the fracture surfaces at tensile test temperatures for the BM and WM.

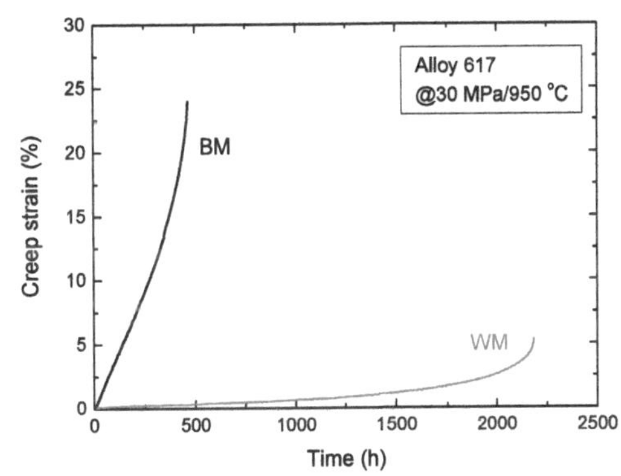


Fig. 10. Comparison of the creep curves of the BM and WM of Alloy 617 tested under 30 MPa at 950°C.

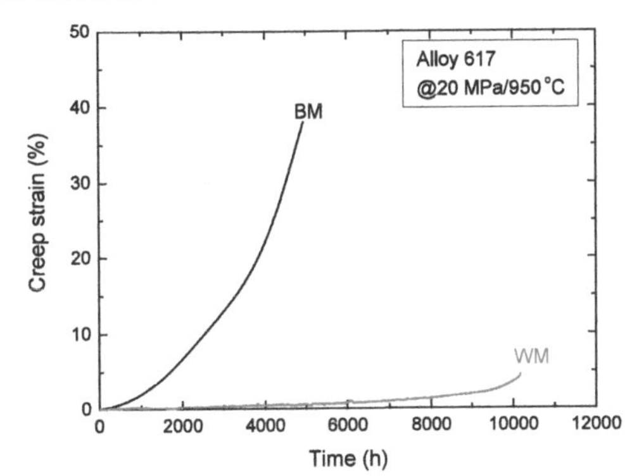


Fig. 11. Comparison of the creep curves of the BM and WM of Alloy 617 tested under 20 MPa at 950°C.

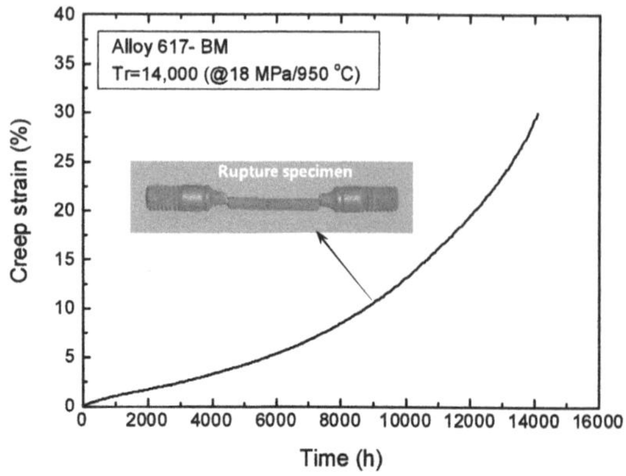


Fig. 12. Typical creep curve of the BM of Alloy 617 tested under 18 MPa at 950°C [$t_r$ = 14,000 h (1.6 years)].

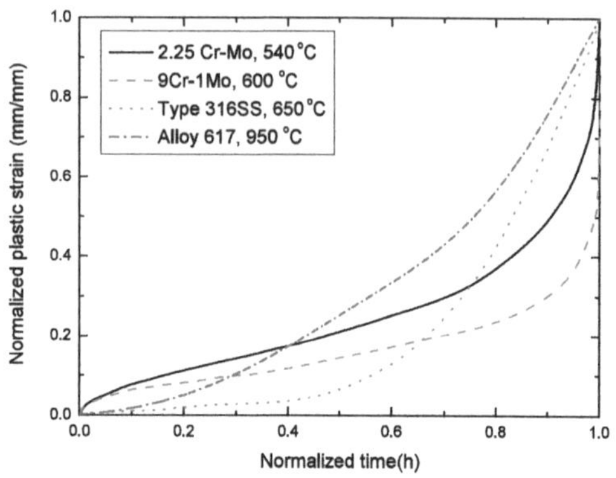


Fig. 13. Comparison of the normalized creep curves for typical heat resistant steels.

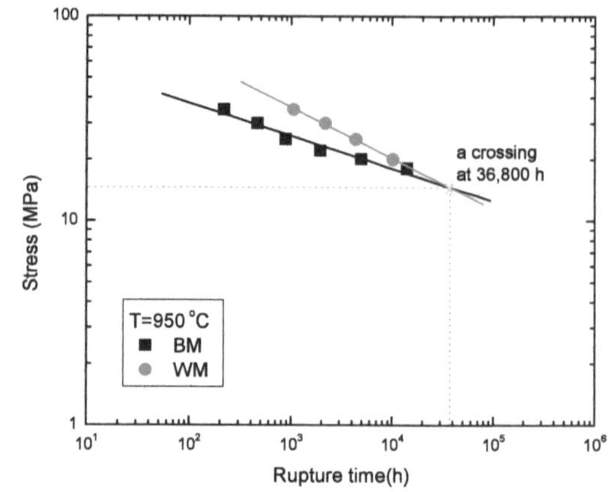


Fig. 14. Plot of log stress versus log rupture time obtained for the BM and WM at 950°C.

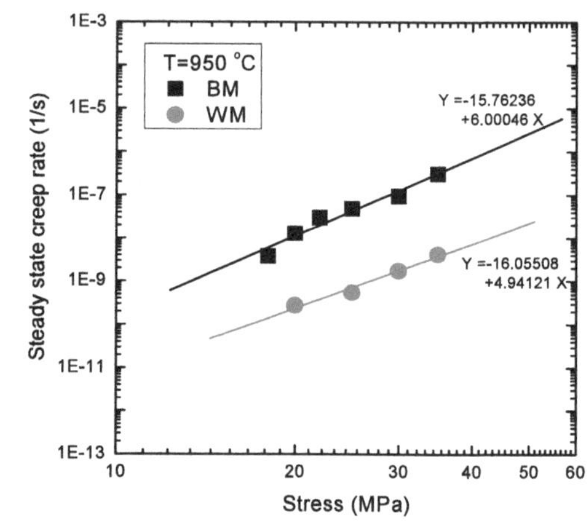


Fig. 15. Plot of the steady-state creep rate versus applied stress for the BM and WM tested at 950°C.

gimes. Compared with the BM, the WM had a longer creep rupture time, but had a lower creep rupture elongation. Creep deformation appeared to develop dominantly from the cavity formation and growth in the creep process, as shown in the ruptured specimen in Fig. 12. The cavities were uniformly distributed over the total gage section of the specimen, and they occurred because of the accumulated damage induced during the creep process under the high-temperature and stress conditions. Cracks were initiated by linking and incorporation with the micro voids, and then developed along the voids. Finally failure occurred because of the interconnections of the voids rather than the necking failure in the tertiary creep region. Owing to the results, it is identified that the Alloy 617 WM has a small reduction in its area.

The creep curve of Alloy 617 appears somewhat different from that of typical heat-resistant steels, such as 2.25Cr-1Mo steel, 9Cr-1Mo steel, and austenitic stainless steel (type 316LN), as shown in Fig. 13. The creep curves of the low and high-Cr ferritic steels plot are distinct primary, secondary, and tertiary creep stages (called "textbook creep curve"). However, Alloy 617 does not exhibit a textbook creep curve, as shown in Figs. 12 and 13, but a predominantly accelerated creep rate from the start of the tertiary creep stage in short time ranges.

Fig. 14 shows the plot of the log stress versus log rupture time obtained for the BM and WM at 950°C. The WM had a higher creep strength (or longer creep rupture life) than the BM. However, as the creep life increased, the two fitting lines between the BM and WM crossed at 36,800 h. After 36,800 h, we estimate that the creep life of the WM is shorter than that of the BM. We thus believe that longer creep tests will be needed to investigate the long-term creep life of the WM.

Fig. 15 shows the comparative plot of the steady-state creep rate as a function of the applied stress for the BM and WM tested at 950°C. The steady-state creep rate $\dot{\varepsilon}_s$ is described by the Norton's power-law function of applied stress as

$$\dot{\varepsilon}_s = A\sigma^n \quad (1)$$

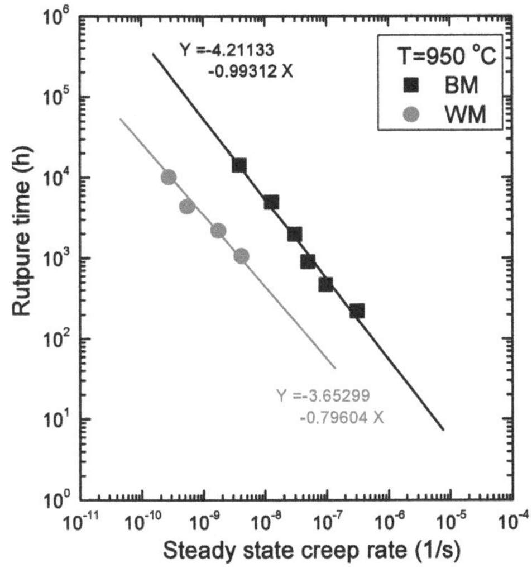


Fig. 16. Plot of the M-G relationship for the BM and WM at 950°C.

where σ is the applied stress, and $n$ is the stress exponent. The value of $A$ and $n$ can be obtained from Fig. 15. A small different slope for the two lines was determined from the data sets obtained at different applied stresses for the BM and WM. The $A$ and $n$ values were determined as $4.82 \times 10^{-20}$ ($MPa^{-n}\ h^{-1}$) and 6.0 for the BM, respectively, and $2.77 \times 10^{-20}$ ($MPa^{-n}\ h^{-1}$) and 4.9 for the WM, respectively. However, these values indicate that the stress exponent for the creep was almost the same, and the same creep mechanism is regarded as operating in both the BM and WM tested at 950°C in this study. Further, the value of n = 4.9 to 6.0 obtained for the BM and WM was reasonable because it was similar to the value of n = 3 to 6 reported for single phase materials [13]. Quantitatively, the WM was 1.75 times lower in the creep strain rate than the BM, although the rupture data were taken for only four tests in this study. It is believed that the longer rupture life of the WM was due to the lower creep rate.

Fig. 16 shows a plot of a Monkman-Grant (M-G) relationship of the log-rupture time versus log-creep strain rate for the BM and WM of Alloy 617. The M-G relationship can be expressed by [14, 15]

$$\log t_r + m \log \dot{\varepsilon}_s = K \tag{2}$$

where $t_r$ is the creep rupture life, and $m$ and $K$ are the material constants. Eq. (2) shows that the creep rupture life is in inverse proportion to the creep strain rate. If the constants for the materials are known, we can predict the other from one. The M-G function can be reasonably used when the relationship follows well the fact that cavity formation and growth are on the basis of the creep process [14].

As described above, because Alloy 617 revealed creep deformation of the cavity formation and growth in the creep process, it is regarded to be fairly adaptable to the BM and WM of Alloy 617. A small different slope for the two lines was determined for the BM and WM. The $m$ and $K$ values were 0.99 and 4.21 for the BM, and 0.80 and 3.65 for the WM. The WM line was shifted to the left side with one order of the creep strain rate.

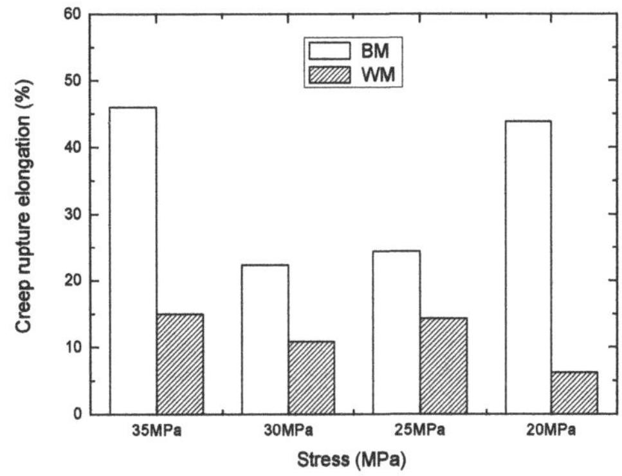


(a) Creep rupture elongation (%)

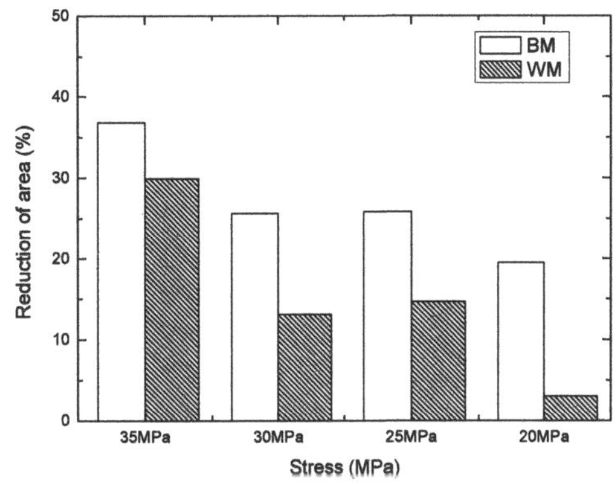


(b) Reduction of area (%)

Fig. 17. Comparison of creep rupture elongation and reduction in area for the BM and WM tested at 950°C.

Fig. 17 shows a comparison of the creep rupture elongation and reduction in the area for the BM and WM tested at 950°C. The WM of the Alloy 617 had a significant reduction in the rupture elongation and reduction in the area compared with the BM. This result is similar to the results in the heat-resistant steel in which the ductility of welds at elevated temperature was generally low. This result should be carefully treated because a large reduction in the rupture ductility of the WM of Alloy 617 may cause a brittle fracture.

Accordingly, during the current test period at the present test conditions, the WM of Alloy 617 revealed a higher creep life (or strength) owing to the lower creep strain rate. However, the creep behavior for a longer creep life of over $10^4$ h has not been found by other researchers until now. Further the longer rupture data are insufficient, and they are not available. We thus emphasize that a number of the longer creep tests will be needed to investigate the long-term creep life of the WM.

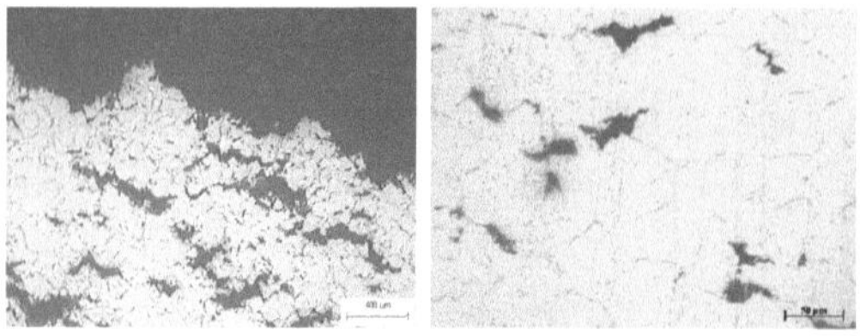

(a) Low magnification (b) High magnification

Fig. 18. Typical OM structures of the BM tested under 20 MPa at 950°C ($t_r$ = 4,943 h).

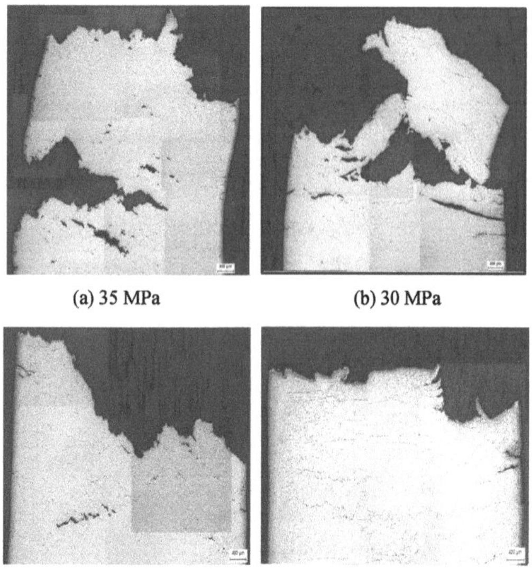

(a) 35 MPa (b) 30 MPa

(c) 25 MPa (d) 20 MPa

Fig. 19. Crept specimens with applied stresses for the WM.

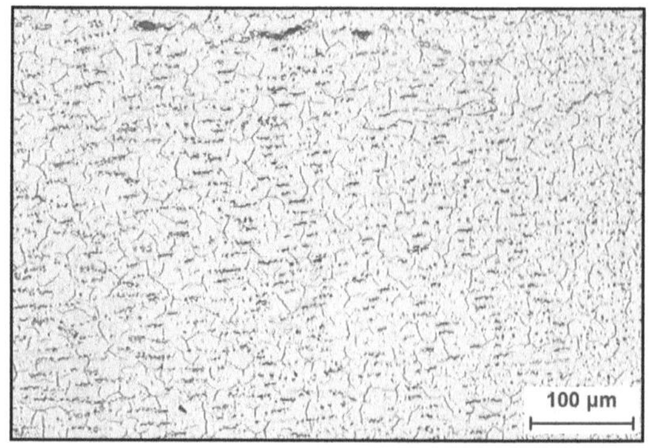


Fig. 20. Typical OM structure of the WM tested under 25 MPa at 950°C.

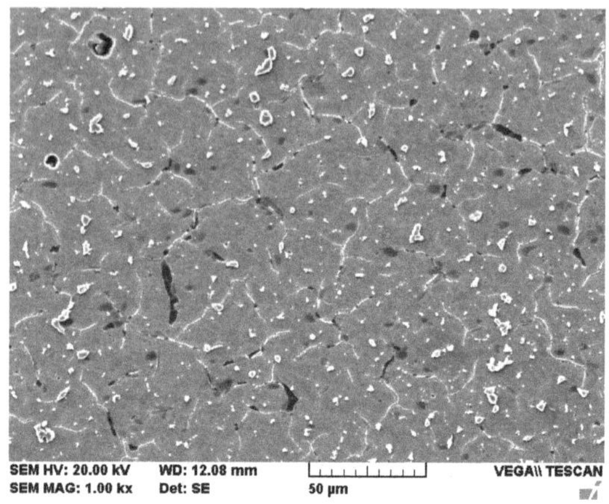


Fig. 21. SEM image showing the precipitates, cavities, and voids of the WM tested under 25 MPa at 950°C.

### 3.3 Fracture micrographs

Fig. 18 shows the OM structures of the BM crept under 20 MPa at 950°C. In the low (Fig. 18(a)) and high (Fig. 18(b)) magnifications, the cavities and cracks are formed along the grain boundary perpendicular to the applied stress and widely distributed in the overall gage section. A creep failure occurred from the interconnections of the cavities and cracks formed during the creep at the present test conditions. The failure was dominant in the intergranular fracture. In the WM, similar failure occurred along the cracks developed perpendicular to the applied stress, as shown in Fig. 19. Many minor cracks were widely distributed in the overall gage section. The failure obviously occurred because of the cracks formed by cavity formation and growth mechanism. At high stresses (Figs. 19(a) and (b)), the failures occurred irregularly on some large cracks. Otherwise, at low stresses (Figs. 19(c) and (d)), they smoothly occurred on one fracture plane, at a low stress of 20 MPa.

Fig. 20 shows a typical OM structure for the WM specimen crept at 25 MPa. In the matrix, many minor cavities and some creep cracks existed along the GBs. However, the image was not clear for the cracks and the precipitates formed in the matrix to be observed. To observe them more clearly, a SEM micrograph was taken.

Fig. 21 shows a typical SEM micrograph of the WM tested under 25 MPa at 950°C, which shows that the precipitates are formed along the GBs, and the cavities and voids are generated at the site of the GBs. We assumed that they played a role as nucleation site that caused cavities and voids during the creep process. We have identified that the precipitates were significantly distributed at a longer rupture time (a low-stress condition) compared with the short rupture time (a high stress condition). Generally, creep properties are closely related to the amount, shape, and size of the distribution of the precipitates.

Fig. 22 shows the SEM micrographs of the fracture surfaces for the BM and WM specimens tested at identical conditions of 20 MPa at 950°C. The low magnification is for the left - hand side, and the high magnification is for the right-hand side. The BM and WM showed combined fracture modes consisting of a dimple rupture and intergranular cracking. However,

(a) BM (20 MPa at 950°C)

(b) WM (20 MPa at 950°C)

Fig. 22. SEM fracture surfaces for the BM and WM tested under 20 MPa at 950°C.

the intergranular cracking is more evident in the WM specimens, which is consistent with the observation of lower creep ductility in the WM than that in the BM. The WM specimen shows a solidification fracture surface formed for solidification of the welds. The BM specimen shows a more ductile fracture surface than the WM does. In addition, oxides of $Cr_2O_3$ formed in the creep tests are observed in the fracture surfaces, as shown in the high magnification photographs for the BM and WM. The amount of oxides increased with the increase in the rupture time.

## 4. Conclusions

The high-temperature tensile and creep properties of Alloy 617 BM and WM, fabricated by the GTAW process, have been investigated by a series of tensile and creep tests, and the properties of the BM and WM were compared. Their fracture morphologies were also examined from the fracture specimens. The following conclusions were drawn:

The WM of Alloy 617 had a higher YS value than the BM, but had a lower UTS value. The WM was significantly reduced in terms of the tensile elongation compared with the BM. The creep curves of the BM and WM appeared to be somewhat different from those of typical heat-resistant steel, and they did not show a textbook creep behavior. The WM showed a longer rupture life, lower creep strain rate, and lower rupture ductility than the BM. The longer rupture life was due to the lower creep strain rate in the WM. As the creep rupture time approached 36,800 h, we estimated that the creep life of the WM was the same as that of the BM; however, after 36,800 h, the WM was worse than the BM. We therefore believe that longer creep tests will be needed to investigate the long-term creep life of the WM. In the micrograph observations, the BM and WM revealed a similar behavior in forming the cracks and cavities along the GB perpendicular to the applied stress, and the cracks and cavities were widely distributed in the overall gage section. The creep failure mode of the BM and WM was obviously an intergranular cracking of the cavity formation and growth mechanisms. However, intergranular cracking was more evident in the WM, which showed a lower rupture ductility than the BM. In addition, the BM revealed a more ductile fracture surface than the WM.

## Acknowledgment

This work was performed under the long-term Nuclear R & D Program sponsored by the Korean Ministry of Science and Technology.

## References

[1] W. G. Kim, S. N. Yin, J. Y. Park, S. D. Hong and Y. W. Kim, An improved methodology for determining tensile design strengths of Alloy 617, *Journal of Mechanical Science and Technology*, 26 (2012) 379-387 .

[2] W. G. Kim, S. N. Yin, Y. W. Kim and J. H. Chang, Creep properties and creep life prediction of Alloy 617 for a very high temperature gas-cooled reactor, *Proc. of ASME PVP 2008 Conf.*, Chicago, USA, July (2008) Paper PVP 2008-61356.

[3] S. J. Dewson and X. Li, Selection criteria for the high temperature reactor intermediate heat exchanger, *Proc. of ICAPP 05*, Seoul, Korea, May (2005) Paper 5333.

[4] W. G. Kim, S. N. Yin, W. Y. Kim and J. H. Chang, Creep characterization of a Ni-based Hastelloy-X alloy by using theta projection method, *Engineering Fracture Mechanics*, 75 (2008) 4985-4995.

[5] W. G. Kim, S. N. Yin, G. G. Lee, Y. W. Kim and S. J. Kim, Creep oxidation behavior and creep strength prediction for Alloy 617, *Int. J. of Pressure Vessels and Piping*, 87 (2010) 289-295.

[6] J. M. Corum and J. J. Blass, *Rules for design of alloy 617 nuclear components to very high temperatures*, Attachment 2, Draft Code Case for Alloy 617, USA (1988).

[7] W. G. Kim, S. N. Yin, W. S. Ryu and J. H. Chang, Analysis of the creep rupture data of Alloy 617 for a high temperature gas cooled reactor, *The 8th Int. Conf. on Creep and Fatigue at Elevated Temperatures*, Texas, USA, July (2007) Paper PVP 2007-26834.

[8] J. H. Chang, Y. W. Kim, K. Y. Lee, Y. W. Lee, W. J. Lee, J. M. Noh, M. W. Kim, H. S. Lim, Y. J. Shin, K. K. Bae and K. D. Jung, A study of a nuclear hydrogen production demonstration plant, *Nuclear Engineering and Technology*, 39 (2) (2007) 111-122.

[9] H. Y. Lee, Y. W. Kim and K. N. Song, Preliminary application of the draft code case for alloy 617 for a high temperature component, *J. of Mechanical Science and Technology*, 22 (2008) 856-863.

[10] J. Y. Park, D. W. Kim, D. J. Kim, E. S. Kim, H. P. Kim, S. H. Kim, W. J. Kim, W. G. Kim, W. S. Ryu and Y. D. Kim,

Development of materials for a high temperature gas cooled reactor in Korea, *The 4th International Topical Meeting on High Temperature Reactor Technology-HTR 2008*, Washington, USA (2008) Paper HTR 2008-58149.
[11] W. G. Kim and W. S. Ryu, Creep rupture properties of type 316LN stainless steel welded by the SAW process, *Solid State Phenomena,* 119 (2007) 91-97.
[12] M. D. Mathew, G. Sasikala, S. L. Mannan and P. Rodriguez, A comparative study of the creep rupture properties of type 316 stainless steel base and weld metals, *J. of engineering Materials and Technology*, 115 (1993) 163-170.
[13] H. E. Evans, *Mechanism of creep fracture*, Elsvier Applied Science Publishers LTD, London and New York, UK and USA (1984).
[14] W. G. Kim, S. H. Kim and W. S. Ryu, Evaluation of Monkman-Grant parameters for type 316LN and modified 9Cr-1Mo stainless steels, *J. of Mechanical Science and Technology,* 16 (11) (2002) 1420-1427.
[15] G. E. Deter and D. Bacon, *Mechanical metallurgy*, Third Ed. McGraw-Hill Book Company, London, UK (1988).

**Woo-Gon Kim** obtained the Ph.D. at the Department of Mechanical Engineering, Chungbuk National University in 1998. He is currently a Principal Researcher with Korea Atomic Energy Research Institute. His specialty is mechanical assessment, analysis, modeling, and database establishment of high-tempera- ture nuclear materials for Gen-IV reactor systems and time-dependent creep and crack growth behaviors at elevated temperature.

Springer

KSME

Journal of Mechanical Science and Technology 26 (2) (2012) 379~387
www.springerlink.com/content/1738-494x
DOI 10.1007/s12206-011-1024-5

# An improved methodology for determining tensile design strengths of Alloy 617 †

Woo-Gon Kim*, Song-Nan Yin, Ji-Yeon Park, Sung-Deok Hong and Yong-Wan Kim

*Korea Atomic Energy Research Institute, 989-111, Deadeok-daero Yuseong-gu, Daejeon, 305-353, Korea*

(Manuscript Received May 11, 2011; Revised September 28, 2011; Accepted October 5, 2011)

## Abstract

This paper presents an improved methodology for determining high-temperature tensile design strengths of Alloy 617, which is regarded as one of main structural materials for very high temperature reactor (VHTR) system. In establishing time-independent allowable stress values, an existing ASME standard procedure is preliminarily analyzed and their limitations are pointed out. Then, an improved methodology, which has a consistent and quantifiable design margin at low and high temperatures for tensile strengths, is proposed and compared with the ASME method. To find suitable curves of temperature trend to the tensile strength data, three fitting methods are demonstrated, and a statistical technique is adopted for design use. The results will be utilized to reasonably determine the tensile design strengths of Alloy 617 for application in the VHTR system.

*Keywords*: Yield strength; Ultimate tensile strength; Alloy 617; Very high temperature reactor

## 1. Introduction

The very high temperature reactor (VHTR) is one of the most promising Generation-IV reactor types for the economic production of electricity and hydrogen. Its major structures, such as the reactor internals, reactor pressure vessel (RPV), piping, hot gas ducts (HGD), and intermediate heat exchangers (IHX), are designed for a life span of 60 years operating at 950°C and 3~8 MPa in helium impurities, as shown in Fig. 1 [1-3]. The HGD and IHX are key components, and their materials have potential for use in nickel-based superalloys such as Alloy 617, Haynes 230, Hastelloy-X and Alloy 800H. Among these, Alloy 617 is a prime candidate material due to its superior creep resistance above 800°C compared with other potential superalloys [4-6]. In present, KAERI (Korea Atomic Energy Research Institute) has been developing the nuclear hydrogen development and demonstration (NHDD) plant, which has a capacity of 200MW$_{th}$ in thermal and core outlet temperature of 950°C. This project is being carried out on a government funded project called "Development of Key Technologies for Nuclear Hydrogen" since 2006 [7-9].

For application of Alloy 617 in VHTR systems, a great deal of yield strength (YS) and ultimate tensile strength (UTS) data is required in higher temperatures beyond and around 1000 K because their strength data are used to set the time-independent primary stress limits for various high-temperature

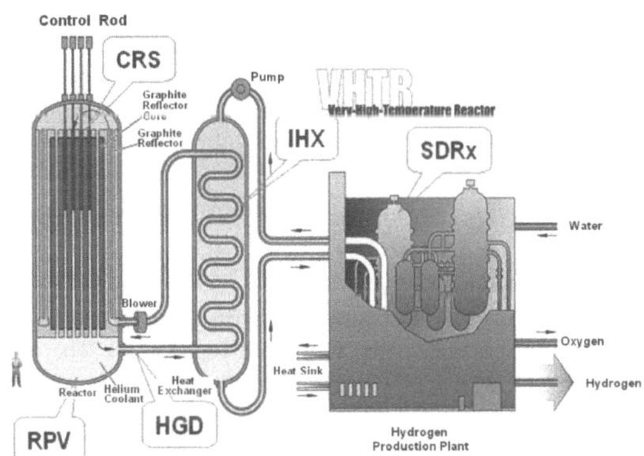


Fig. 1. Main components of VHTR system for nuclear hydrogen production.

design codes. Although the tensile strength data for Alloy 617 are available in the ASME draft code case [10], the data are still insufficient. The ASME code case is also incomplete for use at high temperatures beyond 950°C. Furthermore, the code development work for Alloy 617 was suspended at the end of the 1980s due to a lack of support and interest. Under these circumstances, Sham et al. [11] recently pointed out that the standard ASME method was not suitable for obtaining the design tensile strengths for the Ni-based Alloy 617 and Haynes 230 (Alloy 230), because the temperature trend curves for the tensile strengths did not allow a quantifiable design margin over a full range of low to high temperatures. Therefore, it is necessary to develop an improved method compared

† This paper was recommended for publication in revised form by Editor Jai Hak Park
* Corresponding author. Tel.: +82 42 868 2493, Fax.: +82 42 868 8549
E-mail address: wgkim@kaeri.re.kr

with the existing ASME standard method for the YS and UTS.

In this study, the ASME standard procedure was preliminarily analyzed, and then an improved methodology compared with the ASME procedure was proposed by using three fitting methods; single polynomials, 2-step exponentials, and 2-step polynomials. Using the statistical distributions for the YS and UTS data, the 95% lower bound curves for design use were determined over the full temperature ranges.

## 2. Review of ASME procedures

A procedure for establishing allowable stress for the high-temperature YS and UTS values is given in Appendix 2 of the ASME Boiler and Pressure Vessel Code, Section II, Part D [12]. The YS at a high temperature is based on minimum values, and the UTS at a high temperature based on average values. However, the minimum and average values for the YS and UTS are not interpreted statistically in the ASME code. The ASME procedures for determining the high-temperature tensile strengths for design use are given as follows.

For determining the minimum YS at high temperatures, the YS data for each heat are normalized by room temperature YS for that heat, as a ratio of YS at high temperatures to the room temperature strength. The best-fit trend curve, $R_Y(T)$, as a function of temperature, $T$, is then obtained using a least square fit. The trend curve is customarily expressed in the form of a polynomial. The required minimum YS at a temperature, $S_y(T)$, is defined as $S_Y R_Y(T)$ where $S_Y$ is the specification minimum YS at room temperature.

Also, the procedures for establishing the average UTS at high temperatures are similar to those of the minimum YS at high temperatures. The UTS data are normalized by the room temperature UTS, again as a ratio of UTS at high temperatures to room temperature strength. A best-fit trend curve, $R_T(T)$, is then obtained using a least square fit. Also, the trend curve is customarily expressed in the form of a polynomial. The required average UTS at high temperatures, $S_u(T)$, is set equal to $1.1 \times S_T R_T(T)$ where $S_T$ is the specification minimum UTS at room temperature. These procedures for determining the minimum YS and average UTS values at high temperatures are applied and analyzed to the tensile data of Alloy 617.

In addition, for the minimum YS and average UTS curves of the temperature trend according to the ASME Section II procedures, Sham et al. [11] proposed the following mathematical expressions. They used $\hat{S}_y(T)$ as the best-fit curve for the YS data. The vertical distance between the $\hat{S}_y(T)$ and $S_y(T)$ curves can be determined as,

$$\hat{S}_y(T)-S_y(T)=\hat{S}_y(T)-S_yR_y(T)\approx\hat{S}_y(T)-S_Y\frac{\hat{S}_y(T)}{\hat{S}_y(T_R)} \quad (1)$$

where $T_R$ is room temperature. An approximation was made in equating $R_Y(T)$ to the ratio $\hat{S}_y(T)/\hat{S}_y(T_R)$. After re-arranging Eq. (1), the vertical distance between $\hat{S}_y(T)$ and $S_y(T)$ becomes

$$\hat{S}_y(T)-S_y(T)\approx c_y\hat{S}_y(T),\ c_y\equiv\left(1-\frac{S_Y}{\hat{S}_y(T_R)}\right). \quad (2)$$

The constant $c_y$ is independent of temperature. Thus, the vertical distance between $\hat{S}_y(T)$ and $S_y(T)$ is proportional to $\hat{S}_y(T)$, which is the best-fit YS curve.

Also, similarly for the YS, the vertical distance between the best-fit UTS curve, $\hat{S}_u(T)$ can be analyzed. The curve for the average UTS at a temperature, $S_u(T)$ can be given as approximately

$$\hat{S}_u(T)-S_u(T)\approx c_u\hat{S}_u(T),\ c_u\equiv\left(1-\frac{1.1\times S_T}{\hat{S}_u(T_R)}\right). \quad (3)$$

The constant $c_u$ is also independent of temperature. The vertical distance between and $\hat{S}_u(T)$ and $S_u(T)$ is proportional to $\hat{S}_u(T)$, the best-fit UTS curve.

## 3. Result and discussion

### 3.1 Collection of YS and UTS data

Tensile strength data of Alloy 617 tested in air were collected through world-wide literature surveys, manufacturing companies, and a KAERI effort: Sham et al. [11], European Alloy-DB (EC-JRC) data [13], KAERI data [14], Inco Alloy [15], and Haynes Alloy [16]. Here, Sham's data were obtained by digitizing the document data, and the other data were obtained as numerical values. The total amount of data was 177 sets for the YS and UTS with four product forms such as a bar, plate, sheet, and forging.

Fig. 2 shows the plot of the YS and UTS values as a function of temperature for Alloy 617. The YS and UTS values decreased gradually with increasing temperatures. At higher temperature beyond and around 1000 K for the UTS and 1100 K for the YS, the temperature trends revealed more sharp decays in the strength values than in the lower temperatures. When the entire data for the YS and UTS were plotted together, there was a wide separation for the YS and UTS data in the low temperature ranges, and the scatter bands converged and started to overlap in the high temperature ranges, as shown in Fig. 3.

### 3.2 Analysis of ASME method

To estimate the minimum YS and average UTS at high temperatures for Alloy 617, ASME Section II procedures were applied to the tensile strength data. A commonly used single polynomial fit was adopted for the total data in the full temperature ranges. The best-fit trend curves to the normalized YS and UTS data were obtained as the third-order polynomial for the $S_y(T)$, and fourth-order polynomial for the $S_u(T)$, as follows:

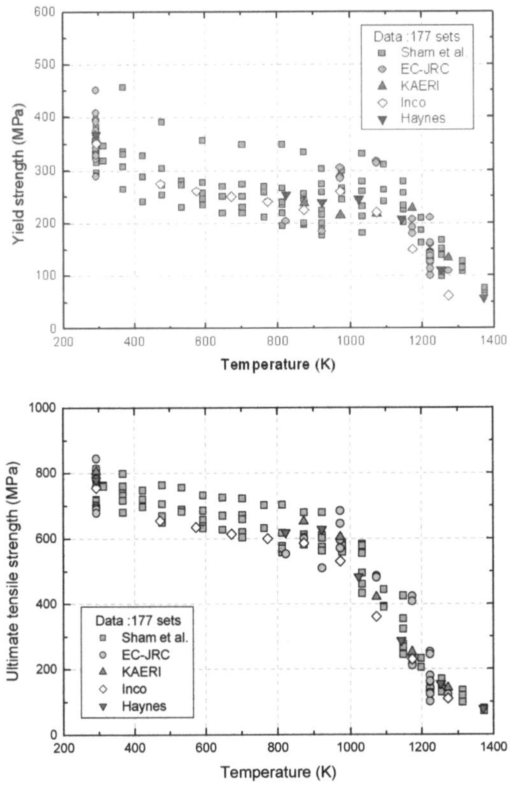


Fig. 2. YS and UTS data collected for Alloy 617.

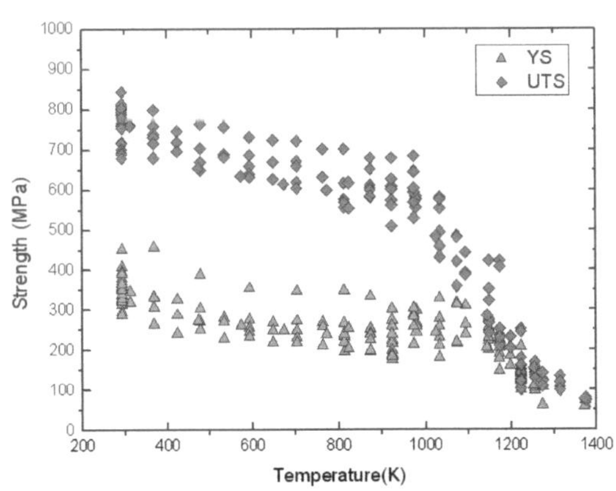


Fig. 3. Total plot of YS and UTS data for Alloy 617.

$$\hat{S}_y(T) = 1.927 - 0.00473\,T + 6.263\text{E}-6\,T^2 - 2.753\text{E-}9\,T^3 \quad (4)$$

$$\hat{S}_u(T) = 2.59 - 0.0103\,T + 2.168\text{E}-5\,T^2 - 1.875\text{E-}8\,T^3 + 5.408\,\text{E}-12\,T^4. \quad (5)$$

However, it is obvious that the best-fit curves obtained for this single polynomial fit were inherent fluctuations due to the third-and fourth-order polynomials over the full temperature ranges.

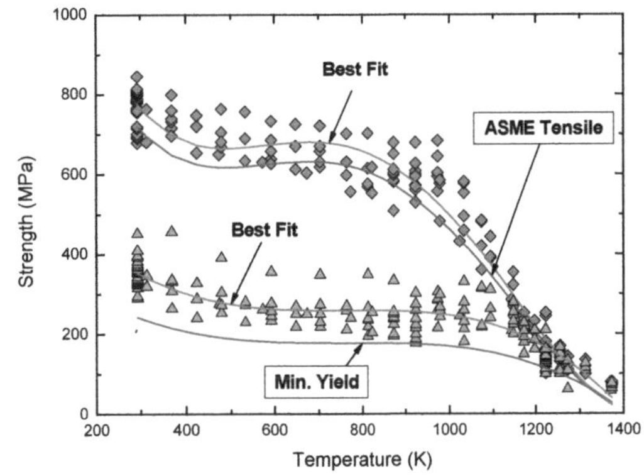


Fig. 4. Minimum YS and average UTS curves for Alloy 617.

As noted in the ASME procedures, using a specific minimum room temperature YS of 240 MPa and a minimum room temperature UTS of 655 MPa [12], the minimum YS and average UTS curves of temperature trend were obtained for Alloy 617, as shown in Fig. 4.

In the lower temperature ranges of $T$<1100 $K$ for the YS and $T$<1000 $K$ for the UTS, the best-fit curves for the YS and UTS did not decrease significantly with increasing temperatures. The minimum YS curve was placed adequately as the low bounds of the tensile data, and the vertical distance between the best-fit and minimum YS curves did not reduce appreciably. However, the average UTS curve was not placed properly as the low bounds of the tensile data. Some of the data were out of the average UTS range due to the fluctuation of the fourth-order polynomial functions.

At the higher temperature range of $T$>1100 $K$ for the YS and $T$>1000 $K$ for the UTS, the YS and UTS values dropped off drastically, and the minimum YS and average UTS curves approached the best-fit YS and UTS curves with increasing temperatures. The vertical distance between the best-fit curves and the minimum YS and average UTS curves was reduced in the higher temperature region. This means that the margin of the minimum YS and average UTS curves relative to the best-fit curve decreases with increasing temperatures, and in particular, the margin of the average UTS curve decreases significantly in the high temperature regime of interest. Consequently, it is identified that the low bound curves of the minimum YS and the average UTS according to the ASME procedures do not allow a consistent and quantifiable design margin in the higher temperature range.

Also, this fact in the higher temperature regime can be expressed from Eqs. (2) and (3). The low bound curves of $S_y(T)$ and $S_u(T)$ for the minimum YS and average UTS depend on the constants $c_y$ and $c_u$, which are independency on temperature. The constants do not have a consistent effect on the $S_y(T)$ and $S_u(T)$ curves within a higher temperature regime. As a result, the two curves of $\hat{S}_y(T)$ and $S_y(T)$ converge more smoothly than those of $\hat{S}_u(T)$ and $S_u(T)$. The reason for this is because the convergence of their curves undermines the ability of $S_y(T)$ and $S_u(T)$ to the lower bound data at a higher temperature regime.

Accordingly, in estimating the time-independent allowable stresses, as discussed above, it is clear that the ASME procedures should be improved for a high temperature range, although they are satisfactory in the low temperature range. The use of the minimum value for the YS and average value for the UTS is not consistent. As a result, the safety margin for the UTS may reduce significantly in the higher temperature regime of interest. Thus, an improved approach to allow for a consistent and quantifiable design margin for low- and high-temperature ranges is proposed herein.

### 3.3 Conservative lower bound curves

To obtain the best-fit curves of temperature trend for the YS and UTS, three fit methods, a single polynomial fit, 2-step exponential fit, and 2-step polynomial fit, are adopted. The 2-step exponential and 2-step polynomial fits are used to separate the fitting temperature ranges for the YS and UTS data into two temperature ranges. As discussed in Fig. 4, the reason for this is that changes in the temperature dependence of the YS and UTS data were evident at around 1100 *K* and 1000 *K*. This approach makes it possible to develop conservative lower bound curves, because the best-fit function can be precisely obtained for the separated temperature ranges, and their curves can be determined through a statistical concept. This statistical approach can define the lower bound with an appropriate level of conservatism for design use.

In addition, in the lower temperature ranges of $T$<1100 *K* for the YS and $T$<1000 *K* for the UTS, the YS and UTS were not reduced and steady regardless of increasing temperatures. This is attributed to a dynamic strain ageing (DSA) phenomenon increasing strengths by obstacles to the movement of dislocations during strain. This DSA phenomenon has been well known to be generated at specific temperature ranges for solid solution alloys like austenitic stainless steels [17]. It is thus believed that the DSA was generated at this temperature range for Alloy 617 which is a solid solution alloy. However, in the higher temperature ranges of $T$>1100 *K* for the YS and $T$>1000 *K* for the UTS, their strengths were rapidly reduced with increasing temperatures. The reason for this is that the density of dislocations was reduced significantly by a crossing slip or a climb of the dislocations at this higher temperature.

A standard deviation (*SD*) was used for comparing the degree of fit between the experimental data and estimated values. The *SD* values are also applied to probabilistically determine the lower bound curves. The standard deviation is defined as

$$SD=\sqrt{\sum_{i=1}^{n}(y_i-\hat{y}_i)^2/(n-n_p)} \tag{6}$$

where $y_i$ is the strength value at the $i^{th}$ data, $\hat{y}_i$ is the mean strength value estimated at the $i^{th}$ data, $n$ is the number of total data, and $n_p$ is the number of undetermined coefficients in the best-fit function. The *SD* values were calculated from the best-fit functions estimated using the three fit methods.

Table 1. Standard normal coverage multiplier in Eq. (7).

| P (x) | z |
|---|---|
| 0.50 | 0.000 |
| 0.75 | 0.674 |
| 0.80 | 0.842 |
| 0.85 | 1.036 |
| 0.90 | 1.282 |
| 0.95 | 1.645 |
| 0.99 | 2.360 |

Also, the $z$ factor in a standard normal distribution was used. The relation equation between $P(x)$ and the $z$ factor in a standard normal distribution is given by

$$z=\ln[4\,P(x)\,(1-P(x)] \tag{7}$$

where $P(x)$ is the probability, and $z$ is the standard normal coverage multiplier. Hence, it indicates that $P(x) \geq 0.5$ is a positive sign (+), and $P(x) \leq 0.5$ is a negative sign (–). For $P(x) = 0.95$, the value of $z$ is 1.645 as listed in Table 1. The plots of 95% lower prediction bound curves were plotted in Figs. 5, 6 and 7, which were determined from each best-fit function for the three fitting methods: the single polynomials, 2-step exponentials and 2-step polynomials.

Fig. 5 shows a comparison for the 95% lower prediction bound curves and minimum YS and average UTS curves for the single polynomial fit method. For the normalized YS data, the best-fit curve was obtained using the third-order polynomials, and for the normalized UTS data, it was obtained using the fourth-order polynomials. Also, the 95% lower prediction bound curves for $S_y$ (YS) and $S_u$ (UTS) were determined by subtracting the amount of $SD \cdot z$ (for $P$=0.95, $z$=1.645) from the functions of the best-fit curves, as given in Eqs. (8) and (9).

$$S_y(T)=1.927-0.00473\,T+6.263\text{E}-6\,T^2-2.753\text{E}-9\,T^3-SD\cdot z \tag{8}$$

$$S_u(T)=2.59-0.0103\,T+2.168\text{E}-5\,T^2-1.875\text{E}-8\,T^3+5.408\text{E-}12\,T^4-SD\cdot z \tag{9}$$

In the above two single polynomials, the *SD* values of 39.8 MPa for $S_y$ and 51.2 MPa for $S_u$, were used respectively, and their results are summarized in Table 1. In higher temperature ranges beyond 1100 K for the YS and 1000 K for the UTS, the minimum YS and average UTS curves which are obtained by the ASME procedures closely approached the best-fit curves with increasing temperatures, particularly, for the average UTS curves.

On the other hand, the 95% lower bound curves allowed a consistent and quantifiable margin over full temperature ranges regardless of whether low or high temperatures. Thus,

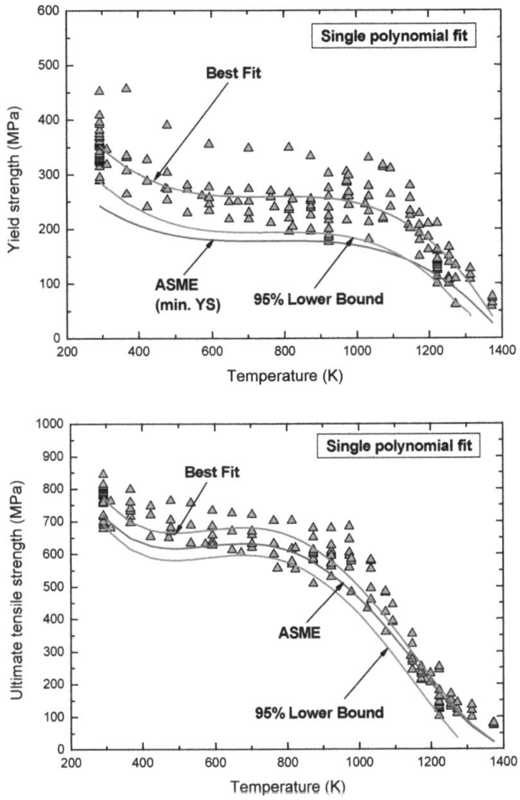


Fig. 5. The 95% lower bound curves of the YS and UTS data from the single polynomial fit.

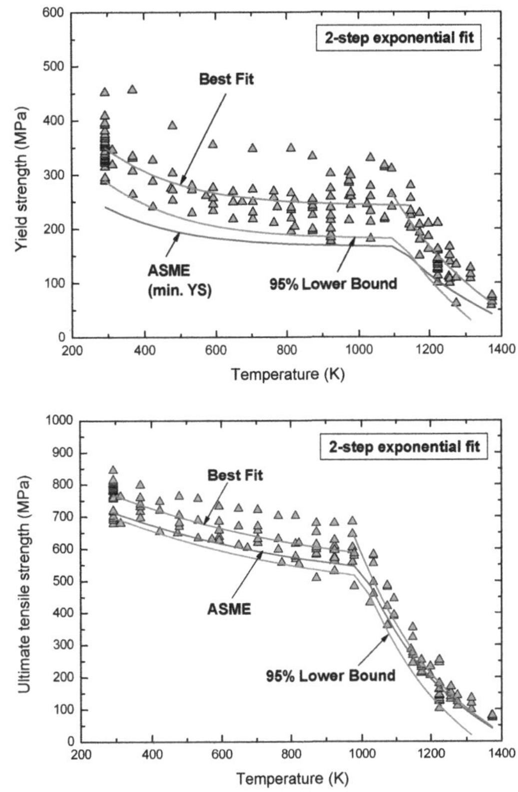


Fig. 6. The 95% lower bound curves of the YS and UTS data from the 2-step exponential fit.

it was clearly found that the 95% lower bound curves from this statistical method are more reasonable than the minimum YS and average UTS curves from the ASME method. However, as discussed in Section 3.2, for a single polynomial fit, the 95% lower bound curves exhibited inherent fluctuations due to the third-order and fourth-order polynomials of the best-fit functions in the full temperature ranges.

Fig. 6 shows the comparison results of the 95% lower prediction bound curves and minimum YS and average UTS curves for the 2-step exponential fit method. For normalized YS and UTS data, the best-fit curves were separately considered at temperatures of 1100 K for $S_y$ and 1000 K for $S_u$, respectively, because the YS and UTS data revealed better regression for the ranges of separated temperatures than in the single polynomial fit. The best-fit curves were obtained using the second-order exponential function at each temperature range. The 95% lower prediction bound curves for $S_y$ and $S_u$ were determined by subtracting the amount of $SD \cdot z$ (for $P$=0.95, $z$=1.645) from each best-fit curve, as given in Eqs. (10)-(12) and (13).

$$S_y(T) = 0.6891 + 1.2775\, exp\,(-0.0048T) - S \cdot z \text{ (for } T<1100\text{K)} \quad (10)$$

$$S_y(T) = -0.2832 + 25.6518\, exp\,(-0.0029\, T) - SD \cdot z \text{ (for } T>1100 \text{ K)} \quad (11)$$

$$S_u(T) = 0.6417 + 0.5623\, exp\,(-0.0016\, T) - SD \cdot z \text{ (for } T<1000 \text{ K)} \quad (12)$$

$$S_u(T) = -0.1999 + 31.8379\, exp\,(-0.0035\, T) - SD \cdot z \text{ (for } T>1000 \text{ K)} \quad (13)$$

In the above 2-step exponentials, the $SD$ values were 37.1 MPa for $S_y$ and 42.5 MPa for $S_u$. The 95% lower bound curves exhibited a reasonable estimation for the YS and UTS in the higher temperature ranges of $T$>1100 K for $S_y$ and $T$>1000 K for $S_u$. The vertical distance between the best-fit and 95% lower-bound curves was not reduced in the higher temperature range, and the 95% lower curves allowed a consistent margin over the full temperature ranges.

Fig. 7 shows the results of the 95% lower prediction bound curves and the minimum YS and average UTS curves for the 2-step polynomial fit method. Similarly, as written for the 2-step exponential fit, for the normalized YS and UTS data, the best-fit curves were obtained using the second-order polynomial functions in the separated temperature ranges at 1100 K of YS and 1000 K for UTS. The 95% lower prediction bound curves for YS and UTS are determined by subtracting the

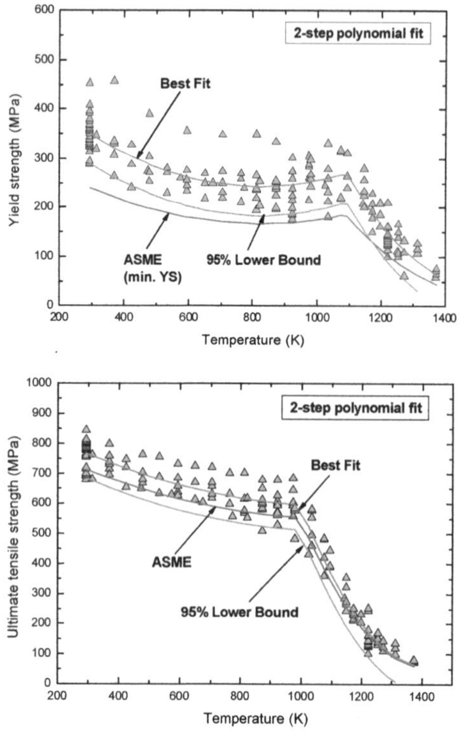


Fig. 7. The 95% lower bound curves of the YS and UTS data from the 2-step polynomial fit.

amount of $SD \cdot z$ (for $P = 0.95$, $z = 1.645$) from each best-fit curve, as given in the following Eqs. (14)-(16) and (17).

$$S_y(T) = 1.4317 - 0.0017\,T + 1.0892\text{E-}6\,T^2 - SD \cdot z \text{ (for } T<1100 \text{ K)} \quad (14)$$

$$S_y(T) = 8.3808 - 0.0108\,T + 3.5689\text{E} - 6\,T^2 - SD \cdot z \text{ (for } T>1100 \text{ K)} \quad (15)$$

$$S_u(T) = 1.1903 - 7.5751\text{E} - 4T + 3.4143\text{E} - 7\,T^2 - SD \cdot z \text{ (for } T<1000 \text{ K)} \quad (16)$$

$$S_u(T) = 7.7667 - 0.0108\,T + 3.8047\text{E} - 6\,T^2 - S \cdot z \text{ (for } T>1000 \text{ K)} \quad (17)$$

For the 2-step polynomials, the $SD$ values are 36.3 MPa for $S_y$ and 41.9 MPa for $S_u$. As a similar result obtained in the 2-step polynomial fit, the 95% lower bound curves were also satisfactory because the vertical distance between the best-fit and lower-bound curves was constant in the higher temperatures beyond 1100 K for $S_y$ and 1000 K for $S_u$.

As discussed above, it is here emphasized that the 2-step exponential and 2-step polynomial fits were lower in $SD$ values than the single polynomial fit, which have been commonly used, as shown in Fig. 8. The 2-step polynomial fit was also lower in $SD$ values than in the 2-step exponential fit. It is thus justifiable that the 2-step polynomial fit method becomes a precise method for determining the conservative lower bound curves for YS and UTS. Thus, it is clearly found that the 2-step fit methods were superior in determining the lower bound curves over the single polynomial fit method, and that the 2-step polynomial fit method was better than the 2-step exponential fit method.

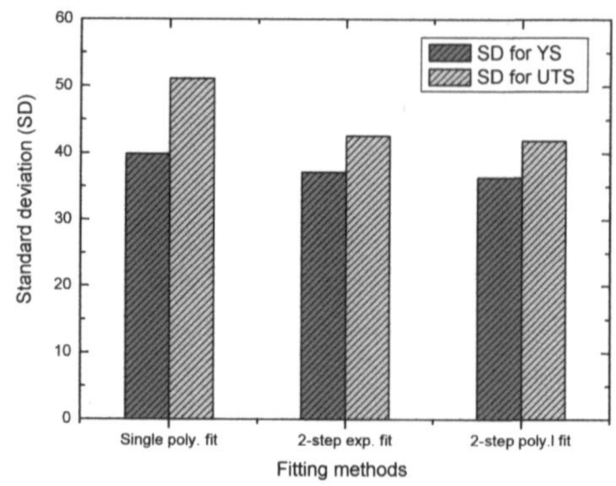


Fig. 8. Standard deviations of the YS and UTS data for the three fit methods.

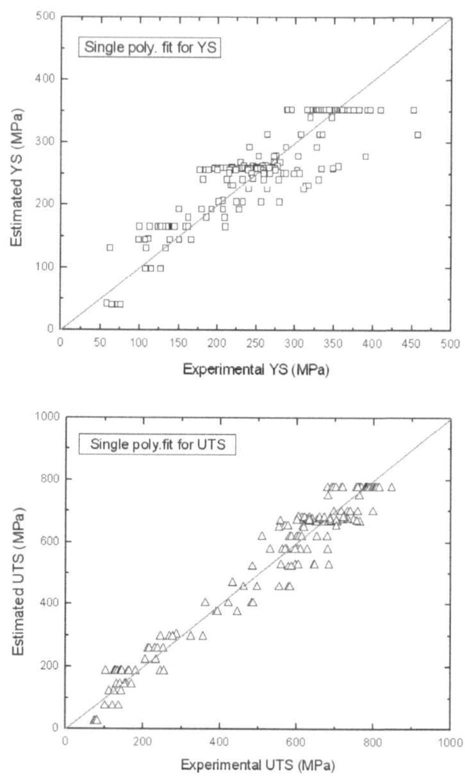


Fig. 9. Estimated vs. experimental YS and UTS for the single polynomial fit.

Figs. 9, 10 and 11 show the relationships between the estimated- and experimental YS and UTS for the three fit methods. The estimated values (y-axis) for YS and UTS were cal-

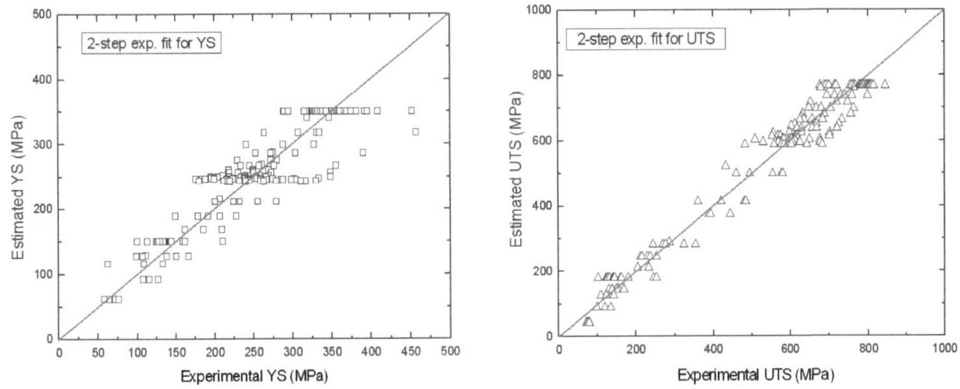


Fig. 10. Estimated vs. experimental YS and UTS for the 2-step exponential fit.

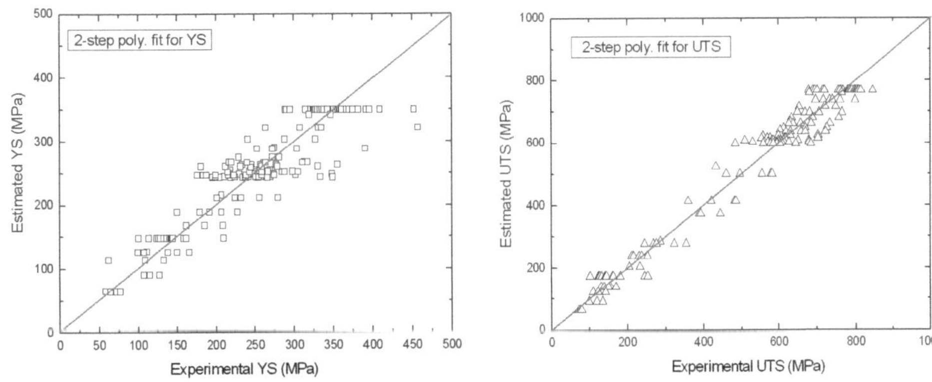


Fig. 11. Estimated vs. experimental YS and UTS for the 2-step polynomial fit.

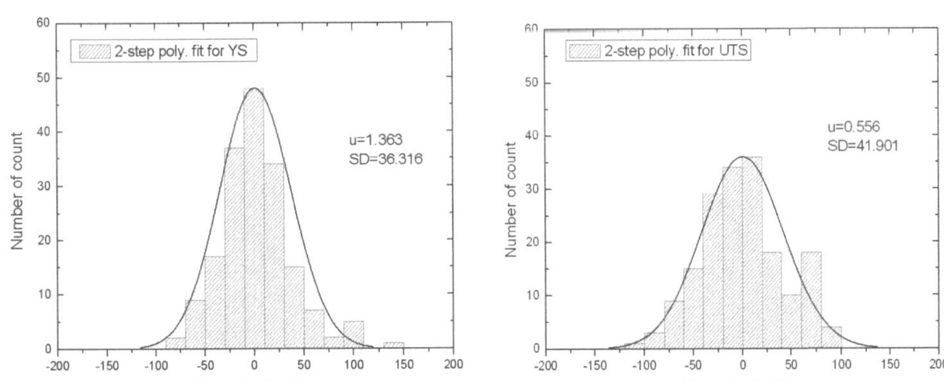


Fig. 12. Normal distributions of the YS and UTS data from the 2-step polynomial fit.

culated by using Eqs. (8)-(17) for different temperature functions. It is emphasized that the 2-step exponential and 2-step polynomial fit revealed good agreement with the linear lines in the low strength regions of interest when compared with the single polynomial fit. The 2-step exponential and 2-step polynomial fits were almost similar. Statistical distributions were also investigated from the YS and UTS data, which were estimated using the best-fit curves for the three fit methods. In total, the YS and UTS data correctly followed the histograms of normal distribution as bell types, as shown in Fig. 12 displayed typically for the 2-step polynomial fit method. Hence, the error of estimation (x-axis) means the difference between

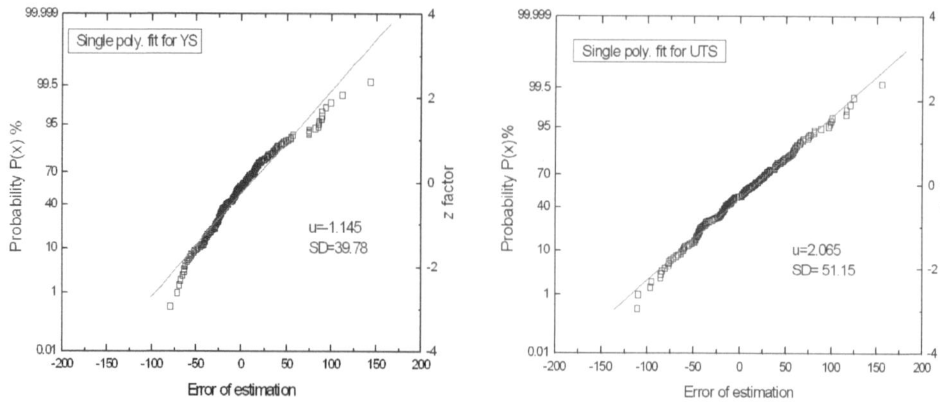


Fig. 13. Normal probability plot of the YS and UTS data from the single polynomial fit.

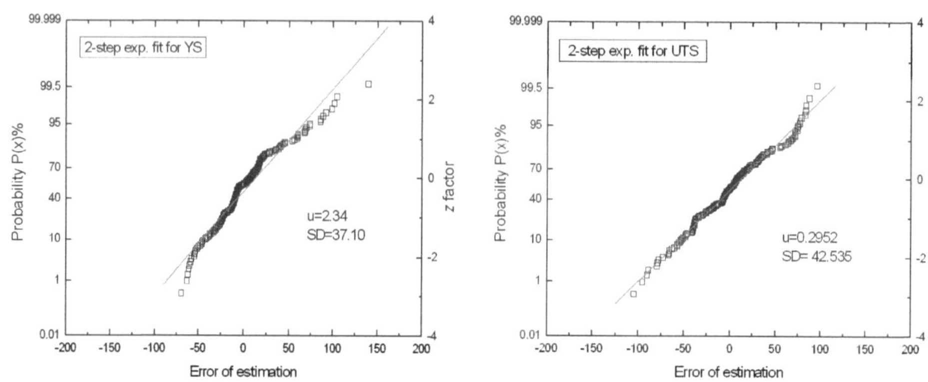


Fig. 14. Normal probability plot of the YS and UTS data from the 2-step exponential fit.

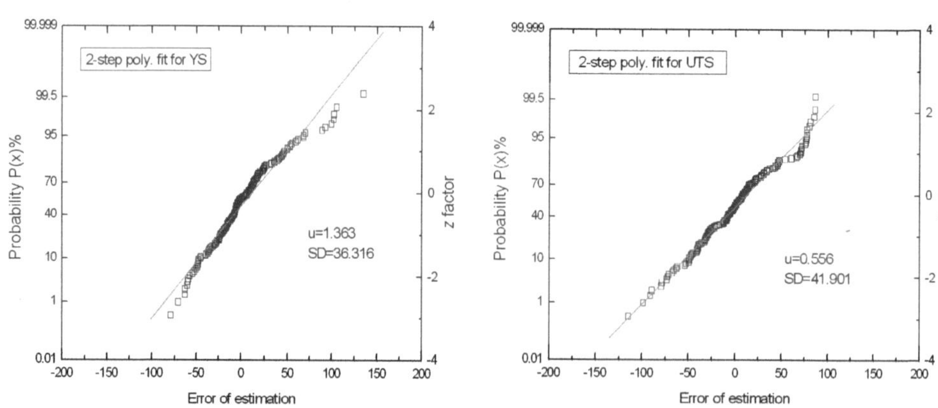


Fig. 15. Normal probability plot of the YS and UTS data from the 2-step polynomial fit.

the experimental data and estimated value.

Figs. 13-15 show the normal probability of the YS and UTS data for the three fit methods. From the normal distributions, the mean (μ) and *SD* for the YS and UTS data were obtained for them. Also, the value of the z factor was estimated from these plots. In the plot of a normal probability for the three methods, it can be clearly seen that the 2-step polynomial fit was lower in *SD* values than the 2-step exponential and single polynomial fits. Using these probabilistic results, the conservative lower-bound curves for the YS and UTS can be probabilistically estimated with a safety margin in the high temperature regime of interest for Alloy 617.

## 4. Conclusions

In establishing the tensile design strengths of temperature trend for Alloy 617, the minimum YS and average UTS curves in the existing ASME procedures were not satisfactory within the higher temperature ranges of $T$>1100 K for $S_y$ and $T$>1000 K for $S_u$, because they did not allow a consistent and quantifiable design margin with increasing temperatures. It appeared that 2-step fit methods were superior in the best-fit curves of temperature trends compared to the single polynomial fit that has been commonly used; the 2-step polynomial fit was particularly better than the 2-step exponential fit. The YS and UTS data correctly followed the histograms of normal distribution. Using this normal probability, the 95% lower bound curves for the three fit methods were probabilistically developed. The results for the 2-step polynomial fit allowed the consistent and quantifiable design margin over the full temperature ranges at the low and high temperatures. The results can be utilized for determining the tensile design strengths with a probabilistic margin at the high temperature regimes of Alloy 617.

## Acknowledgment

This study was supported by the Korean Ministry of Education, Science, and Technology through its National Nuclear R & D Program.

## References

[1] W. G. Kim, S. N. Yin, Y. W. Kim and J. H. Chang, Creep properties and creep life prediction of Alloy 617 for a very high temperature gas-cooled reactor, *Proc. of ASME PVP 2008 Conf.*, Chicago, USA, July (2008) Paper PVP 2008-61356.

[2] S. J. Dewson and X. Li, Selection criteria for the high temperature reactor intermediate heat exchanger, *Proc. of ICAPP 05*, Seoul, Korea, May (2005) Paper 5333.

[3] W. G. Kim, S. N. Yin, W. Y. Kim and J. H. Chang, Creep characterization of a Ni-based Hastelloy-X alloy by using theta projection method, *Engineering Fracture Mechanics*, 75 (2008) 4985-4995.

[4] W. G. Kim, S. N. Yin, G. G. Lee, Y. W. Kim and S. J. Kim, Creep oxidation behavior and creep strength prediction for Alloy 617, *Int. J. of Pressure Vessels and Piping,* 87 (2010) 289-295.

[5] J. M. Corum and J. J. Blass, Rules for design of alloy 617 nuclear components to very high temperatures, Attachment 2, Draft Code Case for Alloy 617, USA (1988).

[6] W. G. Kim, S. N. Yin, W. S. Ryu and J. H. Chang, Analysis of the creep rupture data of Alloy 617 for a high temperature gas cooled reactor, *The 8th Int. Conf. on Creep and Fatigue at Elevated Temperatures*, Texas, USA, July (2007) Paper PVP 2007-26834.

[7] J. H. Chang, Y. W. Kim, K. Y. Lee, Y. W. Lee, W. J. Lee, J. M. Noh, M. W. Kim, H. S. Lim, Y. J. Shin, K. K. Bae and K. D. Jung, A study of a nuclear hydrogen production demonstration plant, *Nuclear Engineering and Technology*, 39 (2) (2007) 111-122.

[8] H. Y. Lee, Y. W. Kim, and K. N. Song, Preliminary application of the draft code case for alloy 617 for a high temperature component, *J. of Mechanical Science and Technology,* 22 (2008) 856-863.

[9] J. Y. Park, D. W. Kim, D. J. Kim, E. S. Kim, H. P. Kim, S. H. Kim, W. J. Kim, W. G. Kim, W. S. Ryu and Y. D. Kim, Development of materials for a high temperature gas cooled reactor in Korea, *The $4^{th}$ International Topical Meeting on High Temperature Reactor Technology-HTR 2008*, Washington, USA (2008) Paper HTR 2008-58149.

[10] Draft Alloy 617 code case, Task Force Very High Temperature Design, ASME (1988).

[11] T. L. Sham, D. R. Eno and K. P. Jensen, Treatment of high temperature tensile data for Alloy 617 and Alloy 230, *Proc. of ASME PVP 2008*. Chicago, USA, July (2008) Paper PVP 2008-61128.

[12] ASME boiler and pressure vessel code, Section II, Part D, ASME, USA, New York (2007).

[13] European Alloy-DB, Licensed CD, European Commission, Joint Research Centre (JRC).

[14] W. G. Kim. D. W. Kim, W. S. Ryu, C. H. Han, J. H. Yoon and J. Chang, Mechanical property and its comparison of superalloys for high temperature gas cooled reactor, *KAERI report*, KAERI/AR-723/2005 (2005).

[15] http:// www. specialmetals.com (2008).

[16] http://www.haynesintl.com (2008).

[17] D. W. Kim, Effect of nitrogen on dynamic strain ageing and high temperature low cycle fatigue properties in type 316L stainless steel, *the Ph.D thesis at KAIST,* Daejeon (1998) 33-38.

**Woo-Gon Kim** received his Ph.D from Chungbuk National University in the Dept. of Mechanical Engineering in 1999. As principal researcher at Korea Atomic Energy Research Institute, his specialty is mechanical assessment, analysis, modeling and database establishment of high temperature nuclear materials for Gen-IV reactor systems, and especially, time-dependent creep and crack growth behaviors at elevated temperature.

*Met. Mater. Int., Vol. 17, No. 3 (2011), pp. 497~504*
doi: 10.1007/s12540-011-0630-1 Published 29 June 2011

# Long-Term Creep Characterization of Gr. 91 Steel by Modified Creep Constitutive Equations

**Woo-Gon Kim***, **Sung-Ho Kim, and Chan-Bock Lee**

Nuclear Materials Research Division, Korea Atomic Energy Research Institute (KAERI), 1045 Daedeokdae-ro, Yuseong-gu, Daejeon 305-353, Korea

(received date: 17 September 2010 / accepted date: 9 March 2011)

This paper focuses on the long-term creep characterization of Gr. 91 steel using creep constitutive equations. The models of three such equations, a combination of power-law form and omega model (CPO), a combination of exponential form and omega model (CEO), and a combination of logarithmic form and omega model (CLO), which are described as sum decaying primary creep and accelerating tertiary creep, are proposed. A series of creep rupture data was obtained through creep tests with various applied loads at 600 °C. On the basis of the creep data, a nonlinear least-square fitting (NLSF) analysis was carried out to provide the best fit with the experimental data in optimizing the parameter constants of an individual equation. The results of the NLSF analysis showed that in the lower stress regions of 160 MPa ($\sigma/\sigma_{ys} < 0.65$), the CEO model showed a match with the experimental creep data comparable to those of the CPO and CLO models; however, in the higher stress regions of 160 MPa ($\sigma/\sigma_y > 0.65$), the CPO model showed better agreement than the other two models. It was found that the CEO model was superior to the CPO and CLO models in the modeling of long-term creep curves. Using the CEO model, the long-term creep curves of Gr. 91 steel were numerically characterized, and its creep life was predicted accurately.

**Keywords:** metals, creep, tensile test, creep test, strain rate

## 1. INTRODUCTION

High-chromium steel is widely used in recent fossil fuel power plants because of its excellent high-temperature strength. In particular, modified 9Cr-1Mo steel (ASME grade 9Cr-1Mo, hereafter G91) is regarded as a promising candidate for structural materials in applications such as steam generators, intermediate heat exchangers (IHX), and the primary pipes used in sodium-cooled fast reactors (SFRs), which is a type of Generation-IV reactor, as shown in Fig. 1. Currently, Korea Atomic Energy Research Institute (KAERI) has been working on a long-term government-funded project to develop a SFR demonstration reactor by 2028. This reactor will have a 600 MWe electricity capacity and a core outlet temperature of 510 °C.

For the design of the SFR components, the creep strain data on the long-term conditions are required to determine the allowable stress, as it is expected that the components may be in use for up to 60 years. It is therefore necessary to estimate their long-term creep behavior by experimental studies, calculations, or both. However, because it is both time-consuming and expensive, there are many difficulties in actually obtaining long-term creep data for more than $10^5$ h (11.4 years) [1-11]. It is therefore necessary to develop a valid constitutive equation by analyzing short-term creep data and then predict the long-term creep life or model its curve using numerical calculations based on the developed creep constitutive equation.

During the past years, various models for describing creep strain-time curves and creep lifetimes have been developed to assess or predict the material behavior of components operating at high temperatures [12-25]. The Larson-Miller (LM) method [12], known as a time-temperature parameter (TTP), has been most commonly used to predict long-term creep life from short-term creep data due to its simplicity and ease of application. However, the LM plot has a basic problem in the prediction because the *C* value for the rupture life in the LM equation changes from a high value in short-term creep to a low value in long-term creep [2,13]. Moreover, the LM plot does not take into account the microstructure changes in creep.

Evans, Brown, and Wilshire [14,21-23] introduced a theta (θ) projection method, as an extrapolation method for long-term creep curves and creep lifetime behaviors. This model describes normal creep curves in terms of a decaying pri-

*Corresponding author: wgkim@kaeri.re.kr

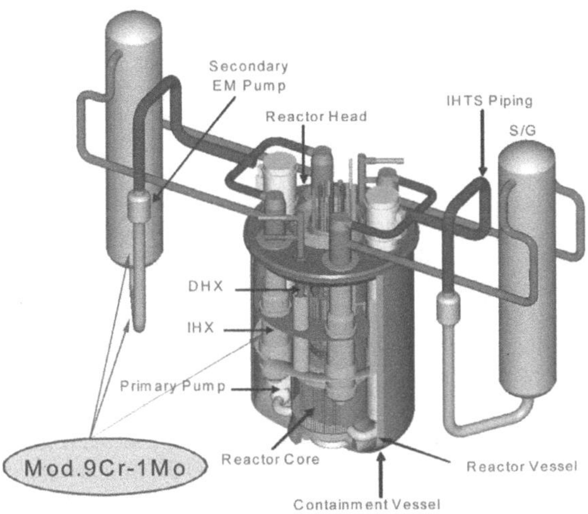


**Fig. 1.** SFR components for a KALIMER-600 system being developed at KAERI.

mary creep and accelerating tertiary creep equations expressed using four θ parameters. A lack of a physical meaning in the model has driven some studies to determine the relationship between the θ parameters and micro-mechanisms of creep, but these approaches have not been fully exploited.

In addition, Prager [15,16] introduced the omega (Ω) method as a good approach for creep life prediction. This model has been known to extend the concepts of the Larson-Miller plot, the Monkman-Grant (M-G) relationship, Robinson, θ projection method, Kachanov's damage mechanics, and Norton's creep law. The Ω method has been successfully applied to various high-temperature components such as process vessels and heater tubes. It is thus necessary to improve the θ and Ω methods for the development of widely applicable creep constitutive equations.

In this paper, models of three modified creep constitutive equations are proposed to characterize long-term creep curves describing up to the tertiary creep curve of G91 steel. The best model is found by using a nonlinear least-square fitting (NLSF) analysis on the experimental creep data. Stress-dependence behavior for the parameter constants is investigated for the best model, and the long-term creep curves are numerically characterized, and the creep life is accurately predicted.

## 2. EXPERIMENTAL PROCEDURES

A commercial grade hot-rolled G91 steel plate was used for the testing material. The heat treatment condition of the steel was normalized and tempered at 1050 °C/1 mim/mm and 770 °C/3 mim/mm, respectively. The plate thickness was 16 mm (0.625 inch). The chemical composition is listed in Table 1, and each element is within the ASTM requirement.

Tensile test specimens were used for a rectangular cross section 2.48 mm thick and 6.33 mm wide, and with a 25 mm gauge length. The tensile tests were conducted using a strain rate of $2 \times 10^{-3}$/s at 600 °C. Creep test specimens were machined to a cylindrical shape with a 30 mm gauge length and 6 mm diameter. All specimens were taken along the rolling direction. The gauge sections of the creep specimens were polished using #1000 grit sandpapers with strokes along the specimen axis. Constant-load creep tests were carried out with various stress levels of 200 MPa, 180 MPa, 160 MPa, 150 MPa, 140 MPa, and 130 MPa at 600 °C. The temperature during the creep tests was controlled within ± 2 °C. Creep strain data were taken automatically at elapsed times using a PC through a high-precision LVDT. The steady-state creep rate was obtained from the experimental creep curves. All of the experimental procedures for the creep tests followed the recommendations of ASTM E139.

## 3. RESULTS AND DISCUSSION

### 3.1. Modified creep constitutive equations

Generally, heat-resistant alloys undergoing creep exhibit creep strain-time curves consisting of primary, secondary, and tertiary creep stages. In the process of designing high-temperature components, total creep curves containing information up to the tertiary stage are required. To facilitate this, there is a need to use a proper creep constitutive equation and to define the parameter constants used in the equation. Thus, various models for creep strain-time curves and creep lifetimes have been developed to provide an overall description of the material behavior of components operating at high temperatures. The θ projection method and the Ω method are well known for modeling or predicting total creep behavior [13-25].

In the θ projection method, a creep curve is expressed as a sum of a decaying primary creep and accelerating tertiary creep, and creep strain, $\varepsilon_c$, can be written as a combination of the two exponential terms :

$$\varepsilon_c = \theta_1[1-\exp(-\theta_2 t)]+\theta_3[\exp(\theta_4 t)-1], \quad (1)$$

where $t$ is time and $\theta_1$, $\theta_2$, $\theta_3$, and $\theta_4$ are parameter constants determined by fitting Eq. 1 to the experimental data. Although the θ projection method was originally developed as an analysis method for constant-stress creep curves, it

**Table 1.** Chemical compositions of G91 steel (wt.%)

| C | Si | Mn | P | S | Ni | Cr | Mo | Cu | V | Al | N | Nb |
|---|---|---|---|---|---|---|---|---|---|---|---|---|
| 0.115 | 0.23 | 0.415 | 0.012 | 0.0014 | 0.22 | 8.9 | 0.869 | 0.038 | 0.194 | 0.020 | 0.0513 | 0.073 |

has been found to be valid for application to constant-load creep curves [17,25].

In addition, the W model in this study was used to assess the creep behavior of G91 steel. This model is based on results showing that a linear relationship between the logarithm of creep rate and true strain exists for Cr-Mo steel. In this model, the creep strain rate of a constant load creep is expressed as follows [6,13,24]:

$$\ln\dot{\varepsilon}_c = \ln\dot{\varepsilon}_o \cdot \Omega + \varepsilon_c \,, \quad (2)$$

where $\dot{\varepsilon}_o$ is the initial creep rate after loading, $\varepsilon_c$ is the creep strain, and the parameter $\Omega$ is a variable which is dependent upon temperature and stress. When the primary creep region is small, $\dot{\varepsilon}_o$ is very close to the minimum creep rate, $\dot{\varepsilon}_{\min}$. The variable W is an implicit function of stress, mechanical damage, and microstructure changes.

Equation 2 can be integrated to give a function containing strain and time. The resultant equation is given by

$$t = \frac{1}{\dot{\varepsilon}_o\Omega}(1 - e^{-\Omega \cdot \varepsilon_c}) \,. \quad (3)$$

From Eq. 3, creep strain can be rewritten in terms of a logarithmic function as

$$\varepsilon_c = -\frac{1}{\Omega}\ln(1 - \dot{\varepsilon}_o\Omega t) \,. \quad (4)$$

For large values of $\Omega \cdot \varepsilon_c$ as creep failure, the exponential term in Eq. 3 is negligible. The rupture time, $t_r$, can be expressed as

$$\frac{1}{t_r} \approx \dot{\varepsilon}_o\Omega \,. \quad (5)$$

The $\Omega$ method can also be used to predict the residual lifetime of a component operating within a creep region.

This $\Omega$ method is fully applicable to various high-temperature components, such as process vessels and heater tubes, because this model extends the concepts of the LM method, M-G relationship, Robinson, $\theta$ projection method, Kachanov's damage mechanics, and Norton's creep law. However, although the $\theta$ and $\Omega$ methods have been used to attempt to predict long-term creep curve behaviors, they are limited in their applicability to creep curve shapes and materials. It is difficult to apply the $\theta$ projection method to dominant curves under a constant load in some cases. Also, the original $\Omega$ method cannot express creep curves with a decaying primary creep. Therefore, the $\theta$ and $\Omega$ methods should be properly modified.

To characterize the total creep curves describing up to a rupture point from a short creep time, in this study, we propose the following models of three modified creep constitutive equations for constant-load creep curves:

$$\varepsilon_c = P_1 t^{P_2} - \frac{1}{P_3}\ln(1 - P_4 t) \,, \quad (6)$$

$$\varepsilon_c = P_1[1 - \exp(-P_2 t)] - \frac{1}{P_3}\ln(1 - P_4 t), \quad (7)$$

$$\varepsilon_c = P_1 \ln(1 + P_2 t)] - \frac{1}{P_3}\ln(1 - P_4 t) \,. \quad (8)$$

The basis of a combination of the power law, the exponential law, and the logarithmic law with the omega model is to be described with the advantages of each combination. Equation 6 expresses a combination of the power-law term describing the primary creep stage and the logarithmic term of the $\Omega$ method describing the tertiary stage. This is hereafter written as "a combination of the power-law term and omega model (CPO)." Next, Eq. 7 expresses a combination of the exponential term describing the primary creep stage and the logarithmic term of the $\Omega$ model describing the tertiary stage. This equation is used to take the logarithmic term in the $\theta$ projection method of Eq. 1 instead of the exponential term as the tertiary creep stage. It is hereafter written as "a combination of exponential term and omega model (CEO)." Finally, Eq. 8 expresses the sum of two logarithmic terms describing the primary creep stage and logarithmic term of the tertiary creep stage. This equation is used to add the primary creep term expressed in terms of the logarithmic term to the original $\Omega$ model. It is hereafter written as "a combination of the logarithmic term and omega model (CLO)."

As described in Eqs. 6, 7, and 8, all three modified equations employ the $\Omega$ method to describe the tertiary creep stage. Also, as in Eq. 4, parameter $P_4$ indicates $1/t_r$. This is given in Eq. 5.

### 3.2. Analysis of creep data

A series of experimental creep curves was obtained using creep tests with various stresses for G91 steel at 600 °C as

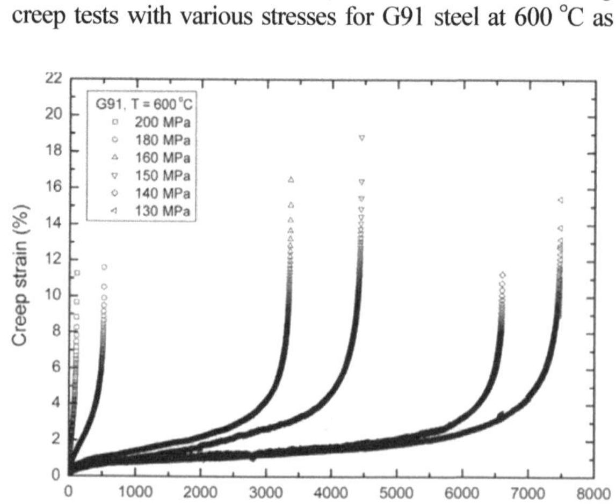


**Fig. 2.** Typical creep strain-time curves with different stress levels for G91 steel at 600 °C.

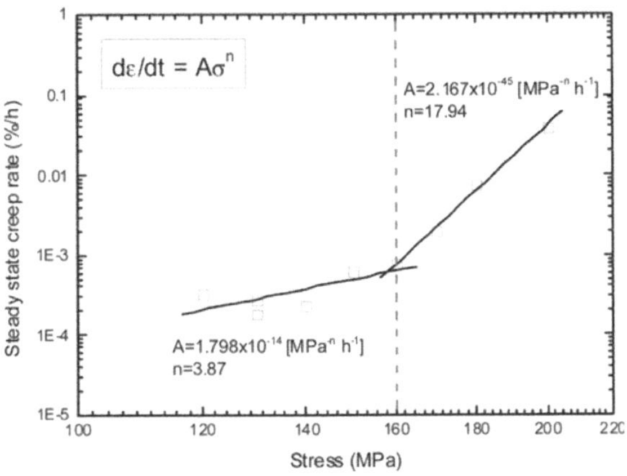


**Fig. 3.** The plot of steady state creep rate vs. stress for G91 steel at 600 °C.

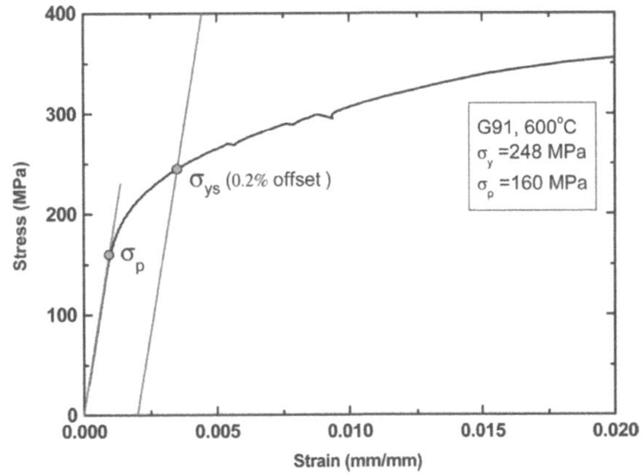


**Fig. 4.** Typical tensile curve obtained for G91 steel at 600 °C.

shown in Fig. 2. The G91 steel clearly exhibited creep strain-time curves for the primary, secondary, and tertiary creep stages. The longest rupture data was obtained for 7,500 h under 130 MPa. The relationship between the steady-state creep rate and stress was plotted in Fig. 3. The linearity for Norton's power rule, $\dot{\varepsilon}_{ss} = A\sigma^n$, was divided into two linear lines at 160 MPa. In the high-stress region (> 160 MPa), the $A$ and $n$ values were $2.167 \times 10^{-45}$ (MPa$^{-n}$ h$^{-1}$) and 17.94, and in the low-stress region (< 160 MPa) the $A$ and $n$ values were $1.798 \times 10^{-14}$ (MPa$^{-n}$ h$^{-1}$) and 3.87.

To investigate the physical meaning of the 160 MPa, tensile tests were conducted for G91 steel at 600 °C. The tensile tests demonstrated that the stress of 160 MPa corresponds to the proportional limit ($\sigma_p$) as shown in Fig. 4. This supports the idea that 160 MPa can be defined as the boundary stress. It seems that the creep fracture mechanism was changed from a ductile fracture into a brittle fracture at the proportional limit of around 160 MPa. The creep strain rate was also different within the range of higher and lower stress at around $\sigma_p$. It is believed that in stress regions higher than $\sigma_p$, a steep reduction in the creep strain rate occurred with the applied stress levels due to a change in microstructure; however, in stress regions lower than $\sigma_p$, a stable reduction in creep strain rate occurred with the applied stress levels due to the stabilizing of the microstructure as shown in Fig. 4.

In a previous study, Kimura *et al.* [3] insisted that the boundary stress corresponds to the elastic limit. In a range lower than 50 % of the 0.2 % offset yield stress ($0.5\sigma_{ys}$), no plastic strain was observed in the tensile tests at a very low strain rate of $5 \times 10^{-5}$ s$^{-1}$. Also, a plastic deformation, which began when the applied stress increased beyond around $0.5\sigma_{ys}$, was identified. Thus, this result is reasonable because the proportional limit value of 160 MPa for G91 steel is similar to the elastic limit value of around $0.5\sigma_{ys}$ for high-chromium ferritic steel (10-12Cr). This proportional limit becomes a very important value for predicting long-term creep strength because changes in the stress dependence of creep rupture strength occur at the proportional limit.

### 3.3. Optimization of parameter constants

To optimize four parameter constants, $P_1$, $P_2$, $P_3$, and $P_4$ in

**Table 2.** Parameter constants optimized by NLSF for three modified constitutive equations

| Models | Parameters | 200 MPa | 180 MPa | 160 MPa | 150 MPa | 140 MPa | 130 MPa |
|---|---|---|---|---|---|---|---|
| CPO model (Eq.6) | P1 | 0.27197 | 1.49E-01 | 0.1770 | 0.13161 | 0.15702 | 0.14212 |
| | P2 | 0.43197 | 0.39037 | 0.2300 | 0.21872 | 0.17449 | 0.13682 |
| | P3 | 0.5583 | 5.94E-01 | 0.7220 | 0.5743 | 0.88506 | 0.87015 |
| | P4 | 8.79E-03 | 1.90E-03 | 0.0003 | 0.00022 | 0.00015 | 0.00014 |
| CEO model (Eq.7) | P1 | 1.12018 | 1.37622 | 0.7195 | 0.52233 | 0.5908 | 0.47909 |
| | P2 | 0.32462 | 0.01043 | 0.0148 | 0.01298 | 0.0059 | 0.01484 |
| | P3 | 0.45124 | 0.5629 | 0.6163 | 0.53453 | 0.8476 | 0.74862 |
| | P4 | 0.00877 | 0.0019 | 0.0003 | 0.00023 | 0.00015 | 0.00014 |
| CLO model (Eq.8) | P1 | 0.24766 | 0.44952 | 0.0808 | 0.08657 | 0.11898 | 0.04618 |
| | P2 | 3.23688 | 0.6817 | 0.70235 | 0.6091 | 7.45E-02 | 0.72264 |
| | P3 | 0.47728 | 0.5928 | 0.62604 | 0.56789 | 8.83E-01 | 0.73698 |
| | P4 | 0.00875 | 0.0019 | 0.0003 | 0.00023 | 1.50E-04 | 0.00013 |

Eqs. 6, 7, and 8, an NLSF analysis of the three modified equations was carried out to provide the best fit with the experimental data. The best fitting curve with the experimental data was generated through iterative NLSF analysis. This work was conducted using a non-linear fitting algorithm based on the function. Also, to evaluate the degree of fitting between the experimental creep data and the calculated values for the three modified equations, the fitting results were compared using the values of standard error, which is commonly used as a statistical parameter. Parameter constants optimized using the NLSF analysis for the three modified constitutive equations are summarized in Table 2. Using these parameter constants, we can numerically plot long-term creep curves describing up to the tertiary creep stage.

Figure 5 shows typical NLSF fitting results of the (a) CPO, (b) CEO, and (c) CLO models with the experimental creep curves for 130 MPa at 600 °C. In each plot, the small figures represent a high magnification to closely observe the result of fitting the primary creep stage to the experimental curve. As shown in the small figures for a primary creep stage of 130 MPa, the CEO model showed a closer agreement than the CPO and CLO models. On the other hand, for a high stress of 200 MPa, the CEO model revealed worse agreement at the primary creep region than the CPO and CLO models as shown in Fig. 6.

Figure 7 shows the comparison results of the standard error values obtained from the NLSF for the three modified equations. In the low-stress regions from 130 MPa to 160 MPa, the CEO model was lower in the standard error values than the CPO and CLO models. This indicates that the CEO model can describe the experimental creep data well because standard error values are low. On the contrary, in high-stress regions from 180 MPa to 200 MPa, the CEO model was higher in standard error values than the CPO and CLO models, and the CPO model was the lowest in the standard error value.

Accordingly, it was determined that the CEO model revealed good agreement with the experimental creep data at low-stress levels of $\sigma/\sigma_{ys} < 0.65$, but the CPO and CLO models showed good agreement with experimental creep data at high-stress levels of $\sigma/\sigma_{ys} > 0.65$. Hence, the $\sigma$ value was 160 MPa, and $\sigma_{ys}$ was 248 MPa for G91 steel at 600 °C. In particular, it appeared that the CPO model showed a good match with the experimental creep curves at a high-stress level of 200 MPa. Therefore, in modeling long-term creep curves of low stress, the CEO model was more useful than the CPO and CLO models. Thus, the CEO model proposed in this study can be applied to G91 steel.

### 3.4. Stress dependence of the parameter constants

As discussed above, the best equation was derived from the CEO model. Therefore, to model long-term creep curves for G91 steel, the stress dependence of four parameter con-

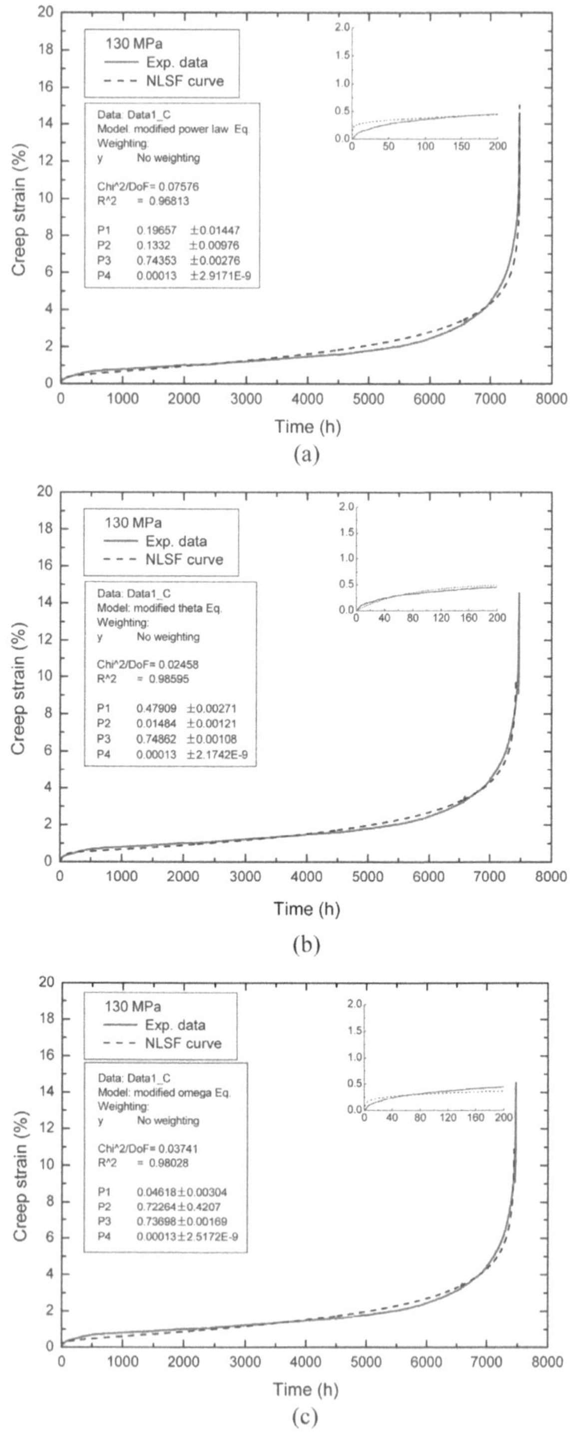


**Fig. 5.** Typical NLSF results on the CPO, CEO and CLO models for experimental creep curves of 130 MPa at 600 °C: (a) CPO model, (b) CEO model, and (c) CLO model.

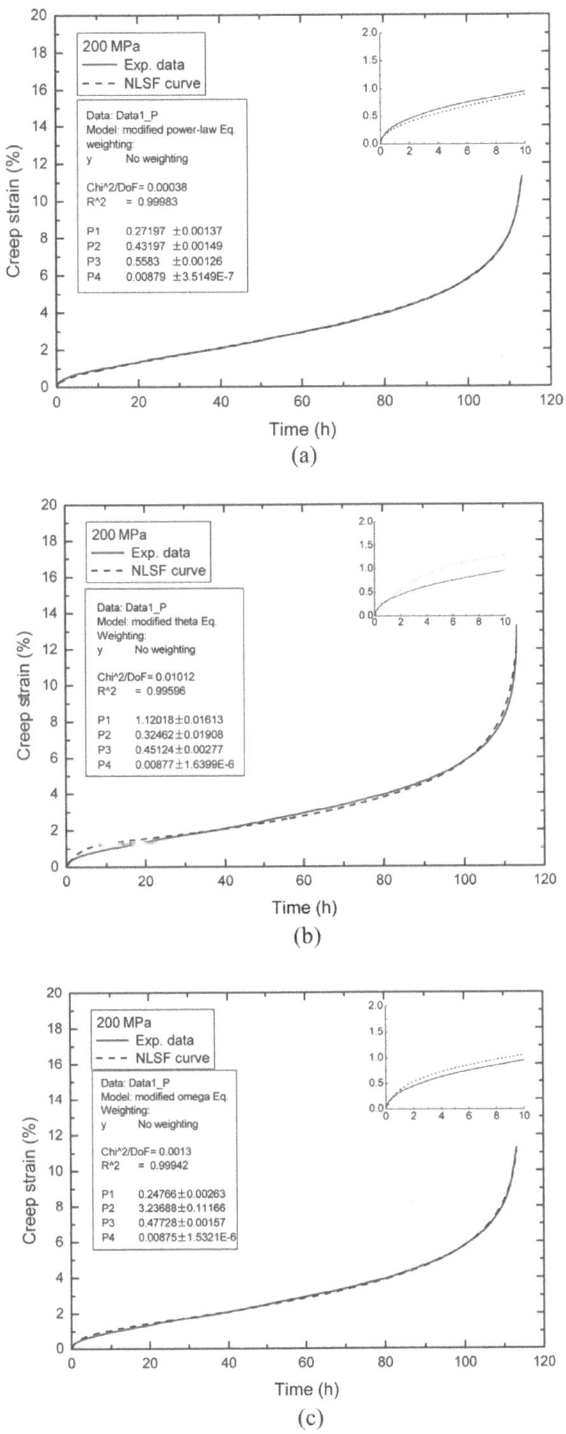


**Fig. 6.** Typical NLSF results on the CPO, CEO and CLO models for experimental creep curves of 130 MPa at 600 °C: (a) CPO model, (b) CEO model, and (c) CLO model.

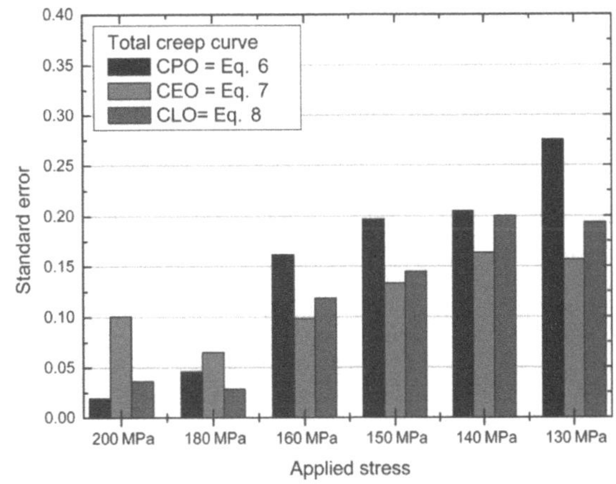


**Fig. 7.** Comparison of standard error values for the CPO, CEO and CLO models.

stants for the CEO model was investigated. The stress dependence of the individual parameter constant ($P_i$) was generally assumed to have a linear form:

$$\log(P_i) = a_i + b_i\sigma \quad (i = 1,2,3,4), \tag{9}$$

where $\sigma$ is stress, and $a_i$ and $b_i$ are constants determined by NLSF analysis. The values of the parameter constants of the CEO model can be calculated using Eq. 9, and the creep curves can be drawn using Eq. 7.

Figure 8 shows the behaviors of stress dependence of the $P_1$, $P_2$, $P_3$, and $P_4$ parameters for the CEO model. The relationships between the four parameters and stresses followed the linear relation well. In the case of the $P_4$ parameter, the linearity was divided into two linear lines at 160 MPa. As written in Eq. 5, the $P_4$ parameter indicates the rupture time as $1/t_r \approx \dot{\varepsilon}_o\ \Omega = P_4$. This means that the creep rupture time at 160 MPa changes. This was similar to the relation of steady-state creep vs. stress as discussed in Fig. 3. As a result, the creep rupture life in ranges of stress higher than the proportion limit will be remarkably reduced in comparison with that extrapolated from the long-term data in stress ranges lower than the proportion limit.

### 3.5. Characterization of long-term creep curves and lives

Figure 9 shows the modeling results of long-term creep curves from short-term creep data using the CEO model for G91 steel at 600 °C. The predicted creep curves are well matched with the experimental creep curves, and a long-term creep curve of 17,000 h at 100 MPa was numerically drawn. Also, the long-term rupture life was predicted using the CEO model as shown in Fig. 10. The predicted results almost overlapped the experimental data, and at a proportional limit of about 160 MPa, the creep rupture strength was

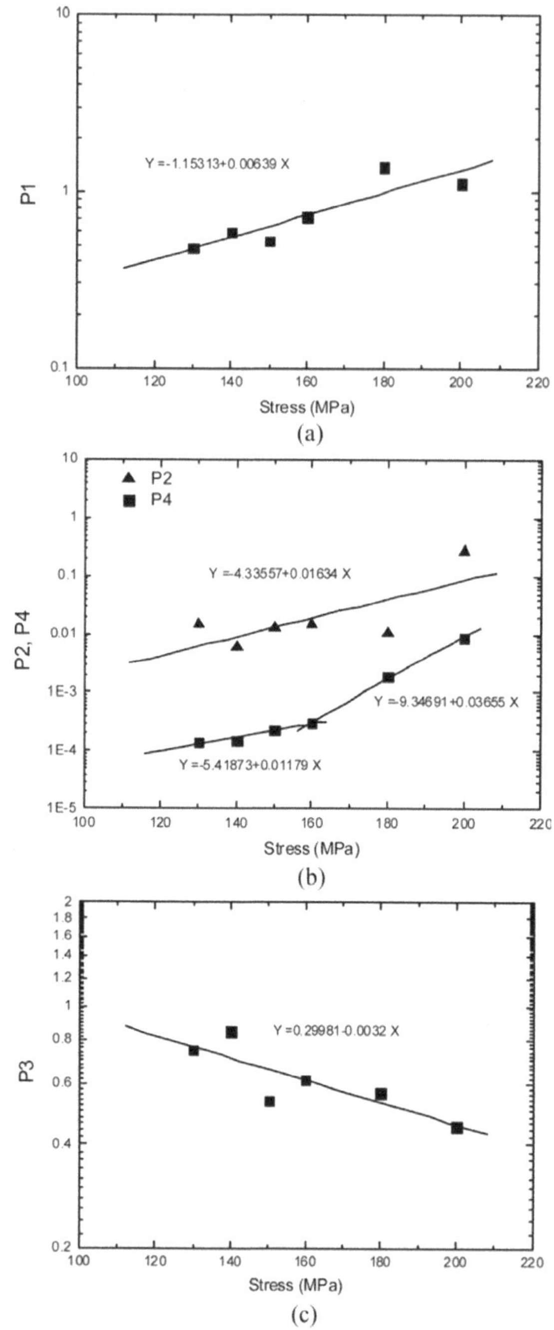


**Fig. 8.** Stress dependence of four parameter constants for the CEO model: (a) CPO model, (b) CEO model, and (c) CLO model.

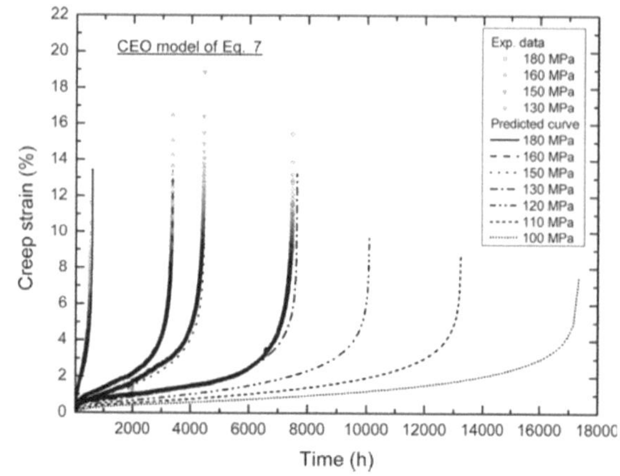


**Fig. 9.** Modeling of total creep curves by the CEO model at 600 °C.

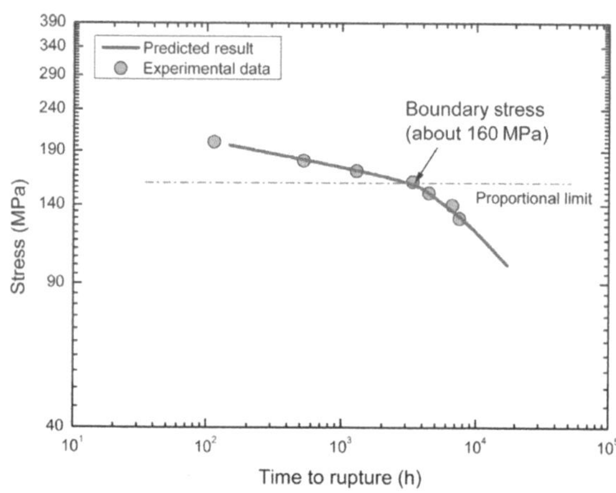


**Fig. 10.** Creep-life prediction of the CEO model for G91 steel at 600 °C.

greatly reduced. The stress value of 160 MPa can be regarded as the boundary stress of G91 steel.

Regarding this boundary stress, it is known that in a stress region lower than the proportion limit, creep rupture life is prolonged by stabilization of the microstructure due to various strengthening factors, such as precipitation strengthening and solid solution strengthening [3]. Thus, creep failure is fractured by a brittleness fracture mechanism. Meanwhile, in a stress region higher than the proportion limit, creep rupture life is reduced by a rapid change in the microstructure due to high external force. Thus, creep failure is fractured by a ductile fracture mechanism. This is because creep deformation mechanisms depend largely on the magnitude of stress.

## 4. CONCLUSIONS

Models of three modified constitutive equations, the CPO, CEO, and CLO, were proposed to accurately characterize total creep curves up to a rupture point. The results demonstrated that in stress regions lower than 160 MPa ($\sigma/\sigma_{ys} < 0.65$), the

CEO model was as well matched with the experimental creep data as the CPO and CLO models, but in stress regions higher than 160 MPa ($\sigma/\sigma_y > 0.65$), the CPO model revealed better agreement than other two models. It was determined that the CEO model was superior in the modeling of long-term creep curves than the CPO and CLO models. Using the CEO model, long-term creep curves for G91 steel were numerically characterized, and the predicted curves almost overlapped with the experimental data. At the proportional limit of about 160 MPa, creep rupture strength was remarkably reduced. The reason for this is that this stress corresponds to the boundary stress of G91 steel. The CEO model can be utilized to numerically predict or model total creep behavior up to the rupture point of G91 steel.

## ACKNOWLEDGMENT

This work was performed under the Long-term Nuclear R&D Program sponsored by the Korean Ministry of Science and Technology.

## REFERENCES

1. W. G. Kim, S. N. Yin, and G. H. Koo, *Met. Mater. Int.* **15**, 727 (2009).
2. W. G. Kim, S. N. Yin, S. H. Kim, W. S. Ryu, C. B. Lee, and S. J. Kim, *Met. Mater. Int.* **15**, 559 (2009).
3. K. Kimura, H. Kushima, K. Sawada, and Y. Toda, *Proc. of CREEP 8*, CREEP 2007-26406, Texas (2007).
4. K. Maruyama and K. Yoshimi, *Proc. of CREEP 8,* CREEP 2007-26150, Texas (2007).
5. S. R. Holdsworth, *Proc. of CREEP 8*, CREEP 2007-26767, Texas (2007).
6. F. Masuyama and T. Tokunaga, *ECCC Creep Conf.*, p. 19-30, Zurich (2009).
7. B. Wilshire and M. R. Bache, *ECCC Creep Conf.*, p. 44-55, Zurich (2009).
8. B. Wilshire and P. J. Scharning, *Int. Mater. Rev.* **53**, 91 (2008).
9. B. Wilshire and P. J. Scharning, *Mater. Sci. Tech.* **24**, 1 (2008).
10. W. G. Kim, S. N. Yin, W. S. Ryu, and J. W. Jang, *Proc. of CREEP 8,* CREEP 2007-26834, Texas (2007).
11. W. G. Kim, S. N. Yin, W. S. Ryu, and J. W. Chang, *Key Eng. Mater.* **326-328**, 1105 (2006).
12. F. R. Larson and J. Miller, *J. Trans. ASME*, **74**, 765 (1952).
13. J. T. Yeom, J. Y. Kim, Y. S. Na, and N. K. Park, *Met. Mater. Int.* **9**, 555(2003).
14. R. W. Evans, J. D. Parker, and B. Wilshire, *Int. J. Press. Vess. Pip.* **50**, 147 (1992).
15. M. Prager, *Development of MPC Omega Method for Life Assessment in the Creep Range*, The Materials Properties Council, INC (1994).
16. M. Prager, *J. Press. Vess. -T. ASME* **122**, 273 (2000).
17. W. G. Kim, S. Y. Yin, Y. W. Kim, and J. W. Chang, *Eng. Fract. Mech.* **75**, 4985 (2008).
18. K. Maruyama, and H. Oikawa, *Scripta metall.* **21**, 233 (1987).
19. H. Wolf, M. D. Mathews, and S. L. Mannan, *Mater Sci. Eng. A* **159**, 199 (1992).
20. A. K. Koul and R. Castillo, *Mater. Sci. Eng. A* **138**, 213 (1991).
21. S. G. R. Brown, R. W. Evans, and B. Wilshire, *Scripta metall.* **20**, 855 (1986).
22. S. G. R. Brown, R. W. Evans, and B. Wilshire, *Int. J. Press. Vess. Pip.* **24**, 251 (1986).
23. S. G. R. Brown, R. W. Evans, and B. Wilshire, *Scripta metall.* **21**, 239 (1987).
24. K. S. Park, D. S. Bae, S. K. Lee, G. H. Lee, J. H. Kim, and T. Endo, *Met. Mater. Int.* **12**, 385 (2006).
25. Y. Kurata and H. Utsumi, *Acta Metall. Sin.* **11**, 397 (1998).

Nuclear Engineering and Design 241 (2011) 3580–3586

Contents lists available at ScienceDirect

Nuclear Engineering and Design

journal homepage: www.elsevier.com/locate/nucengdes

# Probabilistic assessment of creep crack growth rate for Gr. 91 steel

Woo-Gon Kim[a,*], Jae-Young Park[b], Sung-Deok Hong[a], Seon-Jin Kim[b]

[a] Korea Atomic Energy Research Institute, 1045 Daedeok-daero, Yuseong-gu, Daejeon 305-353, Republic of Korea
[b] Pukyong National University, Busan 608-739, Republic of Korea

ARTICLE INFO

*Article history:*
Received 28 March 2011
Received in revised form 20 June 2011
Accepted 23 June 2011

ABSTRACT

This paper focuses on a probabilistic assessment of creep crack growth rate (CCGR) for Gr. 91 steel which is regarded as one of major structural materials of Gen-IV reactors. A series of creep creak growth (CCG) data was obtained from the CCG tests under various applied loads at 600 °C. Using the experimental CCG data, four methods such as a least square fitting method (LSFM), mean value method (MVM), probabilistic distribution method (PDM), and the Monte Carlo method (MCM) were used to determine the parameters $B$ and $q$ for a power law equation between CCGR and $C^*$ integral. The commonly used LSFM revealed a considerable difference in the CCGR lines compared with the MVM and PDM. The PDM was found to be more useful than the LSFM, because it can assess the CCGR lines from the probabilistic viewpoints. It was verified that the two parameters $B$ and $q$ followed a lognormal distribution well. From the lognormal distribution, a number of random variables for the $B$ and $q$ parameters were successfully generated by the Monte Carlo Simulation (MCS) technique. The CCGR lines for the 10% and 90% probabilities were predicted by the PDM and MCM, and the MCM result was compared with the PDM one.


## 1. Introduction

High-chromium steels are widely used in recent fossil power plants because of their excellent high-temperature strength. In particular, modified 9Cr–1Mo steel (ASME grade 9Cr–1Mo, hereafter Gr. 91) is regarded as a promising candidate for structural materials of Gen-IV reactor types such as steam generators, intermediate heat exchangers and hot pipes in sodium-cooled fast reactors (SFR) and pressure vessels in very high temperature reactors (VHTR). The selection of Gr. 91 steel is based primarily on its low thermal expansion parameter and high resistance to stress corrosion cracking in water/steam systems compared to austenitic stainless steels and its better mechanical properties at elevated temperatures compared to alternate 2.25Cr–1Mo steel (Takahasi et al., 2005; Choudhary et al., 1999; Kim et al., 2001, 2009).

Since the components of Gen-IV reactors will be designed for 40–60 years at elevated temperatures, such operational conditions can generate localized creep damage in the components, propagating cracks and ultimately causing fracture. A significant portion of a component's lifetime will be spent in crack propagation (Nikbin et al., 1986; Webster and Ainsworth 1993; Ganesh Kumar et al., 2010). Accordingly, for design and safety assessment purposes, it is necessary to reasonably establish creep crack growth rate (CCGR) law (or creep crack propagation law) in components that are subjected to creep loading.

Over the last few decades, many studies have been conducted for assessing the integrity of various structures containing defects and cracks (Tan et al., 2001; Shibli and Hamata, 2001; Hamata et al., 2001; Gampe and Seliger, 2001; Dogan (2001)). Certain codes of practice, such as R5, R6, BS 7910, and A16, have been developed to deal with crack initiation and growth in components operating at high temperature plants such as those used in power generation and petro-chemical and nuclear industries. However, all these procedures use a deterministic approach based on certain assumptions and fixed material properties. Many studies have not dealt with probabilistic viewpoints that define the probability of a fracture. In such circumstances, for the last few years, there has been a rising need to develop a quantifiable method based on a statistical and probabilistic approach to reduce overly conservative assessments arising from the use of worst case combinations of upper and/or lower bounds of material property data. However, some studies related to statistical and probabilistic methodologies for assessing creep crack growth data have been reported and as being of interest to other researchers (Wei et al., 1994; Gliniak et al., 1997; Nikbin et al., 2003; Kanto et al., 2010).

This study deals with a probabilistic assessment of the CCGR for Gr. 91 steel. A series of creep crack growth (CCG) data was obtained from CCG tests under different applied loads at 600 °C, and the CCG behavior was characterized in terms of the $C^*$ integral. A number of random variables for $B$ and $q$ were generated using Monte Carlo

* Corresponding author. Tel.: +82 042 868 2493; fax: +82 042 868 8549.
*E-mail address:* wgkim@kaeri.re.kr (W.-G. Kim).

0029-5493/$ – see front matter. 
doi:10.1016/j.nucengdes.2011.06.042

simulation (MCS) technique, and the CCGR lines were predicted with probabilistic viewpoints.

## 2. Experimental procedures

The Gr. 91 steel used in this study was a hot rolled Plate 16 mm in thickness. The conditions of the heat treatment condition were normalized and tempered at 1050 °C/1mim/mm and 770 °C/3mim/mm, respectively. The chemical composition is given in Table 1. To obtain the material data used in the $C^*$ equation, tension and creep tests were conducted at 600 °C. Tension specimens were machined with a rectangular cross section 2 mm thick and 6.25 mm wide with a 25 mm gauge length. The strain rate was conducted with $1 \times 10^{-4}$/s at 600 °C. Creep test specimens were machined to a cylindrical shape with a 30 mm gauge length and 6 mm diameter. All the specimens were taken along the rolling direction. The gauge sections of the creep specimens were polished using #1000-grit sand papers and stroking along the specimen axis.

Constant-load creep tests were carried out at different stress levels at 600 °C. The experimental procedures followed the recommendations of the ASTM standard E139 (ASTM E139, 2007). Creep strain data under elapsed times were taken automatically using a PC through a high precision LVDT. The steady state creep rate was obtained from the experimental creep curves. Material constants of $D_1$, $m$, $A$, and $n$ were obtained from the tension and creep tests at 600 °C.

Creep crack growth tests were carried out at various constant-load ranges of 3800 N–5200 N at 600 °C. The compact tension (CT) specimens had a width ($W$) of 25.4 mm, a thickness ($B$) of 12.7 mm, and side grooves with 10% depth. The initial crack ratio ($a/W$) was about 0.5. The test temperature was constantly maintained to within ±2 °C during the test. The load–line displacement was measured using a linear gauge assembly attached to the specimen. Creep crack length was measured by using the direct current potential drop (DCPD) technique. A current of 8 A was used. The output voltage was recorded by PC and converted to crack length by using Johnson's formula as follows:

$$a = \frac{2W}{\pi}\cos^{-1}\frac{\cosh(\pi y_o/2W)}{\cosh\frac{V}{V_o}\left[\cosh^{-1}\left\{(\cosh(\pi y_o/2W)/\cos(\pi y_o/2W))\right\}\right]} \quad (1)$$

where $a$ = crack length, $a_o$ = initial crack length, $V$ = output voltage, $V_o$ = initial output voltage, $W$ = specimen width, and $y_o$ = a half distance between output potential leads. The CCG testing procedures followed the ASTM standard E1457 (ASTM E 1457, 2007) principally.

After CCG testing, the CT specimens were cooled down in a liquid nitrogen solution and fractured to measure the final crack length. Their microstructures in front of the crack tips were observed using a scanning electron microscope (SEM) after etching for about 20 s in the mixed solution, HF (5%) + $HNO_3$ (95%) + $H_2O$ (90%).

## 3. Results and discussion

### 3.1. Material properties and fracture microstructures

As stated above, the material constants of $D_1$, $m$, $A$, and $n$ were obtained from the tension and creep tests for Gr. 91 steel at 600 °C. The $D_1$ and $m$ values in the Ramberg–Osgood relationships, $\varepsilon_p = D_1(\sigma/\sigma_{ys})^m$ were obtained from the tension tests, where, $\varepsilon_p$ is the plastic strain, $\sigma_{ys}$ is the yield stress, and $m$ is the stress exponent. The $A$ and $n$ values in Norton's law, $\dot{\varepsilon}_{ss} = A\sigma^n$, were obtained from the creep tests, where, $\dot{\varepsilon}_{ss}$ is the steady state creep rate, and $n$ is the creep exponent. The results for four material constants are summarized in Table 2.

A summary of the CCG test conditions and the crack length measurements is given in Table 3. The CCG test time for the eight specimens was maintained for 774 h–3275 h. In Table 3, the $a_{pf}$ is a crack size predicted by Johnson's formula, and the $p_{mf}$ is a crack length measured using a nine-point technique for a fractured CT specimen after the CCG tests. The results between the predicted and measured crack lengths were within the error ranges (0.85–1.15) of the requirements in ASTM E1457. It is thus verified that the CCG tests were correctly conducted, as listed in Table 3. Among the tests numbered from 1 to 13, the five tests of 2, 5, 6, 10 and 11 were discarded because they were failed during the tests or not included within the error ranges (0.85–1.15) of ASTM requirement.

Fig. 1 shows a typical SEM micrograph of creep crack propagation for G91-1 (3800N) specimen. Photo (a) represents the total behavior of the creep crack propagation developed by creep damage during the CCG test. Photo (b) represents a high magnification of the front of the crack tip from photo (a). In photo (a), it is well seen that the crack was steadily propagated and that some minor cracks occurred in front of the crack tip. These minor cracks or creep cavities were due to accumulated creep damage, and they initiated crack propagation. As shown in Fig. 2, these cavities or micro cracks were due to the precipitates formed along the grain boundaries in the near region of the crack tips. The Gr. 91 steel also formed a large amount of precipitates at the lath and grain boundaries during the CCG test. The microstructures in the CCG tests were similar to those of the creep tests.

### 3.2. Calculation of the $C^*$

The fracture mechanics parameter, $C^*$-integral has been widely used to characterize creep crack growth rate in metals undergoing a steady state creep. The general form between the creep crack growth rate ($da/dt$) and the $C^*$ parameter can be expressed as follows (Nikbin et al., 1986; Webster and Ainsworth, 1993):

$$\frac{da}{dt} = B[C^*]^q \quad (2)$$

In this expression, the $B$ and $q$ parameters are material constants, which are commonly obtained by a linear regression of the CCGR data. They are related to the intercept and slope, respectively. In order to obtain the $da/dt$ in Eq. (2), the material constants, $D_1$, $m$, $A$, and $n$ were obtained from the tension and creep tests at 600 °C, as described in the above Section 2. Material properties of Table 2 were used to calculate $C^*$. The relationship between $da/dt$ and $C^*$ was obtained for all eight specimens. For the CT specimen, the $C^*$ value was calculated by Eq. (3), and load-line displacement rate ($\dot{V}_c$) due to creep strain was calculated by Eq. (5).

$$C^* = \frac{P\dot{V}_c}{B_N(W-a)}\eta\left(\frac{a}{W}, n\right) \quad (3)$$

$$\eta\left(\frac{a}{W}, n\right) = \frac{1}{(1-a/W)}\frac{n}{n+1}\left(\gamma - \frac{\delta}{n}\right) \quad (4)$$

$$\dot{V}_c = \dot{V} - \frac{\dot{a}B_N}{P}\left(\frac{2K^2}{E'} + (m+1)J_p\right) \quad (5)$$

where $P$ = applied load, $a$ = crack length, $\dot{V}$ = total-line displacement rate, $W$ = specimen width, $B_N$ = net thickness of the specimen, $E'$ = elastic modulus for plane strain, $K$ = stress intensity factor, $\dot{a}$ = crack growth rate, and $m$ = stress exponent in the

**Table 1**
Chemical composition of Gr. 91 steel (wt.%).

| C | Si | Mn | P | S | Ni | Cr | Mo | Cu | V | Al | N | Nb |
|---|---|---|---|---|---|---|---|---|---|---|---|---|
| 0.115 | 0.23 | 0.415 | 0.012 | 0.0014 | 0.22 | 8.9 | 0.869 | 0.038 | 0.194 | 0.020 | 0.0513 | 0.073 |

**Table 2**
Tension and creep properties of Gr. 91 steel at 600 °C.

| E (GPa) | $\sigma_{YS}$ (MPa) | $\sigma_{UTS}$ (MPa) | $\varepsilon_p = D_1(\sigma/\sigma_{ys})^m$ | | $\dot{\varepsilon}_{ss} = A\sigma^n$ | |
|---|---|---|---|---|---|---|
| | | | $D_1$ | m | A | n |
| 164 | 247.5 | 420 | 0.0017 | 6.4 | 1.28E−27 | 9.98 |

**Table 3**
CCG test conditions and crack size measurement of Gr. 91 steel at 600 °C.

| Sp. no. | Test conditions & transition time | | | Results of crack size measurement | | | | |
|---|---|---|---|---|---|---|---|---|
| | Load (N) | Test time (h) | $t_T$ (h) | a (mm) | $a_o$ (mm) | $a_{pf}$ (mm) | $a_{mf}$ (mm) | Error (%) |
| G91-1 | 3800 | 3148.6 | 2051 | 10.13 | 11.60 | 14.67 | 14.92 | −1.729 |
| G91-3 | 4500 | 2541.4 | 1323 | 10.04 | 11.53 | 13.91 | 14.10 | −1.367 |
| G91-4 | 5000 | 1249.8 | 546 | 10.02 | 11.66 | 13.08 | 13.28 | −1.485 |
| G91-7 | 4000 | 3275.6 | 1380 | 10.06 | 11.42 | 13.81 | 14.35 | −3.752 |
| G91-8 | 5000 | 1078.2 | 440 | 10.10 | 11.45 | 13.62 | 13.69 | −0.510 |
| G91-9 | 5000 | 1008.7 | 161 | 10.07 | 11.37 | 13.69 | 13.72 | −0.256 |
| G91-12 | 4800 | 1284.5 | 455 | 10.08 | 11.63 | 17.40 | 17.62 | −1.242 |
| G91-13 | 5200 | 774 | 413 | 9.997 | 11.59 | 17.15 | 17.09 | 0.323 |

a, initial crack size(at sharp notch); $a_o$, pre-cracked size by fatigue; $a_{pf}$, predicted final crack size; $a_{mf}$, measured final crack size, $t_T$, transition time.

Ramberg–Osgood stress vs. strain relationship ($\varepsilon_p = D_1(\sigma/\sigma_{ys})^m$). Also, the $J_p$ contribution for CT specimen is given by

$$J_p = \frac{D_1 h_1(a/W, m)}{(\sigma_{ys}(W-a))^m}\left(\frac{P}{1.455 B_N \alpha}\right)^{m+1} \quad (6)$$

where $h_1$ is a function of $a/W$ and $m$; this value was obtained using the table result of ASTM E1457.

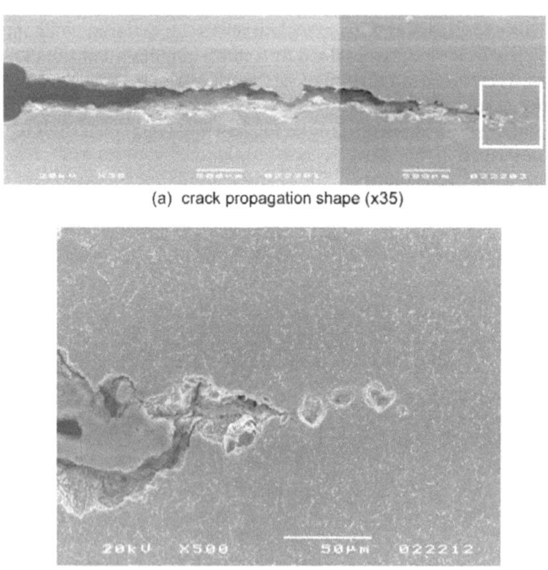

Fig. 1. SEM micrograph showing crack propagation for G91-1 (3800N) specimen (a) crack propagation shape (×35), (b) a high magnification in front of crack tip (×500).

### 3.3. Analysis of the CCGR lines

The $B$ and $q$ parameters are important because the CCGR line depends on them. The $B$ and $q$ values change from material to material and with the regression method used. To logically obtain the CCGR line of Gr. 91 steel, we used three methods; a least square fitting method (LSFM), mean value method (MVM), and probability distribution method (PDM). The $B$ and $q$ values obtained for the three methods are summarized in Table 4. The MVM and PDM were very close in the $B$ and $q$ values, but the LSFM was significantly different. A log–log plot of $da/dt$ vs. $C^*$ in the CCG law is shown in Fig. 3. Detail descriptions on how to obtain the $B$ and $q$ values for each method are given as follows.

First, the LSFM is used to determine the $B$ and $q$ values by taking the best linear regression of the total CCG data. So far, the LSFM has been commonly used to determine the CCGR lines. However, the CCGR line of the LSFM is easily influenced by the number of the

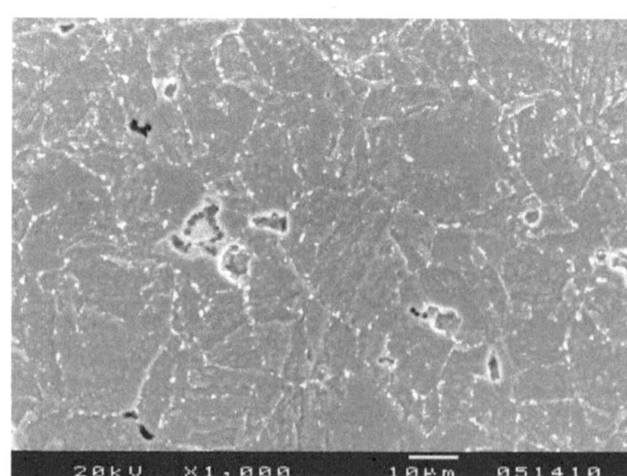

Fig. 2. SEM micrographs showing creep cavities or micro cracks along the lath and grain boundaries for a G91-1 (3800N) specimen.

**Table 4**
Summary of *B* and *q* values obtained from the three methods.

| SP. no. | Applied load (N) | $B$ (×0.01) | $q$ |
|---|---|---|---|
| G91-1 | 3800 | 4.44 | 0.84 |
| G91-3 | 4500 | 3.62 | 0.96 |
| G91-4 | 5000 | 3.43 | 1.05 |
| G91-7 | 4000 | 5.56 | 0.94 |
| G91-8 | 5000 | 3.64 | 1.11 |
| G91-9 | 5000 | 2.63 | 0.95 |
| G91-12 | 4800 | 2.28 | 0.91 |
| G91-13 | 5200 | 1.34 | 0.66 |
| Mean value method (MVM) | | 3.37 | 0.93 |
| Least square fitting method (LSFM) | | 1.89 | 0.77 |
| Lognormal distribution (PDM) | | 3.14 | 0.91 |

CCGR data points regardless of the $C^*$ ranges in the log–log plot of $da/dt$ vs. $C^*$. As a result, the LSFM gave considerably lower values than MVM and PDM in the *B* and *q* values. The CCGR line was not in full accord with the CCG data points, as shown in Fig. 3. In reality, it appeared that the numbers of total CCG data points were different with each test condition. For this reason, the *q* value in the LSFM was the lowest at 0.77, when compared with 0.93 from the MVM and 0.91 from the PDM. Therefore, it is supported that the LSFM is not correct in the linear regression of the CCG data. The CCGR law in the LSFM was obtained as $da/dt = 1.89 \times 10^{-2}[C*]^{0.77}$.

The MVM is then to take just the mean value of the eight values for the *B* and *q* parameters which are obtained from each linear regression to the applied specimens, as listed in Table 4. This method is reasonable because each *B* and *q* value can be determined based on an equal contribution of the individual characters for each applied specimen. The CCGR line in the MVM showed good agreement with the experimental data, as shown in Fig. 3. Accordingly, the CCG law in the MVM was obtained as $da/dt = 3.37 \times 10^{-2}[C*]^{0.93}$.

Finally, the PDM is used to take the probability distribution for the eight *B* and *q* values. As shown in Fig. 4(a) and (b), the *B* and *q* for all the eights specimens followed well the lognormal probability with a linear relation. Thus, the standard deviation ($\sigma_{Le}$) and the mean value ($\mu_{Le}$) were determined from the lognormal distributions for the *B* and *q*. Herein, subscript Le means a natural logarithm. The *B* and *q* values were each obtained at 50% probability. Namely, the *B* value was $\mu_{Le} = -3.4667$, and the *q* value was $\mu_{Le} = -0.0865$. The standard deviation for *B* and *q* was $\sigma_{Le} = 0.1587$ and $\sigma_{Le} = 0.4406$, respectively. Accordingly, the CCGR law in the PDM was obtained as $da/dt = 3.14 \times 10^{-2}[C*]^{0.91}$. From this result, the CCGR line of the PDM closely overlapped that of the MVM, and the PDM can be used as a good method to probabilistically predict the CCGR line.

In addition, as noted above, if using the standard deviation ($\sigma_{Le}$) and mean value ($\mu_{Le}$) obtained from the lognormal distribution for

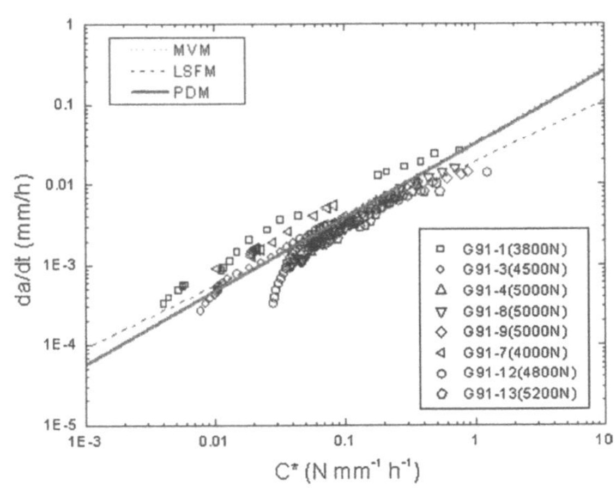


**Fig. 3.** Comparison of the CCGR lines obtained from the three methods.

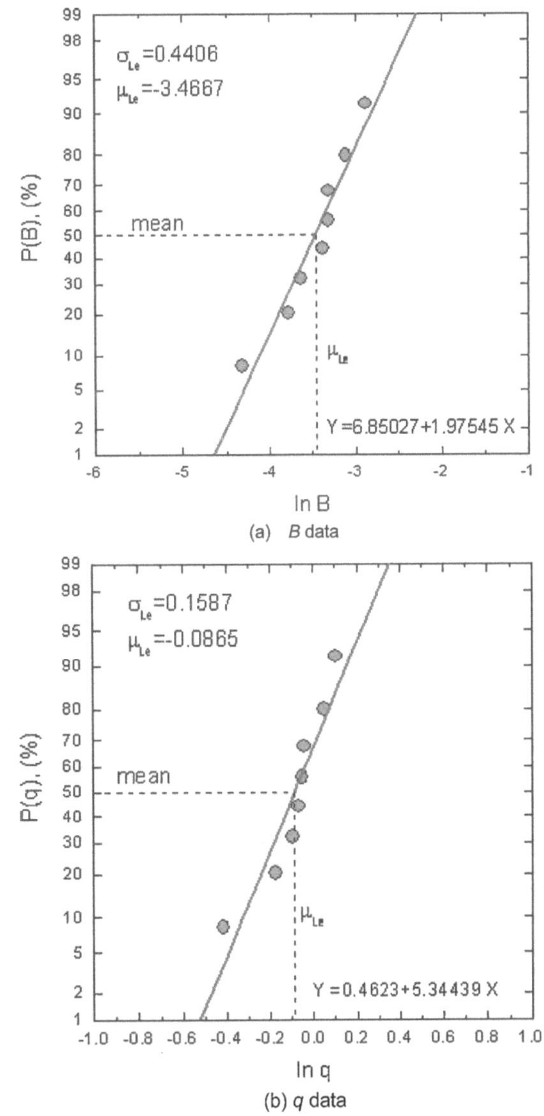


**Fig. 4.** Lognormal distribution of *B* and *q* data.

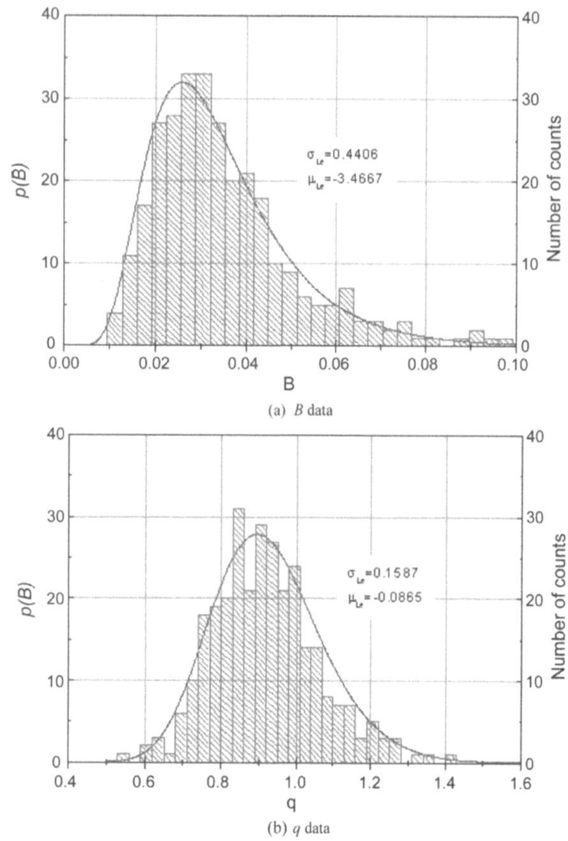


**Fig. 5.** Distribution of PDF for the $B$ and $q$ data.

the $B$ and $q$, the probability density function (PDF) for the $B$ and $q$ can be calculated using Eq. (7).

$$p(x(=B,q)) = \frac{1}{\sqrt{2\pi}\sigma_{\mathrm{Le}}x}\exp\left\{-\frac{(\ln x-\mu_{\mathrm{Le}})^2}{2\sigma_{\mathrm{Le}}{}^2}\right\} \tag{7}$$

The distribution shapes of the $B$ and $q$ data could be obtained, as shown in Fig. 5(a) and (b). From the results, using statistical viewpoints, we can analyze the $B$ and $q$ data in the CCG law of Gr. 91 steel.

### 3.4. Probabilistic prediction of the CCGR line

As described above, it was verified that the $B$ and $q$ parameters of the CCG law followed the lognormal distribution well. Thus, the $B$ and $q$ values at any probability can be calculated from the results obtained in Fig. 4(a) and (b). As an example of application, for 10% and 90% probabilities (the probability variable, $P(B,q)$ is $[10\% \le P(B,q) \le 90\%]$), the probabilistic ranges of the CCGR line were obtained by the PDM, as shown in Figs. 6 and 7.

Fig. 6 shows the results of the CCGR lines for when the $B$ and $q$ ranges were varied to 10% and 90% at $B$ = 50% and $q$ = 50%. An intersection point occurs at $C^*$ = 1 because the value of the slope exponent, $q$, has a constant value at $C^*$ = 1. Also, Fig. 7 shows the results of the probabilistic CCGR lines which were considered for all eight cases; the $B$ and $q$ values were varied to 10%, 50% and 90%. The probabilistic CCGR lines in 90% and 10% were calculated using the lognormal distribution of Fig. 4(a) and (b). The upper prediction line of 90% will be the most conservative prediction, because the fastest CCGR is predicted here. On the contrary, a lower prediction line of 10% will be the most non-conservative prediction because the lowest CCGR is predicted here. This means that by knowing the probability distribution for $B$ and $q$, the CCGR can be predicted probabilistically. Thus, the PDM can be utilized well for assessing the cracked components under creep damage.

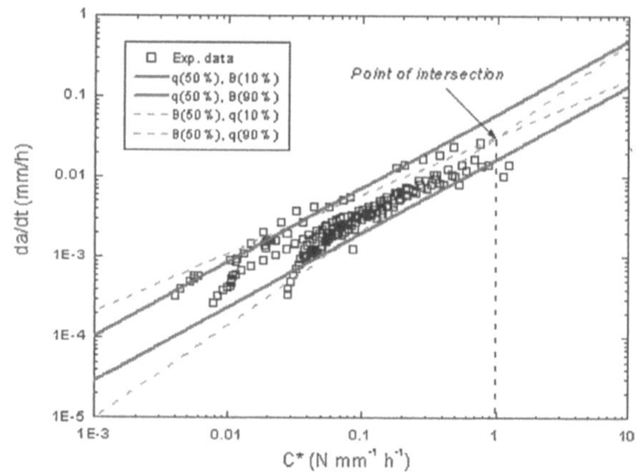


**Fig. 6.** Prediction of CCGR at any probability.

### 3.5. Prediction of CCGR lines by the MCS for B and q data

For problems involving random variables with known or assumed probability distributions, the MCS has been commonly used. The MCS is repeated process of generating deterministic solutions to a given problem; in each solution a particular set of values is generated in accordance with corresponding probability distributions. By repeating the process, a sample of solution, each of which corresponds to a different set of values of the random values, can be obtained. In this study, the MCS was carried out to generate random variables (*i.e.*, random numbers) for the $B$ and $q$ parameters.

Box and Muller (1958) proposed that random variables ($S$) with a standard normal distribution can be represented as

$$S = (-2\ \ln U_1)^{0.5}\cos 2\pi U_2 \tag{8}$$

In this expression, $U_1$ and $U_2$ are standard uniform variables with a uniform probability density function (PDF) between 0 and 1.0. Thus,

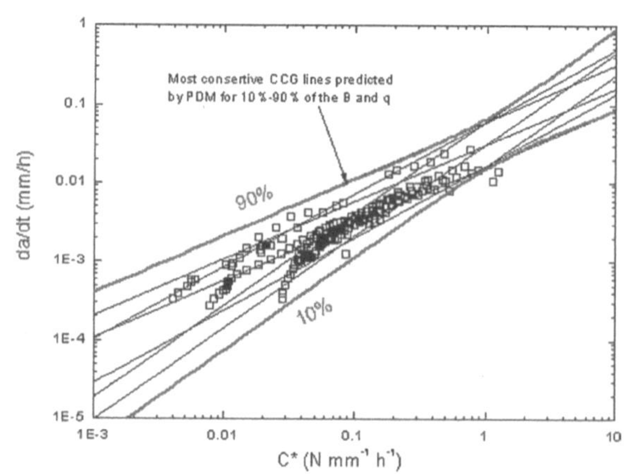


**Fig. 7.** CCGR lines predicted for 10% and 90% of $B$ and $q$.

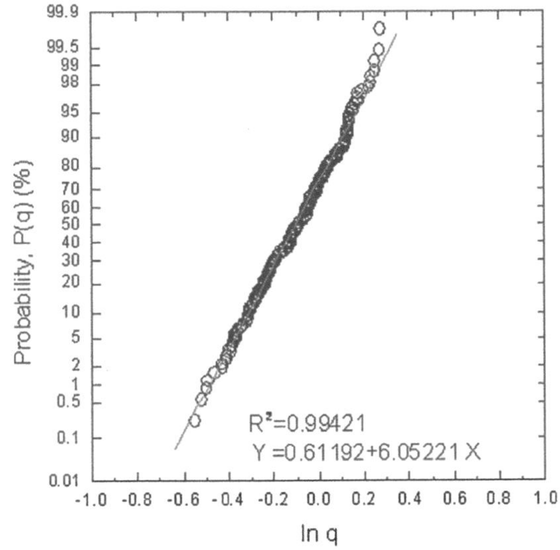


(a) distribution for B

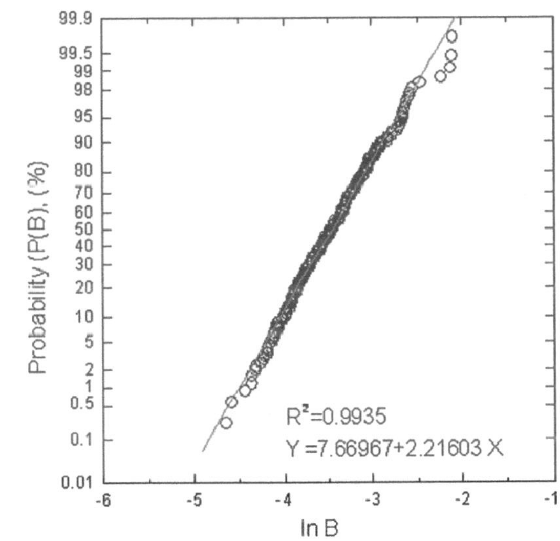


(b) distribution for q

**Fig. 8.** Result of the MCS for the B and q distributions.

random variables ($x$) with a lognormal distribution $N(\mu_{Le}, \sigma_{Le})$ can be given as

$$x = \mu_{Le} + \sigma_{Le} \cdot S = \mu_{Le} + \sigma_{Le}(-2 \ln U_1)^{0.5} \cos 2\pi U_2 \tag{9}$$

For a lognormal random variable, $x'$, the distribution of $x = \ln(x')$ is a normal distribution. Thus, if $x$ is a value generated from Eq. (9),

$$x' = e^x \tag{10}$$

Eq. (9) becomes a random number for lognormal distribution with mean $\mu$ and standard deviation $\sigma$. Eqs. (9) and (10) were used to generate the random values of the $B$ and $q$, which followed a lognormal distribution, as described above.

Fig. 8 shows the MCS results obtained from the lognormal distribution of the $B$ and $q$. Standard uniform variables of the $U_1$ and $U_2$ were generated using Eq. (9). Then, a number of random variables for each $B$ and $q$ were calculated. It is obviously supported that the generated $B$ and $q$ data followed a lognormal distribution because the probability data revealed a linear relationship well in the lognormal plot. Although there were some differences in the linear equations between the PDM and the MCS, the results of the MCS in Fig. 8(a) and (b) were almost similar to those in Fig. 4(a) and (b). It is thus verified that a number of the random variables from only several data sets of the $B$ and $q$ were well realized by the MCS. The CCGR data of 30,000 points were successfully generated by the MCM, as shown in Fig. 9. Using these generated CCGR data, it is possible to probabilistically predict the CCGR lines.

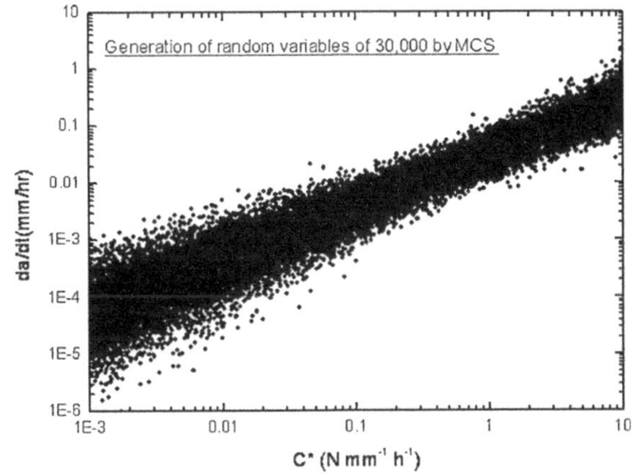


**Fig. 9.** A large amount of CCG data generated by the MCS.

To probabilistically predict the CCGR lines, the probabilistic distributions of $da/dt$ at various $C^*$ values were investigated using the results in Fig. 9. Fig. 10 presents the typical result for the probability distribution of the $da/dt$ at various $C^*$ values: 0.001, 0.01, 0.1, 1, and 10. The relationships between probability and $da/dt$ revealed a lognormal distribution with good linearity.

Fig. 11 shows the CCGR lines predicted using the MCM for the 10% and 90% probabilities. It appeared that all the experimental CCG data were almost included within the range of 10% and 90% probabilities. The upper line of the 90% probability corresponds to the most conservative prediction line, and the lower line of the 10% probability corresponds to the most non-conservative prediction line. Accordingly, if using the MCS, we can assess the CCGR lines with probabilistic viewpoints.

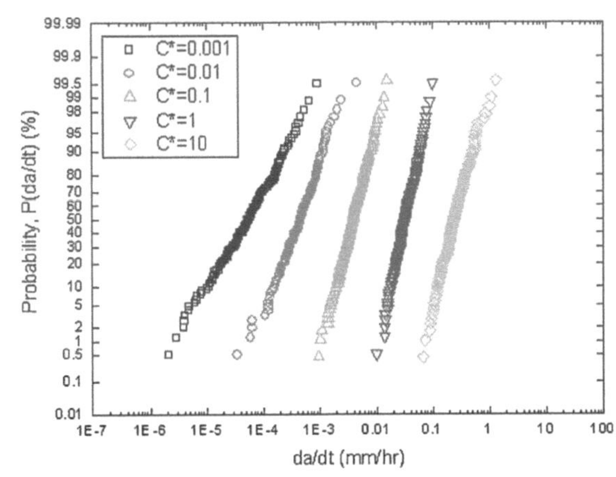


**Fig. 10.** Probability distribution of the CCG data at various $C^*$ ranges.

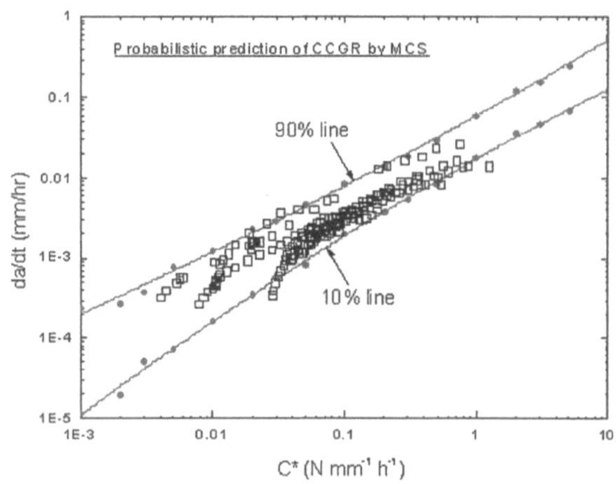


**Fig. 11.** Prediction of the CCGR lines for10% and 90% probabilities.

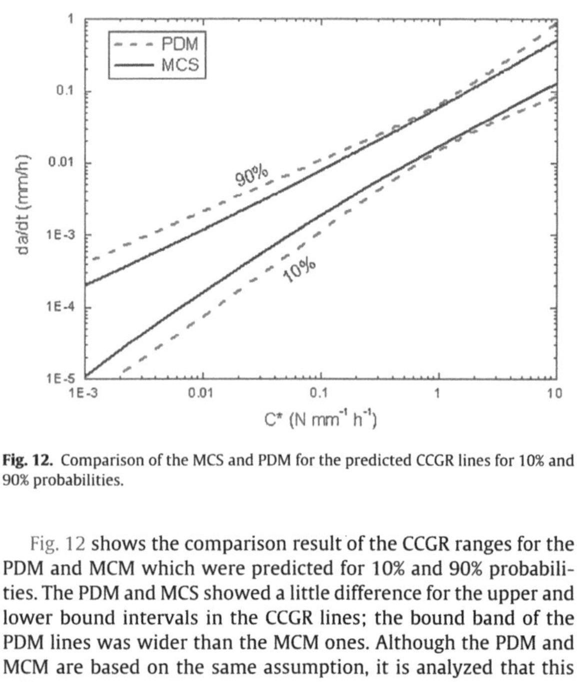


**Fig. 12.** Comparison of the MCS and PDM for the predicted CCGR lines for 10% and 90% probabilities.

Fig. 12 shows the comparison result of the CCGR ranges for the PDM and MCM which were predicted for 10% and 90% probabilities. The PDM and MCS showed a little difference for the upper and lower bound intervals in the CCGR lines; the bound band of the PDM lines was wider than the MCM ones. Although the PDM and MCM are based on the same assumption, it is analyzed that this difference were attributed to the differences in the linear equations for the $B$ and $q$, as shown in Figs. 4 and 8. It is thus expected that the MCM will result in more correct and economic designs than the PDM, because it reduces overly conservative assessments arising from the use of worst case combinations of upper and/or lower bounds of the CCGR data. Also, the MCM would be useful because the CCGR lines can be predicted using only several data for the new and emerging class of high-temperature materials when the CCG data is lacking. Nevertheless, it is believed that the PDM and the MCS are useful to assess the creep crack growth of Gr. 91 steel, because they did not give a significant difference in the predicted CCGR lines, and because they can assess the CCGR with the probabilistic viewpoints.

## 4. Conclusion

This paper has presented four methods to determine the two parameters $B$ and $q$ for a power law equation between creep crack growth rate and the $C^*$ integral. The LSFM revealed a considerable difference in the CCGR lines when compared to the MVM and PDM, because the LSFM is easily influenced by the number of the CCGR data points. It was verified that the $B$ and $q$ parameters followed a lognormal distribution well. From this lognormal distribution, a number of random variables for the $B$ and $q$ were successfully generated by the MCS technique. In the upper and lower bound intervals of the CCGR lines, the MCM was narrower than the PDM. The MCM is expected to achieve the correct and economic designs because it had a good rationale for reasonable conservatism. Nevertheless, it is believed that the PDM and the MCS is useful to assess the creep crack growth of Gr. 91 steel, because they did not give a significant difference in the predicted CCGR lines.

## Acknowledgement

This study was performed under the Long-term Nuclear R&D Program sponsored by the Korean Ministry of Education, Science and Technology.

## References

ASTM E139-07, 2007. Standard test method for conducting creep, creep-rupture and stress rupture tests of metallic materials.
ASTM E1457-07, 2007. Standard test method for measurement of creep crack growth rates in metals.
Box, G.E.P., Muller, M.E., 1958. A note on the generation of random normal deviates. Ann. Math Stat. 29, 610–611.
Choudhary, B.K., Bhanu Sankara Rao, K., Mannan, S.L., 1999. Steady state creep deformation behavior of 9Cr–1Mo ferritic steel forging in quenched and tempered condition. Trans. Indian Inst. Met. 52, 327–336.
Dogan, B., 2001. Creep crack growth of high temperature weldments. Int. J. Pressure Vessel Piping 78, 795–805.
Gampe, U., Seliger, P., 2001. Creep crack growth testing of P91 and P22 pipes bends. Int. J. Pressure Vessel Piping 78, 859–864.
Ganesh Kumar, J., Chowdary, M., Ganesan, V., Paretkar, R.K., Bhanu Sankara Rao, Mathew, M.D., 2010. High temperature design curves for high nitrogen grades of 316LN stainless steel. Nucl. Eng. Des. 240, 1363–1370.
Gliniak, J.T., Harlow, D.G., Delph, T.J., 1997. Probabilistic model for the growth of creep cracks. Eng. Fract. Mech. 57, 25–39.
Hamata, N., Le, Mat., Shibli, I.A., 2001. Creep crack growth of seam-welded P22 and P91 pipes with artificial defects. Part I. Experimental study and post-test metallography. Int. J. Pressure Vessel Piping 8, 819–826.
Kanto, Y., Onizawa, K., Machida, H., Isobe, Y., Yoshimura, S., 2010. Recent Japanese research activities on probabilistic fracture mechanics for pressure vessel and piping of nuclear power plant. Int. J. Pressure Vessel Piping 87, 11–16.
Kim, W.G., Kim, S.H., Ryu, W.S., 2001. Creep characterization of type 316LN and HT9 stainless steels by the K-R creep damage model. KSME Int. J. 15, 1463–1471.
Kim, W.G., Yin, S.N., Koo, G.H., 2009. Generation of isochronous stress–strain curves with a nonlinear least square fitting method for modified 9Cr–1Mo steel. Met. Mater. Int. 15, 727–732.
Nikbin, K.M., Smith, D.J., Webster, G.A., 1986. An engineering approach to the prediction of creep crack growth. J. Eng. Mater. Technol. 108, 186–191.
Nikbin, K.M., Yatomi, M., Wasmer, K., Webster, G.A., 2003. Probabilistic analysis of creep crack initiation and growth in pipe components. Int. J. Pressure Vessel Piping 80, 585–595.
Shibli, I.A., Hamata, N., Le, Mat, 2001. Creep crack growth in P22 and P91 welds-over from SOTA and HIDA projects. Int. J. Pressure Vessels Piping 78, 785–793.
Tan, M., Celard, N.J.C., Nikbin, K.M., Webster, G.A., 2001. Comparison of creep crack initiation and growth in four steels tested in HIDA. Int. J. Pressure Vessel Piping 78, 737–747.
Takahasi, Y., Igari, T., Kawashima, F., Date, S., Isobe, N., Itoh, T., Noguchi, Y., Kobayashi, K., 2005. High-temperature crack growth behavior of high-chromium steels. SMiRT 18, 1904–1915.
Webster, G.A., Ainsworth, R.A., 1993. High Temperature Components Life Assessment. Chapman & Hall, New York.
Wei, R.P., Masser, D., Liu, H., Harlow, D.G., 1994. Probabilistic considerations of creep crack growth. Mater. Sci. Eng. A159, 69–76.

International Journal of Pressure Vessels and Piping 87 (2010) 289–295

Contents lists available at ScienceDirect

International Journal of Pressure Vessels and Piping

journal homepage: www.elsevier.com/locate/ijpvp

# Creep oxidation behaviour and creep strength prediction for Alloy 617

Woo-Gon Kim [a,*], Song-Nan Yin [a], Gyeong-Geon Lee [a], Yong-Wan Kim [a], Seon-Jin Kim [b]

[a] Nuclear Material Research Division, Korea Atomic Energy Research Institute (KAERI), 1045 Daedeokdaero, Yuseong-Gu, Daejeon 305-353, Republic of Korea
[b] Pukyong National University, Department of Mechanical and Automotive Engineering, San 100, Yongdang-Dong, Nam-Gu, Busan 608-739, Republic of Korea

ARTICLE INFO

Article history:
Received 30 November 2008
Accepted 27 January 2010

Keywords:
Alloy 617
Very high temperature gas-cooled reactor (VHTR)
Creep
Creep strength prediction
Larson-Miller (LM) parameter

ABSTRACT

Creep experimental data was obtained by a series of creep tests with different stress levels at 950 °C for Alloy 617. Oxidation behaviour was investigated by observing the microstructures of fractured specimens after the creep tests. Oxidation thickness was measured quantitatively with the creep rupture times, and the oxidation microstructures were represented by a SEM image. In addition, the long-term creep strength for Alloy 617 was predicted by using a multi-constant method with two $C$ instead of the conventional one with a unique $C$ in the Larson-Miller (LM) parameter. For $10^5$ h at 950 °C, the creep strength for the conventional method was 7.2 MPa, but for the multi-constant method it was reduced to 4.7 MPa. The conventional method did not thoroughly match with the creep rupture data, and revealed an overestimation for the prediction of the long-term creep strength. On the other hand, the multi-constant method revealed a good agreement with the creep rupture data, and its method was thus more accurate than the conventional one. This multi-constant analysis can be used to accurately predict the long-term creep rupture of Alloy 617.



## 1. Introduction

A very high temperature gas-cooled reactor (VHTR) is currently the most promising reactor type among the Generation-IV reactors for producing electricity and hydrogen economically. The VHTR structural components like the reactor internals, piping, hot gas ducts, and intermediate heat exchangers (IHX) are designed for a design life of 30–60 years at 950 °C and 3–8 MPa in He impurities. Their components are required to have good mechanical properties, an oxidation resistance, a corrosion resistance, and a phase stability at high temperatures. Among them, their oxidation and creep properties are the most important, because the integrity of the components should be preserved during a prolonged period in a VHTR coolant [1–3].

So far, there have been no materials approved by ASME Section NH Subsection III which is a nuclear code for temperatures reaching 1000 °C. Alloy 617 is considered as a prime candidate material for the VHTR components due to its superior creep resistance above 800 °C when compared to other candidate alloys; Haynes 230, Hastelloy-X and Alloy 800 [4]. Considerable creep data for Alloy 617 is available in the literature, and a draft Alloy 617 code case and ASME Boiler and Pressure Vessel (BPV) Code-Section II [5] have also provided allowable stress values for a $10^5$ h design period at temperatures up to 982 °C. However, the creep and oxidation behaviours for Alloy 617 are not well understood yet, and their long-term creep strength should be characterized for use in a design and analysis [6].

In addition, the Larson-Miller (LM) parameter method known as a time-temperature parameter (TTP) has been the most commonly used to predict the long-term creep life from short-term creep data [7]. TTP constant $C$ in the LM method is unique for a given set of creep rupture data to be analyzed. Temperature dependency of a rupture life, $d\log t_r/d(1/T)$, should not change in the data set. But, this assumption is not always valid, because the $C$ for the rupture life changes from a high value of the short-term creep to a low value of the long-term creep. So, Maruyama et al. [8] have pointed out that a change in $d\log t_r/d(1/T)$ is a major cause of the resultant overestimation, and have recently proposed a multi region analysis for creep rupture data. They also reported that the multi region analysis for $Q$ in the Orr-Sherby-Dorn (OSD) parameter could evaluate the long-term rupture of austenite stainless steels and 9–12% Cr steels accurately. However, an overestimation of the long-term rupture in the LM parameter has not been reported for Alloy 617, and furthermore, to avoid it, a multi-constant method for the $C$ in the LM parameter has not been demonstrated by others.

In this paper, oxidation behaviour was investigated by using a scanning electron microscope (SEM) and an energy dispersive X-ray spectroscopy (EDX) analysis. Also, to accurately predict the long-term creep life for Alloy 617, lots of creep rupture data was

* Corresponding author. Tel.: +82 42 868 2493; fax: +82 42 868 8549.
E-mail address: wgkim@kaeri.re.kr (W.-G. Kim).

0308-0161/$ – see front matter Crown Copyright © 2010 Published by Elsevier Ltd. All rights reserved.
doi:10.1016/j.ijpvp.2010.03.008

**Table 1**
Chemical composition of Alloy 617.

| Element | Ni | C | Fe | Si | Mn | Co | Cr | Ti | P | S | Mo | Al | B | Cu |
|---|---|---|---|---|---|---|---|---|---|---|---|---|---|---|
| Wt.% | Bal. | 0.08 | 1.49 | 0.06 | 0.11 | 11.58 | 22.16 | 0.35 | 0.003 | 0.001 | 9.8 | 1.12 | 0.002 | 0.08 |

collected from world-wide literature surveys and the KAERI experimental data. Using these data, a longer creep life for above $10^5$ h at 950 °C was accurately predicted by using the multi-constant method with two $C$ values in the LM parameter.

## 2. Results and discussion

### 2.1. Experimental creep properties

Commercial grade nickel-based superalloy, Alloy 617 (Inconel 617) was used for this study, which was a hot-rolled plate with a thickness of 15.875 mm (5/8 inch). Chemical composition of the alloy is given in Table 1. Creep specimens were a cylindrical form of a 30 mm gauge length and a 6 mm diameter. Creep tests were conducted with different stress levels, 35 MPa, 30 MPa, 25 MPa, 22 MPa, 20 MPa and 18 MPa at 950 °C. Creep strain data with elapsed times was taken automatically by a personal computer through an extensometer attached to the creep specimens. Creep curves of Alloy 617 were obtained for different stress stresses at 950 °C. Long-time creep curve reaching 14,100 h (1.6 year) was obtained successfully for 18 MPa, as shown in Fig. 1. A minimum creep rate was measured from the obtained creep curves.

Alloy 617 showed little primary creep strain, and a well-defined secondary creep stage was not observed at this temperature condition. The onset of a tertiary creep was unclear, and a tertiary creep stage was initiated from a low strain level. Alloy 617 revealed a sufficient ductility (>30%) in spite of the creep duration of 14,100 h. Also, the relationship between a minimum creep rate and stress showed a good linearity, as shown in Fig. 2. The constants of Norton's power law, $\dot{\varepsilon}_m = A\sigma^n$ were obtained. The $A$ value was $7.41 \times 10^{-14}$ ($\text{MPa}^{-n}\ \text{h}^{-1}$), and the $n$ value was 6.50. Also, the Monkman–Grant (M–G) relationship between the creep rupture time and the minimum creep rate revealed a good linearity, as shown in Fig. 3. In Eq. (1) of the M-G, the $K$ value was −0.63, and the $m$ value was 0.92, respectively.

$$\log t_r + m \log \dot{\varepsilon}_m = K \quad (1)$$

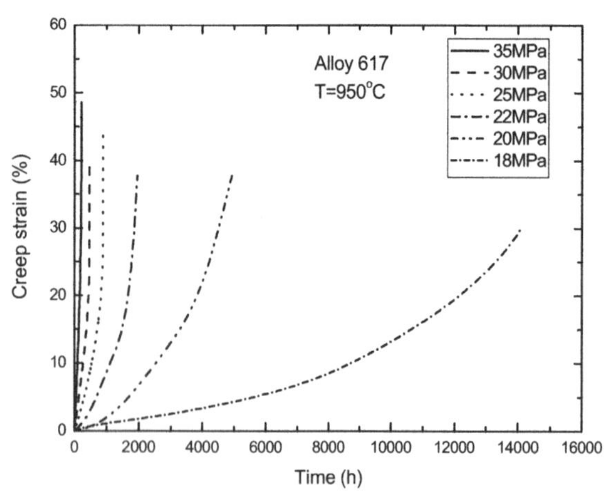


**Fig. 1.** Experimental creep curves obtained with the elapsed times for different stress levels at 950 °C for Alloy 617.

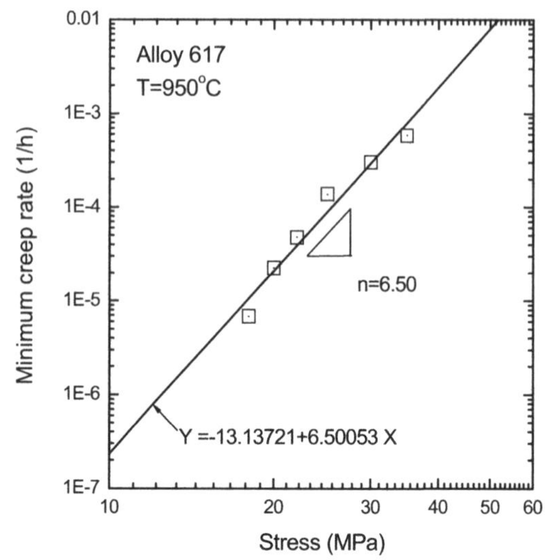


**Fig. 2.** Minimum creep rate vs. stress.

These creep constants of $A$, $n$, $m$, and $K$ can be used as important material constants to characterize the creep behaviour of Alloy 617.

### 2.2. Oxidation behaviour

Fig. 4 shows the SEM image for the oxide layers of the crept specimens with the creep rupture times (or 35 MPa, 30 MPa 25 MPa 22 MPa and 20 MPa) at 950 °C. EDX analysis was performed to

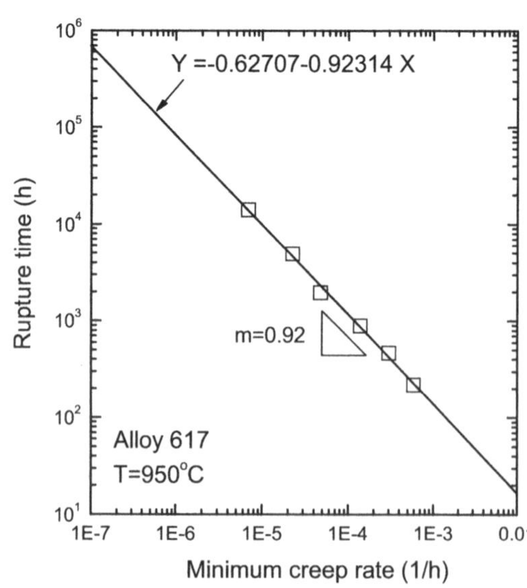


**Fig. 3.** Creep rupture time vs. minimum creep rate.

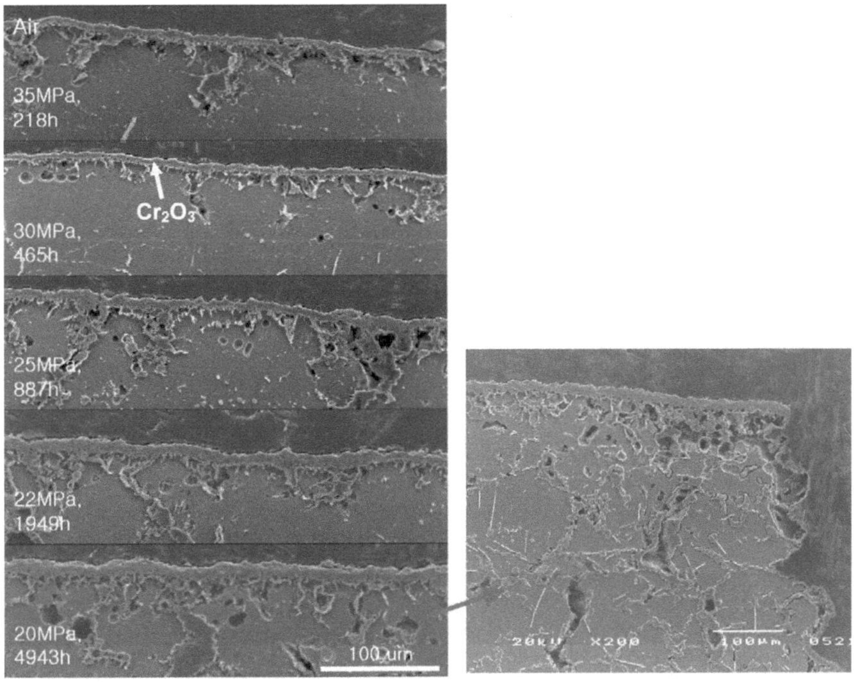


a b

**Fig. 4.** SEM image showing the oxide layers of the crept specimens with creep rupture times at 950 °C (or 35 MPa, 30 MPa 25 MPa 22 MPa and 20 MPa) (a) the oxide layers with time variations (b) at the near fracture region at 20 MPa.

identify the metallic element constituents at different locations. A $Cr_2O_3$ layer was formed on the surface. A thin internal sub layer consisting of (Cr, Al, Ti) oxide was formed with rod shapes just beneath the $Cr_2O_3$ layer. And, a thick carbide-depleted zone was developed by a reaction of the chromia and carbide precipitates below the thin internal sub layer. Low stress specimens (long creep rupture time) formed a heavier oxidation layer and a wider carbide-depleted zone than the high stress ones due to an exposure in air at a high temperature, as shown in Fig. 5. Photo (a) is for the creep rupture time of 465 h (30 MPa @ 950 °C) and photo (b) is for 4943 h (20 MPa @ 950 °C). It appears that the oxide layer was thicker for the specimen with a longer creep rupture time. The thickness of the Cr-oxide layer was measured from a SEM image. The thickness increased with increasing creep rupture times, as shown in Fig. 6. This oxide behaviour was because Alloy 617 is a chromia-forming alloy on the surface due to an exposure at a high temperature during a creep.

In addition, in case of the near creep fracture region, as shown in Fig. 4(b), it is clearly seen that the internal oxides were formed along the creep voids and cavities, which were developed by creep damage, because oxygen can be infiltrated with ease along them. And, because Alloy 617 mainly forms precipitates along a grain boundary during a creep, oxide substances can be formed easily along these grain boundaries. It was identified by EDX

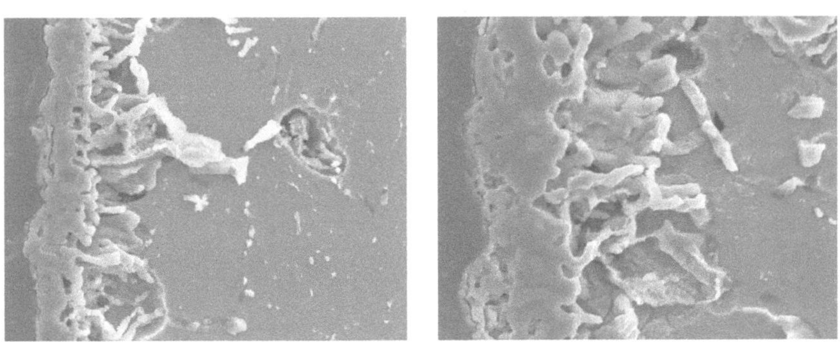

a b

**Fig. 5.** High magnification of the oxide layers for the creep rupture times of (a) 465 h (30 MPa @ 950 °C) (b) 4943 h (20 MPa @ 950 °C).

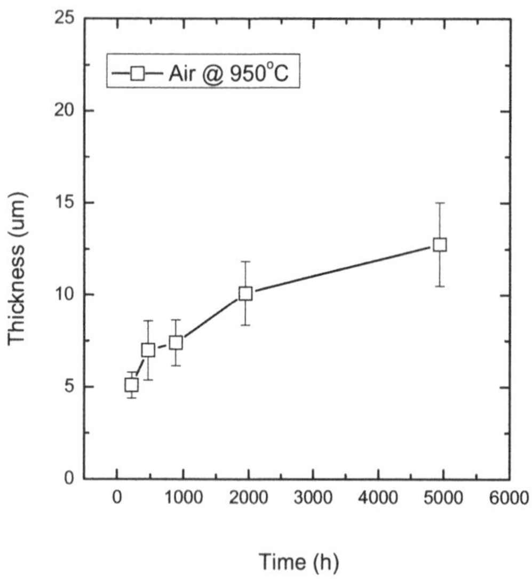


**Fig. 6.** Variations of the oxide layer thickness with the creep rupture times.

analysis that mostly the precipitation of a chromium–molybdenum intermetallic phase was found in the crept specimens. Also, as shown in the SEM fracture morphologies of Fig. 7, the oxide substances on the surface were significantly accumulated like a sand hill in the low stress specimens for a long creep rupture time when compared with the high stress specimens. This is due to an exposure in air during a high temperature creep.

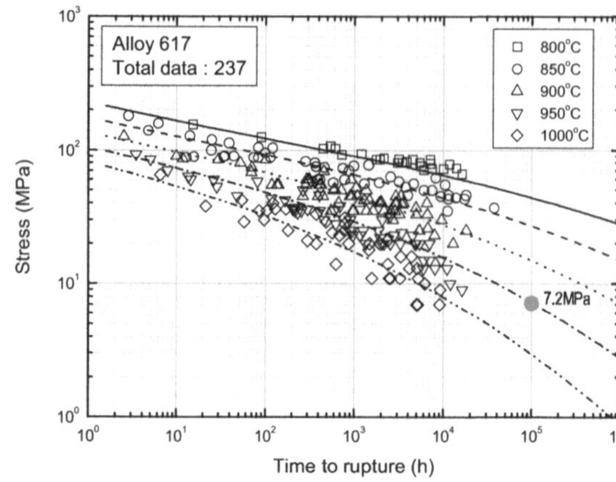


**Fig. 8.** Creep rupture strength predicted by a unique value of $C = 18$ at temperatures of 800–1000 °C of Alloy 617.

### *2.3. Prediction of the long-term creep strength by a conventional method*

A draft Alloy 617 code case has provided allowable stress values for a $10^5$ h design life at temperatures up to 982 °C. However, for an IHX application in a VHTR, because its design life is 30–60 years, a long-term creep strength above $10^5$ h should be predicted. To do this, lots of creep rupture data was collected from world-wide literature surveys for Alloy 617; Schubert et al. [9], ANL (Argonne National Laboratory) reports [10], and KAERI experimental data. The number of total data was 237 points, and the temperature ranges were 800–1000 °C. It was identified that the collected

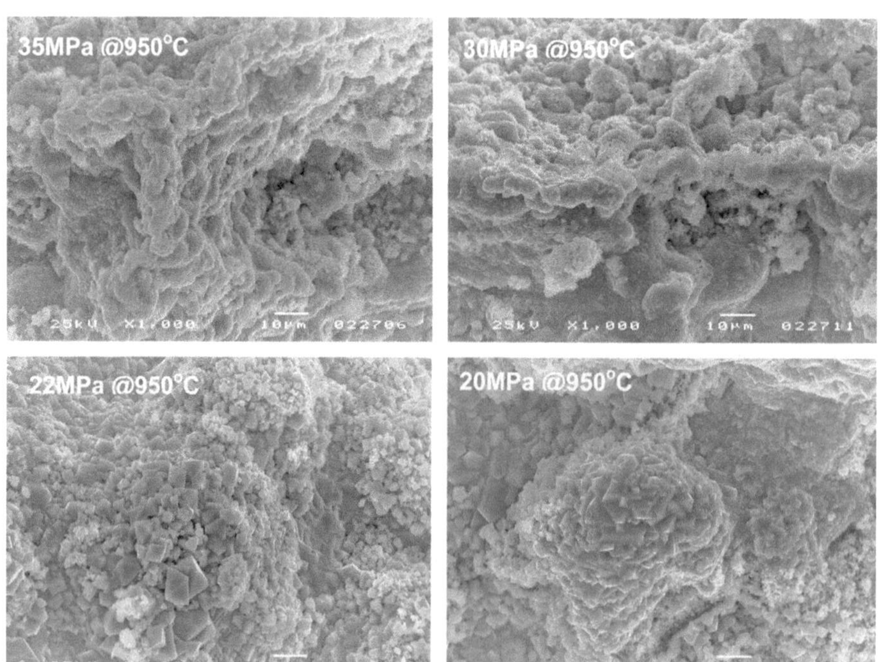


**Fig. 7.** Typical SEM fracture morphologies of the crept specimens with the applied stresses, 35 MPa, 30 MPa, 22 MPa and 20 MPa at 950 °C.

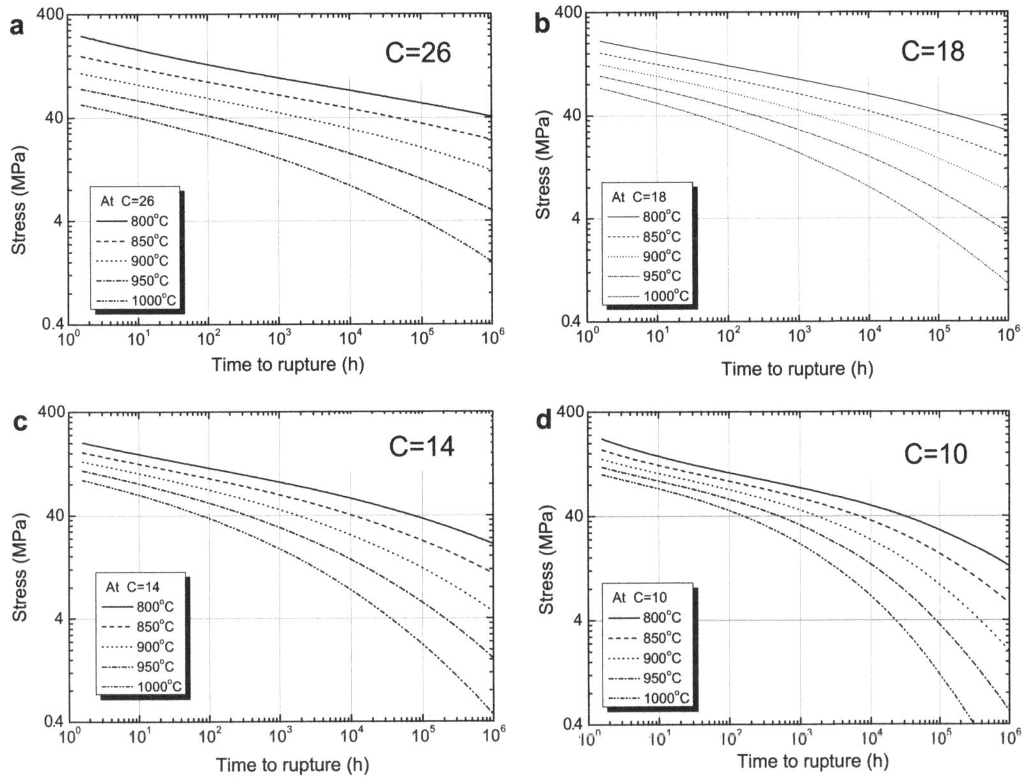


**Fig. 9.** Prediction of the creep rupture strength with the temperatures at $C = 26, 18, 14$ and 10.

rupture data exhibited nearly identical values regardless of the various product forms; such as plates, rods and sheets. When all the creep rupture data were plotted together, they revealed a clear temperature dependency. These creep rupture data were used to predict the long-term creep life of Alloy 617.

To accurately predict the long-term creep strength, the present works have been performed by using two methods; a conventional method using a unique constant and a new one using two constants for the C in the LM parameter. The basic equation of the LM parameter (P) is given by [1],

$$P = (T + 273.15)[\log(t_R + C)] \tag{2}$$

where, T is the absolute temperature, C is a constant, and $t_R$ is the rupture time. A stress function of the x-axis can be expressed by a polynomial equation,

$$P = f(\sigma) = b_o + b_1(\log\sigma) + b_2(\log\sigma)^2 + \cdots b_k(\log\sigma)^k \tag{3}$$

where, $b_k$ is the regression coefficient in the k order, and k is the order of the polynomial equation. The k in the polynomial equation is generally used at under fifth orders of a magnitude.

In the conventional method using a unique value for the C in the LM parameter, to determine an optimal C value, a least square fitting was conducted with various C values for all the creep rupture data. The optimal value was found to be $C = 18$. Third-order polynomial equation was obtained from the master rupture curves of the LM parameter versus the stress. Using the polynomial equation for $C = 18$, the long-term creep strength was predicted for up to $10^6$ h from 800 °C to 1000 °C.

Fig. 8 shows the curves of the creep rupture strength predicted by the conventional method using a unique value of $C = 18$ for 800–1000 °C for Alloy 617. The predicted curves were not bent for a long period of $10^5$–$10^6$ h and did not thoroughly match with the creep rupture data, and they were overestimated as well. For $10^5$ h at 950 °C, the predicted stress was 7.2 MPa. This value was almost similar to 7.9 MPa reported in the ASME BPV Code-Section II. It is obvious that the allowable stress for $10^5$ h for Alloy 617 was overestimated when compared with the experimental creep rupture data. Therefore, both the predicted stress in this study and the allowable stress in ASME should be re-evaluated so as not to be overestimated for a long period above $10^5$ h at 950 °C. For this result, as shown in Fig. 9, it is identified that the application of a unique C value can cause a large error in the predicted stress with to different C values, and especially for a long period of $10^6$ h at high temperatures of 950 and 1000 °C. Their curve shapes for $C = 26$ and $C = 10$ were significantly different. Thus, to accurately predict a long-term stress above $10^5$ h, it is necessary to develop a better method because the conventional method using a unique C value was not proper.

### 2.4. Prediction of the long-term creep strength by a multi C method

As discussed above, to avoid an overestimation of a long-term creep strength for Alloy 617, in the present work, a multi-constant method using two C values was performed instead of the conventional one using a unique C value.

Fig. 10 shows the results predicted for different C values at temperatures of 800, 900, 950 and 1000 °C for Alloy 617. It is clearly

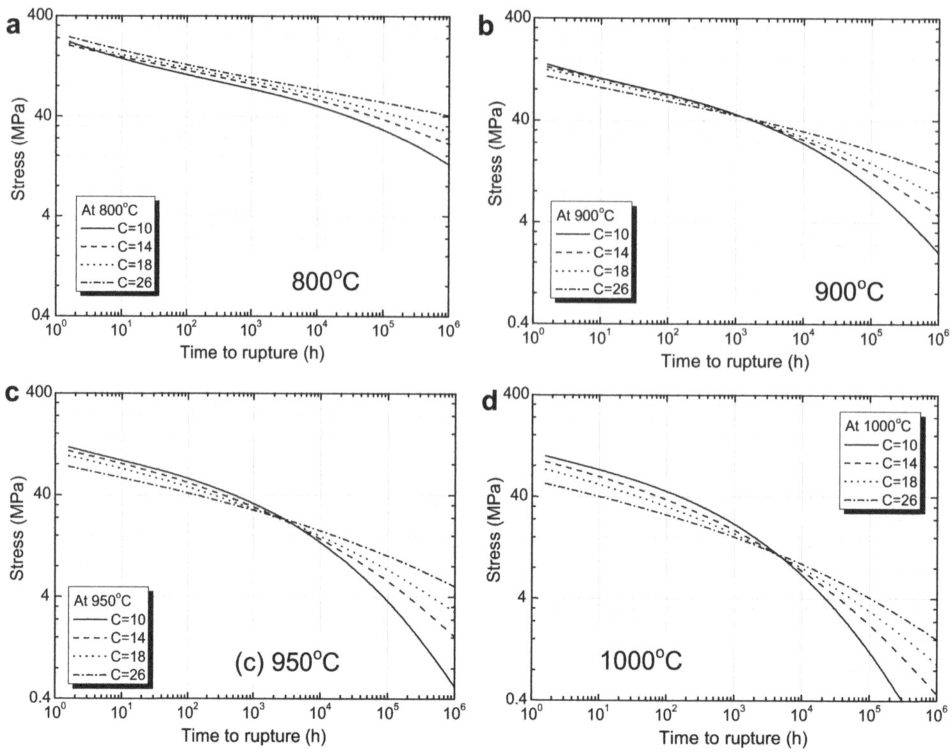


**Fig. 10.** Prediction of the creep strength curves with the $C$ variations at temperatures of 800, 900, 950 and 1000 °C of Alloy 617.

shown that the $C$ values for the predicted curves changed from a high value of a short-term creep to a low value of a long-term creep. Especially, at high temperatures of 950 and 1000 °C, the difference in the predicted stress is large for different $C$ values of $C = 26$ and 10. So, a least square fitting was carried out for the creep rupture data of each temperature. The optimum $C$ value was found to be $C = 12$ for a long-term creep life, and $C = 20$ for a short-term creep life. Using $C = 12$ and $C = 20$, the long-term creep strength for up to $10^6$ h was predicted properly for Alloy 617.

Fig. 11 shows the predicted curves for the long-term creep life by using the multi constants of $C = 12$ and $C = 20$ for 800–1000 °C for Alloy 617. The predicted curves for 900, 950 and 1000 °C were properly bent for the long-term life, and they showed a better agreement with the creep rupture data when compared with the results of the conventional method. For $10^5$ h at 950 °C, the predicted stress was 4.7 MPa. This value was lower when compared with 7.2 MPa of the conventional method. This reason is that the creep fracture mode of Alloy 617 was changed into a brittle fracture

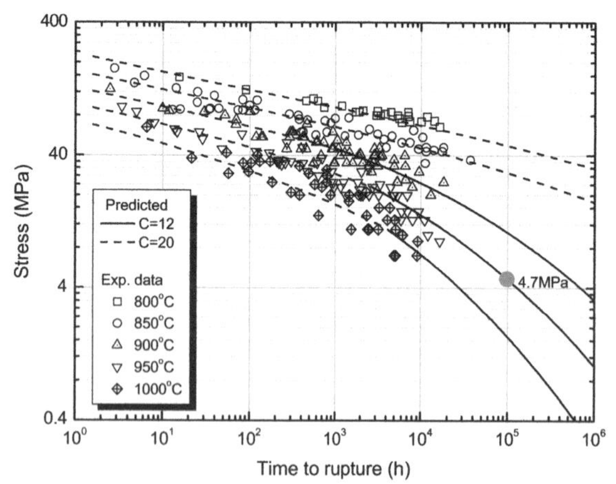


**Fig. 11.** Long-term rupture strength predicted by using multi $C$ methods of $C = 12$ and 20 for Alloy 617.

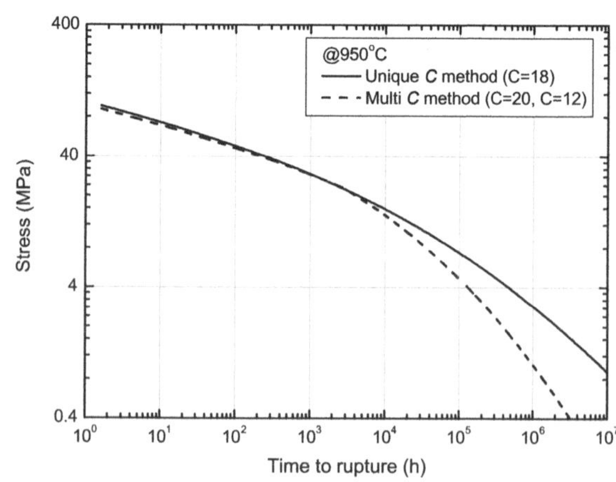


**Fig.12.** Comparison of the creep strength curves predicted by the unique $C$ and multi $C$ methods at 950 °C.

for a long time at a low stress from a ductile fracture for a short rupture time at a high stress. Also, this analysis is reasonable because a set of the creep rupture data is divided into two data sets so that $C$ is unique in the two divided sets.

Fig. 12 shows a comparison of the creep strength curves predicted by the unique $C$ and multi $C$ methods with an application of the LM parameter at 950 °C. The unique $C$ method has a higher value in the predicted curve than the multi $C$ one with $C = 20$ and $C = 10$ for $10^5$ h and $10^6$ h. Especially, the strength gap between the two curves becomes larger as the time is extended from $10^5$ to $10^6$ h. Thus, this multi-constant method can be used to accurately predict the long-term creep strength of Alloy 617 for VHTR components which will be designed for a design life of 30–60 years at 950 °C.

## 3. Conclusions

Experimental creep data was obtained by a series of creep tests with different stress levels at 950 °C for Alloy 617. The creep constants of $A$, $n$, $m$, and $K$ were determined for Alloy 617. A $Cr_2O_3$ layer was formed on the surface of the crept specimens of Alloy 617. A thin internal sub layer oxide was formed with rod shapes just beneath the $Cr_2O_3$ layer. And, a thick carbide-depleted zone was developed by a reaction of the chromia and carbide precipitates below the thin internal sub layer. The low stress specimens with a long creep rupture time formed a heavier oxidation layer and a wider carbide-depleted zone than the high stress ones with a short creep rupture time. In the case of the near creep fracture region, it was observed that the internal oxides were formed along the creep voids and cavities, because oxygen can be infiltrated with ease along them. In addition, to accurately predict the long-term creep strength of Alloy 617 for above 900 °C, a multi-constant method with two $C$ values was adopted instead of a conventional one with a unique $C$ value. The conventional method did not thoroughly match with the creep rupture data, and it showed an overestimation for the prediction of the long-term creep strength. On the other hand, the multi-constant method revealed a good agreement with the creep rupture data, and this method was more accurate than the conventional one. For $10^5$ h at 950 °C, the predicted stress of the multi-constant method was lower with 4.7 MPa than 7.2 MPa of the conventional method. This multi $C$ method can be utilized to accurately predict the long-term creep strength for above $10^5$ h at 950 and 1000 °C for Alloy 617.

## Acknowledgements

This study was supported by the Korean Ministry of Education, Science & Technology through its National Nuclear Technology Program.

## References

[1] Kim W, Yin S, Ryu W, Jang J. Analysis of the creep rupture data of Alloy 617 for a high temperature gas cooled reactor. In: Proceedings of CREEP8, eighth international conference on creep and fatigue at elevated temperature. Texas: ASME PVP2007–26834; 2007.

[2] Kim W, Yin S, Kim Y, Chang J. Creep characterization of a Ni-based Hastelloy-X alloy by using theta projection method. Engineering Fracture Mechanics 2008;75:4985–95.

[3] Kim W, Yin S, Jung I, Kim Y. Modeling of a long-term creep curve of Alloy 617 for a high temperature gas-cooled reactor. Key Engineering Materials 2008;385–387:693–6.

[4] Dewson S and Li X. Selection criteria for the high temperature reactor intermediate heat exchanger. Proceedings of ICAPP 05. Seoul; 2005. Paper No. 5333.

[5] ASME boiler and pressure vessel code, section II, materials part D-properties. New York: ASME; 2004.

[6] Jang C, Lee D, Kim D. Oxidation behavior of an Alloy 617 in very high temperature air and helium environments. Int Jnl of Press Vessels and Piping 2008;85:368–77.

[7] Kim W, Yin S, Ryu W. Application and standard error analysis of the parametric methods for predicting the creep life of type 316LN SS. Key Engineering Materials 2005;297–300:2272–7.

[8] Maruyama K, Yoshimi K. Methodology of creep data analysis for advanced high Cr ferritic steel. In: Proceedings of CREEP8, eight international conference on creep and fatigue at elevated temperature. Texas: ASME PVP2007–26150; 2007.

[9] Schubert F, Bruch U, Cook R, Diehl H, Ennis P, Jakobeit W, et al. Creep rupture behavior of candidate materials for nuclear process heat applications. Nuclear Technology 1984;66:227–40.

[10] Natesan K Purohit A and Tam S. Materials behavior in HTGR environments. Argonne National Laboratory Report: NUREG/CR-6824 ANL-02/37; 2002.

*Met. Mater. Int., Vol. 15, No. 4 (2009), pp. 559~564*
doi: 10.1007/s12540-009-0559-9 Published 26 August 2009

# A Numerical Approach to Determine Creep Constants for Time-Temperature Parametric Methods

**Woo-Gon Kim[1,*], Song-Nan Yin[1], Sung-Ho Kim[1], Woo-Seog Ryu[1], Chan-Bock Lee[1], and Seon-Jin Kim[2]**

[1]1045 Daedeok-Daero, Yuseong-Gu, Daejeon, Nuclear Materials Research Division, Korea Atomic Energy Research Institute (KAERI), 150 Deokjin-dong, Yuseong-gu, Daejeon 305-353, Korea
[2]School of Mechanical Engineering, Pukyong National University, San 100 Yongdang-dong, Nam-gu, Busan 608-739, Korea

(received date: 26 June 2008 / accepted date: 12 February 2009)

This study dealt with how to mathematically obtain creep constants of $C$, $Q/2.3R$, $T_a$ and $t_a$ used in Time-Temperature Parametric (TTP) methods. A number of creep rupture data from 550 °C to 800 °C for Type 316LN stainless steel were collected through a worldwide literature survey and experimental data produced at KAERI. It was found that the creep rupture data revealed a linear relationship when the data were plotted by the semi-logarithmic coordinates of log ($t_r$) versus stress. On the basis of this linear relationship, the creep constants of $C$, $Q/2.3R$, $T_a$ and $t_a$ used in the Larson-Miller (LM), the Orr-Sherby-Dorn (OSD), and the Manson-Haferd (MH) methods were obtained by a new numerical method. This approach can be utilized as a useful method to predict the long-term creep life of Type 316 stainless steel.

**Keywords:** Type 316LN, creep, creep constants, time temperature parameter (TTP)

## 1. INTRODUCTION

Type 316LN Stainless Steels (SS) are major structural materials for currently operating and planned Sodium-cooled Fast Reactors (SFR) because of their good high-temperature mechanical properties, compatibility with liquid sodium coolants, ease of fabrication and weldability, and above all, their commercial availability. Since the SFR components will be designed to be used up to 60 years at the temperature of 550 °C, the design of such high-temperature components requires careful consideration of the time-dependent creep behavior of the material [1,2].

Current design rules of a maximum allowable stress for time-dependent materials are established so as not to exceed the lowest of the following: (i) 100 % of the average stress to produce a creep rate of 0.01 %/1,000 h, (ii) 67 % of the average stress to cause rupture at the end of $10^5$ h, and (iii) 80 % of the minimum stress to cause rupture at the end of $10^5$ h [3]. With respect to the above criterions (ii) and (iii), it is essential to achieve reliable long-term test data beyond $10^5$ h. However, experiments aimed at obtaining long-term creep data (>$10^5$ h) are time-consuming and expensive. Thus, an allowable stress value can be predicted by extrapolating short-term creep data [4].

So far, to accurately extrapolate the creep data to longer times than those for which experimental data is available, various time-temperature parametric methods have been developed. Among them, the Larson-Miller, the Orr-Sherby-Dorn, and the Manson-Harferd methods have typically been used [5-7]. However, these TTP methods have certain inherent restrictions in the application of the TTP data. That is, the isostress curves are frequently not linear, unparallel, or not convergent. Therefore, it is difficult to easily determine the creep constants in the TTP methods. This may also cause errors in the extrapolation of a longer-time rupture. Therefore, a new method is required to conveniently obtain the creep constants in the TTP methods [8].

In this study, a number of creep rupture data for Type 316LN SS were collected through a worldwide literature surveys and experimental effort of KAERI. The linear relationship was identified from the semi-logarithmic coordinates in terms of log ($t_r$) and stress for Type 316LN SS. On the basis of a linear relationship of the creep rupture data for Type 316LN SS, a mathematical method was newly proposed to obtain the creep constants of $C$, $Q/2.3R$, $T_a$ and

*Corresponding author: wgkim@kaeri.re.kr

$t_a$ in the TTP methods. This mathematical result was compared with the conventional methods in determining the values of the TTP constants.

## 2. A CONVENTIONAL METHOD FOR CREEP CONSTANTS

TTP methods depend on the interrelation of the time and temperature in contributing to a creep deformation. Time and temperature variables are combined into one parameter, $P$. The parameter is expressed as a function of a stress,

$$P(t_r, T) = f(\sigma) \tag{1}$$

Each equation for the LM, the OSD, and the MH methods used in this study is given by

$$\text{L-M: } P(t_r, T) = (\log t_r + C)T \tag{2}$$

$$\text{O-S-D: } P(t_r, T) = \log t_r - \frac{Q}{2.3RT} \tag{3}$$

$$\text{M-H: } P(t_r, T) = \frac{\log t_r - \log t_a}{T - T_a} \tag{4}$$

Each parameter can be expressed by a polynomial equation,

$$P = b_o + b_1(X) + b_2(X)^2 + \cdots + b_k(X)^k \tag{5}$$

where, $X = (\log \sigma)$, $b_k$ is the regression coefficient of the $k$ order, and $k$ is the order of the polynomial equation. The $k$ orders in the polynomial equation are generally taken smaller than equal to 5. A plot of the parameter versus the log ($\sigma$) can produce a single curve or line. Longer-time rupture behavior can be predicted by extrapolating the single curve. Until now, in order to determine the constant values for the TTP methods, a conventional method was used, and their values were determined according to the following procedure.

In the LM parameter, $T$ is the absolute temperature, $C$ is a creep constant defined by the intercept on the $1/T = 0$ axis of the isostress lines plotted on the coordinate axes log ($t_r$) versus $1/T$, and $t_r$ is the rupture time, as shown in Fig. 1. In order for the LM parameter to be applied accurately, the isostress data must be linear on coordinate axes log ($t_r$) versus $1/T$, and the families of such lines must converge to the point (0, -$C$). However, practically, the families of the isostress lines do not always converge to the point (0, -$C$). Thus, it is difficult to reasonably determine the creep constant, $C$. So, the $C$ value has been usually determined by taking an average value of the coefficient $b$ in each linear equation (y = ax + b) of the isostress lines that are plotted using the coordinate axes log ($t_r$) versus $1/T$. The constant

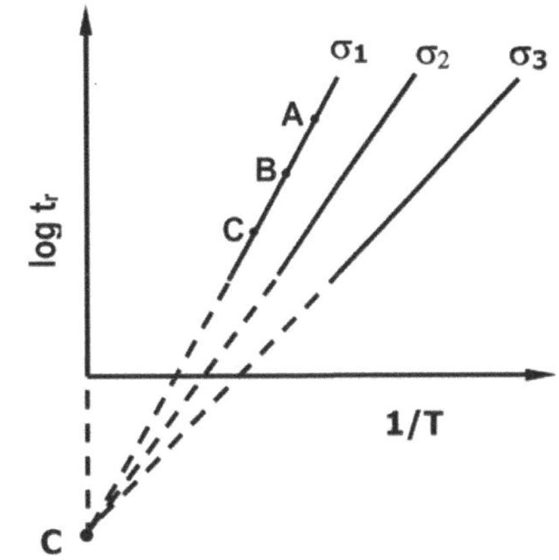


**Fig. 1.** Plot of log ($t_r$) vs. $1/T$ for the LM method.

$C$ value can also be obtained by using the equation,

$$C = \frac{T_1 \log t_{r1} - T_2 \log t_{r2}}{T_2 - T_1} \tag{6}$$

where the rupture times of $t_{r1}$ and $t_{r2}$ result from experiments performed at the same stress and temperatures $T_1$ and $T_2$. For wide temperature ranges, the $C$ can be determined as an average from multiple evaluations of Eq. 6 for different temperature $T_2$.

In the OSD parameter, $Q$ is activation energy and $R$ is universal gas constant. The application of this parameter requires that the isostress data form a linear plot in coordinates log ($t_r$) versus $1/T$ and that the families of such lines are parallel with a common slope $K$ (=$Q/2.3RT$) as shown in Fig. 2. The $K$ value can be obtained as an average value over the individual isostress line slope. It is obvious that if the data fulfill the requirements of the OSD method, they cannot conform to those of the LM procedure.

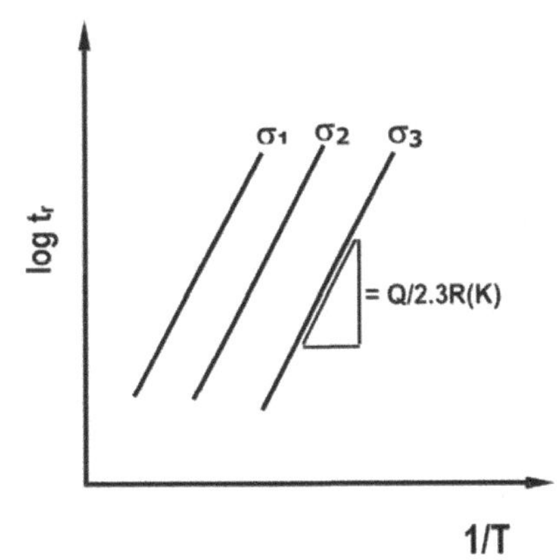


**Fig. 2.** Plot of log ($t_r$) vs. $1/T$ for the OSD method.

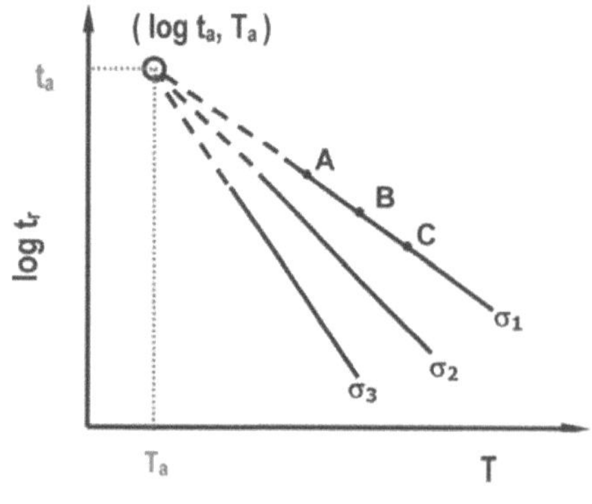


**Fig. 3.** Plot of log ($t_r$) vs. $T$ for the MH method.

Finally, in the MH parameter, $T_a$ and log $t_a$ are constants defined by the point of the intersection of the families of the isostress lines plotted in coordinates log ($t_r$) versus $T$, as shown in Fig. 3. However, practically, the families of the isostress lines do not always converge to the point ($t_a$, $T_a$). Therefore, these two constants are usually determined as an average value of the isostress lines plotted on the coordinates log ($t_r$) versus $T$.

## 3. RESULTS AND DISCUSSION

### 3.1. Linear relationships of creep rupture data

First of all, in order to predict the long-term creep lifeby the LM, the OSD, and the MH methods for Type 316LN SS, it is necessary to determine the constant values of C, Q/2.3R, $t_a$ and $T_a$ used in the TTP methods. To do this, lots of creep rupture data were collected through a worldwide literature survey: Japan [9,10], India [11], the Czech Republic [12], and Korea [13,14]. From this survey, the creep data were selected,

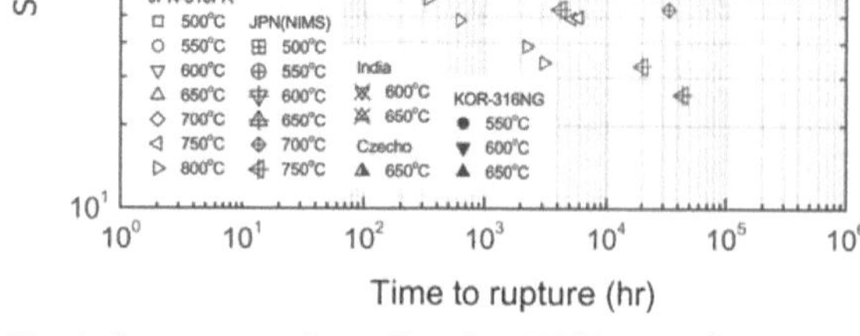


**Fig. 4.** Creep rupture data collected at 550 °C to 800 °C for Type 316LN SS.

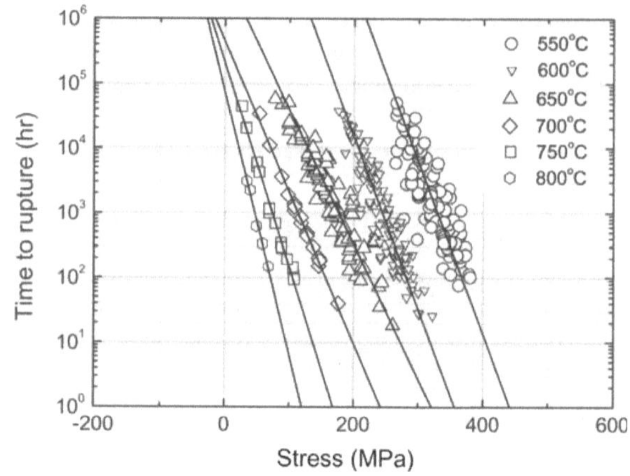


**Fig. 5.** Linear relationships of log ($t_r$) vs. stress for Type 316LN SS.

where the chemical compositions were clearly defined, and the nitrogen content was within the 0.06-0.15 wt.% ranges. The total number of the collected data points was 345, and the temperature ranges were 500-800 °C, as shown in Fig. 4.

Type 316LN SS revealed a good linear relationship when the creep rupture data were plotted in the semi-logarithmic coordinates log ($t_r$) versus the stress and log $t_r$, as shown in Fig. 5. On the basis of this linear relationship, the creep constants of C, Q/2.3R, $t_a$ and $T_a$ were obtained by using a mathematical approach. This approach used the linear relationship of log $t_r$ versus $T$ (or $1/T$) and log $t_r$ versus stress. Mathematical expressions for each parameter were derived in the following sections.

### 3.2. Derivation of the equations for *C* and *K* in the LM and OSD parameters

For the semi-logarithmic plot of log ($t_r$) versus stress, as shown in Fig. 6, the linear relations yield times at the A, B and C points on isothermal lines, $T_1$, $T_2$ and $T_3$ for stress $\sigma_1$.

$$\log t_{r(A)} = a_1\sigma_1 + b_1$$
$$\log t_{r(B)} = a_2\sigma_1 + b_2$$
$$\log t_{r(C)} = a_3\sigma_1 + b_3 \qquad (7)$$

In the case of the LM method, as shown in Fig. 1, the linear relations between the rupture times and corresponding temperatures in the points A, B and C on the isostress lines, $\sigma_1$, $\sigma_2$ and $\sigma_3$ are given by

$$\log t_{r(A)} = K_1(1/T_1) + C$$
$$\log t_{r(B)} = K_1(1/T_2) + C$$
$$\log t_{r(C)} = K_1(1/T_3) + C \qquad (8)$$

By combining Eqs. 7 and 8 for the rupture times, Eq. 9 for $s_1$ is obtained.

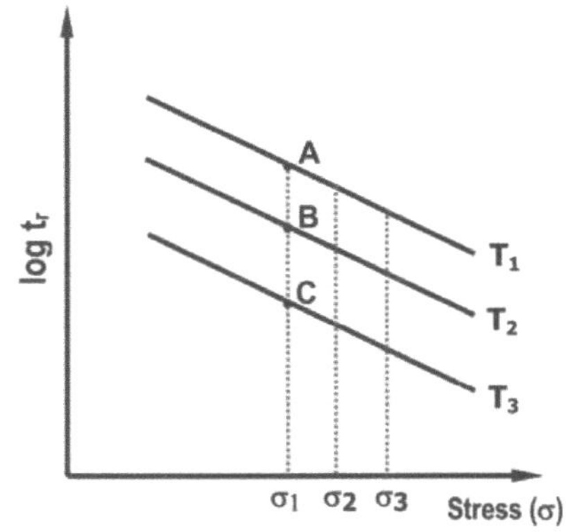


**Fig. 6.** Linear relationships of the isothermal lines of log ($t_r$) vs. stress.

$$a_1\sigma_1 + b_1 = K_1(1/T_1) + C$$
$$a_2\sigma_1 + b_2 = K_1(1/T_2) + C$$
$$a_3\sigma_1 + b_3 = K_1(1/T_3) + C \tag{9}$$

Similarly, the equations for $\sigma_2$, $\sigma_3$, ... $\sigma_n$ can be obtained by using the same approach. By arranging these equations with respect to $K$ for each stress, the equation for a mean slope value, $\overline{K}_i$ can be given by

$$\overline{K}_i = \frac{1}{n}\sum_{j=1}^{n}\frac{(a_j-a_{j+1})\sigma_i+(b_j-b_{j+1})}{(1/T_j-1/T_{j+1})} \tag{10}$$

where $i$ is the stress index, $j$ is the temperature index, and $n$ is the total number of investigated temperatures.

Also, the $C_1$, $C_2$ , $C_3$, ... $C_n$ equations for $\sigma_1$, $\sigma_2$, $\sigma_3$, ...$\sigma_n$ can be obtained from Eq. 9. Eq. 11 can be given for the $C_1$, and the equations for $C_2$, $C_3$, ... $C_n$ are also obtained by the same approach.

$$C_1 = a_1\sigma_1 + b_1 - K_1(1/T_1)$$
$$C_1 = a_2\sigma_1 + b_2 - K_1(1/T_2)$$
$$C_1 = a_3\sigma_1 + b_3 - K_1(1/T_3) \tag{11}$$

Arranging these equations with respect to $C$ for each stress, the $\overline{C}_i$ equation expressing each mean value of $C_i$ can be given by

$$\overline{C}_i = \frac{1}{n}\left[\sum_{j=1}^{n}(a_j\sigma_i+b_j)-\sum_{j=1}^{n}\frac{1}{(1/T_j)}\overline{K}_i\right] \tag{12}$$

A mean value for the number of total stresses can be written by

$$\overline{C} = \frac{1}{m}\sum_{i=1}^{m}\overline{C}_i \tag{13}$$

where, $m$ is the number of the investigated stresses.

In the case of the OSD parameter, the equation for expressing the slopes of the isostress lines, $K(=Q/2.3R)$, is equivalent to Eq. 8. This equation can be obtained with an identical approach to that used for the LM parameter because the plot of log ($t_r$) versus $1/T$ is identical between the LM and the OSD parameters. Also, the $\overline{K}$ equation indicating the number of total stresses can be written by

$$\overline{K} = \frac{1}{m}\sum_{i=1}^{m}\overline{K}_i \tag{14}$$

### 3.3. Derivation of the equations for $t_a$ and $T_a$ in the MH parameter

The equations for $t_a$ and $T_a$ in the MH parameter can be derived by using an identical procedure to that used for the LM parameter. As described above, the times at the A, B and C points on the isothermal lines of $T_1$, $T_2$ and $T_3$ are related to the stress $\sigma_1$ by Eq. 7. Also, as shown in Fig. 3, the rupture times at the A, B and C points are related to $\sigma_1$ by

$$\log t_{r(A)} = K_1(1/T_1) + C$$
$$\log t_{r(B)} = K_1(1/T_2) + C$$
$$\log t_{r(C)} = K_1(1/T_3) + C \tag{15}$$

By combining Eqs. 7 and 15 for the rupture time, and then solving and arranging the $K$ for each stress, $\overline{K}_i$ for $\sigma_i$ is given as

$$\overline{K}_i = \frac{1}{n}\sum_{j=1}^{n}\frac{(a_{j+1}-a_j)\sigma_i+(b_{j+1}-b_j)}{(T_{j+1}-T_j)} \tag{16}$$

The mean value $\overline{C}_i$ of $C_i$ can be obtained as

$$\overline{C}_i = \frac{1}{n}\left[\sum_{j=1}^{n}(a_j\sigma_i+b_j)-\sum_{j=1}^{n}T_j\overline{K}_i\right] \tag{17}$$

In addition, the constants $t_a$ and $T_a$ correspond to the x and y coordinates at the intersection of the isostress lines. The linear equations for each isostress line can be written as

$$y = K_1x + C_1$$
$$y = K_2x + C_2$$
$$y = K_3x + C_3$$
$$\cdots\cdots$$
$$y = K_mx + C_m \tag{18}$$

where $x$ indicates $T$, and y indicates log($t_r$). A solution for x and y in Eq. 18 yields the mean values of $x$ ($T_a$) and y ($t_a$) as

$$\bar{x}_a(T_a) = \frac{1}{m}\sum_{j=1}^{m}\frac{C_j-C_{j+1}}{K_{j+1}-K_j} \tag{19}$$

$$\bar{y}_a(t_a) = \frac{1}{m}\sum_{j=1}^{m}\frac{K_{j+1}C_j-K_jC_{j+1}}{K_{j+1}-K_j} \tag{20}$$

The mathematical formulae derived above have no basic physical meaning, but it is possible to obtain the creep con-

**Table 1.** Comparison of the mathematical and conventional methods for the creep constants

| | Temp. (K) | Items | LM | OSD | MH | |
|---|---|---|---|---|---|---|
| | | | $C$ | $K(=Q/2.3R)$ | $T_a$ | $t_a$ |
| Conventional method | 823-1073 | Constants | 24.56 | 25530 | 679.5 | 12.04E9 |
| | | $R^2$ | 0.96659 | 0.97071 | 0.96378 | |
| | | RSE | 0.01316 | 0.01283 | 0.05254 | |
| Mathematical method | 823-1073 | Constants | 28.43 | 29093 | 602.8 | 6.48E11 |
| | | $R^2$ | 0.96956 | 0.97173 | 0.96455 | |
| | | RSE | 0.01201 | 0.01242 | 0.03726 | |

- Calculated stress, $\sigma$=107 MPa, 150 MPa, 228 MPa
- $R^2$: Coefficient of Determination
- RSE: Relative Standard Error

stants with the numerical algorithm, regardless of any stress and temperature condition.

### 3.4. Application of the mathematical method

Equations derived for determining the constants used in the LM, the OSD, and the MH methods were applied to Type 316LN SS. From examining the linear relationship between log ($t_r$) and stress, as shown in Fig. 5, a linear equation for each investigated temperature in the range from 550 °C to 800 °C was obtained. The constants $C$, $K$, $t_a$ and $T_a$ for the LM, the OSD, and the MH methods were obtained numerically, and the results were compared to a conventional procedure.

The results obtained for the three TTP methods are summarized in Table 1. The numerical method had no large differences in the constant values when compared to the conventional approach. The master rupture curves of the mathematical method were in good agreement with the creep rupture data, as shown in Figs. 7, 8, and 9. Each polynomial relation that predicted the creep rupture life was obtained by the least square fitting of the creep rupture data. The second-order polynomial equation in the LM parameter and fourth-order polynomial in the OSD and the MH parameters were obtained because the regression curves were well matched with each experimental data. By using these polynomial relations, long-term creep life was conducted for each TTP method. In addition, in order to verify

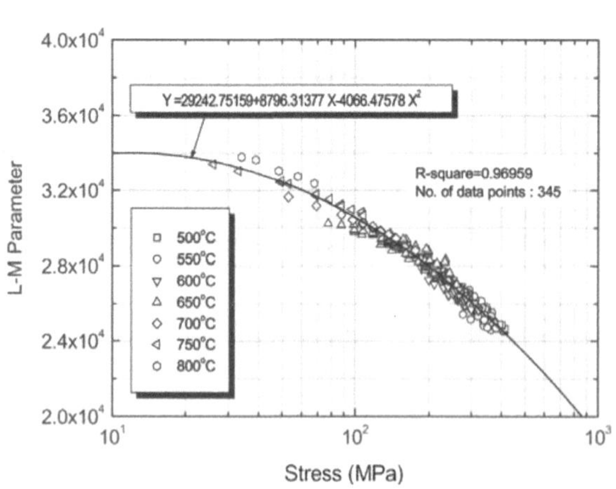


**Fig. 7.** Master rupture curves obtained for the LM parameter.

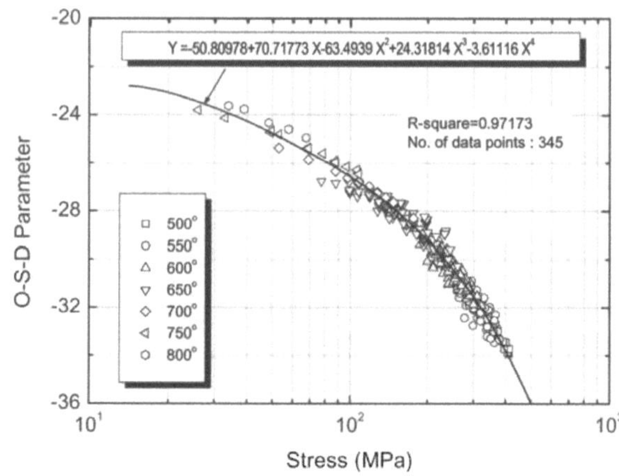


**Fig. 8.** Master rupture curves obtained for the OSD parameter.

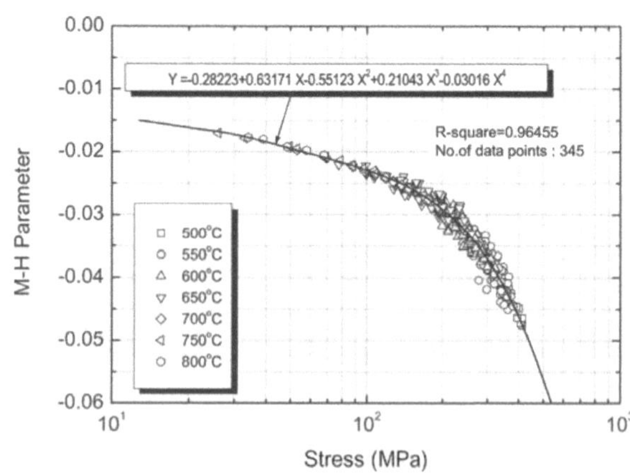


**Fig. 9.** Master rupture curves obtained for the MH parameter.

the quality of the fit, statistical parameters were compared by using both coefficients of determination ($R^2$) and a Relative Standard Error (RSE). As listed in Table 1, it was clear that the mathematical method yielded a better agreement than the conventional one, although the values of $R^2$ and RSE only slightly improved.

In order to make sure this relationship was valid in the other material systems, lots of creep rupture data need to be collected and applied as mentioned above. However, it is not easy to practically obtain the long-term creep rupture data by experimental tests for various high-temperature materials. Therefore, although this study was applied only to Type 316LN SS, it is believed that this mathematical method can be utilized to other materials because the material behavior is generally similar. Furthermore, in determining the creep constants of the LM, OSD and MH methods, it was identified that the mathematical method proposed in this study was not only superior for creep life prediction when compared to the conventional one, but also beneficial for obtaining creep constants by using a numerical program.

## 4. CONCLUSIONS

In order to determine the creep constants for the LM, the OSD and the MH parametric methods, a new approach was demonstrated for Type 316LN SS. A linear relationship between log ($t_r$) and stress was observed. On the basis of this linear relationship, equations were derived to facilitate the determination of the creep constants for the three TTP methods. The results of the mathematical approach were not only superior in predicting the creep life when compared to those of the conventional ones, but also beneficial for obtaining the creep constants by using a numerical program. This new approach can be conveniently used to determine the constants of the TTP methods.

## ACKNOWLEDGMENT

This work was performed under the Long-term Nuclear R&D Program sponsored by the Korean Ministry of Science and Technology.

## REFERENCES

1. W. G. Kim, H. H. Kim, K. B. Yin, and W. S. Ryu, *Mat. Sci. Forum*. **475-479**, 1433 (2005).
2. W. G. Kim, S. N. Yin, W. S. Ryu, S. J. Kim, and W. Yi, *Journal of ASTM International*. STP 1480, **35**, 71 (2006).
3. ASME Boiler and Pressure Vessel Code-Section II Materials-Part D, p. 794-796, ASME New York (2004).
4. W. G. Kim, S. N. Yin, W. S. Ryu, and C. B. Lee, *J. Kor. Inst. Met. & Mater.* **46**, 118 (2008).
5. VAMAS Data Evaluation Committee, Iron and Steel Institute, Japan (1994).
6. I. Le May, *Transactions of the ASME* **101**, 326 (1979).
7. J. T. Yeom, J. Y. Kim, Y. S. Na, and N. K. Park, *Met. Mater. Int* **9**, 555 (2003).
8. K. Maruyama and K. Yoshimi, *Proc.of CREEP8*, CREEP2007-26150, Texas, USA (2007).
9. NRIM Creep Data Sheet, p. 1-12, No. 45 (1997).
10. K. Kazuya, D. Shingo, T. Kenichi, S. Masayuki, K. Nobuchika, and T. Yoshihiko, *ASME PVP*. **391**, 47 (1999).
11. M. D. Mathew, G. Sasikala, K. Bhanu Sankara Rao, S. L. and Mannan, *Mater. Sci. Eng. A* **148**, 253 (1991).
12. M. Liska, V. Vodarek, M. Sobotkova, and J. Sobotka, *Preprints for the 2nd International Conference on High Nitrogen Steels*, p. 78, Aachen, Germany (1990).
13. D. W. Kim, Y. K. Lee, W. G. Kim, and W. S. Ryu, *J. Kor. Inst. Met. & Mater.* **39**, 1241 (2001).
14. W. G. Kim, S. N. Yin, W. S. Ryu, and C.-B. Lee, *J. Kor. Inst. Met. & Mater.* **46**, 118 (2008).

*Met. Mater. Int., Vol. 15, No. 5 (2009), pp. 727~732*
doi: 10.1007/s12540-200-0727-3 Published 26 October 2009

# Generation of Isochronous Stress-Strain Curves with a Nonlinear Least Square Fitting Method for Modified 9Cr-1Mo Steel

**Woo-Gon Kim*, Song-Nan Yin, and Gyeong-Hoi Koo**

Department of Nuclear Materials Technology Development, Korea Atomic Energy Research Institute, 1045 Daeduk-daero, Yuseong-gu, Daejeon 305-353, Korea

(received date: 16 October 2008 / accepted date: 24 March 2009)

This study deals with the generation of isochronous stress-strain curves for modified 9Cr-1Mo steel. Material property data ($s_y$, $D$, and $m$) from tensile tests and strain-time curves from creep tests were obtained at 550 °C. On the basis of the experimental data, the creep curves were characterized by Garofalo's creep model. The three parameters of $P_1$, $P_2$ and $P_3$ in Garofalo's model could be optimized properly by a nonlinear least square fitting analysis. The three parameters showed a good stress dependency with a linear relation. However, the $P_3$ parameter, which represents the steady state creep rate, exhibited two-slope behavior with different stress exponents at a transient stress of about 250 MPa. The results verify that the long-term creep curves for G91 steel can be modeled by Garofalo's model with only several short-term creep data. When the modeled creep curves are used, long-term isochronous curves of up to $10^5$ h can be successfully constructed.

**Keywords:** creep curve, isochronous stress-strain curve (ISSC), modified 9Cr-1Mo (G91) steel, Garofalo's model, nonlinear least square fitting (NLSF)

## 1. INTRODUCTION

Compared with austenitic stainless steels, high chromium ferritic-martensitic (FM) steels have a lower irradiation-induced swelling, a lower thermal stress, a lower induced activation, a lower thermal expansion, a higher thermal conductivity and a higher fatigue strength at a high temperature. Modified 9Cr-1Mo steel (ASTM grade 91, (G91)) has been widely used for boiler components in ultrasupercritical thermal power plants operated at about 600 °C. Furthermore, G91 is a candidate for the structural material of next-generation fission power plants and new-generation nuclear power reactors, particularly as a pressure vessel material of a very high temperature gas-cooled reactor and the nuclear fuel cladding tubes of a sodium-cooled fast reactor [1].

Because these components are designed to be used over a 30 year life span at a high temperature, one of the most critical properties in their design is their time-dependent creep behavior [2]. For the design application of high-temperature materials, the isochronous stress-strain curves (ISSCs) of these components must be such that excessive deformation of the materials is avoided during the intended service life because the designed target life of the materials is based on a specified amount of allowable strain and stress. However, the task of obtaining long-term ISSCs of up to 30 years is very difficult and very time-consuming [3]. Thus, long-term creep curves from short-term creep curves must be modeled accurately to generate long-term ISSCs.

To describe a creep strain-time curve well, Garofalo [4], Kachanov-Rabotnov [5], Blackburn [6], Evans [7], and Maruyama *et al.* [8] have all developed creep constitutive models. Of these, Garofalo's model is the simplest in terms of its various constitutive equation forms; it is also suitable for G91 steel because it is well defined for primary and secondary creep strains. In addition, methods of generating an ISSC for high-temperature materials have been reported by Swindeman [9] for 2.25Cr-1Mo steel and Blackburn [6] for austenitic stainless steels. However, Swindeman's method requires a large amount of creep experimental data because it involves the use of a Larson-Miller analysis for coupling the temperature and stress dependencies of the minimum creep rate. Blackburn's method also has difficulty characterizing several creep parameters because it uses a complicated constitutive equation. Hence, there is a need for a useful method of conveniently generating an ISSC with a minimal amount of creep data.

In this paper, Garofalo's model was selected as a suitable model for characterizing the creep curves of the G91 steel. The nonlinear least square fitting (NLSF) method was used to obtain the $P_1$, $P_2$ and $P_3$ parameters of Garofalo's model. The results of the stress dependencies of these three parame-

*Corresponding author: wgkim@kaeri.re.kr

**Table 1.** Chemical composition of the G91 steel (wt.%)

| C | Si | Mn | P | S | Ni | Cr | Mo | Cu | V | Al | N | Nb |
|---|---|---|---|---|---|---|---|---|---|---|---|---|
| 0.115 | 0.23 | 0.415 | 0.012 | 0.0014 | 0.22 | 8.9 | 0.869 | 0.038 | 0.194 | 0.020 | 0.0513 | 0.073 |

ters were used to model the long-term creep curves and to construct the ISSCs for up to $10^5$ h.

## 2. EXPERIMENTAL METHOD

The testing material was a hot-rolled G91 steel plate of a commercial grade. The heat treatment condition of the steel was normalized and tempered at 1050 °C / 1 min/mm and at 770 °/3 min/mm. The plate thickness was 16 mm. Table 1 shows the chemical composition, and each element is well within the ASTM requirement.

Tensile tests were performed to obtain the material data used in the inelastic constitutive equations. The tensile test specimens were machined with a rectangular cross section with a thickness of 2 mm, a width of 6.25 mm, and a gauge length of a 25 mm. The strain rate in the tensile tests was a slow rate of $1 \times 10^{-4}$/s at 550 °C. The creep test specimens were machined to a cylindrical shape with a gauge length of 30 mm and a diameter of 6 mm. The creep tests were conucted at 550 °C with different stress levels of 300 MPa, 280 MPa, 260 MPa, 250 MPa, 240 MPa, and 220 MPa. The creep tests were stopped when the secondary creep stage was considerably developed. The reason for this cessation is that the full creep curves up to a tertiary creep stage are not needed when the creep curves are modeled to generate an ISSC. The creep strain data with elapsed times was taken automatically by a PC through a high precision extensometer. The steady state creep rate was measured from the experimental creep curves.

## 3. RESULTS AND DISCUSSION

### 3.1. Selection of a suitable creep model for G91 steel

Strain depends on both the stress level and time at a given temperature within a creep range. For a design application, the limitation of a strain may be an important factor for preserving the structural integrity or functional capability of a component. Such strain limitations are more readily incorporated into a design process if they are first translated into design stress limits that preclude any excessive distortion of the intended service [6]. Thus, a design for creep materials should reflect the essential interrelation of the stress, strain, and time. One method of depicting these interrelations is through isochronous curves. In a plot of the stress against strain, isochronous curves present the locus of the total strains that are accumulated when different constant stresses are applied for a fixed time. An ISSC of up to $10^5$ h must be constructed for the design of high-temperature nuclear components. For this reason, the creep curves of each material are properly modeled on a proper constitutive model because

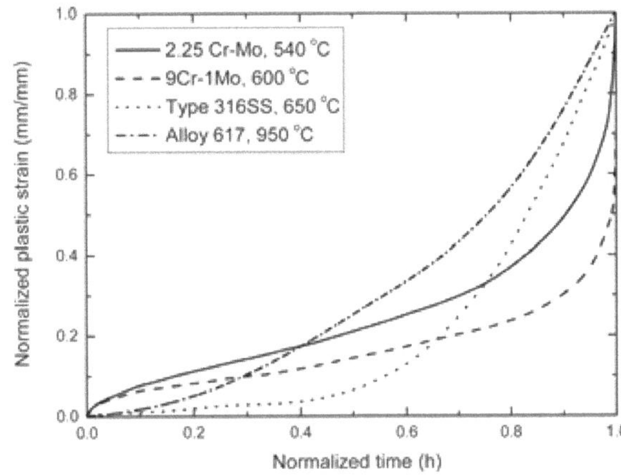


**Fig. 1.** Normalized creep curves for typical high-temperature materials; 2.25Cr-Mo, 9Cr-1Mo, type 316 SS, and alloy 617.

the shapes of the creep curves depend on the materials, temperatures and stresses.

Figure 1 shows the normalized creep curves at different temperatures for typical high-temperature materials, such as 2.25Cr-Mo, 9Cr-1Mo, type 316 SS, and alloy 617. The shapes of the creep curves vary with the type of material used. FM steels such as 2.25Cr-Mo and 9Cr-1Mo steel have a clear primary, secondary and tertiary stage in the creep curves. Most of the creep life belongs to the secondary stage. On the other hand, type 316 and alloy 617 almost fail to generate a primary creep stage, and a secondary creep stage is unclear at high this temperature. At 950 °C, alloy 617 has a higher ductility than the FM steels, its creep strain increases significantly with time variations, and the end point of its secondary creep stage is not clear.

Figure 2 shows the raw data of the creep curves with different stresses for the G91 steel at 550 °C. The creep tests were stopped when the secondary creep stage was considerably developed because the generation of an ISSC obviates the need for full creep curves up to a tertiary creep stage. The secondary creep stage is clearly observable in the creep curve. The applied stress dependency of the steady state creep rate obeys Norton's power law of $\dot{\varepsilon}_s = A\sigma^n$; furthermore, as shown in Fig. 3, at 550 °C it exhibits two-slope behavior with a stress exponent of about 10 for low levels of stress and about 19 for high levels of stress. Choudhary *et al.* [10] reported that this two-slope behavior is closely related to the resisting stress, the transition stress, and the threshold stress of G91 steel. Specifically, at low levels of stress the stress sensitivity is low, whereas at high levels of stress the stress

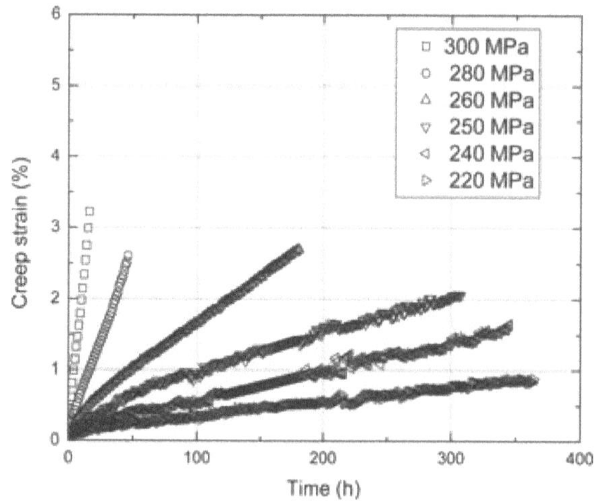


**Fig. 2.** Creep curves obtained with stress levels for the G91 steel at 550 °C.

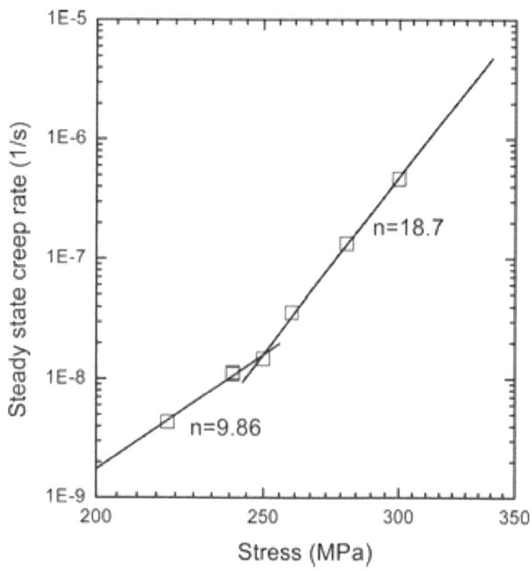


**Fig. 3.** Stress dependency of the steady state creep rate for the G91steel at 550 °C.

sensitivity is high. The resisting stress increases linearly with the applied stress in a low stress region but, in a high stress region, remains either constant (that is, independent of the applied stress) or weakly dependent on the applied stress.

As discussed above, Garofalo's model was selected because of its suitability for modeling the creep curves of G91 steel. The reason for this suitability is that Garofalo's model can describe the creep stage of G91 steel (where a small amount of instantaneous strain is clearly revealed) as well as the primary and secondary creep stages.

Garofalo's model for creep strain-time curves is given as

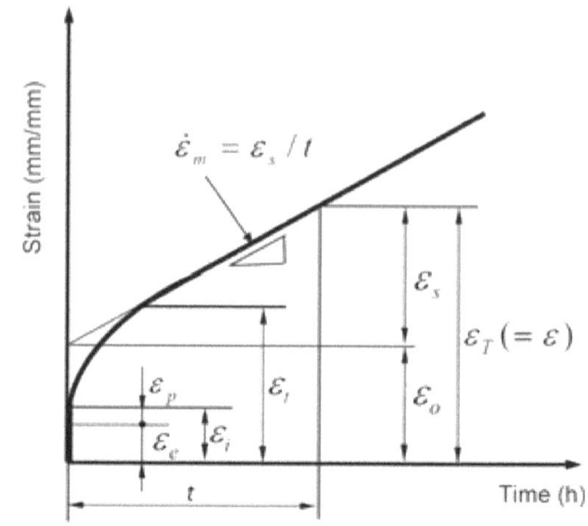


**Fig. 4.** Detail definitions for various creep strains.

follows [4]:

$$\varepsilon = \varepsilon_i + \varepsilon_t[1-\exp(-\beta t)] + \dot{\varepsilon}_m t \tag{1}$$

where $\varepsilon$ is the total strain, $\varepsilon_i$ is the instantaneous elastic-plastic strain, $\varepsilon_t$ is the limiting magnitude of a transient creep strain, $\hat{a}$ is a constant related to the rate of exhaustion of a transient creep, $\dot{\varepsilon}_m$ is the secondary creep rate (or minimum creep rate), and $t$ is the time. Detailed definitions of the various creep strains are given in Fig. 4. The initial strain effect in the modeling process of a creep strain can be avoided if Eq. 1 is rewritten as Eq. 3, where $\varepsilon_c$ is a pure creep strain, and $P_1$, $P_2$, and $P_3$ are the equation parameters corresponding to $\varepsilon_t$, $\beta$ and $\dot{\varepsilon}_m$, respectively.

In addition, the stress dependence of each parameter is considered as a linear form of Eq. 4. From this relation, each value of $P_i$ at any given low stress can be calculated and predicted as follows:

$$\varepsilon_c(=\varepsilon-\varepsilon_i) = \varepsilon_t[1-\exp(-\beta t)] + \dot{\varepsilon}_m t \tag{2}$$

$$\varepsilon_c = P_1[1-\exp(-P_2 t)] + P_3 t \tag{3}$$

$$\log P_i = a_i + b_i\sigma \tag{4}$$

The long-term creep curves of G91 steel can be modeled by Eq. 4 for a linear relation between each parameter and stress.

### 3.2. Determination of the instantaneous elastic-plastic strain ($\varepsilon_i$) value

The basic equations for a time-independent elastic plus plastic strain are given by

$$\varepsilon_i = \frac{\sigma}{E} \quad \text{for} \quad \sigma \le \sigma_p \tag{5}$$

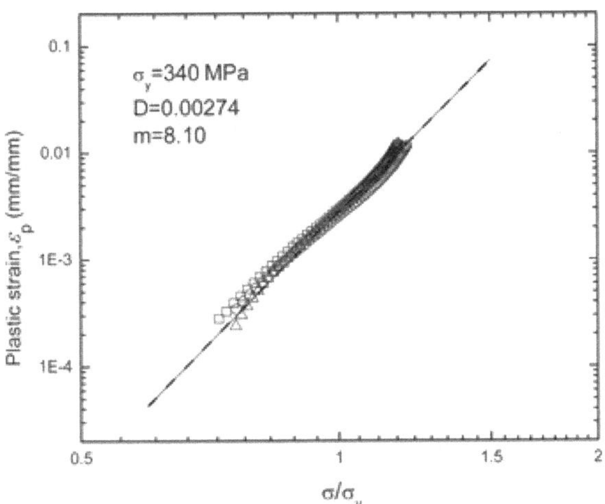


**Fig. 5.** Material constants $D$ and $m$ of the G91 steel at 550 °C.

$$\varepsilon_i = \varepsilon_e + \varepsilon_p = \frac{\sigma}{E} + D\left[\frac{\sigma}{\sigma_y}\right]^m \quad \text{for} \quad \sigma > \sigma_p \tag{6}$$

where $E$, $\sigma_y$, $D$ and $m$ are material constants obtained from a tensile test, $\sigma$ is the true stress, $E$ is the elastic modulus, and $\sigma_p$ is the true stress at a proportional elastic limit. Eq. 5 is used for $\sigma \leq \sigma_p$ with an applied creep stress, and Eq. 6 is used for $\sigma > \sigma_p$ with an applied creep stress. The $D$ and $m$ constants can be determined from the slope and intercept of the plots of log ($\sigma/\sigma_y$) vs. log ($\varepsilon_p$). Note that Eq. 6 is derived in terms of a true stress and a true strain, and that all the equations for a creep strain are in engineering strain units. Thus, when significant plastic strains are involved, the stress and strain values from Eq. 6 must be converted to engineering units before $\varepsilon_i$ is combined with the creep strains.

As shown in Fig. 5, the least square method was used for three specimens to characterize the material constants of $D$ and $m$ in the tensile tests. The average values of the G91 steel were 0.00274 for $D$ and 8.1 for $m$. In addition, $\sigma_y$ was 340 MPa and $\sigma_p$ was 230 MPa. The elastic modulus of the 9Cr-Mo steel, as shown in Fig. 6, was also obtained by referring to the ASME Boiler and Pressure Vessel Code, Section II, Materials-Parts D. The elastic modulus was 173 GPa at 550 °C. This value was used to calculate the instantaneous elastic-plastic strain ($\varepsilon_i$) value in Eqs. 5 and 6.

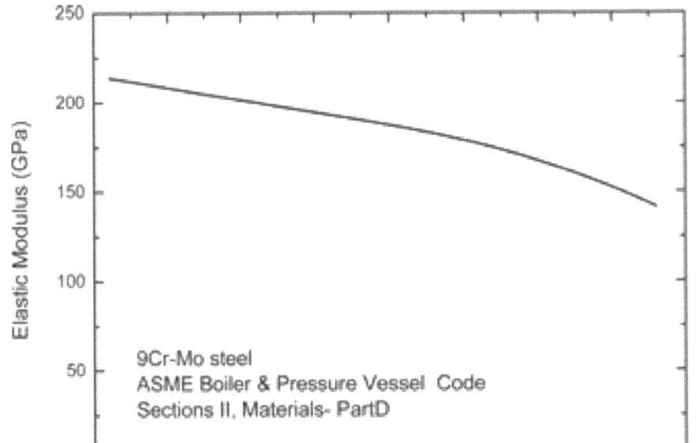


**Fig. 6.** Elastic modulus with temperatures for the G91 steel.

### 3.3. Creep curve modeling and calculated isochronous curves

A number of computing processes for the NLSF analysis were conducted for the experimental creep curves so that an optimum value of the three parameters in Eq. 3 could be obtained. Through an iterative NLSF analysis, an optimized fitting curve was obtained for the experimental creep curve. The Nonlinear Curve Fit function of the Advanced Fitting Tool of the software program Origin 7.5 was used for the NLSF analysis.

Figure 7 shows the NLSF results for the experimental creep curves of the G91 steel at 550 °C. The solid lines indicate the NLSF curves based on Garofalo's creep model. The values of $P_1$, $P_2$ and $P_3$ were properly optimized because Garofalo's equation is simple. The NLSF curves closely match the experimental creep data. Hence, the NLSF analysis is suitable for modeling the creep curve of the G91 steel. The long-term creep curves at any low stress level must then be modeled. To that end, the stress dependency of each $P_i$ parameter should be investigated at a given stress level. Eq. 4 can be used to determine whether the parameters $P_1$, $P_2$ and $P_3$ depend on the level of stress. In Fig. 8 the three parameters tend to increase in an almost linear manner as the stress increases. In the case of $P_3$, which represents the steady state creep rate, the linear relation is separated into two parts

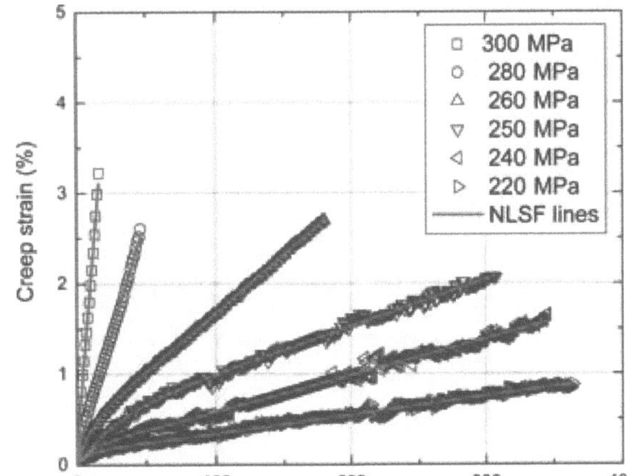


**Fig. 7.** The NLSF results of the creep curves by the Garofalo's model.

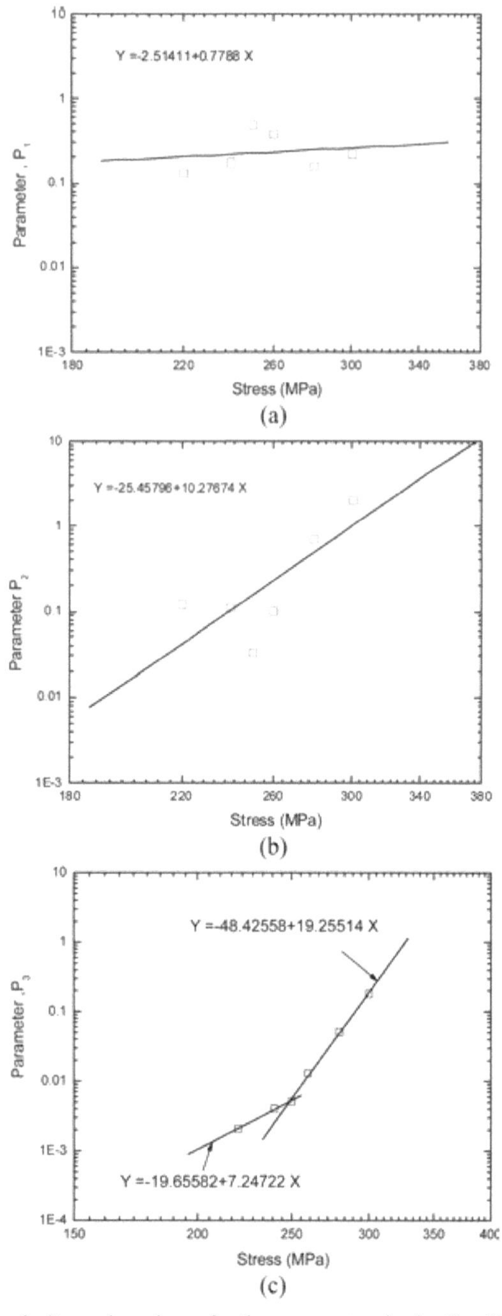


**Fig. 8.** Stress dependency for three parameters in the Garofalo's model: (a) $P_1$ parameter, (b) $P_2$ parameter, and (c) $P_3$ parameter.

at a transient stress of about 250 MPa. This two-slope behavior is almost the same as the result of Fig. 3, where the apparent stress exponent, $n$, is of 7.2 at low levels of stress and 19.2 at high levels of stress.

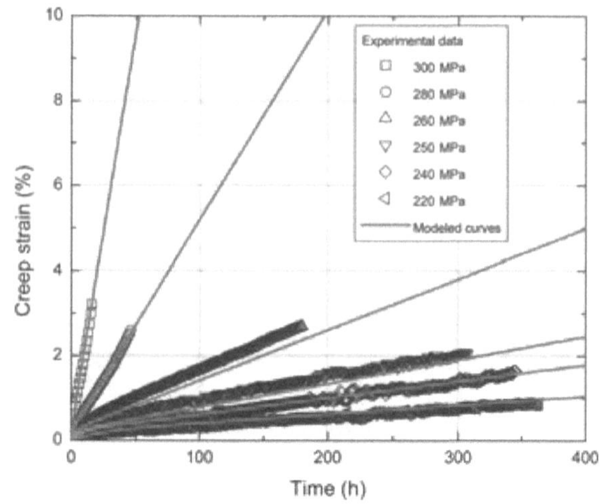


**Fig. 9.** Comparison of modeled curves and experimental creep data.

**Table 2.** The $a$ and $b$ coefficients obtained for each $P_i$ parameter

| Parameters | Log ($P_i$) = $a_i + b_i\sigma$ |
|---|---|
| $P_1$ | Y =-2.5141 + 0.7788 X |
| $P_2$ | Y =-25.4579 +10.2767 X |
| $P_3$ | Y =-48.4255 +19.2551X (at high stress) |
| | Y =-19.6558 + 7.2472 X (at low stress) |

Figure 9 compares the modeled creep curves and the experimental creep data. The modeled curves coincide well with the experimental creep data. Table 2 gives the coefficients $a$ and $b$ for the linear equations of the three parameters. With these linear equations, the long-term creep curves at any low stress levels can be modeled, and the isochronous curves at any given time can then be calculated with Eq. 1. Figure 10 plots the results of the isochronous curves obtained from 1 h to $10^5$ h at 550 °C. Consequently, when the NLSF

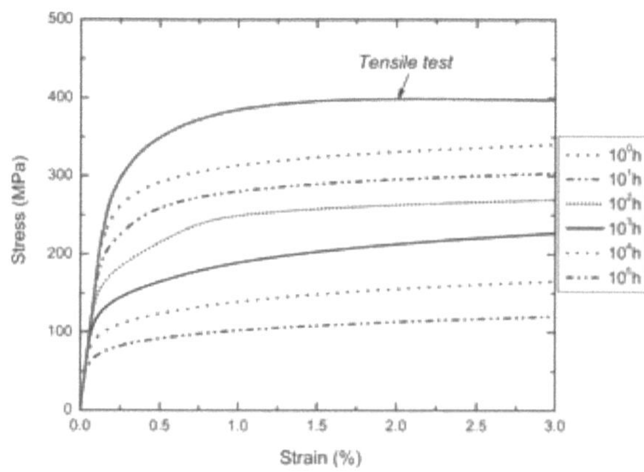


**Fig. 10.** The ISSC results calculated up to $10^5$ hours for the G91 steel at 550 °C.

method is applied to Garofalo's simple model, the isochronous curves for the G91 steel can be easily obtained with a small amount of creep data.

## 4. CONCLUSION

The creep curves of the G91 steel clearly reveal a small amount of instantaneous strain as well as primary, secondary and tertiary creep stages. Garofalo's model, which conforms well to the instantaneous strain stage and the primary and secondary creep stages, is suitable for modeling the creep curves of the G91 steel. The $P_1$, $P_2$, and $P_3$ parameters are properly optimized by the NLSF method, and the stress dependency of the three parameters is a linear relation. However, $P_3$ parameter, which represents the steady state creep rate, exhibits two-slope behavior with different stress exponents at a transient stress of about 250 MPa. The curves modeled on the basis of the NLSF analysis coincide well with the experimental creep data, and the long-term creep curves of the G91 steel are modeled well on the basis of Garofalo's model. The modeled creep curves were used to successfully construct long-term isochronous curves for up to $10^5$ h.

## ACKNOWLEDGMENT

This work was performed as part of a nuclear R&D program sponsored by the Korean Ministry of Education, Science and Technology.

## REFERENCES

1. S. G. Hong, W. B. Lee, and C. G. Park, *J. Nucl. Mater.* **288**, 202 (2001).
2. J. T. Yeom, J. Y. Kim, Y. S. Na, and N. K. Park, *Met. Mater. Int.* **9**, 555 (2003).
3. W. G. Kim and W. S. Ryu, *Key Eng. Mater.* **261-263**, 1103 (2005).
4. F. Garofalo, *Fundamentals of Creep and Creep Rupture in Metals*, p. 10-27, Macmillan Co., New York (1966).
5. R. K. Penny and D. L. Marriott, *Design for Creep, 2nd ed.*, p. 24-27, Chapman & Hall, London (1995).
6. L. D. Blackburn, *Isochronous Stress Strain Curves for Austenitic Steels*, p. 15-48, ASME, New York (1972).
7. R. W. Evans and B. Wilshier, *Creep of Metals and Alloys*, p. 197-256, Institute of Metals, London (1985).
8. K. Maruyama and H. Oikawa, *Scripta metal.* **21**, 233 (1987).
9. R. W. Swindeman, *ASME Pressure Vessel and Piping Conf.*, ASME, Fiorida, USA (1997).
10. B. K. Choudhary, K. Bhanu Sankara Rao, and S. L. Manan, *Trans. Indian Inst. Met.* **52**, 327 (1999).

Materials Science and Engineering A 483–484 (2008) 495–497

MATERIALS SCIENCE & ENGINEERING A

www.elsevier.com/locate/msea

# Tension and creep design stresses of the "Hastelloy-X" alloy for high-temperature gas cooled reactors

Woo-Gon Kim [a,*], Song-Nan Yin [b], Woo-Seog Ryu [a], Jong-Hwa Chang [a], Seon-Jin Kim [c]

[a] Nuclear Materials Technology Development Division, Korea Atomic Energy Research Institute (KAERI), 150 Deogjin-dong, Yuseong-gu, Daejeon, 305-353, Korea

[b] School of Mechanical Engineering, Soong-sil University, Seoul 156-743, Korea

[c] School of Mechanical Engineering, Pukyong National University, Busan 608-737, Korea

Received 6 June 2006; received in revised form 9 October 2006; accepted 3 December 2006

**Abstract**

To provide the design data for the Hastelloy-X alloy, the tensile and creep test data was obtained experimentally, and a number of creep data was also collected through literature surveys. All of the experimental and collected data was combined to determine a minimum design value for a tensile stress and to predict a long-term life for a creep rupture time. The design yield and ultimate tensile stress curves for the Hastelloy-X alloy were determined by a best fitting all of the data up to 1000 °C. Creep master curves based on the Larson–Miller parameter were obtained for the standard deviations of $1\sigma$, $2\sigma$ and $3\sigma$. The creep-rupture life for up to $10^5$ h was predicted for the Hastelloy-X alloy.



*Keywords:* Hastelloy-X Alloy; Creep; Creep life; High temperature gas cooled reactor; Larson–Miller parameter

## 1. Introduction

The Hastelloy-X alloy is considered as a candidate material for the high-temperature gas cooled reactor (HTGR) components, such as the hot-gas ducts and intermediate heat exchanger (IHX) tube to produce hydrogen using the processed nuclear heat, because it has a good oxidation resistance at temperatures up to 1200 °C and moderately good strength at temperatures up to 870 °C, especially, in a helium environment [1]. However, high-temperature tensile and creep data up to 1000 °C is not sufficient to establish the design stresses for the Hastelloy-X alloy, especially, to predict a long-term creep life reaching $10^5$ h (about 11 years), a lot of creep data should be obtained from experiments and worldwide literature surveys, or other country's databases.

In this paper, the tensile and creep experimental data for the Hastelloy-X alloy were obtained from the tensile tests from 600 °C to 1000 °C and the creep tests at 950 °C. The experimental creep data were combined with the data obtained from the literature surveys; European Alloy DB [2], published data in the literatures [3,4], and manufactures' data sheets [5,6]. Using these data, the minimum design tensile value was obtained at temperature up to 1000 °C, and the creep master curves to predict a long-term creep life for up to $10^5$ h were obtained by the Larson–Miller (LM) parameter.

## 2. Results and discussion

Fig. 1 shows the results of the minimum design stress curves for $S_y$ and $S_u$, which were obtained using the tensile data collected from Haynes Co. [5], European Alloy DB [2], and the experimental data in this study for the Hastelloy-X alloy. Our experimental tensile values showed nearly similar ones to the collected data at the same temperature condition, although the data was collected from different countries. The $S_y$ curve of the yield stress was obtained with a minimum value determined by a lower prediction line of 95% (95% LPL) at each temperature, by referring to ASME Code Case N47-17 [7]. Also, the $S_u$ curve of the ultimate tensile stress was obtained at the same manner, as described above for $S_y$. The LPL line of the minimum value is to assure a statistical reliability of 95% for one datum point at each temperature. In Fig. 1(b), the range of the error bars indicates a

* Corresponding author. Tel.: +82 42 868 2493; fax: +82 42 868 8549.
*E-mail address:* wgkim@kaeri.re.kr (W.-G. Kim).

0921-5093/$ – see front matter © 2007 Elsevier B.V. All rights reserved.
doi:10.1016/j.msea.2006.12.184

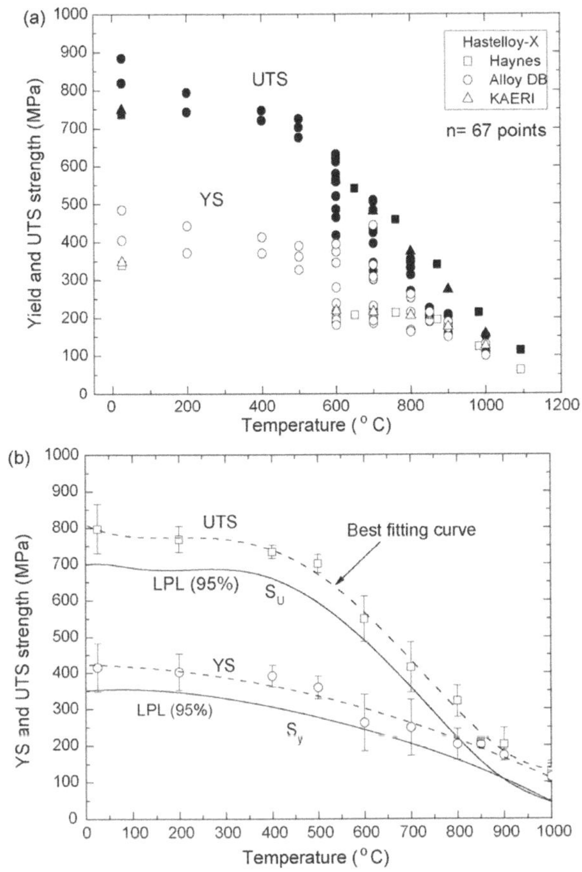


Fig. 1. The YS and UTS data collected for the Hastelloy-X alloy. (a) is all the data collected from Haynes Co., Alloy DB and KAERI, and (b) is the LPL of 95% obtained from the best fitting curve.

standard deviation ($1\sigma$) which was calculated from all the data points at each temperature.

Fig. 2 shows the results of the steady-state creep rate (SSCR) and the stresses tested for the Hastelloy-X alloy. The Hastelloy-X alloy used in this study is a commercial type of a hot-rolled plate with a thickness of 19 mm. The creep tests were conducted with different stresses from 14 MPa to 35 MPa at 950 °C. The relationship between the SSCR and the stress showed a good linearity. Creep constants for Norton's law were obtained; the $A$ value was $2.57 \times 10^{-13}$ ($\mathrm{MPa}^{-n}\,\mathrm{s}^{-1}$) and the $n$ value was 4.15.

In addition, the Monkman–Grant (M–G) relationship is useful to perform the prediction of a lifetime of high-temperature structural components. The relationship between the SSCR and the rupture time is expressed as $\log t_\mathrm{R} + m \log \dot{\varepsilon}_\mathrm{sscr} = C$, where $C$ and $m$ are constants, and $t_\mathrm{R}$ is time to rupture. The relation for the Hastelloy-X alloy showed a good linearity. It was found to be $m = 0.81$. Also, the modified M–G one is expressed as $\log t_\mathrm{R}/\varepsilon_\mathrm{R} + m' \log \dot{\varepsilon}_\mathrm{sscr} = C'$, where $\varepsilon_\mathrm{R}$ is the elongation at rupture. The creep data showed a good linearity, as shown in Fig. 3. It was found $m'$ to be 1.02, which is close to unity. Good agreement of the modified relation was due to the linear relationship

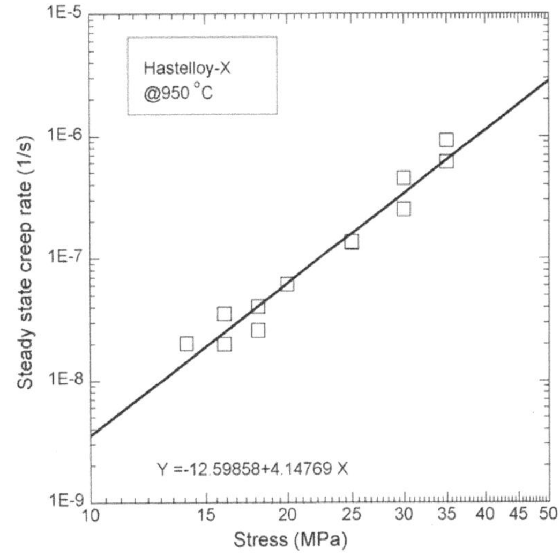


Fig. 2. A linear relationship between the SSCR and the stresses.

of the elongation to the rupture time. It means that creep deformation of the Hastelloy-X alloy progressed by the formation of cracks and cavities.

A long-term creep life was estimated by using the L–M parameter, as follows:

$$P = T(\log t_\mathrm{R} + C) = f(\sigma). \quad (1)$$

Hence, $t_\mathrm{R}$ is the time at rupture. The $f(\sigma)$ is a stress function and the $C$ value used was 20. All of the creep data was combined with the data collected from the literature surveys; European Alloy DB [2], published data of the literatures [3,4], manufactures' data sheets [5,6], and the experimental data of

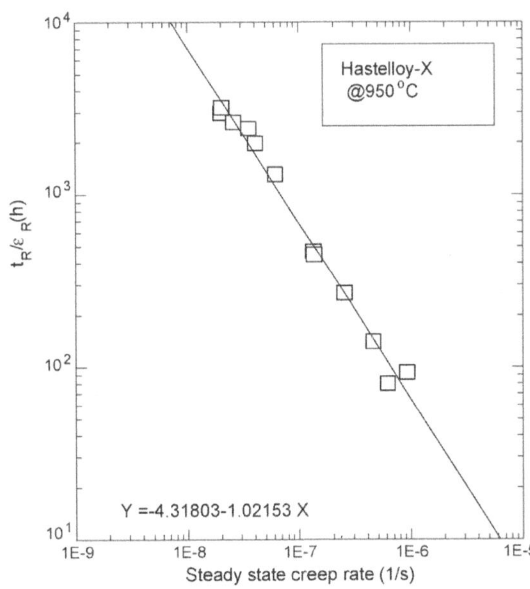


Fig. 3. Plot of the modified M-G relation.

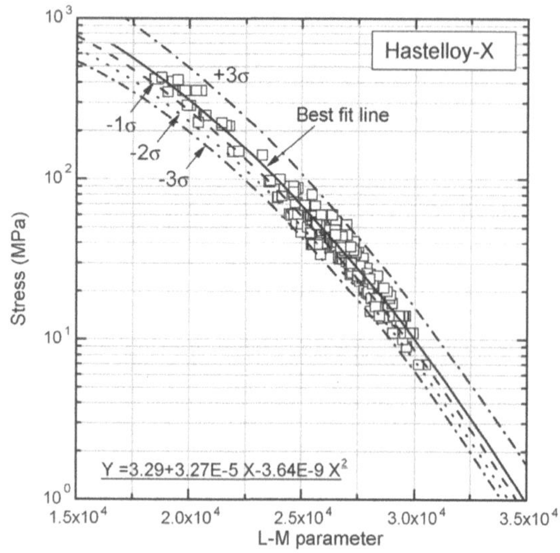


Fig. 4. Creep master curves for the 1σ, 2σ and 3σ standard deviations.

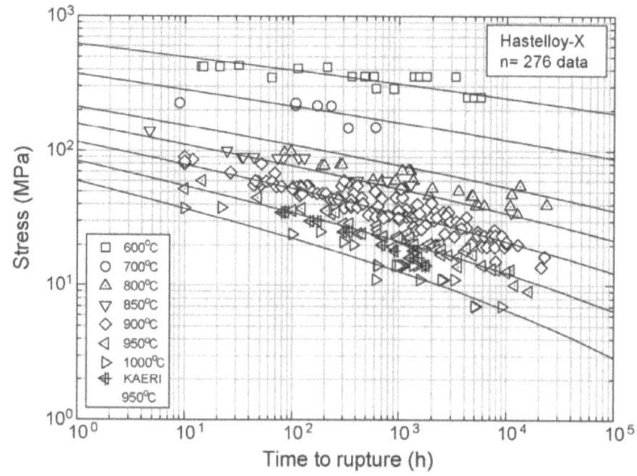


Fig. 5. Prediction of a creep life up to $10^5$ h with the temperatures for the Hastelloy-X alloy.

this study. Although both the experimental data and the data collected in literature were different at experimental conditions, all of the data nearly conformed to each temperature.

Fig. 4 shows the creep master curve calculated by the L–M parameter by using all of the creep rupture data. A polynomial equation of the second order was obtained from a best fitting of the creep data, and each curve was obtained for the standard deviations of $1\sigma$, $2\sigma$ and $3\sigma$. Long-term creep life for up to $10^5$ h for the Hastelloy-X alloy was extrapolated by the L–M parameter at each temperature, as shown in Fig. 5.

## 3. Conclusion

A number of tension and creep data for the Hastelloy-X alloy was provided through available literature surveys and experiments. The values of the minimum design stresses for $S_y$ and $S_u$ were obtained up to 1000 °C. The creep constants in the Norton's law and the M-G relationship were investigated to be $A = 2.57 \times 10^{-13}$ (MPa$^{-n}$ s$^{-1}$), $n = 4.15$, $m = 0.81$ and $m' = 1.02$. The creep master curves based on the Larson–Miller parameter were obtained for the standard deviations of $1\sigma$, $2\sigma$ and $3\sigma$, respectively. A long-term creep life up to $10^5$ h was predicted for a temperature up to 1000 °C. This result can be utilized for estimating the long-term creep life or the creep-rupture strength for the creep design of the Hastelloy-X alloy.

## References

[1] K. Natesan, A. Purohit, S.W. Tam, Materials Behavior in HTGR Environments, NUREG/CR-6824, ANL-02/37, 2003, 19–26.
[2] European Alloy-DB, Licensed CD, European Commission-Joint Research Centre (EC-JRC).
[3] Y. Muto, H. Nakajima, M. Eto, Nucl. Eng. Des. 144 (1993) 305–315.
[4] K. Hada, I. Nishiguchi, Y. Muto, H. Tsuji, Nucl. Eng. Des. 132 (1991) 1–11.
[5] Haynes International, www.haynesintl.com, Hastelloy-X Alloy, H-3009A (1997).
[6] Special Metals, www.specialmetals.com, Inconel Alloy HX, SMC-007 (2004).
[7] H. Yoshizu, Y. Monma, E. Baba, Y. Kurata, H. Nakajima, Study of Creep and Rupture Behavior for a Ni-base Heat Resistant Alloy–Improved for High Temperature Strength Data of Hastelloy XR-II, JAERI-M 93-231, 1993, pp. 59–96.

Engineering Fracture Mechanics 75 (2008) 4985–4995

Contents lists available at ScienceDirect

Engineering Fracture Mechanics

journal homepage: www.elsevier.com/locate/engfracmech

# Creep characterization of a Ni-based Hastelloy-X alloy by using theta projection method

Woo-Gon Kim *, Song-Nan Yin, Yong-Wan Kim, Jong-Hwa Chang

*Korea Atomic Energy Research Institute, 150 Deokjin-dong, Yuseong-gu, Daejeon 305-353, Republic of Korea*

ARTICLE INFO

*Article history:*
Available online 1 July 2008

*Keywords:*
Creep
Hastelloy-X alloy
Theta ($\theta$) projection method
Nonlinear least square fitting
Creep rate

ABSTRACT

The $\theta$ projection method was applied to characterize the creep behavior of the Hastelloy-X alloy at 950 °C. Four $\theta$ parameters were established by a nonlinear least square fitting (NLSF) to the creep curves. In the NLSF of the full creep curves, the $\theta_1$ and $\theta_2$ parameters were not defined with a large error, but the $\theta_3$ and $\theta_4$ parameters were defined well without an error. An optimum cutoff strain range for defining the four $\theta$ parameters was found to be a 3% strain. Four $\theta$ parameters revealed a good linearity as a function of stress. The predicted minimum creep rate showed a good agreement with the experimental data. At 950 °C of the Hastelloy-X alloy, the creep curves, the creep rate, and the time to reach a limiting strain were estimated with a wide range of stresses.



## 1. Introduction

Since very high temperature gas-cooled reactor (VHTR) components are designed for a lifetime above 30 years at 950 °C, one of the most important properties of their metallic components is the time-dependent creep behavior [1,2]. The criteria for setting allowable design stress are based on the ASME nuclear Code NH [3]. The maximum allowable value is established not to exceed the lowest of the following: (i) 100% of the average stress to produce 1% total strain (elastic, plastic and creep), (ii) 80% of the minimum stress to cause a tertiary creep, and (iii) 67% of the minimum stress to cause rupture. To obtain this creep data, it is necessary to conduct long-term creep tests. However, it is not practical and economical to obtain these creep properties directly by long term experiments, because the service lives of such structural materials are generally longer than ten years. It is thus necessary to apply a realistic method characterizing a long-term creep behavior from short term tests.

Over the past several decades, to describe the creep curves accurately, various equations have been developed by Kachanov–Rabotnov (K–R) [4], Garofalo [5], Davies [6], Blackburn [7], Evans et al. [8] and Maruyama et al. [9,10]. Among them, the $\theta$ projection method proposed by Evans et al. [8] is being widely used with advantages and flexibility being appreciated more. The $\theta$ projection method can describe the full creep curve and the dependency of the curve shape on the test conditions accurately, because the creep equation can model the decelerating primary and the accelerating tertiary stages of a creep, and imply that any constant secondary creep region is an inflection on the curves where the effects of a strain hardening and strain softening process are equal. Originally, the $\theta$ projection method was developed by using constant-stress creep data for ferritic steels, but it has also been found to be valid with the constant-load creep data of the 316 austenite steel [11]. However, in the case of IN-100 and IN-738 alloys, Koul and Castillo [12] pointed out that the method considerably underestimates the rupture lives. Thus, it is necessary to investigate its validity for nickel-based Hastelloy-X alloy, which is being considered as a candidate alloy for the VHTR components, such as the hot gas ducts and intermediate heat exchanger (IHX).

* Corresponding author. Tel.: +82 042 868 249; fax: +82 042 868 8549.
*E-mail address:* wgkim@kaeri.re.kr (W.-G. Kim).

0013-7944/$ - see front matter © 2008 Elsevier Ltd. All rights reserved.
doi:10.1016/j.engfracmech.2008.06.017

In this paper, the creep properties for the Hastelloy-X alloy were characterized by the $\theta$ projection method. The four $\theta$ parameters were determined accurately by a nonlinear least square fitting (NLSF) analysis. Creep curves, creep rate, and the time to reach a limiting strain under a given stress condition were predicted with the viewpoint of a design usage.

## 2. Theta projection method

Brown et al. [13–16] have introduced the $\theta$ projection concept to express a creep curve accurately. The equation of a creep strain $\varepsilon_c$ with time $t$ is given by following form,

$$\varepsilon_c = \varepsilon_t - \varepsilon_0 = \theta_1(1 - e^{-\theta_2 t}) + \theta_3(e^{\theta_4 t} - 1) \tag{1}$$

where $\varepsilon_t$ is the total creep strain at time $t$ and $\varepsilon_0$ is the instantaneous strain on loading. The four $\theta$ values are parameters to be fixed by the least square regression analysis. The second and last terms are a decaying creep in the primary stage and an accelerating creep in the tertiary stages. The terms $\theta_1$ and $\theta_3$ are scaling parameters for describing the primary and tertiary stages with respect to a strain, while the terms $\theta_2$ and $\theta_4$ are rate parameters for characterizing the curvatures of the primary and tertiary stages.

In addition, each $\theta$ value is written as function of a stress and temperature:

$$\log\theta_i = a_i + b_i\sigma + c_iT + d_i\sigma T \tag{2}$$

where $\sigma$ is stress, $T$ is the temperature and $i$ = 1, 2, 3, 4. The log $\theta_i$ values can be derived from a testing vary approximately linearly as functions of a stress and temperature. If the $\theta_i$ functions is defined on the basis of short term tests at a high temperature, their values for long term creep data can be predicted.

A creep rate is usually used as an important parameter to predict the creep life of high temperature materials. The creep rate value can be determined by differentiating Eq. (1) with respect to a time:

$$\dot{\varepsilon} = \frac{d\varepsilon_c}{dt} = \theta_1\theta_2 e^{(-\theta_2 t)} + \theta_3\theta_4 e^{(\theta_4 t)} \tag{3}$$

The time at a minimum creep rate, $\dot{\varepsilon}_m$ can be obtained by differentiating Eq. (3) with respect to a time and setting it equal to zero. The time at a minimum creep rate can be calculated by

$$t_m = \frac{1}{\theta_2 + \theta_4} \ln\frac{\theta_1(\theta_2)^2}{\theta_3(\theta_4)^2} \tag{4}$$

If the four $\theta$ values are known for any set of testing condition, the $t_m$ is obtained by Eq. (4) and the minimum creep rate is obtained by Eq. (3). Hence, in Eq. (3), time $t$ is substituted by $t_m$. This provides a useful check on the accuracy of the creep curve shape description since the calculated value can be directly compared to those observed experimentally. Furthermore, if the variations of $\theta$ values with a stress and temperature are obtained, a minimum creep rate at any extended stress condition can be predicted by Eqs. (3) and (4).

## 3. Results and discussion

### 3.1. Creep properties

The Hastelloy-X alloy used in this study is a commercial type hot-rolled plate with a thickness of 19 mm. The chemical composition is given in Table 1. Creep specimens were used with a cylindrical form of 30 mm in gauge length and 6 mm in diameter. Constant-load creep tests were conducted with different stress levels, 35 MPa, 30 MPa, 25 MPa, 20 MPa, 18 MPa, 16 MPa, and 14 MPa at 950 °C. The values for the minimum creep rate, the rupture time and rupture strain were obtained from these tests.

Fig. 1 shows the experimental creep strain–time curves for different stress levels at 950 °C. The Hastelloy-X alloy has a high rupture ductility of 50% to 100%. The creep rupture strain was decreased with a linear relation as the rupture time was increased, as shown in Fig. 2. The primary creep strain was significantly low, and the secondary creep range was unclear. The tertiary creep strain was initiated at a lower strain level, and this tertiary creep governed the total creep deformation.

**Table 1**
Chemical composition of the Hastelloy-X (wt.%)

| Ni | C | Mn | Si | Cr | Co | Cu | Mo |
|---|---|---|---|---|---|---|---|
| 48.04 | 0.082 | 0.82 | 0.42 | 21.91 | 0.79 | 0.13 | 8.65 |
| W | Fe | B | Al | Ti | P | N | S |
| 0.44 | 19.00 | 0.0020 | 0.17 | 0.007 | 0.013 | 0.015 | 0.0003 |

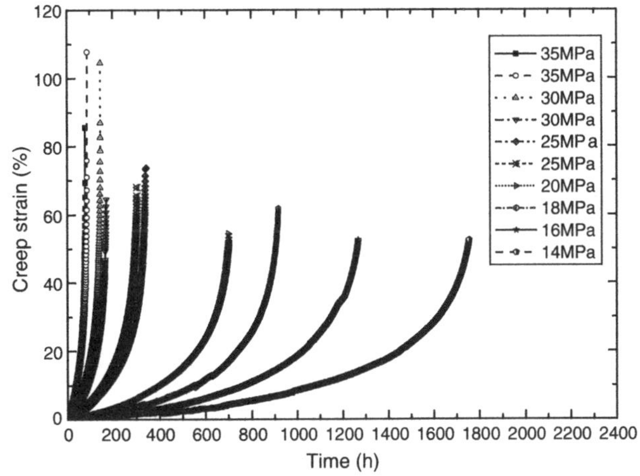


**Fig. 1.** Creep strain–time curves obtained with different stresses for the Hastelloy-X at 950 °C.

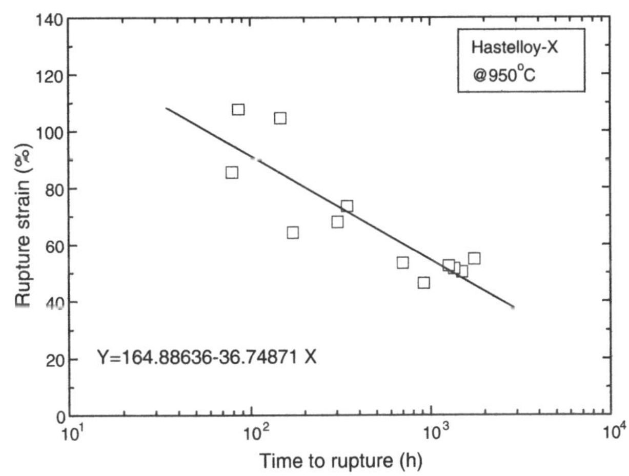


**Fig. 2.** Rupture strain vs. time to rupture with different stresses for the Hastelloy-X at 950 °C.

Fig. 3 shows the results of typical creep rate obtained with different stresses for the Hastelloy-X alloy at 950 °C. The values of the minimum creep rates were determined by a minimum value at the point of an inflection of the creep curve. The relationship between the minimum creep rate and the stress revealed a good linearity, as shown in Fig. 4. From the results, the creep constants for Norton's power law, $\dot{\varepsilon}_m = A\sigma^n$ can be obtained for the Hastelloy-X alloy. The $A$ value was $7.53 \times 10^{-11}$ ($MPa^{-n}\,h^{-1}$) and the $n$ value of the slope was 4.78.

### 3.2. NLSF analysis for creep curves

To obtain an optimum value of the four $\theta$ parameters in Eq. (1), a number of computing processes of the NLSF analysis were carried out for full creep curves. Through the iterative NLSF analysis, a new fitting curve was generated closely to experimental creep curve. A calculation method was conducted by using the "Nonlinear Curve Fit" function of the "Advanced Fitting Tool" in Origin 7.5 version.

Fig. 5 shows a typical result of the NLSF at 35 MPa for the Hastelloy-X at 950 °C. For a description of the creep curve at a stress level, one set of the four $\theta$ values was obtained. The fitted curve closely follows the experimental data. However, the $\theta_1$ and $\theta_2$ parameters for describing the primary creep could not be optimized well and the parameters generated a large error during the fitting process for the full creep curves. When the parameter value (PV) is compared to the error values, the PV is

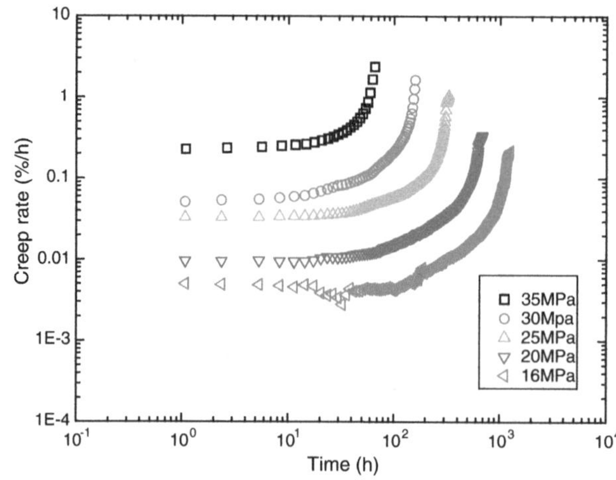


**Fig. 3.** Typical creep rate obtained with different stresses for the Hastelloy-X at 950 °C.

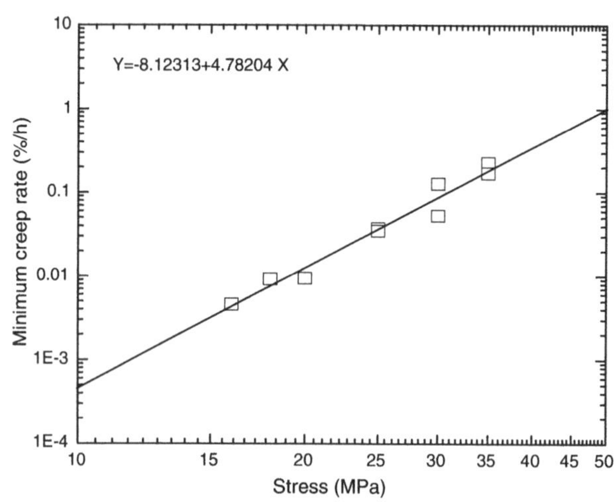


**Fig. 4.** Relationship between the minimum creep rate and the stress for the Hastelloy-X at 950 °C.

smaller than the error values, as underlined in Fig 5. The reason for this is because the primary creep strain was significantly low when compared to the tertiary creep strain up to rupture. Hence, the PV means the value for a maximum likelihood estimates determined from the NLSF of the full creep curve, and also the error value means the value for a standard deviation of the PV. That is, as mentioned in the results of the creep properties, this large error is attributed to the high rupture ductility reaching 50% to 100% in the Hastelloy-X alloy. Also, because the primary range is neglected in a regression process, the $\theta_1$ and $\theta_2$ values are not defined by large errors.

On the other hand, the $\theta_3$ and $\theta_4$ values for describing the tertiary range were accurately achieved through a regression process. It was investigated to see whether the PV was much larger than the error one. From the results, it is necessary to determine a suitable cutoff strain for fitting the creep curves. To investigate an optimum cutoff strain, the NLSF fitting analysis was performed with various cutoff strains for all the creep curves. It was found to be within a 3% cutoff strain. Therefore, in this study, when determining the four $\theta$ parameters of Eq. (1), we discarded the creep strain data after a 3% cut-off strain and used the low strain data within a 3% strain range.

Fig. 6 shows typical results of the NLSF curve at the 3% cutoff strain for different stresses: (a) 35 MPa, (b) 25 MPa and (c) 16 MPa at 950 °C. Each fitted curve showed a good agreement with the experimental data. All of the four $\theta$ values were accurately obtained because the PV was much larger than the error value. Especially, the fitted curve at a low strain level of the primary creep stage coincided well with the experimental data. Thus, the 3% cutoff strain criterion is reasonable.

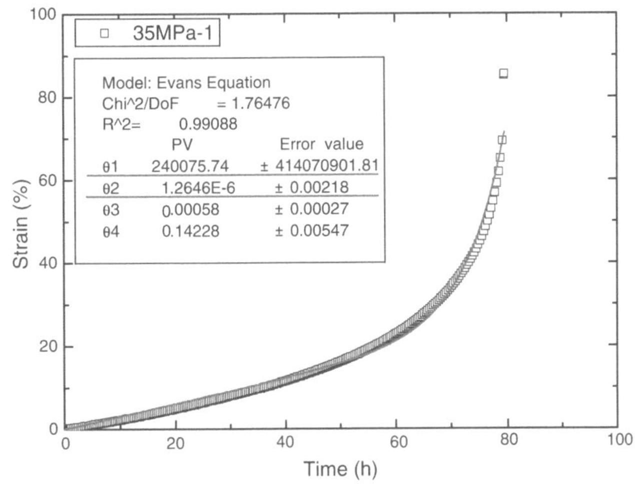


**Fig. 5.** Results of the NLSF for a full creep curve up to rupture at 35 MPa at 950 °C.

### 3.3. Modeling of creep curves

Figs. 7 and 8 show the stress dependency of the parameters $\theta_1$, $\theta_2$, $\theta_3$ and $\theta_4$ in Eq. (1) for constant stress tests carried out for the Hastelloy-X alloy at 950 °C. The four $\theta$ parameters revealed a stress dependency with a good linearity. Based on Eq. (2), for each $\theta$ parameter, the values of the coefficients $a$ and $b$ were calculated, and these values are given in Table 2. The linear relationships indicates that the $\theta$ projection analysis permits both an interpolation and extrapolation of the creep data, allowing short-term test results to be used to predict long-term creep curves. Using this linear relationship, in present study, the creep curves for the Hastelloy-X alloy were predicted to a low stress level of 8 MPa. A detail summary of each $\theta$ value calculated with the stresses is listed in Table 3.

Fig. 9 shows the result of predicted curve to experimental creep data. The predicted curve coincided well with the experimental data, and also the creep curve was modeled to the low stress level of 8 MPa. From this result, it is obvious that this analysis was performed well without an error in determining the four $\theta$ values.

### 3.4. Prediction of time to reach a limiting strain

Once the shape of the creep curve is known by a mathematical expression using Eq. (1), we can calculate the time to reach any creep strain. The time to reach a limiting strain is important from a design viewpoint. As for an example, for a design of time-dependent materials, ASME nuclear Code NH requires information on the time (or stress) to produce a 1% strain. For this purpose, it will be based on a precise description of the shape of the long-term creep curves at various stresses and temperatures.

The time to attain any creep strain ($\varepsilon^*$) at any stress and temperature condition can be obtained simply by solving the following Eq. (5) numerically:

$$\theta_1(1 - e^{-\theta_2 t}) + \theta_3(e^{\theta_4 t} - 1) - \varepsilon^* = 0 \tag{5}$$

where $\varepsilon^*$ is at any specified strain which will be required for a design purpose.

Fig. 10 shows the result of such calculation at any creep strains for the Hastelloy-X alloy at 950 °C. The curves represent for the strains from 0.008 to 0.16. Also, the time to reach a 1% creep strain was predicted for different stresses, as shown in Fig. 11. Thus, because the $\theta$ projection method provides the contours of the time to attain any specified strain (or time to rupture) in log stress vs. log time space, it will be utilized to accurately predict a creep strain and a creep life data required for design of high temperature plant.

### 3.5. Prediction of creep rate

Fig. 12 shows the results of a minimum creep rate predicted by the $\theta$ projection method for the Hastelloy-X alloy at 950 °C. The values of the minimum creep rates were calculated by using Eq. (3). Circle symbols indicate experimental data points, a solid line indicates the result of the $\theta$ analysis, and a dashed line indicates the prediction of Norton's law. The $\theta$ analysis shows a close agreement with the experimental data. From an inspection of Fig. 12, the slope at the low stresses and high stresses is different, because the line shows a nonlinear curve. The stress exponent, $n$ in Norton's law of $\dot{\varepsilon}_m = A\sigma^n$ is

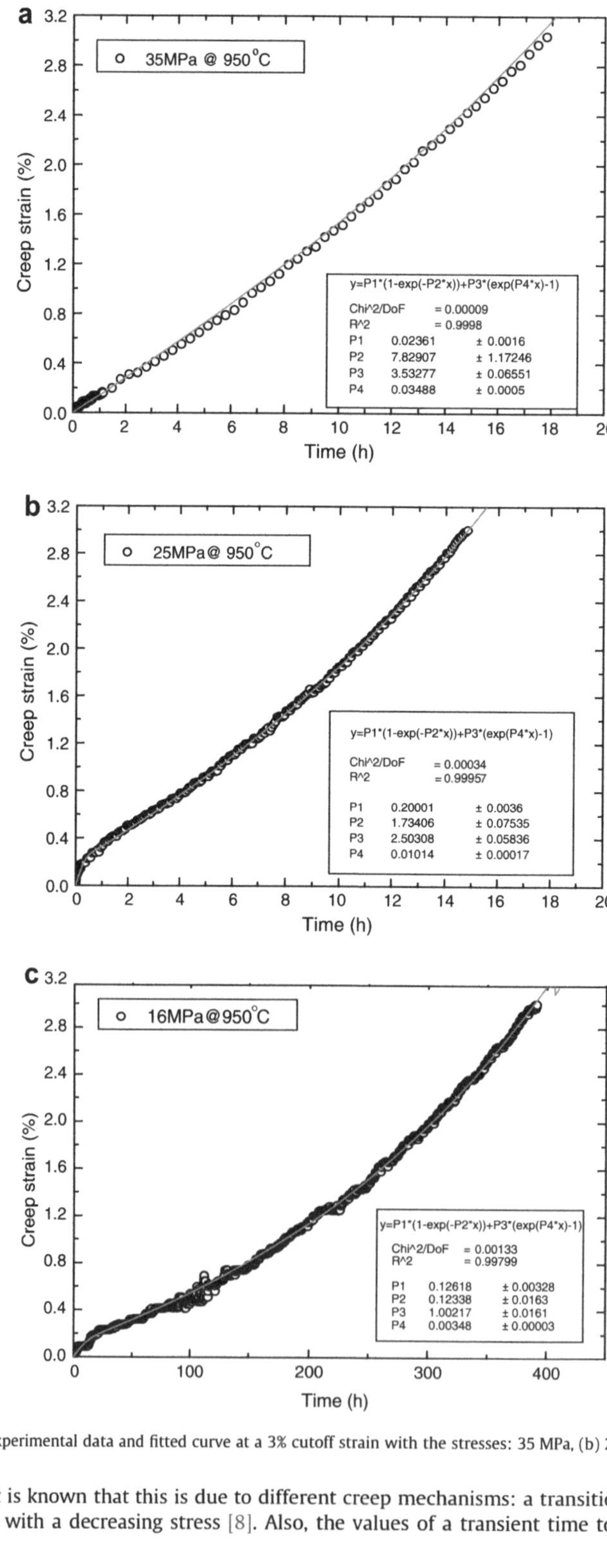


**Fig. 6.** Typical results of the experimental data and fitted curve at a 3% cutoff strain with the stresses: 35 MPa, (b) 25 MPa and (c) 16 MPa at 950 °C.

different in two regions. It is known that this is due to different creep mechanisms: a transition occurs from a dislocation creep to a diffusion creep with a decreasing stress [8]. Also, the values of a transient time to reach the minimum creep

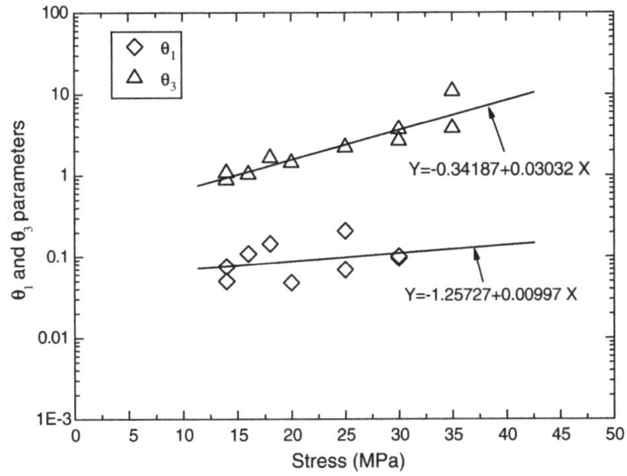


**Fig. 7.** The stress dependence of the parameters $\theta_1$ and $\theta_3$ in Eq. (1) for constant stress tests carried out for the Hastelloy-X alloy at 950 °C.

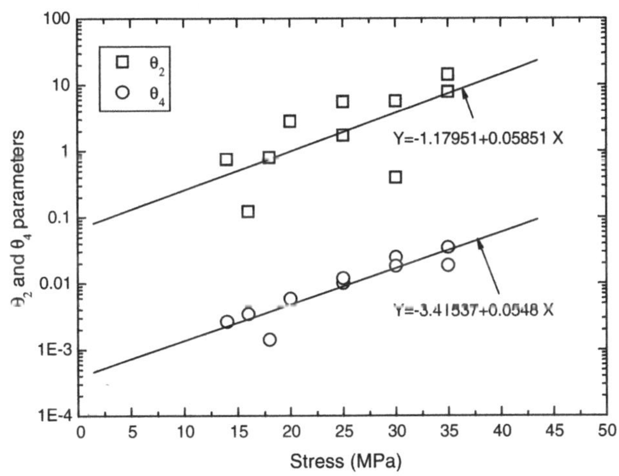


**Fig. 8.** The stress dependence of the parameters $\theta_2$ and $\theta_4$ in Eq. (1) for constant stress tests carried out for the Hastelloy-X alloy at 950 °C.

**Table 2**
The $a$ and $b$ coefficients for each $\theta$ parameter

| Parameters | $\text{Log}(\theta_i) = a_i + b_i\sigma$ |
|---|---|
| $\theta_1$ | Y = −1.25727 + 0.00997 *X* |
| $\theta_2$ | Y = −1.17951 + 0.05851 *X* |
| $\theta_3$ | Y = −0.34187 + 0.03032 *X* |
| $\theta_4$ | Y = −3.41537 + 0.0548 *X* |

rate, $t_m$, were estimated by Eq. (4), as shown in Fig. 13. The transient time exponentially decreased with increasing the stress.

Fig. 14 shows the result of the creep rate curves predicted by the $\theta$ projection method as a function of the time for different stresses at 950 °C. The creep rate curves numerically can be constructed for a wide range of stresses by using Eq. (3). Also, the log $\dot{\varepsilon}$ vs. $\varepsilon$ curve was calculated using Eqs. (1) and (3) with the coefficients from Table 2 and the result is plotted in Fig. 15. As seen in Fig. 15, the strain rate decreases steeply at the primary stage of a low strain, after a transient region, it

**Table 3**
Calculated results of four $\theta$ values with the stresses

| Stress (MPa) | $\theta_1$ | $\theta_2$ | $\theta_3$ | $\theta_4$ |
|---|---|---|---|---|
| 20 | 0.09164 | 0.978791 | 1.838781 | 0.004793 |
| 18 | 0.08752 | 0.747601 | 1.599153 | 0.003724 |
| 16 | 0.0836 | 0.571018 | 1.390753 | 0.002894 |
| 14 | 0.07985 | 0.436144 | 1.209511 | 0.002248 |
| 12 | 0.07626 | 0.333127 | 1.051889 | 0.001747 |
| 10 | 0.07284 | 0.254443 | 0.914808 | 0.001357 |
| 8 | 0.06957 | 0.194343 | 0.795591 | 0.001054 |

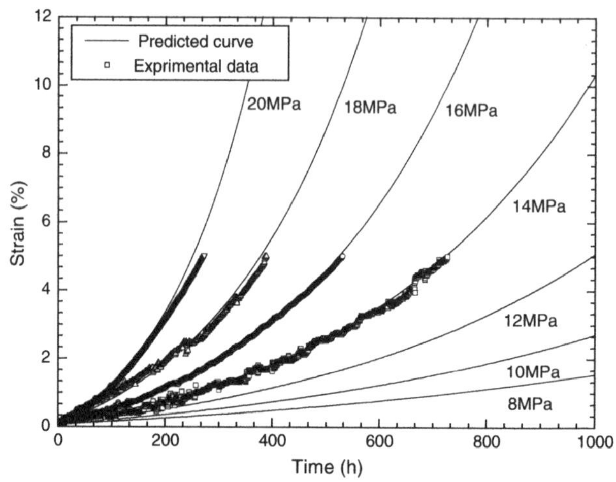


**Fig. 9.** Modeling of the creep curves with different stress levels for the Hastelloy-X alloy at 950 °C.

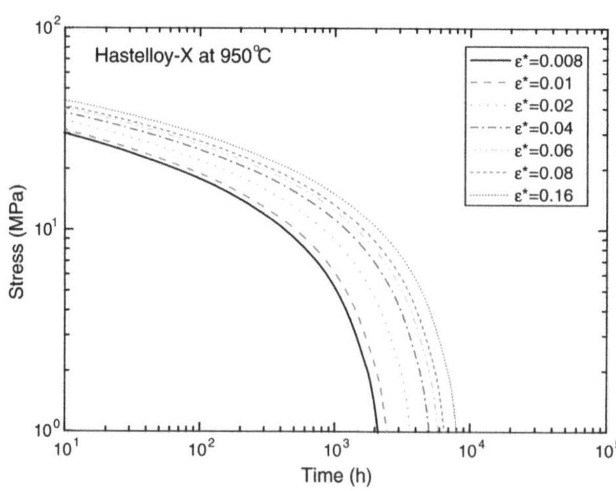


**Fig. 10.** Calculated isostrain data for the Hastelloy-X alloy at 950 °C.

slightly increases with strain. By using the results of Figs. 14 and 15, some information on the instantaneous creep resistance of the Hastelloy-X alloy at any particular time or strain can be achieved conveniently without any experimentation. The behavior of the creep strain rate at any strain and time can be numerically estimated for a wide range of stresses.

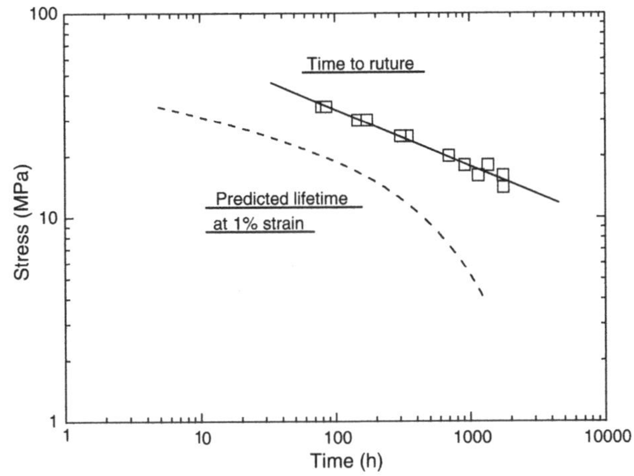


**Fig. 11.** Plot of stress–time obtained at 1% strain for the Hastelloy-X alloy at 950 °C.

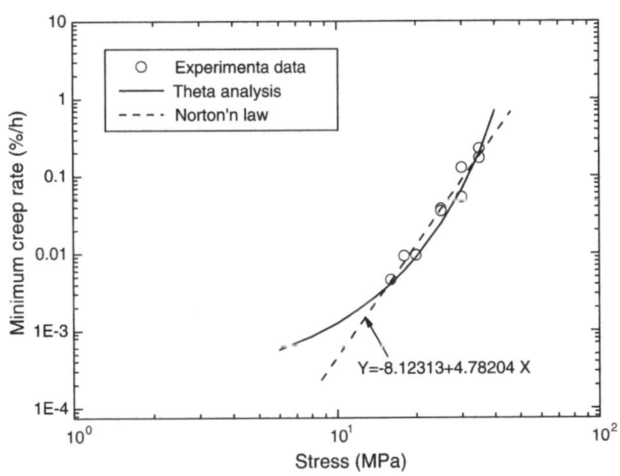


**Fig. 12.** Results of the minimum creep rate predicted by $\theta$ projection method for the Hastelloy-X alloy at 950 °C.

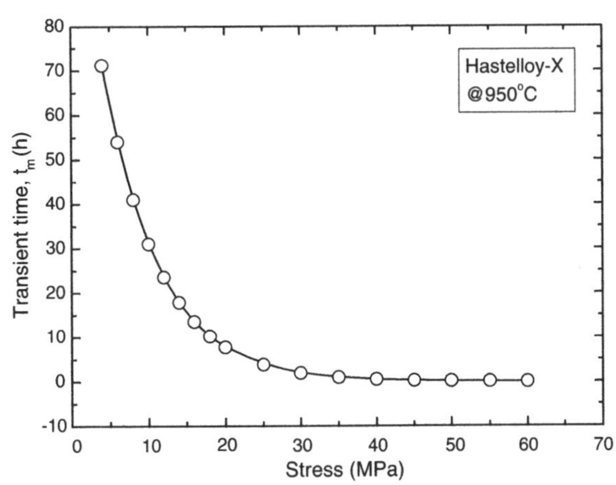


**Fig. 13.** Variations of transient time to reach the minimum creep rate with the stresses.

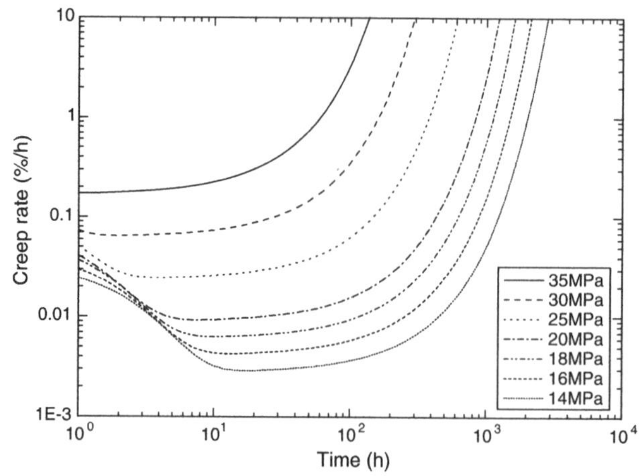


**Fig. 14.** Creep rate curves predicted by the $\theta$ projection method as a function of time for different stresses at 950 °C.

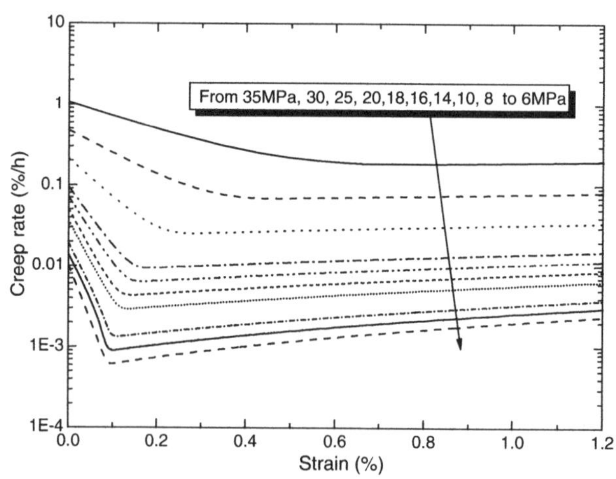


**Fig. 15.** Creep rate behavior predicted by the $\theta$ projection method as a function of strain for different stresses at 950 °C.

## 4. Conclusions

The $\theta$ projection method was applied to characterize the creep behavior for the Hastelloy-X alloy which is being considered as on of candidate materials for the VHTR components. In the NLSF of the full creep curves, the $\theta_1$ and $\theta_2$ parameters could not be optimized well with a large error during the fitting of the full creep curves. This reason is that the primary creep strain is significantly low when compared to the tertiary creep strain up to rupture. On the other hand, the $\theta_3$ and $\theta_4$ parameters were optimized well without an error to the creep curves. An optimum strain range for the four $\theta$ parameters was found to be within a 3% cutoff strain. The predicted creep curves at a 3% cutoff strain coincided well with experimental ones. Four $\theta$ parameters revealed the stress dependency with a good linearity, and the creep curves were modeled for the low stress levels. Predicted minimum creep rate showed a good agreement with the experimental data. The creep rate curves with time and strain were constructed numerically for a wide range of stresses. This result can be utilized for a design purpose to predict the creep behavior for the Hastelloy-X alloy at 950 °C.

## Acknowledgement

This work was performed under the Long-term Nuclear R&D Program sponsored by the Korean Ministry of Science and Technology.

## References


[1] Kim WG, Yin SN, Ryu WS, Jang JH. Analysis of the creep rupture data of alloy 617 for a high temperature gas cooled reactor. In: Proceedings of ASME PVP2007/CREEP8, Taxas, USA; 2007.

[2] Kim WG, Yin SN, Ryu WS. Creep properties of Hastelloy-X alloy for the high temperature gas-cooled reactor. Key Engng Mater 2006;326–328:1105–8.

[3] ASME B&PV Code Sections III Division 1 Subsection NH. Class 1 components in elevated temperature service. ASME, New York; 2004.

[4] Penny RK, Marriott DL. Design for creep. London: Chapman & Hall; 1995.

[5] Garofalo F. Fundamentals of creep and creep rupture in metals. New York: Macmillan Co.; 1966.

[6] Davies PW, Evans WJ, Williams KR, Wilshire B. An equation to represent strain/time relationship during high temperature creep. Scripta Metallurgica 1969;3(9):671–4.

[7] Blackburn LD. Isochronous stress–strain curves for austenitic steels. In: Smith GV, editor. The generation of isochronous stress–strain curves. New York: ASME; 1972. p. 14.

[8] Evans RW, Parker JD, Wilshire B. The theta projection concept – a model-based approach to design and life extension of engineering plant. Int J Pres Ves Piping 1992;50(1–3):147–60.

[9] Maruyama K, Harada C, Oikawa H. A strain–time equation applicable up to tertiary creep stage. J Soc Mater Sci Japan 1985;34:1289–95.

[10] Maruyama K, Oikawa H. Comments on exponential descriptions of normal creep curves by SG R Brown, RW Evans and B Wilshire. Scripta Metallurgica 1987;21(2):233–7.

[11] Wolf H, Mathews MD, Mannan SL. Prediction of creep parameters of type 316 stainless steel under service conditions using the $\theta$-projection concept. Mater Sci Engng A 1992;159:199–204.

[12] Koul AK, Castillo R. A critical assessment of $\theta$-projection concept for creep life prediction of nickel-based superalloy components. Mater Sci Engng A 1991;138:213–9.

[13] Brown SGR, Evans RW, Wilshire B. Creep strain and creep life prediction for the cast nickel-based superalloy IN-100. Mater Sci Engng A 1986;84:147–56.

[14] Brown SGR, Evans RW, Wilshire B. Exponential descriptions of normal creep curves. Scripta Metallurgica 1986;20(6):855–60.

[15] Brown SGR, Evans RW, Wilshire B. A comparison of extrapolation techniques for long-term creep strain and creep life prediction based on equations designed to represent creep curve shape. Int J Pres Ves Piping 1986;24(3):251–68.

[16] Brown SGR, Evans RW, Wilshire B. Alternative methods of parameter estimation in exponential descriptions of normal creep curves. Scripta Metallurgica 1987;21(2):239–42.

*Journal of ASTM International*, January 2006, Vol. 3, No. 1
Paper ID JAI13375
Available online at www.astm.org

*Woo-Gon Kim,*[1] *Song-Nam Yoon,*[2] *Woo-Seog Ryu,*[1] *Seon-Jin Kim,*[3] *and Won Yi*[2]

# Probabilistic Analysis of the Creep Crack Growth Rate of Type 316LN Stainless Steel by the Monte Carlo Simulation

**ABSTRACT:** This paper presents a probabilistic analysis for evaluating the creep crack growth rate (CCGR) of type 316LN stainless steel. The CCGR data was obtained from the creep crack growth tests, which are conducted under various applied loads at 600°C. The crack growth rate was characterized as a function of the $C^*$ fracture parameter. In order to logically obtain the $B$ and $q$ values in the CCGR equation of $\dot{a} = B(C^*)^q$, three methods of the least square fitting method (LSFM), a mean value method (MVM), and a probabilistic distribution method (PDM) were adopted. Also, using the Monte Carlo simulation, a number of random variables was generated, and the CCGR lines were predicted probabilistically. The three methods did not show a large difference in the CCGR lines, but the PDM was most useful because the CCGR line can be evaluated with a probabilistic reliability. Both the $B$ and $q$ coefficients followed a lognormal distribution, even though the $B$ ones were a little scattered for the points of the data. In the case of a standard deviation of $1\sigma$ for the probability variables, $P$ $(B,q)$, the results of the MCS and the PDM for the distribution ranges of the CCGR lines were compared and discussed.

**KEYWORDS:** creep crack growth rate, Type 316LN SS, probability, Monte Carlo simulation

## Introduction

Type 316LN stainless steel (SS) is widely used as the structural material for liquid metal reactor (LMR) components, as a consequence of its good high temperature mechanical properties, compatibility with the coolant liquid sodium, and adequate weldability. Their components may be subjected to a non-uniform stress and temperature distribution during service. These conditions generate localized creep damage and propagate the cracks and ultimately cause a fracture. A significant portion of the component's life can be spent in a crack propagation. Most GEN-IV reactors are also operated at the high temperature creep range above 500°C [1–3]. Thus, it is very important to evaluate the creep crack growth rate (CCGR) from a design concern and in predicting the residual life of the components during the high temperature service [4–6]. In addition, for ensuring the safety and integrity of the components, it is necessary to assess the CCGR from the probabilistic viewpoint which defines the probability of a fracture. However, until now, the design and/or evaluation of the components have been conducted mostly by deterministic fracture mechanics (DFM). This method may bring about an overly conservative evaluation by applying highly upper limit values in the material properties. So the crack propagation is reasonable to evaluate probabilistically using probabilistic fracture mechanics

Manuscript received 7 April 2005; accepted for publication 22 June 2005; published January 2006. Presented at ASTM Symposium on Fatigue and Fracture Mechanics: 35th Volume on 18-20 May 2005 in Reno, NV.

[1] Principal Researcher, Dept. of Nuclear Material Technology Development, Korea Atomic Energy Research Institute (KAERI), 150 Deogjin-dong,Yuseong-gu, Daejeon, 305-353, Korea (South).

[2] Graduate Student and Professor, respectively, Soong-sil University, 1-1 Sangdo-dong, Dongjak-gu, Seoul, 156-743, Korea.

[3] Professor, Pukyong National University, 599-1 Daeyon-dong, Nam-gu, Busan, 608-737, Korea.

(PFM).

In this paper, the CCGR data were dealt with from probabilistic viewpoints in order to logically evaluate the CCGR in type 316LN SS. The creep crack growth data were obtained from several experimental CCGR tests under different applied loads at 600°C and characterized in terms of the $C^*$ parameter. The CCGR equations were compared with the least square fitting method (LSFM), a mean value method (MVM), and a probability distribution method (PDM). Also, using the Monte Carlo simulation (MCS), a number of random variables was generated, and the CCGR lines were predicted probabilistically. For an application of the PDM and the MCS, in the case of a standard deviation of $1\sigma$ for the probability variables, $P$ ($B$, $q$ ), the CCGR lines were predicted and discussed.

## Experimental Method

Chemical compositions of type 316LN SS are given in Table 1. Laboratory ingots were prepared by a vacuum induction melting (VIM) and were homogenized at 1150°C for 2 h in an argon atmosphere and shaped to 15 mm plates by the hot rolling processes. The type 316LN plates were solution annealed for 1 h at 1100°C and water quenched.

Tension testing specimens were machined to be cylindrical with a 25.4 mm gauge length and 6.35 mm in diameter. And a creep test specimen was machined to be cylindrical with a 30 mm gage length and 6 mm diameter. All the specimens were taken along the rolling direction. The gage sections of the creep specimens were polished using #1000 grit sand papers with strokes along the specimen axis. In order to obtain the material properties of the type 316LN SS, tension and creep tests were conducted at 600°C. Creep tests were carried out using dead weight machines with a lever ratio of 20/1. The experimental procedures for the creep tests followed the recommendations of the ASTM standard E 139. Creep data were automatically collected through a personal computer. The mechanical properties of $D_1$, $m$, $A$, and $n$, which are obtained from the tension, and creep tests at 600°C are listed in Table 2.

Creep crack growth tests were carried out at a constant load with applied load ranges of 5500 N to 7000 N at 600°C. Compact tension (CT) specimens were a width ($W$) of 25.4 mm, a thickness ($B$) of 12.7 mm, and side grooves of a 10 % depth. Initial crack ratio ($a/W$) was about 0.5. Load-line deflection was measured using a linear gauge assembly attached to the specimen, and the crack length was determined by the direct current potential drop (DCPD) method. Crack extension data were continuously collected using a data acquisition system. The test temperature was constantly maintained to within 2°C during the test. The test was continued until a sufficient crack growth occurred. After testing, the CT specimens were cooled down in a liquid nitrogen solution and fractured to measure the final crack length.

The crack length was calculated using the Johnson's formula from the results of the DCPD. All the experimental procedures followed the ASTM standard E 1457. The results of the crack size measurement in each condition are given in Table 3. As seen in Table 2, the results between the predicted and measured crack sizes are well within the 5 % of the requirements in the ASTM E 1457 [7].

TABLE 1—*Chemical composition of type 316LN SS.*

| Fe | C | Si | Mn | P | S | Cr | Ni | Mo | N | B |
|---|---|---|---|---|---|---|---|---|---|---|
| Bal. | 0.022 | 0.70 | 1.01 | 0.029 | 0.005 | 17.15 | 12.30 | 2.35 | 0.094 | 0.003 |

TABLE 2—*Material properties at 600°C of type 316LN SS.*

| E (GPa) | $\sigma_{YS}$ (MPa) | $\sigma_{UTS}$ (MPa) | Plastic constants, $\varepsilon_p = D_1\sigma^m$ | | Creep constants, $\dot{\varepsilon}_{ss} = A\sigma^n$ | |
|---|---|---|---|---|---|---|
| | | | $D_1$ | m | A | n |
| 149 | 104 | 420 | 6E-4 | 2.54 | 6.35E-33 | 11.8 |

TABLE 3—*CCGR test conditions and crack size measurement at 600°C of type 316LN SS.*

| S.P –I.D. | Test conditions & transition time | | | | Results of crack size measurement | | | | |
|---|---|---|---|---|---|---|---|---|---|
| | Load (N) | Applied K ($MPa\sqrt{m}$) | Test time (hr) | $t_T$ (hr) | a (mm) | $a_o$ (mm) | $a_{pf}$ (mm) | $a_{mf}$ (mm) | Error (%) |
| K1 | 7000 | 34.9 | 526.3 | 245 | 10.19 | 12.11 | 13.52 | 13.41 | 0.80 |
| K2 | 6500 | 32.7 | 667.9 | 221 | 10.29 | 12.18 | 14.52 | 14.65 | -0.88 |
| K3 | 6000 | 30.0 | 884.9 | 428 | 10.18 | 12.13 | 13.04 | 13.33 | -2.26 |
| K5 | 5500 | 28.5 | 1284.0 | 423 | 10.59 | 12.43 | 14.90 | 15.00 | -0.64 |
| K6 | 6000 | 29.8 | 749.8 | 245 | 10.28 | 12.08 | 22.50 | 23.64 | -4.81 |
| K7 | 6000 | 29.9 | 811 | 301 | 10.12 | 12.02 | 25.27 | 24.68 | -2.31 |
| KR | 6500 | 31.9 | 662.6 | 331 | 10.13 | 12.03 | 14.06 | 14.28 | 1.52 |

- a : initial crack size(at sharp notch), $a_o$: pre-cracked size by fatigue
- $a_{pf}$ : predicted final crack size, $a_{mf}$ : measured final crack size
- $t_T$ : transition time

## Results and Discussion

### *Calculation of the $C^*$ values*

Fracture mechanics parameter, $C^*$-integral has been widely used to characterize the CCGR in metals undergoing a steady state creep. The general form between the CCGR (*da/dt*) and the $C^*$ can be expressed as the following [8, 9]:

$$da/dt = B[C^*]^q \tag{1}$$

In this expression, the *B* and *q* coefficients are material constants, which are generally obtained from a regression line of the CCGR data. They are related to the intercept and slope, respectively, of the *da/dt* versus $C^*$ relationship on a log-log plot [5,10]. For calculating the *da/dt* in Eq 1, the material constants, $D_1$, *m*, *A*, and *n* were obtained from the tension and creep tests at 600°C, as described in Section 2. Material properties of Table 2 were used to calculate the $C^*$ values. The relationship between the *da/dt* and the $C^*$ was obtained for all the seven samples. Calculating procedures of the $C^*$ values followed the ASTM E 1457.

*The CCGR Lines for Regression Methods*

The values of the $B$ and $q$ coefficients are very important because the CCGR behavior depends on them. Their values change from material to material and may change with regression methods. Therefore, to logically obtain the CCGR line of the type 316LN SS, the three methods of the LSFM, the MVM, and the PDM are applied and compared. Here, the MVM is to take a mean value which is obtained by the total sum of each $B$ and $q$ value of the CCGR lines. The PDM is obtained by a statistical distribution of the CCGR data. Figure 1 shows the results of the CCGR lines obtained from the LSFM, the MVM, and the PDM, respectively. The three methods are found to have a little difference in the CCGR lines.

At first, the CCGR equation in the LSFM was obtained as

$$da/dt = 3.84\times10^{-2}[C^*]^{0.89} \tag{2}$$

It is shown that the slope, $q$ of a regression line was the lower than the other two methods and was not in full accord with the experimental data points. Although the LSFM has been generally used as a good method for obtaining the best fitting of the CCGR data so far, it obtains only the best regression line of the y-axis for all the data points, regardless of the $C^*$ ranges of the x-axis for each applied condition. It has been known that the LSFM estimates an independent regression for the data of the x-axis and of the dependent one for the data of the y-axis [11]. In reality, each range of the $C^*$ values for the CCGR lines are different with the applied load conditions. The numbers of data points is also different with the $C^*$ values. Owing to this reason, the $q$ value in the LSFM is low with 0.89. As a consequence, the LSFM may generate some error in the regression line.

Secondly, the conception of a mean value was adopted for obtaining the CCGR line. The MVM line showed a good agreement with the experimental data, as shown in Fig. 1. Because the MVM line takes a mean value of each $B$ and $q$ of all the CCGR lines, this method equally reflects the individual characters of each experimental line. The $q$ value in the MVM was 0.93 higher than that in the PDM, and the CCGR equation in the MVM was obtained as

$$da/dt = 5.32\times10^{-2}[C^*]^{0.93} \tag{3}$$

Finally, the conception of a probability distribution was applied to obtain the CCGR line from a probabilistic viewpoint. For using this method, it is necessary to investigate the probability distribution of the $B$ and $q$ coefficients. Figures 2 and 3 show the results of the $B$ and $q$ distributions obtained from seven experimental CCGR data. Probability distribution of lognormal distribution on paper follows a linear relationship, even though the $B$ distribution was a little scattered for the data points. Thus, the $B$ and $q$ data are shown to follow a lognormal distribution. Standard deviation, $\sigma_{L10}$ (subscript L means log) and mean value, $\mu_{L10}$ for the lognormal distributions of the $B$ and $q$ can be obtained using the results of Figs. 2 and 3. The standard deviation for the $q$ is low with $\sigma_{L10} = 0.056$, while the one for the $B$ is a little bit high with $\sigma_{L10} = 0.324$. The CCGR equation in the PDM was obtained as

$$da/dt = 4.64\times10^{-2}[C^*]^{0.92} \tag{4}$$

The CCGR line of the PDM shows a good result when compared with that of the LSFM because the PDM line closely approaches the MVM one. Thus, the PDM can be used as a reasonable method for estimating the CCGR of high temperature materials. Especially using this method, we can predict the CCGR with a probabilistic reliability. For the usability of the three methods, each method may have an advantage or disadvantage. The MVM is not convenient because each CCGR line should be obtained for each applied condition.

Meanwhile, the LSFM is very convenient because the CCGR line is easily obtained by a fitting at a time for all the data points. The PDM is superior from the probabilistic viewpoints to the other two methods.

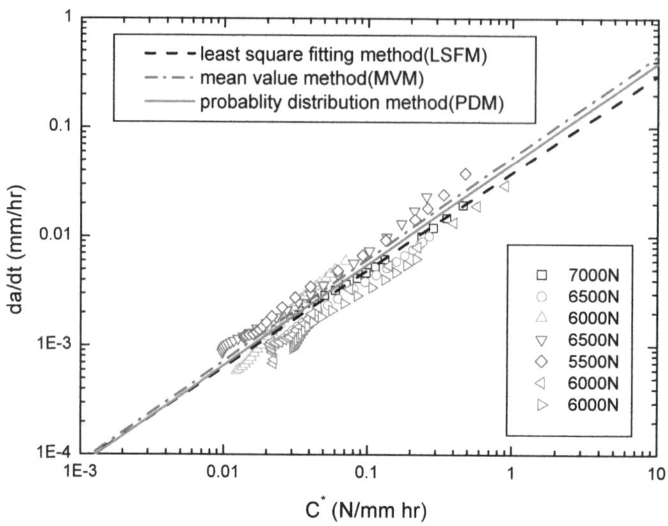


FIG. 1—*Comparison of the CCGR lines for the three methods.*

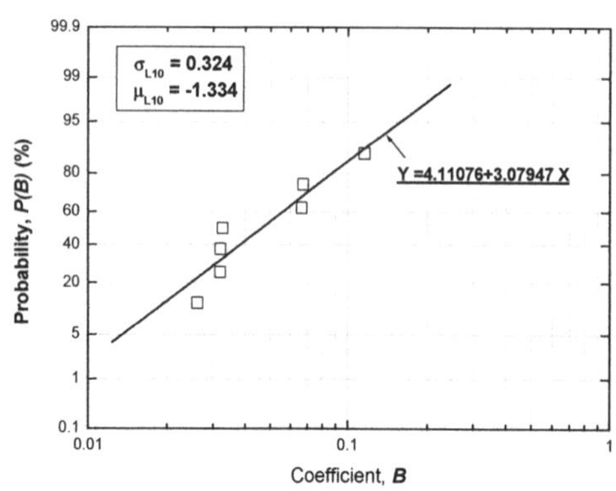


FIG. 2—*Lognormal distribution of the B data.*

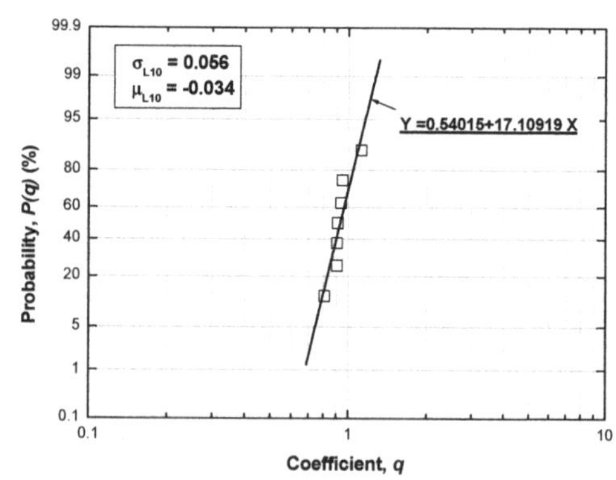


FIG. 3—*Lognormal distribution of the q data.*

*Probabilistic Prediction of the CCGR Line*

By knowing the probabilistic distributions of the $B$ and $q$, the CCGR line can be predicted probabilistically. As described above, the $B$ and $q$ data followed a lognormal distribution. Thus, the $B$ and $q$ values at any probability can be calculated from Figs. 2 and 3. As an example for an application, in the case of $1\sigma$ in which the probability variable, $P(B, q)$, is [$15.87\ \% \le P(B, q) \le 84.13\ \%$], the distribution ranges of the CCGR line are plotted in Figs. 4 and 5.

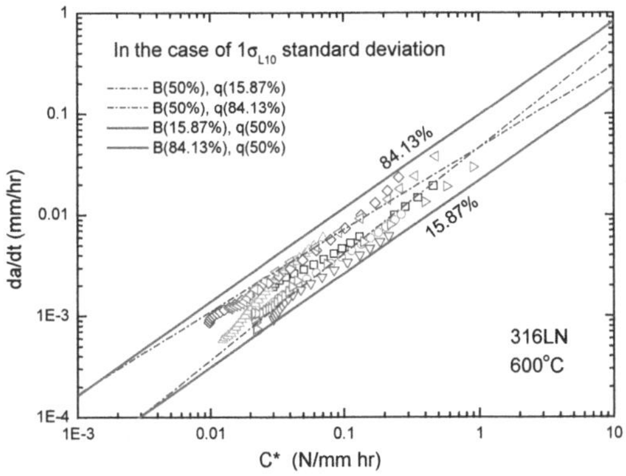


FIG. 4—*Probability line in the 1$\sigma$ of the B.*

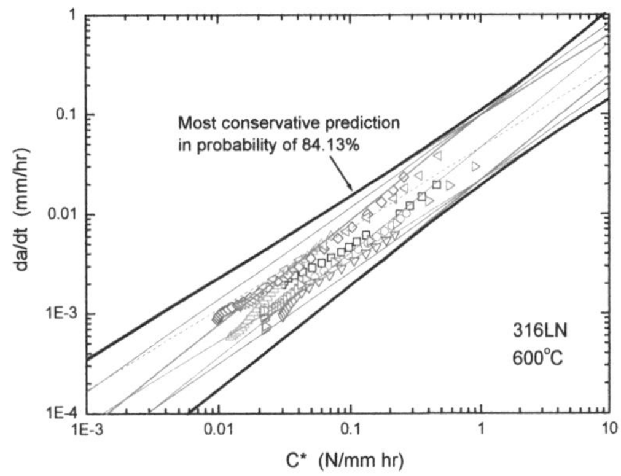


FIG. 5—*Probability line in the 1$\sigma$ of the B and q.*

Figure 4 shows the results of the CCGR lines in the case that the $B$ and $q$ ranges are changed with 1$\sigma$ for $B$ and $q$ of 50 %, respectively. In the 1$\sigma$ range of $B$ (15.87–84.13 %) for $q$ of 50 %, all data are included. Meanwhile, in the 1$\sigma$ range of $q$ (15.87–84.13 %) for $B$ of 50 %, all the data are not included perfectly. Thus, the $B$ and $q$ coefficients should be considered with the wider probability to predict more conservatively the CCGR line.

Figure 5 shows the results of the CCGR lines which are considered for all the eight cases. The values of the $B$ and $q$ are freely changed with 15.87 %, 50 %, and 84.13 %, respectively. Probability lines of 84.13 % and 15.87 % were calculated using the results of the lognormal distribution as shown in Figs. 2 and 3, respectively. In this case, the upper and lower prediction band is wider than that of Fig. 4. The upper prediction line of 84.13 % will be the most conservative prediction, because the fastest CCGR is predicted here. On the contrary, the lower prediction line of 15.87 % will be the most non-conservative prediction, because the lowest CCGR is predicted here. It indicates that by knowing the probability distribution for the $B$ and $q$, the CCGR can be predicted probabilistically. The PDM can be utilized usefully for predicting a design life and a remaining life of cracked components under a high temperature creep condition.

*Monte Carlo Simulation (MCS) for the B and q*

For engineering purposes, a simulation may be applied to predict the performance and response of a system. For problems involving random variables with known or assumed probability distributions, the MCS has been commonly used. In this study, the MCS is applied to generate the appropriate values of random variables (i.e., random numbers) for the $B$ and $q$ coefficients. Box and Muller [12] have proposed that random variables ($S$) with a standard normal distribution can be represented as

$$S = (-2 \ln U_1)^{0.5} \cos 2\pi U_2 \tag{5}$$

In this expression, the $U_1$ and $U_2$ are standard uniform variables, which have a uniform probability density function (PDF) between 0 and 1.0. Thus, random variables ($x$) with a normal distribution $N(\mu, SD)$ can be given as [13]

$$x = \mu + SD \cdot S = \mu + SD(-2\ln U_1)^{0.5} \cos 2\pi U_2 \tag{6}$$

For a lognormal random variable $x'$, the distribution of $x = \log(x')$ is a normal distribution. Thus, if $x$ is a value generated from Eq 6

$$x' = 10^{x} \quad (7)$$

Equation 7 becomes a random number for the lognormal distribution with a mean $\mu$ and standard deviation *SD*. Equations 6 and 7 were used to generate the values of the $B$ and $q$, which were shown to be a lognormal distribution, as shown above.

Figures 6 and 7 show the results of the MCS for the lognormal distribution of the $B$ and $q$. Standard uniform variables of the $U_1$ and $U_2$ were generated using Eq 6. The numbers of the random variables were 100 for each $B$ and $q$. The $B$ and $q$ data follows the lognormal distribution because the data plots a linear relationship. The $B$ and $q$ values are almost similar to those of the experimental data obtained in Figs. 2 and 3. It means that the random variables of the $B$ and $q$ were well realized by the MCS.

Figure 8 is the typical results of the MCS showing the variations of the $da/dt$ at various $C^*$ values. The probability distribution of the $da/dt$ was calculated using Eq 1 from the $B$ and $q$ distributions of Figs. 6 and 7. Probability distributions of 50 numbers were generated for the wide ranges of the $C^*$ values from 0.001 to 100. From these distributions, the CCGR lines can be predicted probabilistically. The results are shown in Figs. 9 and 10, respectively. Figure 9 presents the results of the probability line obtained from the MCS for the 1$\sigma$ of the $B$ value, as shown in Fig. 4. However, as clearly shown in Fig. 9, in the lower and upper values of the $C^*$, several data points are out of the boundaries. So the $q$ coefficient should be considered probabilistically.

Figure 10 presents the results of the probability line obtained from the MCS in the 1$\sigma$ of the $B$ and $q$ ones, as noted in Fig. 5. The probability lines of 84.13 % and 15.87 % were calculated using the results of the MCS obtained in Figs. 6 and 7, respectively. In this case, the band of the upper and lower prediction lines is wider than that of Fig. 9. All the data are included in the 1$\sigma$ range of 15.87 % and 84.13 %. The upper prediction line of 84.13 % will be the most conservative prediction. Meanwhile, the lower prediction line of 15.87 % will be the most non-conservative prediction. The bands of the upper and lower lines in the PDM and the MCS are a little different, as shown in Figs. 11 and 12. The comparative results are discussed below.

Figs. 11 and 12 show the comparative results of the PDM and the MCS for the predicted CCGR lines. In the 1$\sigma$ of the $B$ value and in the 1$\sigma$ of the $B$ and $q$ ones, the two methods showed a little difference in the bands of the predicted CCGR lines. It is found that the PDM line is wider than the MCS one. It means that the PDM line was faster than the MCS one, because the upper prediction line is only considered in the design of the CCGR. Thus, the PDM will make a more conservative design than the MCS. Contrarily, the MCS will make an economic and reasonable design. However, the PDM and the MCS do not generate a large difference in the predicted lines, so both methods are used for predicting the creep crack growth of type 316LN SS. It is also regarded that the MCS of this study were successfully performed without an error.

Consequently, by using the lognormal distribution investigated for type 316LN SS, a number of random variables, $P(B, q)$ with the various $C^*$ values can be generated by the MCS. Then, the CCGR lines can be predicted with a probabilistic reliability by using the generated random variables.

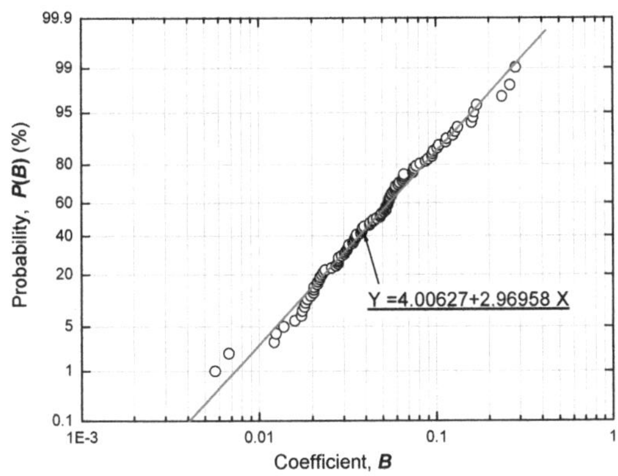


FIG. 6—*Result of the MCS for the B distribution.*

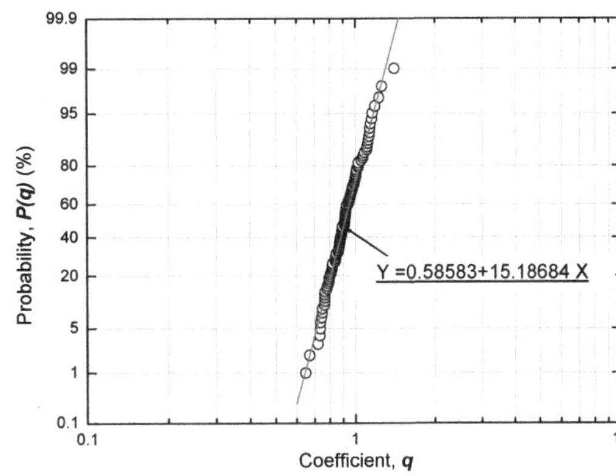


FIG. 7—*Result of the MCS for the q distribution.*

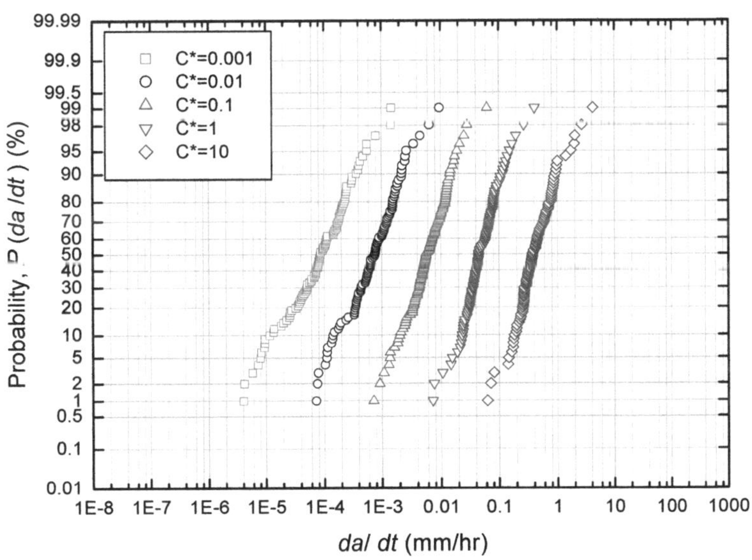


FIG. 8—*The MCS of the CCGR with the* $C^*$ *values.*

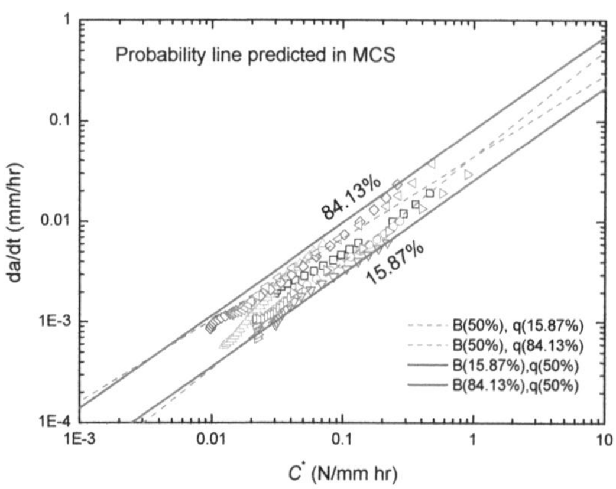


FIG. 9—*Probability line obtained in the MCS.*

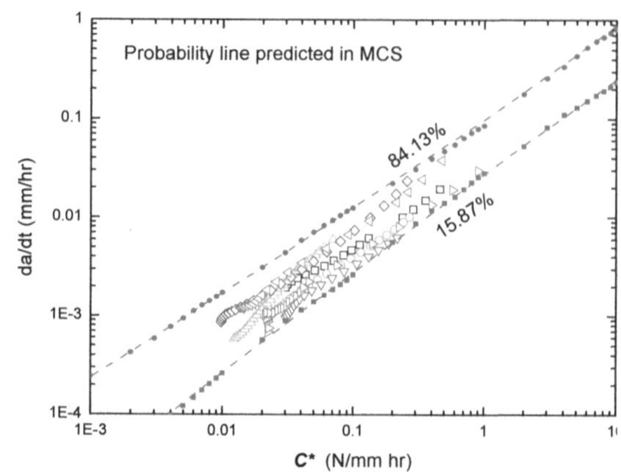


FIG. 10—*Probability line obtained in the MCS in the 1σ of the B in the 1σ of the B and q.*

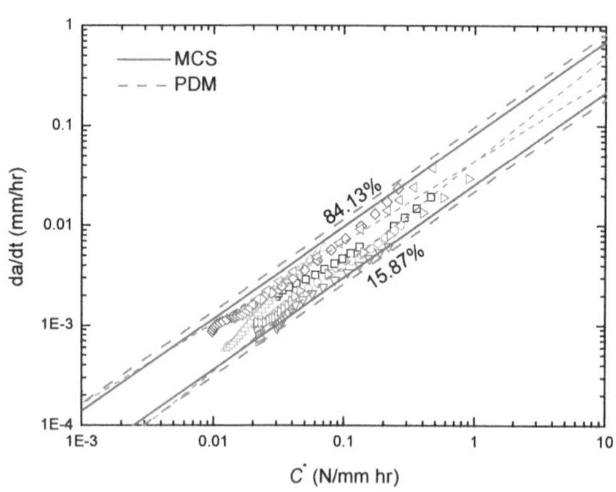


FIG. 11—*Comparison of CCGR lines for MCS and MCS PDM in the 1σ of the B.*

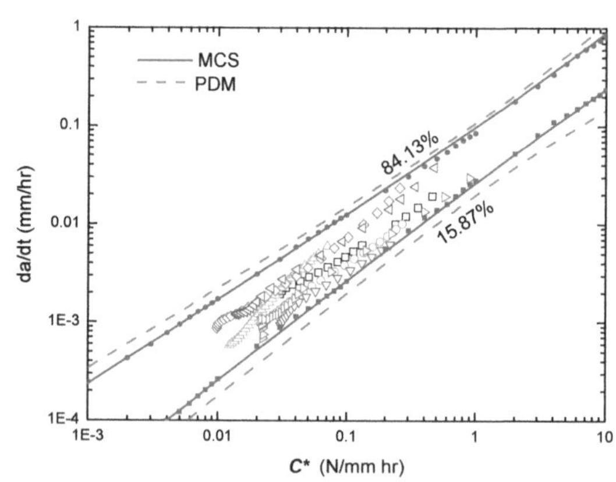


FIG. 12—*Comparison of CCGR lines for and PDM in the 1σ of the B and q.*

## Conclusions

In order to logically obtain the creep crack growth rate (CCGR) of type 316LN stainless steel, the CCGR lines were obtained by the three methods of the least square fitting method (LSFM), the mean value method (MVM), and the probability distribution method (PDM), and the CCGR lines were analyzed probabilistically by the Monte Carlo simulation (MCS). The three methods did not show a significant difference in the CCGR lines, but, above all, the PDM was most useful because the CCGR line can be predicted probabilistically. It was found that both the $B$ and $q$ coefficients of the CCGR data of type 316LN stainless steel followed a lognormal distribution, even though the $B$ ones were a little scattered in the CCGR data. By using the lognormal distribution, the CCGR lines were predicted with a probabilistic reliability. In addition, by applying the Monte Carlo simulation, a number of random variables for the $B$ and $q$ was generated, and the CCGR lines were successfully predicted without an error. As an example for application, the results of the CCGR line between the PDM and the MCS, in the case of the 1σ of the *B* value and of the *B* and *q* ones, did not show a significant difference. It is thus regarded that the MCS is useful in the probabilistic evaluation of the CCGR for type 316LN stainless steel, and especially for new and emerging class of high-temperature materials where the CCGR data is lacking.

*Acknowledgments*

This work was performed under the Long-term Nuclear R&D Program sponsored by the Korean Ministry of Science and Technology.

**References**

[1] Kim, D.W., Lee, Y.K., Kim, W.G. and Ryu, W.S., "Effect of Nitrogen on Creep Properties of Type 316L(N) Stainless Steel," *Journal of Korean Institute Metal & Material*, Vol. 39, 2001, pp.1241-1248.
[2] Kim, W.G., Kim, S.H. and Ryu, W.S., "Creep Characterization of Type 316LN and HT9 Stainless Steels by the K-R Creep Damage Model," *KSME International Journal*, Vol. 15, 2001, pp. 1463-1471.
[3] Jang, J., Ryu, W.S. and Hong, J.H., "Material-Related Issues for Gen IV Supercritical Water-Cooled Reactor," *Proc. of Sixteenth Conference on Mechanical Behaviors of Materials*, Cheju, Korea, Nov. 8, 2002, pp. 425-431.
[4] Nikbin, K.M., Smith, D.J. and Webster, G. A., "An Engineering Approach to the Prediction of Creep Crack Growth," *Transactions of the ASME*, Vol. 108, 1986, pp. 186-191.
[5] Saxena, A., "Nonlinear Fracture Mechanics for Engineers," *CRC Press*, New York, 1997, pp. 363-377.
[6] Yagi, K., Tabuchi, M., Kubo, K., Kim, J.J. and Huk Y.H., "The Influence of Fracture Mechanisms on the Creep Crack Growth Behavior of 316 Stainless Steel," *Engineering Fracture Mechanics*, Vol. 57, 1997, pp. 463-473.
[7] ASTM E1457-00, "Standard Test Method for Measurement of Creep Crack Growth Rates in Metals," ASTM Standard, 2000.
[8] Anderson, T.L., "Fracture Mechanics," Second Edition, *CRC Press*, Boca Raton, Ann Arbor, London, Tokyo, 1995, pp. 226-228.
[9] Ryu, W.S., Kim, W.G. and Lee, K.Y., "Evaluation of Creep Crack Growth Behavior of Type 316LN Stainless Steel, SMiRT-17, Paper No. F05-4, 2003.
[10] Kim, W.G., Kim, H.H., Yoon, K.B., Ryu, W.S., "Application of Creep Ductility Model for Evaluating Creep Crack Growth Rate of Type 316SS Series," *Materials Science Forum*, Vols. 475-479, 2005, pp. 1433-1436.
[11] Microcal Software, Inc, Origin User Manual, Version 5, 1997, pp. 460-463.
[12] Box, G.E.P. and Muller, M.E. "A Note on the Generation of Random Normal Deviates," *Annals of Math. Stat.*, Vol. 29, 1958, pp. 610-611.
[13]Yatomi, M., "Factors Affecting the Failure of Cracked Components at Elevated Temperature," Ph.D. Thesis, Imperial College, 2003, pp. 58-59.

*Key Engineering Materials Vols. 326-328 (2006) pp. 1105-1108*
*online at http://www.scientific.net*
*© (2006) Trans Tech Publications, Switzerland*

# Creep Properties of Hastelloy-X Alloy for the High Temperature Gas-Cooled Reactor

Woo Gon Kim[1, a], Song Nan Yin[2,b], Woo Seog Ryu[3,c] and Jong Hwa Chang[4,d]

[1,3,4] KAERI, 150 Deogjin-dong, Yuseong-gu, Daejeon, 305-353, Korea

[2] Soongsil Univ., 1-1 Sangdo-dong, Dongjak-gu, Seoul, 156-743, Korea

[a]wgkim@kaeri.re.kr, [b]yjsongnan@yahoo.co.kr, [c]wsryu@kaeri.re.kr, [d]jhchang@kaeri.re.kr

**Keywords:** Hastelloy-X, Creep, Creep Rupture Time, High Temperature Gas-Cooled Reactor (HTGR), Larson-Miller Parameter (LMP), Standard Deviation (SD)

**Abstract.** The creep properties for the Hastelloy-X alloy which is one of candidate alloys for a high temperature gas-cooled reactor are presented. The creep data was obtained with different stresses at 950°C, and a number of the creep data was collected through literature surveys. All of the creep data were combined together to obtain the creep constants and to predict a long-term creep life. In the Norton's creep law and the Monkman-Grant relationship, the creep constants, $A$, $n$, $m$, and $m'$ were obtained. Creep master curves based on the Larson-Miller parameter were presented for the standard deviations of 1σ, 2σ and 3σ. Creep life at each temperature was predicted for a longer-time rupture above $10^5$ hours. Failure probability was also estimated by a statistical process of all the creep rupture data.

## Introduction

Since high temperature gas-cooled reactor (HTGR) components are designed for a lifetime of 40 years at 950°C, one of the most important properties in the metallic components is the creep behavior. The Hastelloy-X (hereafter HX) is considered as a candidate alloy of the HTGR components, such as hot gas ducts, intermediate heat exchanger (IHX) tube, steam reformer tubes, and immersion heaters, to produce hydrogen using processed nuclear heat [1].

The HX has a good oxidation resistance at temperatures up to 1200°C and moderately good strength properties at temperatures up to 870°C, especially, superior corrosion and oxidation properties in the helium environment of the HTGR. Several studies for the HX have been conducted in the U.S. and in Japan to evaluate the creep and the stress rupture properties at temperatures above 650°C [2-4]. However, because longer-time rupture data reaching $10^5$ hours is still insufficient, many creep data for the alloy must be obtained from experimentations, literature surveys, and other country databases. Finally, a material database for high-temperature design code should be provided.

In this paper, the creep properties for the HX were investigated from the creep tests at 950°C, and their creep constants were obtained. To predict longer-time rupture, the experimental data have been combined with the data of the literature surveys; European Alloy DB [5], published literatures [3, 6], and manufactures' data sheets [7, 8]. Using this data, the creep master curves based on the Larson-Miller (LM) parameter were obtained for the standard deviations of 1σ, 2σ and 3σ. Also, the failure probability was estimated by a statistical process of all the creep data.

## Experimental Procedures

The HX used in this study is a commercial type hot-rolled plate with a thickness of 19 mm. Chemical composition is given in Table 1. Creep specimens were taken along the rolling direction and machined into a cylindrical form with a 30 mm gauge length and 6 mm diameter. Creep tests were conducted by using constant-load machines in air, and creep stresses were applied from 14MPa to 35MPa. Test temperature was maintained constantly within ± 2°C at 950°C during the period of the test. Before starting the creep test, all the specimens were held at the test temperature for 1 hr. Creep strain data with time variations were automatically measured by using an extensometer gage.

Table 1 Chemical composition of the Hastelloy-X (wt.%)

| C | Mn | P | S | Si | Ni | Cr | Mo | Co | Cu | W | N | Fe | B | Ti | Al |
|---|---|---|---|---|---|---|---|---|---|---|---|---|---|---|---|
| 0.082 | 0.32 | 0.013 | 0.003 | 0.42 | 48.04 | 21.91 | 8.65 | 0.79 | 0.13 | 0.44 | 0.015 | 19.00 | 0.002 | 0.007 | 0.17 |

## Results and Discussion

**Creep properties.** Fig. 1 shows the results of the typical creep curves obtained with different stresses for the HX at 950°C. The creep curves at this temperature and stress conditions did not generate the primary creep strain and the steady state creep rate (SSCR) at a secondary creep range was unclear. As the creep stress decreases, the rupture elongation decreased with a linear relation. Fig. 2 shows the relation between the SSCR and the stresses. The HX showed a good linearity, as shown in Fig. 2. From the results, creep constants for Norton's power law of $\dot{\varepsilon}_m = A\sigma^n$ can be obtained for the HX. Where $A$ is a constant and $n$ is a stress exponent, and $\dot{\varepsilon}_m$ is the minimum creep rate. The $A$ value was obtained as $2.57\text{x}10^{-13}$ ($\text{MPa}^{-n}\ \text{s}^{-1}$) and the $n$ value was obtained as 4.15. The $n$ value is similar to the result of 4.2 to 5.7 reported for Hastelloy-XR at 800 to 1000°C [9].

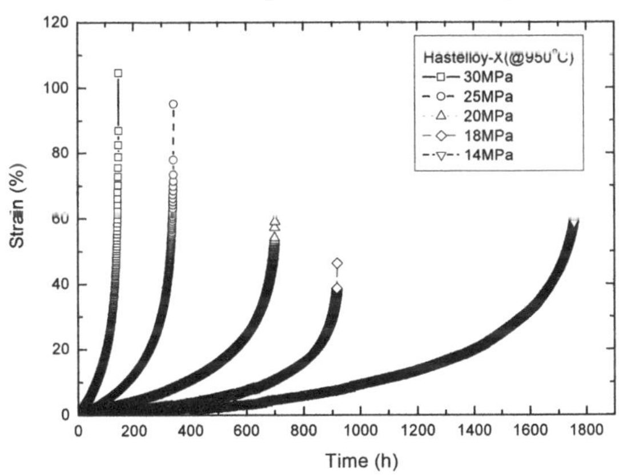

Fig. 1 Creep curves for the HX at 950°C

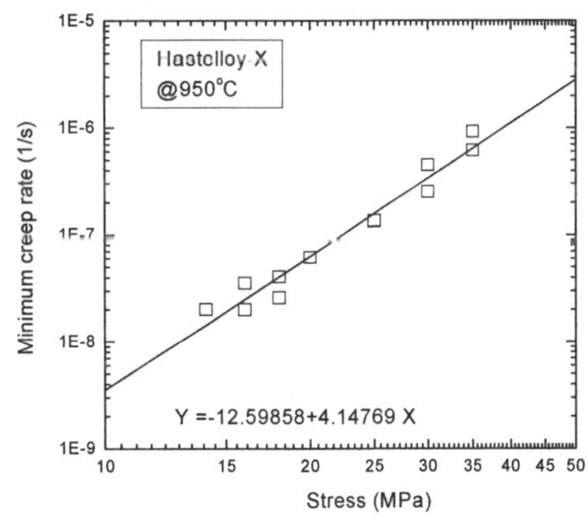

Fig. 2 Minimum creep rate vs. stress

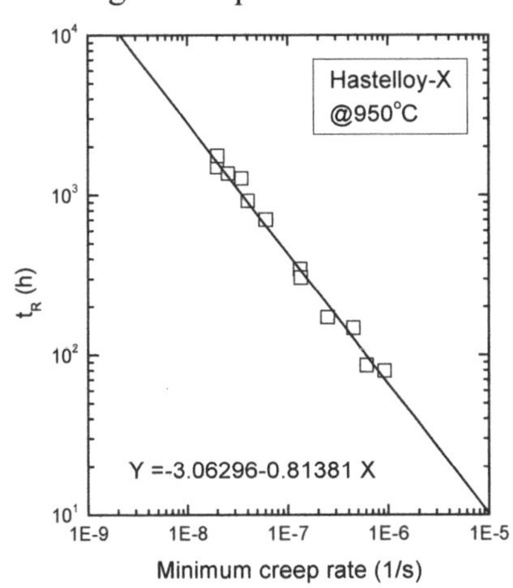

Fig. 3 Monkman-Grant relationship at 950°C

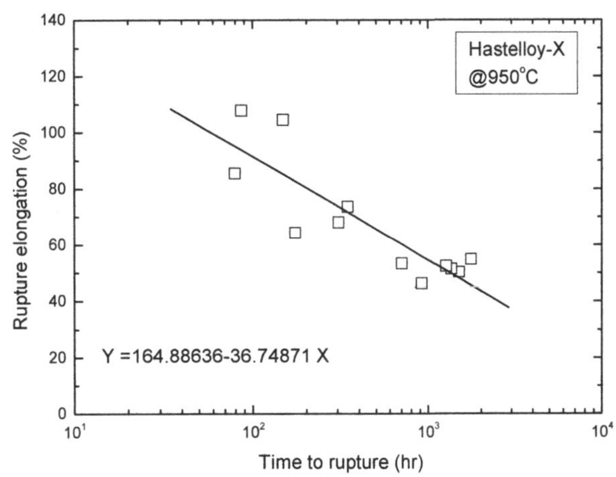

Fig. 4 Rupture elongation vs. time to rupture

In addition, the Monkman-Grant (M-G) relationship is useful to perform the prediction of a life of high-temperature structural components. The relationship between the minimum creep rate and the rupture time is expressed as Eq. (1).

$$\log t_R + m \log \dot{\varepsilon}_m = C\,, \tag{1}$$

where $C$ and $m$ are constants, and $t_r$ is the time to rupture. The HX followed well the M-G relationship with a good linearity, as shown in Fig. 3. The $m$ value was obtained as 0.81. Also, the modified M-G one is expressed as Eq. (2).

$$\log t_R / \varepsilon_R + m' \log \dot{\varepsilon}_m = C'\,, \tag{2}$$

where $\varepsilon_R$ is the rupture elongation. All the creep data showed a good linearity. The $m'$ value was obtained as 1.02, which is close to unity. Good agreement of the modified M-G one is because the relationship between the elongation and the rupture time was linear, as shown in Fig. 4. It means that the creep deformation of the HX alloy was progressed by the formation of cracks and cavities.

**Creep life prediction.** The experimental creep data was combined with the data collected from different countries; European Alloy DB, published literatures, and manufactures' data sheets. Creep life was estimated by using the LM parameter, as written in Eq. (3),

$$P = T(\log t_R \cdot \alpha_R + C) = f(\sigma)\,. \tag{3}$$

Hence, $\alpha_R (= t_R / t_r)$ is a time coefficient, and $t_R$ is the experimental rupture time and $t_r$ the estimated one. The $f(\sigma)$ is a stress function of the polynomial equation. The $C$ value was used as 20.

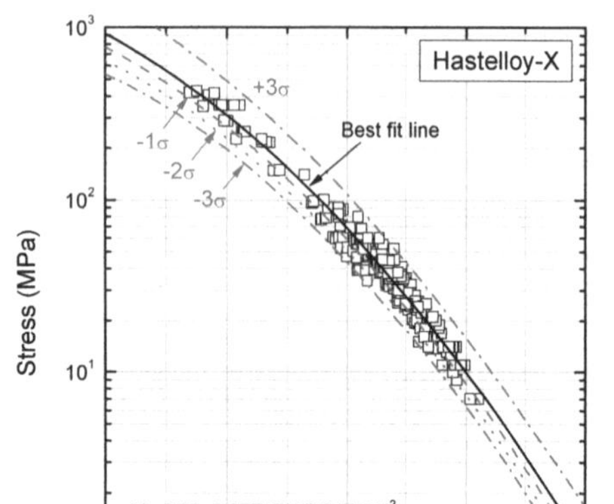


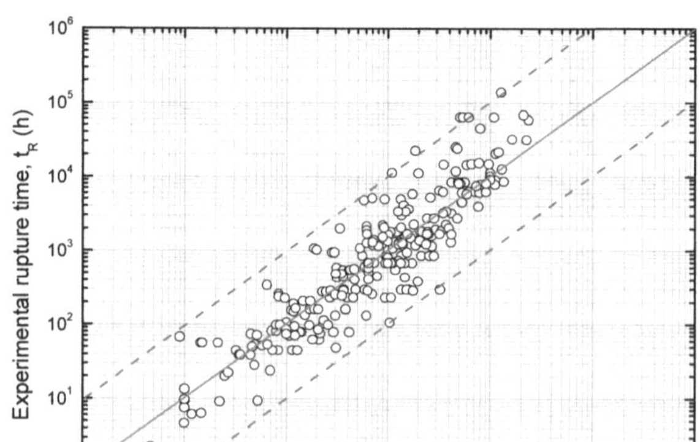


Fig. 5 Creep master curve of the HX Fig. 6 Correlation of the experimental and estimated results

Fig. 5 shows the results of the creep master curves calculated by the LM parameter by using all the rupture data. Each curve presents the standard deviations of 1σ, 2σ and 3σ. Almost all of the data is included in 3σ, which has a reliability of 99.7%. Polynomial equation of two orders was obtained by the best fitting of the total data. Relationship between the experimental and estimated rupture times is presented in Fig. 6. It is identified that a scattering band of all the data is well included in about a factor 10 from the average line. Thus, a safe design can be established by applying a minus 10 value from the average line of the estimated rupture time.

Fig. 7 shows the prediction curves obtained from the LM parameter at each temperature. Creep rupture time up to $10^5$ hours was extrapolated for the HX. Each prediction curve presents a good conformation of the experimental data. However, in the case of a longer-time prediction up to and over $10^5$ hours, because the strength of the materials can be deteriorated due to the thermal-mechanical damage at high temperature, the stability of the metallurgical structure can not be assured. Thus, this result may not proper in predicting a longer-time rupture over $10^5$ hours, especially at high temperature above 900°C.

Fig. 8 shows the failure probability with the time coefficient of $\alpha_R (= t_R / t_r)$. The $\alpha_R$ value was calculated by using Eq. (3). Total data of 276 points were also calculated by a statistical process. The creep rupture data followed well the lognormal distribution with a good linearity. Using this result, the failure probability for an individual data set was estimated probabilistically.

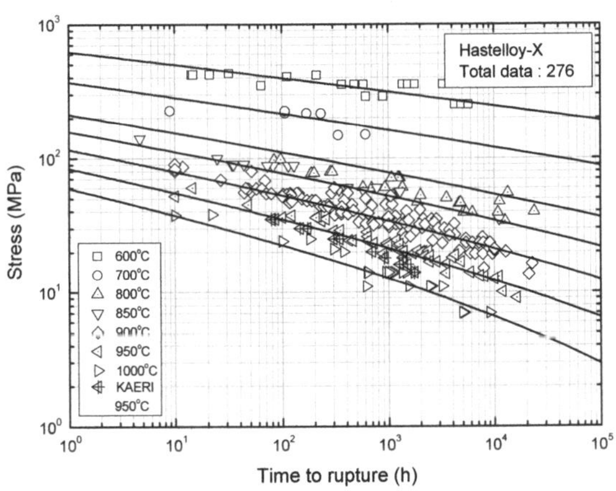


Fig. 7 Creep-life prediction curve at each temperature by the LM parameter

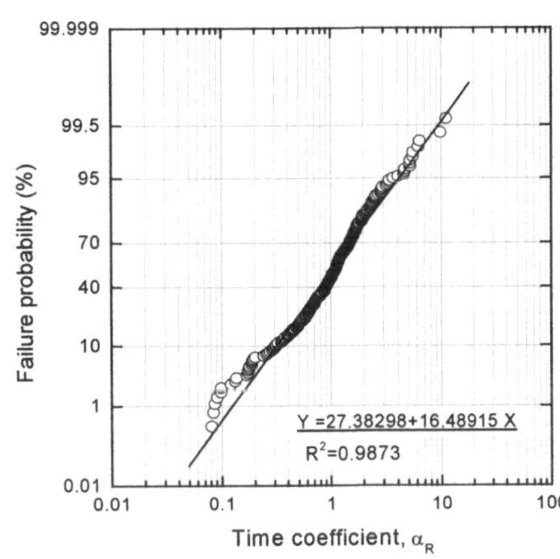


Fig. 8 Failure probability for all data

## Conclusion

The creep properties for the HX alloy were obtained from the creep tests at 950°C, and a number of the creep data was collected through available literature surveys to predict long-term creep lifetime. The creep constants of *A*, *n*, *m* and *m'* used in the Norton's law and the M-G relationship were determined for the HX alloy. The regression curves based on the LM parameter were plotted for the 1σ, 2σ and 3σ. All of the creep data were included in 3σ. The creep rupture life was estimated up to $10^6$ hours for the HX alloy. The results of failure probability of the creep data in the HX alloy well followed the lognormal distribution.

## References

[1] K. Hada, I. Nishiguchi, Y. Muto and H. Tsuji: Nuclear Eng. and Design, Vol. 132(1991), p.1.
[2] ANL Report, Materials Behavior in HTGR Environments, NUREG/CR-6824 ANL-02/37(2002).
[3] Y. Muto, H. Nakajima and M. Eto: Nuclear Engineering Design, 144 (1993), p.305.
[4] T. Nakanishi and H. Kawakami: Nuclear Technology, Vol. 66 (1984), p.273.
[5] European Alloy-DB, Licensed CD, European -Joint Research Centre (EU-JRC).
[6] JAERI Report, JAERI-M-89-005 (1989), p.32.
[7] Haynes International, www.haynesintl.com, Hastelloy-X Alloy, H-3009A (1997).
[8] Special Metals, www.specialmetals.com, Inconel Alloy HX, SMC-007(2004).
[9] Y. Kurata and H. Nakajima: Journal Nuclear Materials, Vol. 228(1996), p.176.

*Key Engineering Materials Vols. 326-328 (2006) pp. 1313-1316*
*online at http://www.scientific.net*
*© (2006) Trans Tech Publications, Switzerland*

# Creep-Life Prediction of Type 316LN Stainless Steel by Minimum Commitment Method

Woo Gon Kim[1, a], Song Nan Yin[2,b], Woo Seog Ryu[3,c] and Won Yi[4,d]

[1,3] KAERI, 150 Deogjin-dong, Yuseong-gu, Daejeon, 305-353, Korea

[2,4] Soongsil Univ., 1-1 Sangdo-dong, Dongjak-gu, Seoul, 156-743, Korea

[a]wgkim@kaeri.re.kr, [b]yjsongnan@yahoo.co.kr, [c]wsryu@kaeri.re.kr, [d]yiwon@hanmail.net

**Keywords:** Minimum Commitment Method (MCM), Type 316LN Stainless Steel, Creep, Creep-Life Prediction, Time-Temperature Parameter (TTP), Coefficient of Determination (COD)

**Abstract.** This paper presents the results of the Minimum Commitment Method (MCM) applied to predict the creep rupture life of type 316LN SS. Constant $A$, and the function of $P(T)$ and $G(\sigma)$ being used in the MCM equation were determined. To determine a proper value of the constant $A$, a focal point method and a trial and error one were adopted, respectively. It was found to be $A$=-0.02~-0.05 for type 316LN SS. Each prediction curve with the $A$ values were presented up to $10^6$ hours and compared to the experimental data at each temperature. Using the short-term creep rupture data for under 2,000 hours, a long-time rupture reaching up to $10^6$ hours was predicted by the MCM.

## Introduction

Type 316LN stainless steel (SS) is a candidate material for the major structural components of a liquid metal reactor (LMR) because of its good high-temperature mechanical properties. Since the LMR components are used for 20 to 30 years at a high temperature up to 550°C, one of the most important properties of components is the creep behavior [1, 2]. To predict accurately the long-term creep rupture of the components, it is essential to achieve reliable long-term test data beyond their design life. But, it is difficult to actually obtain long-term creep data because it is time-consuming work and it is an expensive process.

So far, a variety of time-temperature parameters (TTPs) have been developed to predict a long-time rupture from shorter-time tests at higher temperatures. Among them, the Larson-Miller, the Orr-Sherby-Dorn, the Manson-Harferd and the Manson-Succop parameters have been typically used [3, 4]. However, none of these parameters has an overwhelming preference, and they have certain inherent restrictions imposed on the data in the application of the TTPs parameters. That is, a linearity of the isostress curves, and a parallelism or a convergence of the families of such curves are conditions imposed on the data. The data do not always conform to these restrictions. It may cause large errors in the extrapolation of a longer-time rupture. Meanwhile, it has been reported that the Minimum Commitment Method (MCM) proposed by Manson and Ensign [5,6] has a greater flexibility for a creep rupture analysis. It can be shown that this method not only includes all the commonly used TTPs, but that it has no restrictions imposed on the data in the common TTPs. Thus, it is believed that the MCM will be useful as another approach. But, until now, the applicability of the MCM has not been investigated for type 316LN SS because of insufficient creep data.

In this paper, the MCM was applied to predict a long-term rupture of type 316LN SS. For this purpose, a number of creep rupture data was collected through literature surveys and the experimental data of KAERI. Using the short-term rupture data for under 2,000 hour, longer-time rupture above $10^5$ hours was predicted by the MCM at temperatures of 550°C to 800°C.

## Result and Discussion

**Creep data collection.** Fig. 1 shows the creep-rupture data of the type 316LN SS collected from worldwide literature surveys [7]; Japan, India, Czechoslovakia and Korea. This data was selected because it has some reliability among the many creep data reported for type 316LN SS, which are clearly defined by chemical compositions. Especially, the *nitrogen content* is within the 0.06~0.15wt.% ranges. The total number of data is 345 points, and the temperature range is from 500 to 800°C. All of the creep rupture data is filed by digitizing it to apply the MCM.

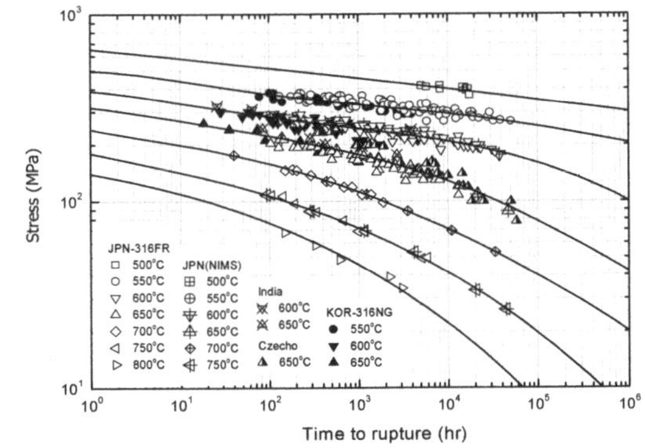


Fig. 1 Plot of creep rupture data obtained for type 316LN SS

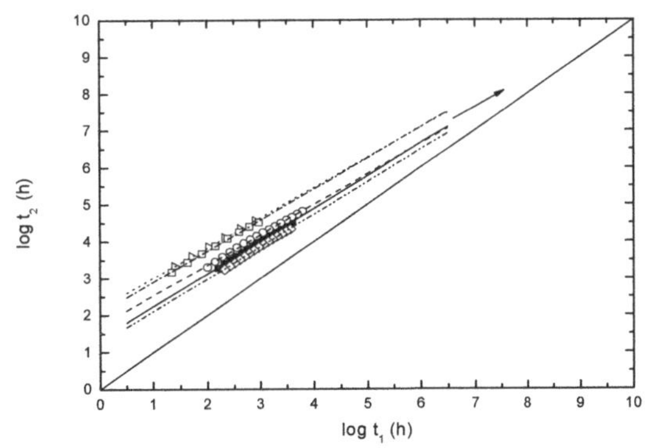


Fig. 2 Plot to determine A from focal point of such lines along the 45° line

**Determination of constant *A*.** The MCM equation has the form as

$$\log t + AP(T)\log t + P(T) = G(\log\sigma), \tag{1}$$

where $t$ is the rupture time and $A$ is a material constant relating to a structural stability, $P(T)$ is a temperature function, and $G(\sigma)$ is a stress function. The value of stability factor $A$ is a single constant. If the $A$ and the $P$ function are known, the $G$ function can be determined simply. But, to solve Eq. (1) is complicated by a nonlinear form, resulting from the product of an unknown $A$ and the $P$ function.

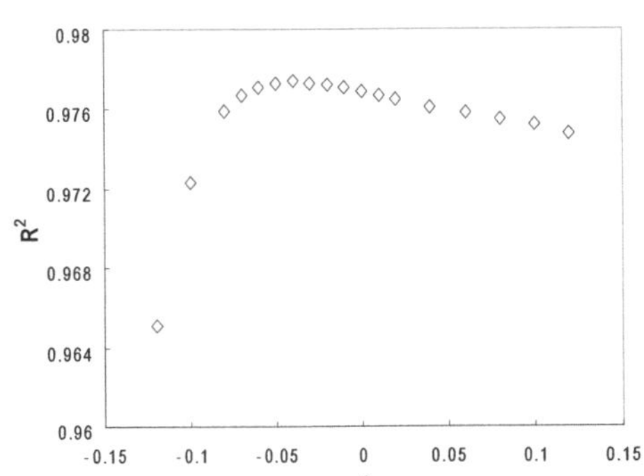


Fig. 3 Coefficient of a determination with A values

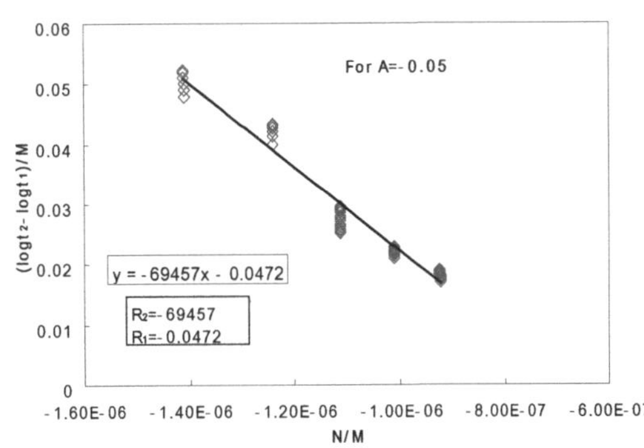


Fig. 4 Determination of $R_1$ and $R_2$ in $A$=-0.05

In order to determine the $A$ value, two methods were demonstrated in this study. Firstly, the "focal point method" was adopted. Log $t_1$ and log $t_2$ were taken from a point intersected with isothermal lines at equal stress levels in a typical log $\sigma$–log $t$ plot. Then, the $A$ value can be obtained at a convergent point of each extended isothermal line. But, the data did not create a convergent point for the extended isothermal lines. So, the $A$ value in this study was determined by taking a reciprocal of the mean slope of each isothermal line between log $t_1$ and log $t_2$, as shown in Fig. 2. The $A$ value for the focal point method was obtained with -0.089. Secondly, the "trial and error method" was adopted to determine the optimum value of the $A$. A series of $A$ values are chosen from a wide range of -0.15 to 0.15. This value was referred to the reported results that for steels and superalloys the optimum value was -0.05 to -0.10 and for aluminum alloy it was 0 to 0.1 [6]. From a series of the $A$ values, by using Eq. (1), a polynomial equation is obtained by best fitting the data in the plot of the log $\sigma$ and the $G$ function. Then, a proper value for the $A$ was determined by using a statistical parameter of coefficient of the determination (COD), $R^2$, indicating whether the polynomial equation is a good or bad conformation of the data. In the case of a high COD value, it becomes a good match to the data.

Fig. 3 presents the results of the COD values with respect to the $A$ values obtained by the trial and error method. Optimum values for $A$ were found to be -0.02 to -0.05. On the other hand, the $A$ value of -0.089 obtained from the focal point method is not proper because the COD value was low.

**Determination of the *P* and G function.** The $P$ function has the form as in Eq. (2) [6],

$$P(T) = R_1(T - T_m) + R_2\left(\frac{1}{T} - \frac{1}{T_m}\right) \tag{2}$$

where $R_1$ and $R_2$ are material constants, and $T_m$ is the absolute temperature in the midrange of the experimental data. In this case, we have used 923K (650°C) as the midrange between 550°C and 800°C. Substituting Eq. (2) into Eq. (1) and by re-arranging it, we can get Eq. (3) as

$$\frac{(\log t_2 - \log t_1)}{M} = \frac{N}{M}R_2 + R_1 . \tag{3}$$

Hence,
$$M = A\left[(T_1 - T_m)\log(t_1) + (T_2 - T_m)\log(t_2)\right] + (T_1 - T_2) \tag{4}$$

$$N = A\left[\left(\frac{1}{T_1} - \frac{1}{T_m}\right)\log(t_1) + \left(\frac{1}{T_2} - \frac{1}{T_m}\right)\right]\log(t_2) + \left(\frac{1}{T_1} - \frac{1}{T_2}\right) . \tag{5}$$

Since the $M$ and $N$ are known at each point, then we can plot (log $t_2$-log $t_1$)/$M$ vs. $N/M$. The $R_1$ becomes the intercept and the $R_2$ is the slope of the resulting straight line. Fig. 4 shows the result of the analysis for type 316LN SS. The $R_1$ and $R_2$ values were obtained by a linear equation. In the case of $A$=-0.05, the $R_1$ value was -0.0472 and the $R_2$ value was -69,457. The plot of the $P$ function vs. $T$ is shown in Fig. 6. The $A$ and the $P$ function were obtained by using the data for under 2,000 hours.

In addition, the $G$ function was determined by using the known $A$ and the $P$ function from above. For every available datum point, all the terms on the left side of Eq. (1) are known since log $t$, $A$, and $P$ for the temperature are known. Thus, the $G$ can be calculated for each datum point. Then, these values can be plotted against the values of log$\sigma$ to obtain the curve. Fig. 6 shows the typical curve for type 316LN SS by using only the short-time data for under 2,000 hours in $A$=-0.05. The polynomial equation of three orders was obtained and it showed a good match with $R^2$=0.98 for the data.

**Creep-life prediction with the *A* values.** Figs. 7 and 8 present typical results of the creep-life prediction with respect to the $A$ values at 650°C and 750°C. Long-time rupture was predicted up to $10^6$ hours for each temperature from 550°C to 800°C. For the values of $A$=-0.02~-0.05, each curve showed a good match with the data without any differences in all the temperature ranges, but for $A$=-0.07 at 750°C, the curve did not show a good match with the data of a longer-time rupture, as shown in Fig. 8. This is because the COD value indicating an agreement to the data was low. Optimum values of the $A$ values for type 316LN SS were found to be in the ranges of -0.02 to -0.05. Using the results, it is believed that the creep life can be predicted by a long-time rupture reaching $10^6$ hours from the short-time data for type 316LN SS.

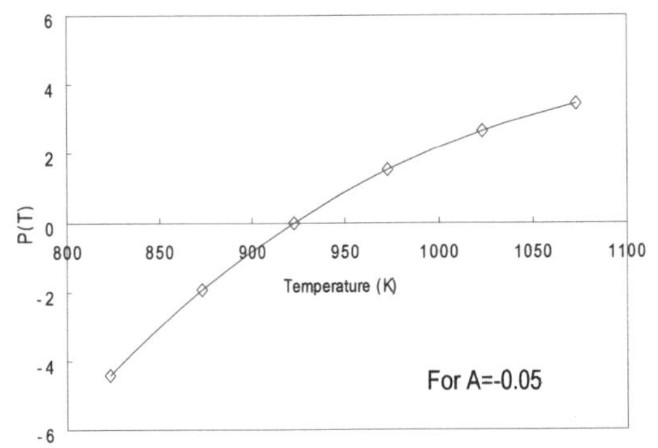


Fig. 5 Results of the $P(T)$ function at $A$=-0.05

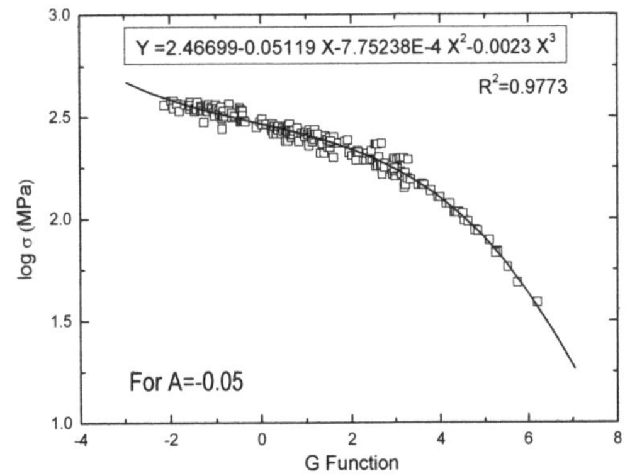


Fig. 6 Regression curve at $A$=-0.05

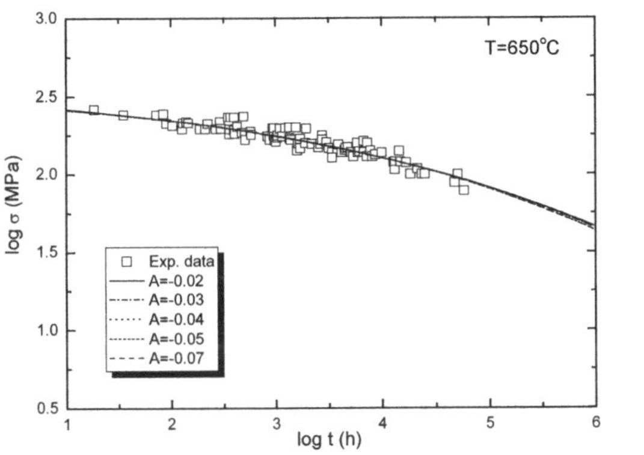


Fig. 7 Prediction curves with $A$ values at 650°C

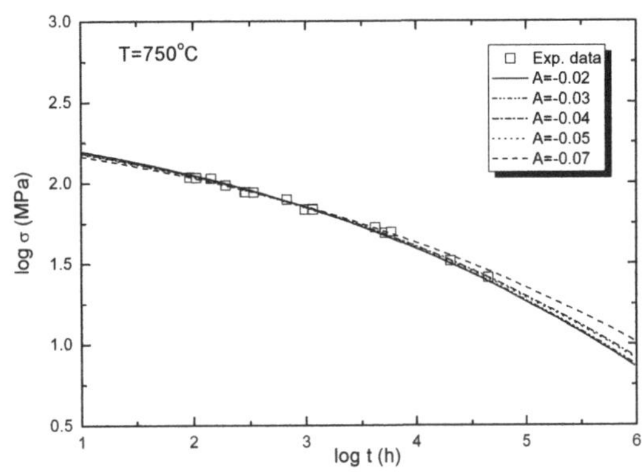


Fig. 8 Prediction curves with $A$ values at 750°C

## Conclusion

The MCM was applied to predict the creep rupture life of type 316LN SS. Constant $A$, $P(T)$ and $G(\sigma)$ functions being used in the MCM equation were determined. Optimum values of the $A$ was found to be from -0.02 to -0.05 for type 316LN SS. The prediction curve for $A$=-0.02~-0.05 showed a good match with the experimental data without any differences in all the temperature ranges, but the curve for $A$=-0.07 at 750°C did not show a good match with the data of a longer-time rupture. This is because the COD value indicating an agreement with the data was low. Long-time rupture reaching $10^6$ hours from the short-time data of under 2,000 hours was predicted by the MCM.

## References

[1] W.G. Kim, S.H. Kim and W. S. Ryu: KSME Int. J., Vol.16 (2002), p. 420.
[2] W.G. Kim, S.H. Kim and W. S. Ryu: KSME Int. J., Vol.15 (2001), p.1463.
[3] VAMAS Data Evaluation Committee: *Iron and Steel Institute of Japan* (1994).
[4] I. Le May: Transactions of the ASME, Vol. 101 (1979), p. 326.
[5] S.S. Manson and C.R. Ensign: NASA TM X-52999 (1971), p. 1.
[6] S.S. Manson and C.R. Ensign: MPC-7 ASME New York, (1978), p. 299.
[7] W.G. Kim, S.N. Yin and W.S. Ryu: Key Engineering Materials, Vol. 297~300(2005), p.2272.

*Key Engineering Materials Vols. 297-300 (2005) pp. 2272-2277*
*online at http://www.scientific.net*
*© (2005) Trans Tech Publications, Switzerland*

# Application and Standard Error Analysis of the Parametric Methods for Predicting the Creep Life of Type 316LN SS

Woo Gon Kim[1,a], Song Nam Yoon[2,b] and Woo Seog Ryu[1,c]

[1]KAERI, Daejeon, Korea

[2]Soongsil Univ., Seoul, Korea

[a]wgkim@kaeri.re.kr, [b]ex-yjsongnam@kaeri.re.kr, [c]wsryu@kaeri.re.kr

**Keywords:** Type 316LN Stainless Steel, Creep, Creep Life, Liquid Metal Reactor (LMR), Time-Temperature Parameter (TTP), Standard Error of Estimate (SEE)

**Abstract.** To predict the creep-rupture life of type 316LN stainless steels which are major structural components of liquid metal reactors, a number of creep-rupture data were collected through literature survey and experimental data of KAERI. Using the data, the creep-rupture life was analyzed by means of the Larson-Miller, the Orr-Sherby-Dorn and the Manson-Haferd parametric methods. Polynomial equations for predicting the creep life were obtained. In order to analyze the acceptance and use of the parametric methods, standard error values were accurately investigated by statistical process of the creep data. As for the results, the three parametric methods are found to be favorable in predicting the creep life of type 316LN stainless steel. Each method did not generate a large error in the standard error of the estimate with variations of the temperatures, but the Orr-Sherby-Dorn and the Manson-Haferd methods showed a better agreement than the Larson-Miller one. Especially, at higher the 700°C, the Manson-Haferd method conformed well to the experimental data. The reason is because the Manson-Haferd method includes two constants of $t_a$ and $T_a$.

## Introduction

Type 316LN stainless steels (SS) are candidate materials for major structural components of a liquid metal reactor (LMR) because of its good high-temperature mechanical properties. Since the LMR components are used for 20 to 30years at a high temperature up to 550°C, one of the most critical factors of the components is creep behavior [1, 2]. For a creep rupture of these components, it is essential to achieve reliable long-term test data beyond its design life. However, it is difficult to actually obtain long-term creep data because of time-consuming work and it is an expensive process. So, a creep design has been carried out by an extrapolation of short-term creep data to approximate the long-term behavior. This method will be very useful if the creep rupture can be predicted well with a sufficient accuracy for design purposes.

So far, a variety of parameters have been developed for an accurate extrapolation of creep data to longer times than those for which experimental data is available. Among them, the Larson-Miller (L-M), the Orr-Sherby-Dorn (O-S-D), the Manson-Harferd (M-H), the Manson-Succop (M-S), and the Goldhoff-Sherby (G-S) parametric methods have been used typically [3, 4]. They are basically established by a time-temperature parameter (TTP). A TTP method is to provide a modification to the time scale for a creep deformation such that all the isothermal plots of the creep strain are coincident or a rupture occurs at the same temperature for a specified stress level. TTP method depends on an interrelation of the time and temperature in contributing to a creep deformation. That is, among the three variables of time, temperature and stress in a creep, the time and temperature variables are combined into one parameter, and then it is expressed as a function of a stress. Thus, a plot of log (stress) versus the parameter produces a single curve or line. Longer-time rupture behavior can be predicted by an extrapolation of the single curve. However, the TTP methods have been found to be applicable to particular materials and data sets, providing the best fit in particular cases. They have merits and demerits in the use of different material cases, respectively. In addition, in the case that a material is damaged due to the thermal-mechanical effect during a long period at a high temperature, stability of the metallurgical structure is not assured. Thus, because none of the parameters are perfect for predicting a longer-time rupture from the results of the shorter time tests, it is necessary to analyze acceptance and use of the TTP parametric methods.

In this paper, a number of creep-rupture data for type 316LN SS are collected through a literature survey and produced at KAERI. These data are applied to the typical L-M, the O-S-D and the M-H parametric methods. Polynomial equations for predicting the creep life are obtained for each parametric method. An acceptance of each method is accurately investigated by analyzing the standard error of the estimate (SEE) through a statistical process of the creep data with different temperature ranges.

**Time-Temperature Parametric (TTP) methods**

TTP methods depend on the interrelation of the time and temperature in contributing to the creep deformation. Time and temperature variables are combined into one parameter, *P*. The parameter is expressed as a function of a stress as in Eq. (1). A plot of the log (stress) versus the parameter produces a single curve or line. Longer-time rupture behavior can be predicted by an extrapolation of the single curve.

$$P\,(t_r, T) = f(\sigma)\,. \tag{1}$$

Among the TTP methods, the L-M, the O-S-D, the M-H, the M-S and the G-S methods have been commonly used [5]. In this study, the L-M, the O-S-D and the M-H methods were typically applied because the G-S method is similar to the L-M one, and the M-S method is similar to the M-H one. Each parametric equation can be written in the following forms,

$$\text{L-M}: \; P(t_r, T) = (\log t_r + C)T\;, \tag{2}$$

$$\text{O-S-D}: \; P(t_r, T) = \log t_r - \frac{Q}{2.3RT}\;, \tag{3}$$

$$\text{M-H}: \; P(t_r, T) = \frac{\log t_r - \log t_a}{T - T_a}\;. \tag{4}$$

where, in the L-M parameter, $T$ is the absolute temperature, $C$ is a constant defined by the intercept on the $1/T = 0$ axis of the isostress lines plotted on the coordinate axes log $t_r$ versus $1/T$, and $t_r$ is the time to rupture. In order that the parameter can be applied accurately, the isostress data must be linear on coordinate axes log $t_r$ versus $1/T$ and the families of such lines must converge to the point (0, -C). In the O-S-D parameter, $Q$ is the activation energy and $R$ is the universal gas constants. The application of this parameter requires that the isostress data is linear on the coordinate axes of log $t_r$ versus $1/T$ and that the families of such lines are parallel with a common slope $B$ ($=Q/2.3RT$). If the data clearly conform to the requirements of the O-S-D parameter, they cannot conform to those of the L-M parameter above. In the M-H parameter, $T_a$ and log $t_a$ are material constants defined by the point of the intersection of the families of the isostress lines plotted on the coordinate axes of log $t_r$ versus $T$.

In addition, a stress function of the x-axis can be expressed by a polynomial equation, as follows,

$$f(\sigma) = b_o + b_1(\log\sigma) + b_2(\log\sigma)^2 + \cdots b_k(\log\sigma)^k \tag{5}$$

where, $b_k$ is the regression coefficient of the $k$ order, and $k$ is the order of polynomial equation. The $k$ in the polynomial equation is generally used under five orders.

## Results and Discussion

**Creep-life prediction of type 316LN SS.** Fig. 1 shows the creep-rupture data of the type 316LN SS collected from worldwide literature surveys; Japan [6, 7], India [8], Czechoslovakia [9] and Korea [10]. These creep data are selected to have some reliability among the many creep data reported for type 316LN SS, which are clearly defined by chemical compositions. Especially, the *nitrogen content* is within the 0.06-0.15wt.% ranges. The total number of data is 345 points, and the temperature range is 500-800°C, as shown in Fig. 1. All of the data are filed by digitizing them to apply parametric methods. These creep data are used as basic data for obtaining the polynomial equations and for analyzing the standard error values by means of the L-M, the O-S-D and the M-H methods with different temperatures of type 316LN SS.

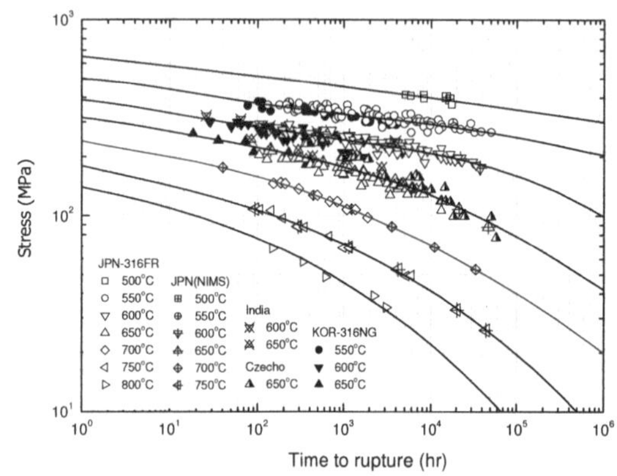


Fig. 1 Stress vs. time to rupture of type 316LN SS

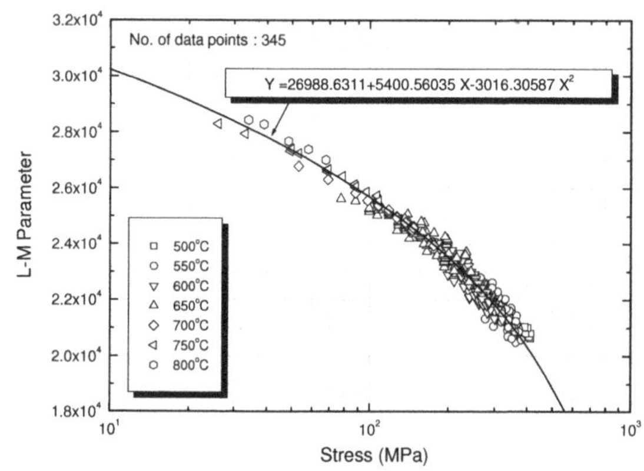


Fig. 2 Regression line of the L-M parameter

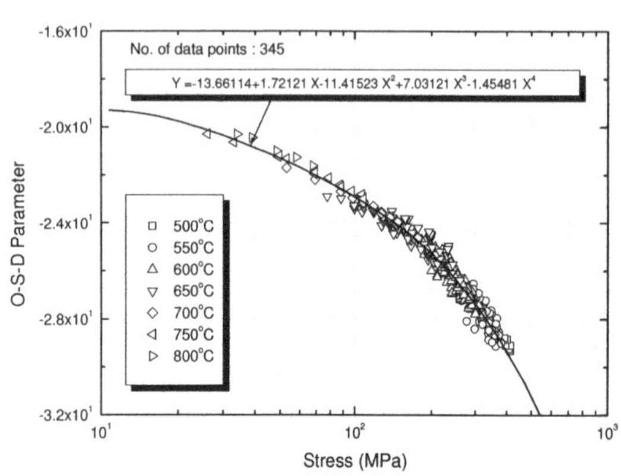


Fig. 3 Regression line of the O-S-D parameter

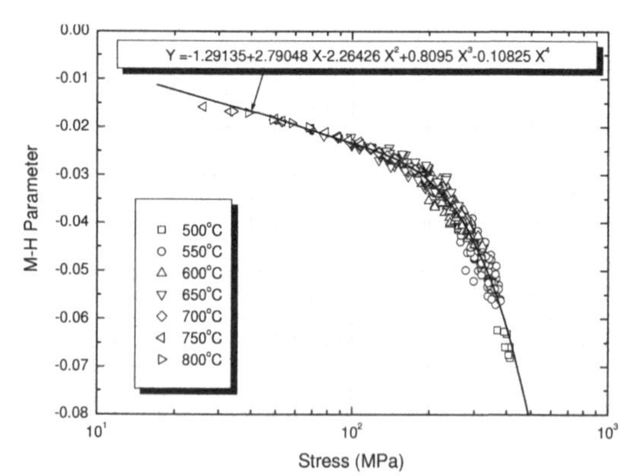


Fig. 4 Regression line of the M-H parameter

In order to predict the creep life by the L-M, the O-S-D and the M-H methods, first of all, it is required to determine the constant values of $C$, $B$, $t_a$ and $T_a$ used in each TTP equation. In this study, their values for each method were obtained as follows. In the L-M method, because the families of the isostress lines were not converged to the point (0, -C), the $C$ value was obtained by taking an average value of the coefficient $b$ in each linear equation (y=ax+b) of the isostress lines which are plotted on coordinate axes log $t_r$ versus $1/T$. Also, in the O-S-D parameter, the isostress lines showed a good parallel relation with the same slope $B$ $(=Q/2{,}3RT)$ to each other. Thus, the $B$ value was obtained by taking a slope of the straight line. In the M-H parameter, $T_a$ and log $t_a$ values were obtained taking an intersecting point of the isostress lines plotted on the coordinate axes of log $t_r$ versus $T$. But, the families of the isostress lines were not converged to an intersecting point. So, each liner equation (y=ax+b) of the isostress lines was solved for x and y, respectively, and its value was finally obtained. Hence, the x-axis corresponded to $T_a$, and the y-axis corresponded to $t_a$. Using the obtained constants and regression coefficients, the polynomial equations for predicting the creep life were obtained by a best fitting of the creep data.

Fig. 2, Fig. 3 and Fig. 4 show the polynomial equations obtained for the L-M, the O-S-D and the M-H methods, respectively. The L-M method was obtained with second-order polynomial equation to the data, and both the O-S-D and the M-H methods are obtained with quadratic-order polynomial equations for the data. It is obvious that the polynomial equations for each method conformed favorably to the experimental data. The summary of constants and regression coefficients for each parameter are given in Table 1.

Table 1 Summary of constants, regression coefficients and RSE values for three parameters

| Parameters | Constants | Regression coefficients | | | | | RSE |
|---|---|---|---|---|---|---|---|
| | | $b_o$ | $b_1$ | $b_2$ | $b_3$ | $b_4$ | |
| L-M [1)] | C (= 23.03) | 26988.6 | 5400.5 | -3016.3 | | | 0.01316 |
| O-S-D [2)] | Q/2.3R(=25530) | -13.66 | 1.72 | -11.41 | 7.03 | -1.45 | 0.01283 |
| M-H [3)] | $T_a$(=679.5) $t_a$(=12.0475E9) | -1.29 | 2.79 | -2.26 | 0.80 | -0.10 | 0.05254 |

[1)] $Y(P_{L-M}) = T(\log t_r + C)$, $Y = b_o+b_1X+b_2X^2$
[2)] $Y(P_{O-S-D})=\log t_r - Q/2.3RT$, $Y = b_o+b_1X+b_2X^2 +b_3X^3 +b_4X^4$
[3)] $Y(P_{M-H}) = (\log t_r-\log t_a)/ (T-T_a)$, $Y = b_o+b_1X+b_2X^2 +b_3X^3 +b_4X^4$

**Standard error analysis of each parametric method.** The polynomial equations predicted by the three parametric methods was analyzed and compared by means of relative standard error (RSE). It is defined by Eq. (6),

$$RSE = \sqrt{\sum_{i=1}^{i=n} \left( \frac{Y_i - \hat{Y}_i}{\hat{Y}_i} \right)^2 /(n-n_p)} \qquad (6)$$

where, $Y_i$ is log $t_r$ at $i$ data, $\hat{Y}_i$ is the average value estimated at $i$ data, and $n$ is the total number of creep data. $n_p$ is the number of undetermined coefficients included in the regression equation. The L-M and O-S-D methods become $n_p = k+2$, and the MH method becomes $n_p = k+3$. The summary of the RSE values for each parameter is given in Table 1. The RSE values for the three methods have very low values such as 0.01316, 0.01283 and 0.05254, and there is not a large difference between them, as given in Table 1. It is found that the three parametric methods can be used without a large error to predict the creep life of type 316LN SS. In addition, to compare some of the error values for the regression equations of the three methods, the SEE value is analyzed. It is defined by Eq. (7),

$$SEE = \sqrt{\sum_{i=1}^{i=n} (Y_i - \hat{Y}_i)^2 /(n-n_p)} \quad . \qquad (7)$$

The SEE values for the three methods were obtained with variations of the temperatures, 550, 600, 650 and 700°C. The results for them are presented in Fig. 5, Fig. 6, Fig. 7 and Fig. 8. As a whole, the SEE values do not generate a large difference between the three parametric methods, because they are used on the basis of the TTP relationship. However, the M-H method at 700°C showed a significantly low value in the SEE value comparing with the L-M and the O-S-D methods. The reason for this is because the M-H method includes the constants of $t_a$ as well as $T_a$, Thus, it is believed that the M-H method has a wider acceptance than the L-M and the O-S-D ones.

In addition, as shown in Fig. 5, Fig. 6, Fig. 7 and Fig. 8, the regression lines of the O-S-D and the M-H methods show better fitting lines to the data points than that of the L-M one. The O-S-D method conformed better in all the temperature ranges than the L-M one. This reason for this is because the families of the isostress lines, which are plotted on the coordinate axes log $t_r$ versus $1/T$, have good parallel relationships with respect to each slope $B$. On the contrary, in the case of the L-M method, because the families of the isostress lines were not converged to one point, the regression line did not conform any more favorably the data points than those of the O-S-D and the M-H ones. The SEE values are summarized in Table 2.

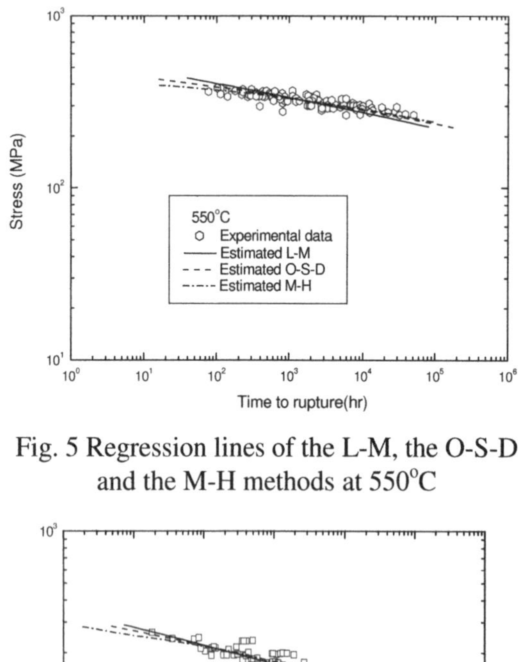

Fig. 5 Regression lines of the L-M, the O-S-D and the M-H methods at 550°C

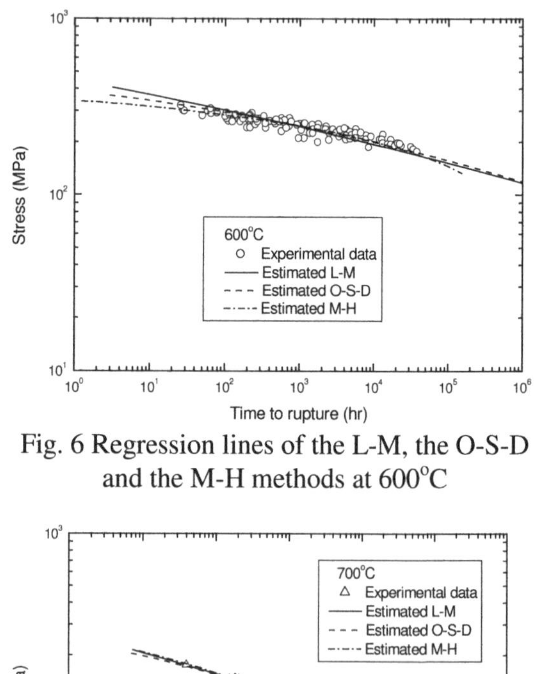

Fig. 6 Regression lines of the L-M, the O-S-D and the M-H methods at 600°C

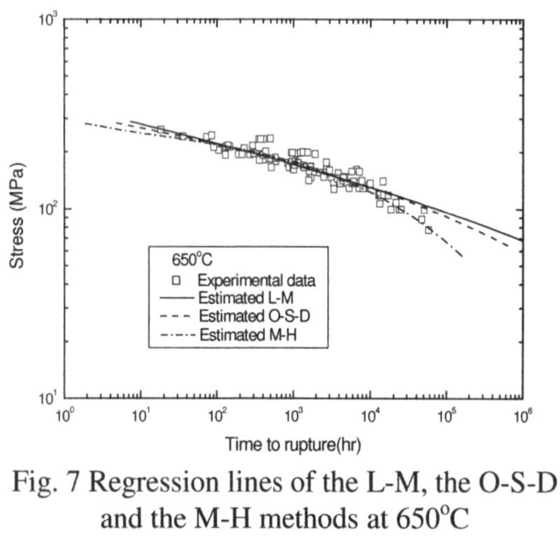

Fig. 7 Regression lines of the L-M, the O-S-D and the M-H methods at 650°C

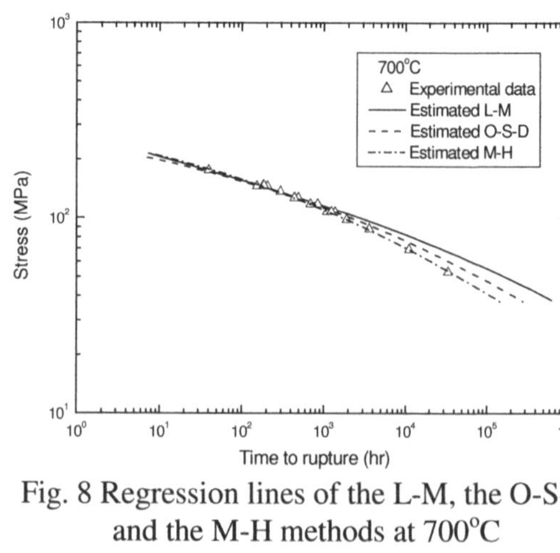

Fig. 8 Regression lines of the L-M, the O-S-D and the M-H methods at 700°C

The SEE values are mainly generated by (i) an error due to a formula for the temperature dependency, (ii) an error due to a formula for the stress dependency and (iii) an error due to the data itself. It is thus important to reduce some of the errors generated in the formula process of (i) and (ii) above. In addition, the error may depend on whether the creep-rupture data for one material of the same composition or different materials of several compositions. Generally, for one material of same composition, if SEE >0.2, it becomes a bad acceptance, and if SEE<0.2, it becomes a favorable acceptance [3]. The *SEE* results obtained in this study showed slightly higher values than 0.2. This reason is because the creep data was collected from several countries and the *nitrogen contents* were as wide as 0.06-0.15wt. %.

Table 2 Summary of the SEE values with different temperatures of each parameter

| Temp.( °C) | SEE values for log $t_r$ | | |
|---|---|---|---|
| | L-M parameter | O-S-D parameter | M-H parameter |
| 550 | 0.3823 | 0.3795 | 0.4066 |
| 600 | 0.3443 | 0.3280 | 0.3340 |
| 650 | 0.3649 | 0.3629 | 0.3640 |
| 700 | 0.2212 | 0.1251 | 0.0636 |

**Summary**

Creep-rupture life for type 316LN SS was predicted by means of the L-M, the Q-S-D and the M-H parametric methods, and the polynomial equations for predicting its life were obtained. Each parametric method did not generate large *RSE* values, and they conformed favorably to the experimental data even although a number of creep data were collected from several countries. However, the O-S-D and M-H methods showed a better agreement than the L-M one. Especially, it was found that the M-H method at 700°C was lower in the SEE values than the L-M and the O-S-D ones. This reason is because the M-H method includes the constants of $t_a$ as well as $T_a$. Thus, it is believed that the M-H method has a wider acceptance in the temperature ranges than the L-M and the O-S-D ones.

**References**

[1] W.G. Kim, S.H. Kim and W.S. Ryu: KSME Int. J. Vol. 16 (2002), p. 420

[2] W.G. Kim, S.H. Kim and W.S. Ryu: KSME Int. J. Vol. 15 (2001), p. 1463

[3] VAMAS Data Evaluation Committee: *Iron and Steel Institute of Japan* (1994)

[4] I. Le May: Transactions of the ASME Vol. 101 (1979), p. 326

[5] W.E. White and I. Le May: J. of Eng. Mater. and Tech. Vol. 100 (1978), p. 333

[6] NRIM: NRIM Creep Data Sheet No. 45 (1997)

[7] K. Kazuya, D. Shingo, T. Kenichi, S. Masayuki, K. Nobuchika and T. Yoshihiko: ASME, PVP Vol. 391 (1999), p. 47

[8] M.D. Mathew, G. Sasikala, K. Bhanu Sankara Rao and S.L. Mannan: Materials Science and Engineering, A148 (1991), p. 253

[9] M. Liska, V. Vodarek, M. Sobotkova and J. Sobotka: High Nitrogen Steels (1990), p. 78

[10] D.W. Kim, Y.K. Lee, W.G. Kim and W.S. Ryu: J. Kor. Inst. Met. & Mater. Vol. 39 (2001), p. 1241

*Materials Science Forum Vols. 475-479 (2005) pp. 1433-1436*
*online at http://www.scientific.net*
*© (2005) Trans Tech Publications, Switzerland*

# Application of Creep Ductility Model for Evaluating Creep Crack Growth Rate of Type 316SS Series

Woo Gon Kim[1,a], Hyun Hie Kim[2,b], Kee Bong Yoon[3,c] and Woo Seog Ryu[4,d]

[1,4] KAERI, 150 Deokjin-Dong, Yuseong-Gu, Daejeon, 305-353, Korea

[2,3] Chung-Ang University, 221 Huksuk-Dong, Dongjak-Gu, Seoul, 156-756, Korea

[a]wgkim@kaeri.re.kr, [b]ewtecdx@wm.cau.ac.kr, [c]kbyoon@cau.ac.kr, [d]wsryu@kaeri.re.kr

**Keywords:** Creep, Creep Crack Growth Rate (CCGR), $C^*$, Creep Ductility Model, Type 316 Stainless Steel, Plane Stress, Plane Strain.

**Abstract.** This paper is to evaluate the creep crack growth rate (CCGR) of the type 316SS series: 316SS, 316FR and 316LN, and to apply a creep ductility model. A number of the data are collected through wide literature surveys and experiment, and evaluated by the $C^*$ parameter. The results of the CCGR data were nearly matched with a small scattering band regardless of the different applied stresses, temperatures and test specimens configuration. In the CCGR, type 316FR and 316LN steels were slower than type 316SS. Type 316SS showed a better agreement in the application of the creep ductility model than the type 316FR and 316LN steels.

## Introduction

The type 316 stainless steel (SS) series has been widely used as the major structural components material of liquid metal reactor (LMR). Especially, type 316LN stainless steel containing both low carbon (0.03C wt.%) and an appropriate amount of nitrogen (0.1Nwt.%) is a candidate material for the structural components in the LMR, because the 'LN' grade is superior in creep resistance to the 'L' grade of low carbon [1,2]. The structural components of the LMR are operated up to approximately 550°C, the components are often subjected to non-uniform stress and temperature distribution during service. These conditions generate localized creep damage and propagate the cracks and ultimately cause a fracture. A significant portion of the component life can be spent in crack propagation. Thus, a creep crack growth rate (CCGR) is important in a design concern and in predicting the residual life of the components during service.

In this paper, to evaluate the CCGR of the type 316SS series: 316SS, 316FR and 316LN, a number of CCGR data were collected through wide literature surveys and experiment. Using the collected data, the CCGR equations were proposed for the each steel, and all the data are applied to the creep ductile model and its applicability are evaluated in engineering viewpoints.

## Creep ductility model (NSW model)

Nikbin, Smith and Webster (NSW) have developed the creep ductility model (hereafter NSW model), which the CCGR can be simply predicted by using uniaxial creep ductility data [3]. The creep crack growth rate is predicted as

$$\dot{a} = \frac{(n+1)}{\varepsilon_f^*}(A r_c)^{1/(n+1)}\left(\frac{C^*}{I_n}\right)^{n/(n+1)} . \tag{1}$$

Where $\varepsilon_f^*$ is the creep ductility and its value is determined with readuction of area, and $r_c$ is the size of the process zone and its value is determined with grain size of materials. $I_n$ is the nondimensional function of $n$, and its value is determined approximately as 3 for the plane stress and as 4 for the plane strain. $A$ and $n$ are the coefficient and exponent of the Norton's rule. Reducing Eq. 1, for the plane stress conditions, it is simply written as $\dot{a} = 3C^{*0.85}/\varepsilon_f^*$, and for the plane strain conditions, it is $\dot{a} = 150C^{*0.85}/\varepsilon_f^*$. In units, $\dot{a}$ is in mm/h, $\varepsilon_f^*$ is strain as a fraction and $C^*$ is in MJ/m$^2$. The $\dot{a}$ is very sensitive to creep ductility. The model has been demonstrated for experimental evidence on various materials. In this study, the model is applied to the type 316SS series, which is used widely as nuclear structural components.

**Creep crack growth rates for type 316SS series**

When secondary creep doninates, the general form between the CCGR ($\dot{a}$) and $C^*$ is expressed as $\dot{a} = D[C^*]^q$[3]. $D$ and $q$ are material constants. The CCGR data wer collected from wide literature surveys, reported by Yagi[4], Merckling[5], Radharkrishnam[6], Hyde[7], Nakayama[8], Fookes[9], Tans[10] and Kim[11]. Nakayama's data are for type 316FR developed for a Fast Breeder Reactor in Japan. Kim's data are for type 316LN produced for a LMR in Korea. Table 1 shows the results of typical tensile and creep properties for type 316SS and 316LN. Type 316LN was superior in creep properties to type 316SS. The $D$ and $q$ values are summarized in Table 3. Here, $D_a$ and $q_a$ indicate the average values. The $q$ value did not show a large difference in the 0.85~0.90 ranges regardless of different temperatures, stresses and test specimens. The reason is because the $C^*$ parameter can well reflect the material constsnts, $A$ and $n$.

Table 1 Typical tensile and creep properties in the type 316SS and 316LN

| Materials | Temp. (°C) | Tensile properties | | | Creep properties | |
|---|---|---|---|---|---|---|
| | | YS (MPa) | UTS (MPa) | Elongation (%) | Applied stress (MPa) | Rupture time (hr) |
| 316SS | 600 | 110 | 350 | 45.2 | 210 MPa | 120 |
| 316LN | 600 | 131 | 418 | 45.7 | 210 MPa | 1105 |

Table 2 Summaries of the constants, $D$ and $q$ and the test conditions of the type 316SS series

| Materials | Data source | Temp.(°C) | Material constants of CCGR Eq. | | | | Specimen thickness(mm) | Specimen geometry |
|---|---|---|---|---|---|---|---|---|
| | | | D | $D_a$ | Q | $q_a$ | | |
| 316SS | Yagi (97) | 550-800 | 4.409x10$^{-8}$ | 4.840x10$^{-8}$ | 0.851 | 0.846 | 12.7 | CT |
| | Radhakrishnan(90) | 600-800 | 1.909x10$^{-8}$ | | 0.985 | | 1 | SENT |
| | Hyde(88) | 600 | 4.828x10$^{-8}$ | | 0.903 | | 16, 15 | CT |
| 316FR | Nakayama(98) | 550-650 | 3.800x10$^{-8}$ | 7.158x10$^{-8}$ | 0.790 | 0.713 | 1.5, 7.62, 5 | WCCP,CT,CCP |
| 316LN | Tan(01) | 650 | 1.379x10$^{-8}$ | | 0.865 | | - | CT |
| | Fookes(98) | 600 | 3.839x10$^{-8}$ | | 0.970 | | 4 | SENT |
| | Merckling(98) | 550-650 | 5.275x10$^{-8}$ | | 0.882 | | 25, 19 | CT, CT |
| | Kim(02) | 600 | 1.342x10$^{-7}$ | | 0.641 | | 12.7 | CT |

Note : CT : Compact Tension, SENT : Single Edge Notched Tension, CCP : Center Cracked Plate, WCCP : Wide Center Cracked Plate

Fig. 1 shows the results of creep crack growth rates ($da/dt$) vs. $C^*$ in type 316SS reported from Yagi, Radharkrishnam and Hyde. The $D$ and $q$ values were obtained by the best fitting of the data. When all the data points are overlapped together, the average values are as $q_a$ = 0.846 and $D_a$ = 4.840x10$^{-8}$. It is identified that all the data points are placed within a small scatter band. For conservatively predicting a result, it is reasonable to use an upper regression line[12]. Thus, using the 95% upper prediction limit(UPL) line, the CCGR Eq. for type 316SS is obtained as

$$da/dt = 1.5314\times10^{-7}[C^*]^{0.854} \quad \text{(m/h)}. \tag{2}$$

Fig. 2 shows the results of $da/dt$ vs. $C^*$ in type 316FR and 316LN reported from Nakayama, Tan, Fookes, Merckling, and Kim. The $q$ value was about in the 0.641~0.97 ranges. the $q_a$ and $D_a$ values were 0.713 and 7.158x10$^{-8}$, respectively. As in type 316SS. The data of type 316SS are located in a higher position than those of type 316FR and 316LN. Type 316SS is faster in the CCGR than type 316FR and 316LN. From the 95% UPL line, the CCGR Eq. of type 316FR and 316LN is obtained as

$$da/dt = 1.6726\times^{-7}\,[C^*]^{0.718} \quad \text{(m/h)}. \tag{3}$$

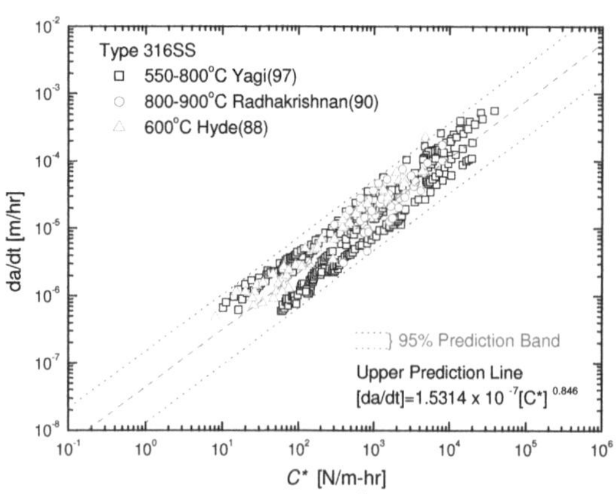


Fig. 1 *da/dt* vs. $C^*$ for type 316SS

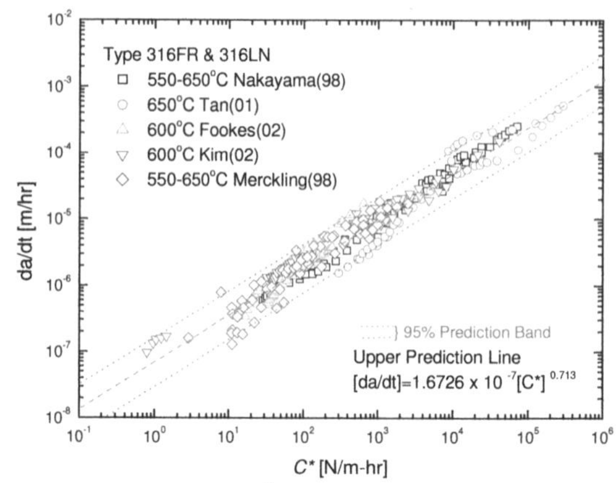


Fig. 2 *da/dt* vs. $C^*$ for type 316FR and 316LN

**Application to the creep ductility model**

Using the NSW model, the relationship between *da/dt* and $C^*$ can be obtained for the type 316SS series. Uniaxial creep strain is converted into the reduction of the area and then used with the true strain. When Yagi's and Tabuchi's data for type 316SS are used, the reduction of area was in the 20-70% ranges. Converting this value into the true strain, it is corresponded to 0.22-1.39 ranges. Thus, the $\varepsilon_f^*$ value can be obtained using uniaxial creep strain, $\varepsilon_f$.

Figs. 3 and 4 show the results in the application of the NSW model to type 316SS and 316FR/316LN, respectively. Dotted line indicates the NSW model and solid line indicates the experimental data fitted by the 95% prediction. The 95% lower prediction limit (LPL) line of the NSW model is the plane stress condition with a reduction of the area of 75% (1.39 strain). Here, if an appropriate creep strain from the LPL line is taken, the UPL line can be obtained. Nikbin and Tan reported on the UPL line with $\varepsilon_f/50$ for creep brittle materials and the UPL line with $\varepsilon_f/30$ for creep ductile materials. Because type 316SS corresponds to the creep ductile materials, the UPL line in this study is calculated at 30 times to the plane stress. As presented in Fig. 3, the NSW model for type 316SS well is matched with the experimental data, and the line slope is nealy the same. But, the UPL line of the NSW model evaluates very conservativly to the experimental data. Thus, when taking $\varepsilon_f/10$ ($\varepsilon_f = 0.2$ strain), the NSW model is well matched to the experimental data. The CCGR data fot type 316SS are closely distributed for the plain stress condition.

On the contrary, for type 316 FR and 316LN, the NSW model shows some differences in the line slope of the experimental data. As in type 316SS, the model evaluates conservatively with the increasing $C^*$ values. Type 316FR and 316LN data are more closely distributed for the plain stress condition than type 316SS. The CCGR of type 316SS was faster than that of type 316FR and 316LN. The reason is because type 316FR and 316LN were superior in creep resistance and in tensile properties to type 316SS as seen in Table 1. Superior mechanical properties in 316LN are due to the nitrogen addition, and the effects of nitrogen on creep properties were presented in previous studies[2, 13]. In addition, whether the plane stress or plane strain conditions are determined is dependent upon the specimen dimensions and the creep deformation which accompanies the fracture. Generally, a low ductility, large sample size and the presence of side grooves favor the plane strain situations[3].

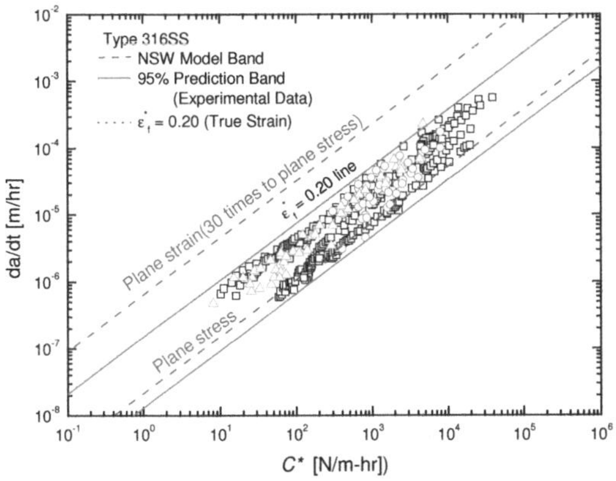


Fig. 3 Application to type 316SS

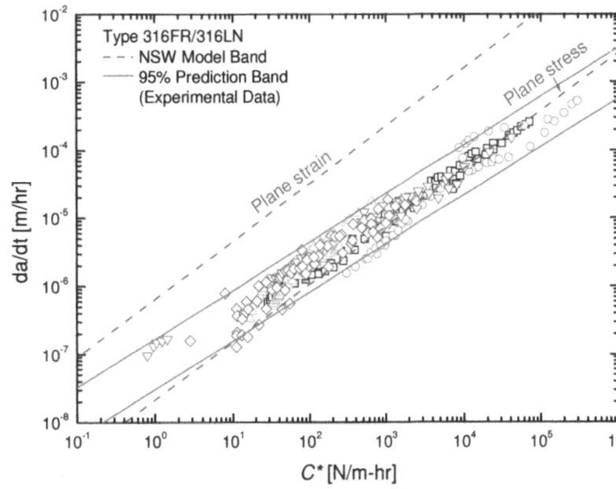


Fig. 4 Application to type 316FR and 316LN

## Conclusion

The CCGR equations were obtained for type 316SS, 316FR and 316LN, respectively. All the CCGR of the type 316SS series showed a small scattering band regardless of applied stresses, temperatures and specimen configurations. The reason for this is because the $C^*$ parameter reflects well creep material constsnts, $A$ and $n$. Type 316SS was higher in the CCGR than type 316FR and 316LN. All the CCGR data of the type 316SS series were closely distributed for the plain stress condition. This is attributed to the solid-solution strengthing due to the nitrogen addition. In the application of the creep ductility model, type 316SS showed a better agreement than the type 316FR and 316LN.

## Acknowledgement

This work is a part of the Nuclear Material Technology Development Project, which has been financially supported by the Korean Ministry of Science and Technology.

## References

[1] J.T. Yeom, J.Y. Kim, Y.S. Na and N.K. Park: *Met. Mater. - Int.* Vol.9 (2003), p. 555.

[2] W.G. Kim, S.H. Kim and W.S. Ryu: *KSME Int. J.*, Vol.15 (2001), p. 1463.

[3] K.M. Nikbin, D.J. Smith and G.A. Webster: *Trans. of the ASME*, Vol.108 (1986), p. 186.

[4] K. Yagi, M. Tabuchi, K. Kubo, J.J. Kim and Y.H. Huh: *Eng. Fract. Mech.*, Vol. 57 (1997), p. 463.

[5] G. Merckling, C. Guardamagna and L. Villa: *Materials at High Temp.*, Vol. 15 (1998), p. 219.

[6] V.M. Radhakrishnan and M. Kamaraj: *High Temperature Technology*, Vol.8 (1990), p. 219.

[7] T.H. Hyde: *High Temperature Technology*, Vol.6 (1988), p. 50.

[8] Y. Nakayama, N. Miura, Y. Takahashi and Y. Toya: *ASME PVP*, Vol. 365 (1998), p. 191.

[9] A.J. Fookes, S.X.Li and D.J. Smith: *Materials at High Temp.*, Vol. 15 (1998), p.187.

[10] M. Tan, N.J.C. Celard, K.M. Nikbin and G.A. Webster: *ASME PVP*, Vol. 78 (2001), p. 737.

[11] W.G. Kim, W.S. Ryu, K.Y. Lee and K.B.Yoon, *16th Conf. on Mech. of Mater.* (2002), p. 45.

[12] K.Y. Lee, U.B. Back and K.B.Yoon: *J. of the KIIS*, Vol.17 (2002), p. 5.

[13] D.W. Kim, Y.K. Lee, W.G. Kim and W.S. Ryu, *J. Kor. Inst. Met & Mater.*, Vol. 39 (2001), p.1241.

*Key Engineering Materials Vols. 261-263 (2004) pp. 1103-1108*
*online at http://www.scientific.net*


# Usefulness of the Creep Work-Time Relation for Stress Intensity Limits ($S_t$)

Woo Gon Kim and Woo Seog Ryu

Department of Nuclear Materials Development Technology, KAERI
150, Deogjin-dong Yuseong-gu, Daejeon, 305-353, Korea

**Keywords**: Creep, Isochronous Stress-Strain Curve, Type 316 Stainless Steel, Stress Intensity Limit, Strain Energy, Creep Work

**ABSTRACT**

The usefulness of the creep work ($W_c$) and time ($t$) relation, defined as $W_c\ t^{\,p} = B$, was investigated for use in determining the stress intensity limit of the creep components. The creep tests to obtain the creep work were conducted for type 316SS at 593°C and the data of the isochronous stress-strain curve was collected from the ASME BPV NH. The result of the relation showed good agreement with the ISSC one. However, it was found that the linear fitting prediction generated a little error for the $10^5$ hr. The error value was below 7.0% in a 1.0% strain and 8.0%~9.0% in a 1.0%~2.0% strain. From the viewpoint of engineering design, the error for $10^5$ hr is not large and is regarded as conservative. It is thus believed that the creep work-time relation can be used as a useful method for estimating the stress intensity limits of creep materials.

## 1. INTRODUCTION

Since the structural components of a liquid metal reactor are operated at high temperatures up to 550°C, one of the most critical factors for design is the time-dependant creep behavior. These creep components are designed with a target life based on a specified amount of allowable strain and stress [1-3]. The basic criteria for establishing the allowable stress values of the materials are well described in the ASME BPV (boiler-pressure vessel) NH and its Code Section-II-PART D [4,5].

At low temperatures below the creep range, the maximum allowable stress value, $S_m$ is lowest in the following: (i) 1/3.5 of the specified minimum tensile strength at room temperature; (ii) 1/3.5 of the tensile strength at temperature; (iii) 2/3 of the specified minimum yield strength at room temperature; (iv) 2/3 of the yield strength at temperature. At elevated temperatures in the creep range, the temperature and time-dependent stress intensity limit, $S_t$ for all the materials is defined not to exceed the lowest of the following: (i) 100% of the average stress to produce a creep rate of 0.01%/ $10^3$ hr (1%/ $10^5$ hr); (ii) 67% of the average stress to cause a rupture at the end of $10^5$ hr; and (iii) 80% of the minimum stress to cause a rupture at the end of $10^5$hr. The allowable limit of the general primary membrane stress intensity, $S_{mt}$ to be used as a reference for stress calculations for the service life is the lowest of the two stress intensity values, $S_m$ in a time-independent material and $S_t$ in a time-dependent material [5-7].

Among the above three conditions, the first condition (i) is generally used for design and isochronous stress-strain curve (hereafter ISSC) which is needed to avoid excessive deformation of the materials during the intended service life. For a relatively short service life of a few thousand hours, the generation of the ISSC can be easily obtained by taking a constant time through a family of creep curves. On the contrary, for a long service life over 30 to 40 years such as in nuclear power plants, it is not only very difficult but also very time-consuming work. To obtain 1% strain for $10^5$ hr (11.4 year), such as in the first condition (i), it is not easy. So, the generation of the ISSC requires extrapolation of the short-term data within $10^4$ hr [8]. At present, the ISSC data is only available in the ASME BPV NH [5,9]. So, several approaches have been employed to generate the ISSC. As a method, the relationship between the creep work and time, based on the strain energy (SE) concept done during time $t$, causes a concern for engineers because the relationship is too convenient for use. But, its usefulness is unclear for a long-term prediction up to $10^5$ hr such as in nuclear power plants.

The purpose of this paper is to identify the usefulness of the creep work and time relation. For this purpose, the ISSC data for a commercial type 316 stainless steel (hereafter SS) was collected from the ASME BPV NH, and a series of creep tests were conducted to obtain the creep work as a function of time. The stress intensity values obtained from the relation were compared with the ISSC ones, and its usefulness is discussed for the long-term calculations up to $10^5$ hr.

## 2. EXPERIMENTAL PROCEDURE

The testing materials were chosen from a commercial type 316 SS (C316SS) to utilize the ISSC data presented in ASME BPV NH [5,8]. Chemical compositions of the material are given in Table 1. Creep specimens were machined to be cylindrical with a 30 mm gauge length and 6 mm diameter. During the heating of the specimens in the furnace, the initial load with 10% of the test load was constantly applied to the specimens. The creep tests were carried out using constant-load machines with a lever ratio of 20/1. A linear variable differential transformer (LVDT) was used for measuring the strain with the elapsed time. All experimental procedures referred to the ASTM standard E139 [10]. The creep test was conducted at 593°C(1100°F) with various stresses: 135MPa, 140MPa, 145MPa, 150MPa, and 160MPa. Temperature was maintained constantly within ± 2°C during the period of the test. The time in generating a strain was measured with each stress level. Stress intensity values were determined using the experimental data of the log $W_c$ - log $t$ plot, and the creep work as a function of time was extrapolated up to $10^5$ hr for each stress level.

Table 1. Chemical compositions of commercial type 316 SS (wt.%)

| Material | Fe | C | Si | Mn | P | S | Cr | Ni | Mo |
|---|---|---|---|---|---|---|---|---|---|
| C316SS | Bal. | 0.50 | 0.58 | 1.26 | 0.032 | 0.001 | 16.77 | 10.75 | 2.06 |

## 3. RELATIONSHIPS BETWEEN CREEP WORK AND TIME

The area under the ISSC is the stored energy of the creep, $W$ done during time, $t$. Stored energy of the creep is evaluated by counting the number of squares for various durations of time. It was empirically found to follow the functional relation of the following Eq. (1) [6,11],

$$W t^m = C. \quad (1)$$

The $C$ and $m$ are constants depending on the strain level and temperature, and the $m$ indicates the slope values in the log $W$-log $t$ plot. If the $W$-$t$ relation accurately follows the linear relation with a double logarithmic plot in the relations up to $10^5$ hr at the given strain levels and temperatures, the $C$ or $m$ values can be used for the estimation.

Using the similarity of $W$ and $t$, we are able to generate the ISSC from a few short-time creep tests. Considering that a material has undergone creep deformation at a stress, $\sigma$ to a strain, $\varepsilon$ at time, $t$. The creep work can be written as $W_c = \sigma \cdot \varepsilon$. The ISSC for time $t$ reaches the point ($\sigma$, $\varepsilon$), and the area under the ISSC, $W$ is as defined above,

$$W = \int_o^{\varepsilon} \sigma \, d\varepsilon = \sigma \varepsilon - \int_o^{\sigma} \varepsilon \, d\sigma = W_c - \int_o^{\sigma} \varepsilon \, d\sigma. \qquad (2)$$

The $W_c$ is the creep work, and the $W$ is the stored energy. The second term on the right hand side of Eq. (2) is the dissipated component of the creep work, $W_d$ as the complementary energy. The $W_d$ is corresponding to the area above the ISSC. For a linearly elastic material, $W=W_d$, but the $W$ and $W_d$ for a nonlinear elastic material are different. The SI unit of strain energy density (strain energy per unit volume) is the joules per cubic meter (J/m$^3$).

Assuming that a Hollomon's Eq., $\sigma = K\varepsilon^n$ is valid for the ISSC, the $n$ is the strain-hardening exponent of the ISSC [12]. By differentiating Hollomon's Eq., which makes Eq. (3). Substituting it into Eq. (2), and then simplifying it, Eq. (5) can be obtained, as follows

$$W = K \cdot n \cdot \varepsilon^{(n-1)} \, d\varepsilon \qquad (3)$$

$$W = \sigma \, \varepsilon - [K \cdot n / (n+1)] \, \varepsilon^{(n+1)} \qquad (4)$$

$$W = \sigma \varepsilon / (n+1) = W_c / (n+1). \qquad (5)$$

As seen in Eq. (5), during a creep experiment at an applied stress $\sigma$, the $W_c$ for deformation up to a strain $\varepsilon$ is proportional to the $W$ corresponding to an ISSC of the time $t$. The time is taken to attain the strain level under the applied stress $\sigma$. Consequently, the $W_c$ can be also described as a time-related Eq. (6), which is similar to Eq. (1).

$$W_c t^m = B \qquad (6)$$

Hence, $B = (n+1)\, C$, and $W_c = \sigma \varepsilon = W(n+1)$. Using Eq. (6), the stress values can be calculated from the creep work obtained from a strain and a time. This method allows us to simply determine stress intensity values with only a few short-term creep tests.

## 4. RESULTS AND DISCUSSION

### 4.1. The ISSC data of type 316 SS

The ISSC can provide the designer with information regarding the total strain caused by stress under elevated temperature conditions. At present, the ISSC data is available in the ASME BPV NH for class 1 components; 304SS, 316SS, alloy 800H and 2.25Cr-1Mo [9]. Fig. 1 shows typical ISSC graphs collected in the ASME BPV NH for type 316 SS at 593°C (1100°F) [5,9]. This data is later used as comparative data for evaluating the creep work and time relation. Fig. 2 shows the ISSC graphs for $10^5$ hr in the 426°C~815°C ranges for type 316 SS. All the graphs were obtained by converting the ISSC data of the ASME BPV NH. Each curve significantly shows a temperature dependency. The data for $10^5$ hr is especially important for estimating the total strain caused by a stress under each temperature condition, because the elevated-temperature components such as in nuclear power plants are designed with respect to a target life of 30 to 40 years.

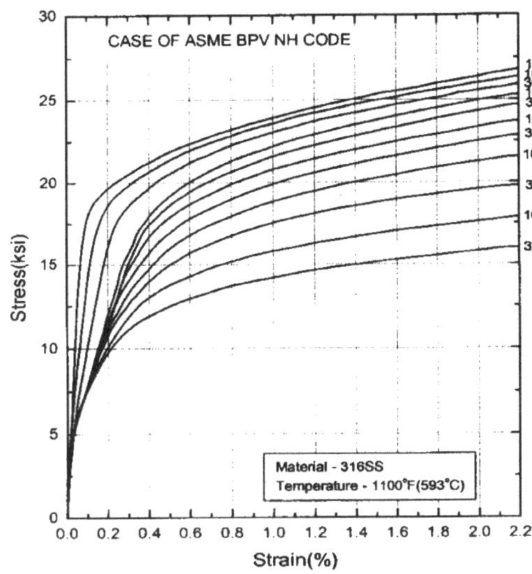


Fig.1. Average ISSC of type 316SS at 593°C

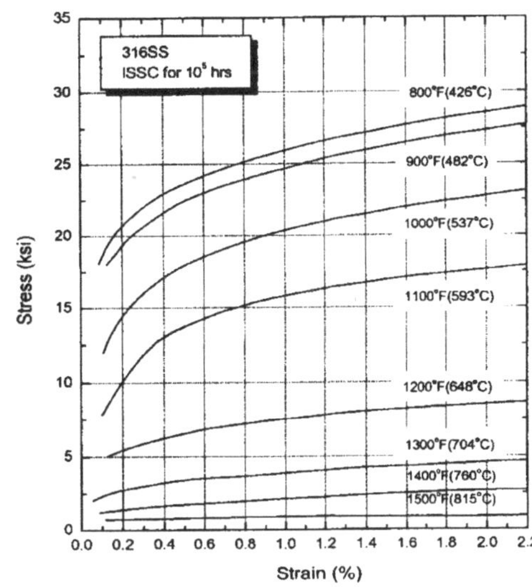


Fig. 2. Average ISSC with temperatures in $10^5$ hr

## 4.2. Application of creep work and time relation

Fig. 3 shows the creep strain-time curves obtained with different stresses for a commercial type 316SS at 593°C. The creep curves show well the stress dependency; creep strains increase by increasing the applied stresses. Also, the time to attain each strain, 0.6%, 0.8%, 1.0% and 1.2%, is longer as the stress levels are lower. The value of the creep work was calculated as a function of time from the relationship between the stress and strain, and the value of the stress intensity was finally obtained by converting the creep work.

Fig. 4 shows the log $W_c$-log $t$ plot with strain variations for the experimental and the ISSC results. Close symbols indicate the experimental results based on the strain energy concept. Open symbols indicate the ISSC results. The data was linearly extrapolated up to $10^5$ hr for each strain.

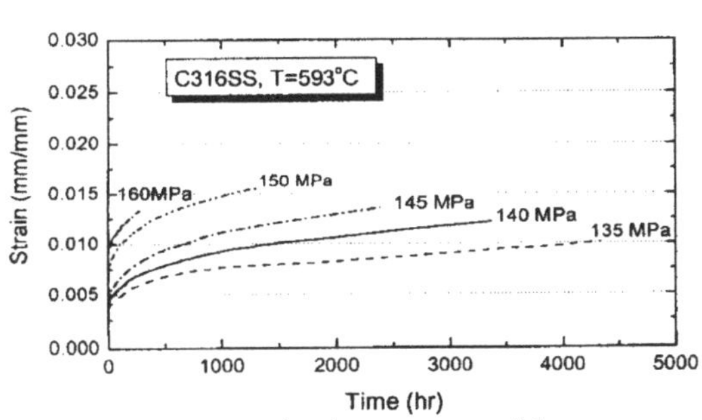


Fig. 3. Creep strain-time curves with stresses

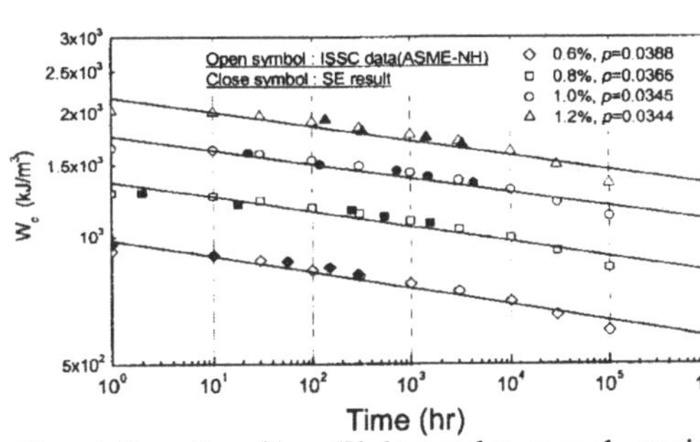


Fig. 4. Results of log $W_c$-log $t$ plot at each strain

The log $W_c$-log $t$ plot follows well a linear relation in a double logarithmic plot. The SE results show a good match for the line with the ISSC ones up to $10^4$ hr. The $B$ and $p$ values are obtained and summarized in Table 2. The $p$ value is determined from the slope of the straight line from a linear prediction up to $10^5$ hr. Here, it is noted that the linear fitted line has a little error at $10^5$ hr, as listed in Table 2. The error percentage is about 7.0% up to 1.0% strain and 8.0% ~ 9.0% in the 1.0% ~ 2.0% strain ranges. And the error reduces with the decreasing strains. It is believed that the errors at $10^5$ hr are due to the thermal damage generated in high-temperature creep deformation. However, from the viewpoint of creep design, it is thought that the creep work and time relation can be used for estimating stress intensity, because most creep structures are practically used for elastic regions in the low strain ranges. The 6.9% error at the 1.0 % strain is conservative for long-term predictions up to $10^5$ hr. Thus, the relation is useful for estimating stress intensity for time-dependant creep materials. In particular, it is useful for newly developed materials.

Table 2. The values of $B$, $P$ and $\sigma$ determined from linear prediction for the ISSCs in $10^5$ hr

| Strain (%) | $P$ | $B$ (KJ/m$^3$) | $\sigma$ in $10^5$ hr (MPa) | | |
|---|---|---|---|---|---|
| | | | Predicted by SE | ISSC data | Error (%) |
| 0.6 | 0.0388 | 978.58 | 104.33 | 98.58 | 5.5 |
| 0.8 | 0.0365 | 1362.23 | 111.85 | 104.78 | 6.7 |
| 1.0 | 0.0345 | 1764.67 | 118.62 | 110.92 | 6.9 |
| 1.2 | 0.03448 | 2167.45 | 121.44 | 112.37 | 8.0 |
| 1.4 | 0.0344 | 2588.13 | 124.41 | 114.44 | 8.7 |
| 1.6 | 0.0340 | 3003.82 | 126.92 | 117.19 | 8.3 |
| 2.0 | 0.0338 | 3895.93 | 132.00 | 121.33 | 8.7 |

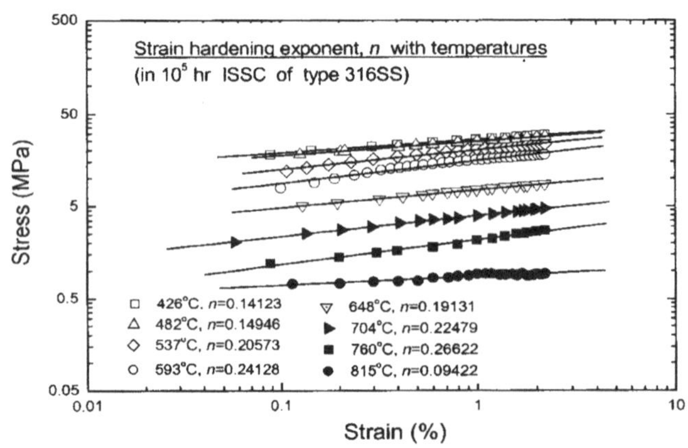


Fig. 5. The $n$ values determined for various temperatures in $10^5$ hr of type 316SS

In addition, in order to conveniently use Eq. (5), it is necessary to obtain the strain-hardening exponent, $n$ for the ISSC. By knowing the $n$ value, the work or stored energy during creep can be easily calculated from the proportion relation of $W/W_c$. Fig. 5 shows the $n$ values determined with different temperatures at $10^5$ hr of the type 316SS. The results were obtained by converting the ISSC obtained in Fig. 2. The $n$ value in the Hollomon Eq., $\sigma = K\varepsilon^n$ is determined from the slope of the log$\sigma$-log$\varepsilon$ plot. The plastic strain range in this calculation was used for the 0.1%~2.2% ranges. The $n$ value is found to vary from 0.10~0.30 for the type 316SS. Tendency of the $n$ variations is increased with temperature in all the temperature ranges except 815°C. Also, the gaps of the $n$ value are small with the temperature variations. In the case of the tension test of austenitic stainless steel, it was reported that $n$ = a 0.45~0.55 range[12]. It is identified that the $n$ values of the ISSCs become much lower than those of the tension test. This is attributed to a basic difference between the time-independent tension and time-dependant creep. As a matter of fact, it is difficult to obtain the $n$ for the ISSC because a lot of long-term creep data is required to make the ISSC.

## 5. CONCLUSIONS

The relation of the creep work ($W_c$) and time ($t$) showed a good agreement to the ISSC results presented in the ASME BPV NH. The relation showed a good linear result with the ASME NH data, and the $B$ and $p$ values were obtained from the linear prediction. However, it was found that the linear fitting prediction for $10^5$ hr had some errors in the $B$, $P$ and stress values. The error value was below 7.0% in a 1.0% strain and 8.0%~9.0% in a 1.0%~2.0% strain. The reason for it is attributed to the thermal damage with a long-term creep deformation. From the viewpoint of engineering design, because the error for $10^5$ hr is not large and is regarded as conservative, it is thus believed that the creep work-time relation can be used as a useful method for estimating the stress intensity for creep materials and particularly for newly developed materials. Strain-hardening exponent, $n$ was found to vary from 0.10~0.30 for the ISSC of the type 316SS in the 425°C~815°C ranges. The $n$ values did not have any large differences with the temperature variations.

## ACKNOWLEDGEMENTS

This work was performed under the Long-term Nuclear R&D Program sponsored by the Korean Ministry of Science and Technology.

## REFERENCES

1. W.G. Kim, S.H. Kim and W.S. Ryu, J. International KSME, **16** (2001) p.1420
2. D.W. Kim, Y.K. Lee, W.G. Kim and W. S. Ryu, J. Kor. Inst. Mater, **39** (2001) p.1241
3. W.G. Kim, S.H. Kim and W.S. Ryu, J. International KSME, **15** (2001) p.1463
4. ASME BPV Code, Section II-Part (D), Appendix I, New York (2001) p.762
5. ASME BPV Code, Rules for Construction of Nuclear Facility Components-Class 1 Component in Elevated Temperature Service Section III, Division 1-Subsection NH, New York (2001) p.26
6. K.G. Samuel and P. Rodriguez, Int. Journal Pressure Vessels and Piping, **78** (1998) p.939
7. R.K. Penny and D.L. Marriott, Design for Creep, Chapman & Hall, London (1995) p.355
8. L.D. Blackburn, Isochronous Stress Strain Curves for Austenitic Steel, ASME (1972) p.14
9. ASME BPV N-47-33, Class 1 Component in Elevated Temperature Service Section III, Division 1, New York (1996) p.132
10. ASTM Standard E139, (1995) p.257
11. A.C. Ugral and S.K. Fenster, Advanced Strength and Applied Elasticity, Prentice Hall PTR, New Jersey (1994) p.77
12. R.W. Hertzberg, Deformation and Fracture Mechanics of Engineering Materials, John Wiley & Sons, New York (1976) p.175

# 저 자 약 력

| | |
|---|---|
| 현) | 한국에너지기술단(KETG) 수석 컨설턴트 |
| 1999.08. | 충북대학교 기계공학과 박사과정 졸업(공학박사) |
| 1981~2021 | 한국원자력연구원(책임연구원) 정년 퇴직 |
| 1998~2021 | 대한기계학회 신뢰성부문 이사 및 부회장, 학회 평의원 |
| 2009~2021 | 중소기업기술개발지원 평가위원 |
| 2008.08. | IAEA 기술 전문가 활동 |
| 2020.12.17 | 대한기계학회 기술상(제130호 2020-285) |
| 2013.12.09 | 대한기계학회 신뢰 기술상(제1호 2013-413) |
| 2006.01.02 | 연구업적 논문상(한국원자력연구원, 제2941호) |
| 2006.01.02 | KM 지식인상(한국원자력연구원, 제2994호) |
| 1990.12.27 | 연구개발상(과학기술처장관 제198호) |
| 1989.03.29 | 근무유공상(한국원자력연구소) |

**[ 저자 연락처 ]**

e-mail : smartgon_1@naver.com or wgkim100@gmail.com

# 고온 원자력 재료의 크리프 수명 평가 및 예측

# Creep Life Assessment and Prediction of High-Temperature Nuclear Materials


2021년 12월 10일 초판 인쇄

2021년 12월 15일 인쇄 발행

저　　　자 김우곤

발 행 인 송기수

발 행 처 도서출판 GS인터비전

편 집 처 도서출판 GS인터비전

편 집 인 공예서

표지디자인 디자인붐

인 쇄 처 트윈벨미디어

등 록 번 호 제 25100-2016-000050호

I S B N 979-11-5576-378-0(93550)

주　　소 서울 은평구 증산로 15길 69 2층

전　　화 02-976-7898, 02-3272-7898.

팩　　스 02-6468-7898

홈페이지 gsintervision.co.kr

E-Mail gsinter7@gmail.com

정 가 25,000원